T0171775

# Patentism Replacing Capitalism

Samuel Meng

# Patentism Replacing Capitalism

## A Prediction from Logical Economics

Samuel Meng
Griffith University
Brisbane, QLD, Australia

ISBN 978-3-030-12249-2      ISBN 978-3-030-12247-8    (eBook)
https://doi.org/10.1007/978-3-030-12247-8

Library of Congress Control Number: 2018968268

This Palgrave Macmillan imprint is published by the registered company Springer Nature Switzerland AG
The registered company address is: Gewerbestrasse 11, 6330 Cham, Switzerland

*I dedicate this book to my father, and to the memory of my mother.*

# Preface

The purpose of this book is threefold: providing an explanation and a solution to the phenomenon of economic recession/stagnation, foreshadowing the coming societal changes, and redirecting economic research from robotic data mining back to rigorous logical thinking. At this very beginning, I shall make a claim that this book will help society avoid economic stagnation and bring the world one step closer towards a knowledge economy. Is it true or not? You can judge after reading. Or time will tell eventually.

The world has experienced significant social changes over thousands of years. In what direction is society heading? Will socialism replace capitalism or will capitalism survive even after the implication of the 2008 global financial crisis? Based on economic knowledge and careful logical reasoning, this book demonstrates that the recurrence of economic recessions spells the end of capitalism and the dawn of patentism.

The book highlights the vital role of innovation in economic growth. It is widely appreciated that innovation is extremely important for economic growth and for our daily life, but it appears that the mechanism by which innovation contributes to economic growth is little understood. The limited understanding of this mechanism greatly constrains the development of both innovations and the economy. This book will reveal the mechanism and dynamics between innovation and economic development, as well as their implications for society.

The revolutionary ideas expressed in this book have evolved from scrutiny of old ideas, so the new ideas at first sight appear unremarkable except that they are not harmonious with any existing ideas. This reflects the author's assessment of the current state of schools of economic thought: all ideas have contained only some elements of truth. By emphasizing logical thinking and theoretical contribution, the author contests the trend of empiricalization—a trend which makes economics a one-dimensional discipline that focuses on empirical relationships between economic variables. Why has the author not followed the popular empirical approach that has become so fashionable? Because he firmly believes that the current trend is flawed and is harmful to scientific research.

Since the book challenges the current trend, it is certain that the book will be criticized and the ideas presented will not be accepted immediately. It is of no concern whether the book is controversial or popular, rather, the author's interest is to generate a substantial positive change to improve the living standards of humankind. However, a real change can be made only when the policy makers and the majority of the public appreciate the solution provided. From this point of view, any comments or arguments from readers are welcome.

This book is written for both economists and general readers. While the chapters in this book have strong logical links, each chapter is an independent piece, so the reader can begin with any chapter. The historical review and discussion from Chapters 1 to 3 is to expose the flaws in various economic ideas. To get the main message of the book, a speedy reader can start at Sect. 4.3 in Chapter 4. The book is written in a way that most contents can be understood by the lay reader with no economics training although some parts of the book require economics knowledge. To make a more rigorous argument, a few models and a number of equations are used in some sections, but the ideas and implications of the models are fully explained in the text. The advanced sections are marked with * and most mathematical contents are moved to appendixes. If the reader is not interested in economics or mathematics, all mathematical contents and all appendixes can be safely ignored.

Interestingly enough, a non-economist might have a much better overall understanding of this book than the majority of current professionals in economics. The truth this book tries to reveal is rather simple: the features of an economic recession (e.g. stagnation of sales, unemployment and unutilized capital) indicate an overproduction of old products (market saturation) and an undersupply of new products (e.g. a cure for cancer, tourism travel to the Moon or Mars), which results from the scarcity of inventions. The shortcomings in the current patent laws are responsible for invention scarcity. If the patent law can be successfully revised to encourage enough inventions, the supply of new products can satisfy demand, create jobs and utilize capital. As such, economic recessions can be avoided. This truth is very clear to me and is understandable to most non-economists. However, it is not so easy to convince majority of economists of this truth.

The minds of most economists today are complicated, or even distorted, by the conflicting theories in macroeconomics and by the still fashionable statistical or empirical approach in the economics profession. These prejudgements tend to prevent economists from accepting new ideas. If one's mind is already set by certain theories, it is suggested that he/she ignores all theories before reading this book. If it is really hard to ignore the existing theories (this is very likely for a believer of current econometrics, certain schools of economic thought and the balanced approach to the patent system), it is suggested that he/she focuses on the validity of the theory in this book: Are the axioms and assumptions plausible? Are the models and reasoning valid? Does this book have an element of truth despite its not being consistent with the known theories? If one does not agree with the assumptions and reasoning in this book, do continue to read but, in the meantime, form a concrete argument and put it aside for a discussion later. In this way, the reader can truly follow the logic of the arguments made in this book. After reading, you can compare the arguments herein with the existing theories and make a renewed judgement. It is unlikely that one would readily discard his/her lifetime belief and immediately embrace the ideas in this book, but the author hopes that the book can provoke the reader to reconsider his/her old belief and as necessary provide a counter-argument. Through debate, the subtle points involved may become more apparent. All arguments the reader reaches are welcome, but the arguments should be based directly on logical reasoning (induction and/or deduction) rather than based on what most people believe in or who in history has said what. Nobody is right about everything and popular opinions are not necessarily closer to the truth.

Three old and widely accepted ideas or trends need to be corrected: (1) the trend of empiricalization or econometricalization of recent decades, which downplays the importance of any theory and logical reasoning in general; (2) the various conflicting macroeconomic theories that have some elements of truth but fail to reveal the major mechanism in an economy; and (3) the balanced approach and anti-patent ideology shown in the heated patent debate, in which the participants are driven more by emotion and enthusiasm than by rational thinking. Based on the feedback from my economist friends and colleagues, journal editors and referees, the author has included and discussed these issues and many relevant arguments. In fact, a substantial effort in this book has been made to refute some erroneous but popularly accepted ideas. Even so, the book might not have included all such ideas.

Given the dominance of empirical research in economics research, the author has had to expose the fallacies of the approach. Empiricalization has been embraced by both orthodox and heterodox economists. For orthodox economists, empirical studies are in line with the belief that theories have to be examined and supported by data. For heterodox economists, the empirical approach fits their realism. It is common wisdom that the current economic discipline is an empirical one—the vast majority of economists are using empirical data to calibrate or estimate economic models, simulate economic performance and provide economic forecasts. There is nothing wrong in doing empirical research. In fact, every theory has to be examined in the end by empirical data or reality. However, empirical economic research faces a big problem: unlike most research in natural science, the data used in the social sciences are not scientific experimental data.

Most natural scientists have the luxury of doing experiments under the same conditions, so the accurately measured experimental data can reflect the impact of the variable of interest. Consequently, empirical studies based on scientific experimental data are reliable. Economists, on the other hand, have to rely on detailed microeconomic data or aggregated macroeconomic data which are obtained under remarkably different conditions. As a result, many factors (including some unknown) have influenced the data. Since numerous factors are involved in economic data and the relationships between factors are complex, it is not surprising then that many empirical economic studies have produced conflicting results. This indicates that empirical economic results may be inaccurate and can only be indicative. Econometricians claim that, by introducing a random disturbance into an economic model, they can model economic time series data to provide reasonable projections into the future. The fact that no econometrician foresaw the global financial crisis is a clear rejection of this claim. The truth is that a statistical or econometric model on economic time series data does not satisfy the condition for random experiments—a precondition for this type of approach. Due to the complexity of data, economists should be extremely careful in using modelling approaches and in interpreting econometric modelling results.

Compared with empirical studies, theoretical research is largely neglected in current economics disciplines. This is evidenced by the dominance of empirical research in economic journals. However, the importance and impact of an innovative theory cannot be emphasized enough. This has been demonstrated by numerous examples such as Adam Smith's theory of competitive markets, Charles Darwin's theory of evolution and the separation of the power system proposed by Baron de Montesquieu. Regarding the relationship between empirical and theoretical research, the great examples can be found in physics, for example, Kepler's law of planetary motion and Newton's gravitation law. Kepler's empirical study described the movement of planets in the solar system very well, but it was Newton's gravitation theory that changed our world view, and eventually changed our world dramatically. Based on basic observation and logical reasoning, theories sometimes need to be proposed and subsequently examined by empirical studies later. A typical example is Einstein's theory of relativity. In short, although there are different opinions on the methodology of scientific research, all would agree that empirical studies are necessary but they need to progress to theories, which play a central role for improving understanding and in solving problems. In contrast to the current empirical or statistical approach in economics which tends to belittle logic and theory, the author of this book calls for the return to logical economics.

Some readers may still be in favour of econometrics and hold objection to logical economics. One may argue that, although the econometric approach entails various problems, the approach is at least backed by data. How can the logical approach do better if it does not rely on empirical support? One may further argue that, the return to logical economics is actually a return to the pre-econometric era and that historical regression is simply implausible.

However, the above reasoning is superficial because it simplifies the relationship between the truth (mechanism) and data. Data contain truth or mechanism but data themselves cannot speak. Raw data can be misinterpreted by flawed statistical methods and thus lead to incorrect conclusions. There are numerous examples of this. The typical ones are spurious correlations, which have already been recognized by an econometrician, such as the high correlation between ice cream consumption data and crime data, between rainfall data and GDP data, and between women's ovulatory cycle and their voting preferences. On the other hand, logical economics does not conflict with empirical data. It emphasizes reliable statistical methods and logical reasoning to uncover the main mechanism buried in raw data. As a result, the return to logical economics is not a simple regression to the pre-econometrics era, rather, it is an objection to the flawed methods and misleading interpretation of data in econometrics.

Upon the publication of this book, the author wishes to express his thanks to many people for their direct or indirect contribution. At the risk of incurring criticism, here the author would like to, and should, acknowledge the influence of the economic thought of giants in history, especially Adam Smith, Karl Marx, John Keynes and Joseph Schumpeter. This book essentially extends Smith's theory of the invisible hand to a new market—the patent or innovation market, which is a necessary outcome of the advanced development of civilization. The invisible hand works well only after the necessary legal framework is established. This book demonstrates how to establish a new patent system which enables the invisible hand to promote innovations. The book also explains why it is necessary to have a well-functioning patent market. The thoughts of Marx, Keynes and Schumpeter influenced this book indirectly. These giants have responded to repeated occurrence of economic recessions by proposing different theories and solutions such as communism, government intervention and the innovation by entrepreneurs. Although history has shown that these theories and solutions failed, they are invaluable attempts based on innovative thinking. These attempts are the midwives for the ideas presented in this book. If my solution is proven to be right or is of any use to society, the contribution of the past giants in economics can never be emphasized enough. Without their efforts, I might still be exploring the solutions they believed in.

The author has discussed with and obtained comments from his colleagues, students and friends, including Benjamin Mitra-Kahn, John Qiggin, Malcolm Treadgold, Peter Ellston, Judith McNeill, Mahinda Siriwardana, George Chen, Shawn Leu, Nam Hoang, Callian Fellows, Frank Stilwell and Warren Halloway. Their comments are greatly appreciated. The author would also like to thank the publisher, Palgrave and Macmillan, especially Rachel Sangster and Joseph Johnson, for their patience and excellent publishing services. Last, but most importantly, the author must acknowledge the support from my core and extended family. I thank my wife for her understanding and for her tolerance of my almost workaholic lifestyle, and thank my brothers, sisters and cousins for their emotional and financial support of my overseas study. I am indebted to my parents for their wholehearted support of my pursuit of an ambitious and unconventional goal, even though they did not understand exactly what my goal was (but they trusted me that it was for the good of people), and even though they knew my pursuit meant years and years of separation from them.

Brisbane, Australia                                    Samuel Meng

# Contents

# List of Figures

# List of Tables

# Abstract

Economic recessions have haunted mankind for a long time. While the Great Depression from 1929 to 1939 has become a remote memory, the recent global financial crisis serves as a powerful reminder of the devastating consequence of a global recession. What causes recessions? What would be the best cure? What are the implications for society? The answers provided by Orthodox economists, Keynesian economists, Marxists and the institutionalists are not foolproof. Moreover, although most people have realized the important contribution of technological progress to economic growth, nobody fully understands how technology impacts economic growth and how to speed up technological progress. This book demonstrates that technology affects economic growth from both the demand and supply sides, and that a new patent system can prevent economic stagnation and bring us a step closer to a knowledge economy.

The book begins with a brief review of the history of economics, then questions and rejects the trend of econometricalization or empiricalization in recent decades—economists' complacency with empirical relationships between economic variables. By reviewing the different schools of economic thought and by scrutinizing the limitations of existing theories on business cycles and economic growth, the author forms a new theory to explain cyclic economic growth. According to this theory, innovation scarcity leads to saturation of demand, stagnation of sales and decline in investment. These in turn lead to economic recessions, featuring widespread unemployment and unutilized capital. Firms in recessions are waiting for, or more engaged in, innovation. Successful innovations will eventually lift the economy out of recessions and into economic booms.

This book also demonstrates that the scarcity of innovation results from the flawed design of the patent system. The author suggests a new design for the patent system and envisions that the new design will bring about large economic and societal changes. Under a new patent system, the synergy of the patent and capital markets would ensure that economic recessions can be avoided and that the economy would grow at the highest rate possible. The new patent system would also make inventors the richest class and thus transform our society from capitalism to patentism.

# 1

# Patents and Economics

Both patents and economics have a long history. Documented economic thought in Western civilization can be dated back to Ancient Greek times when Aristotle (384–322 BC) completed his books 'Politics' and 'Ethics'. In eastern culture, economic thought can be dated even further back to the period of Spring-Autumn and Warring States in China, during which a political-economic book 'Guanzhi' was written by Guanzhong (725–645 BC). The history of patents is somewhat shorter. The word 'patent' comes from the Latin word 'litterae patentes', meaning 'open letter'. Patents were initially used by medieval European monarchs to sell monopoly rights over the trade of specific commodities. The use of patents to encourage invention can be dated back to 1474, when the Venetian state government established a statute on patents. The principle that patents should be given only to inventors was laid down by Francis Bacon in 1602 and enacted by the British Parliament in 1623.

© The Author(s) 2019
S. Meng, *Patentism Replacing Capitalism*,
https://doi.org/10.1007/978-3-030-12247-8_1

## 1.1   Linkage Between Patents and Economics

Patents and economics might seem totally unrelated, but they are linked to each other through innovations or inventions (generally speaking, inventions mean significant innovations. However, since they are of the same nature, we use them interchangeably in this book). Nowadays, most people acknowledge the importance of innovations in an economy, but very few realize that innovations are the ultimate source of economic growth. The increased productivity due to process innovations (innovations improving production processes) has been widely studied and confirmed by economists, but little attention has been paid to the role of product innovations (innovations creating new products) in an economy. The importance of product innovations is related to varieties of commodities and thus to consumption ceilings. These ceilings stem from the fact that, although the desire of a human being is unlimited, one's ability to consume one specific type of good or service is limited. For example, a person can drink only a limited amount of beer a day and no one can listen to the same music all day every day (even for a teenage!). Hence, without product innovation, the limited varieties of goods and the limit on consumption of each good lead to a consumption ceiling. As household consumption approaches this ceiling and the economy will stagnate (Some economists might disagree on this. We will discuss this later).

Since innovation is vital to economic growth, the question we need to ask is: do we naturally have enough innovation? The answer is negative for two reasons. One is that innovations are extremely costly and have a high chance of failure. Most innovations need funding and time in order to do experiments and research. More importantly, innovations are full of uncertainty. By definition, innovation means trials or attempts in the hope to come up with some new products or solutions. The innovators have no idea if their attempts would work. Since innovation activity is a trial-and-error process, innovations have a high chance of failing, with many useful innovations succeeding only after numerous failures. Edison's invention of carbon-filament electrical bulb is a good example. The other reason that imitation deters innovation. Generally speaking, innovation takes time and is costly, but imitation of an innovation is generally much easier. This means that innovators spend a lot of their energy and money in creating new products for society, but they get very little reward for their creativity due to imitation. Consequently, people shy away from innovation activities. In short, the high risk of innovation failure and low return on successful innovations greatly deter innovation activity.

There are many ways to encourage innovation and scientific discovery. One is prize money. It is well known that the Nobel Prize has a significant impact on scientific discovery. Other prizes in different countries also have played an important role. For example, the invention of marine clocks was largely the outcome of the Longitude Prize offered by the British government for an invention which could easily determine the longitude of the ship during its long-haul travel at sea. A second way to encourage innovation is for the government to set up or fund research/innovation organizations to conduct innovation activities (i.e. the public produce option). A third method is for the government to provide some funding to private innovation firms in order to reduce their innovation costs. Each method has its own limitations. For example, the prize money can reward only a limited number of innovators (in other words, there is a very low chance for innovators to get the prize, so it is not an incentive for most inventors), the public produce suffers from the shortcoming that government-funded institutes are insensitive to the market potential of their innovation, and the government-subsidized private produce is also inefficient because it is hard (if not impossible) for the government to monitor the performance of the private firms due to information asymmetry.

In summary, these methods have three limitations. First, all methods require extra funding to speed up innovation. Second, all methods need to be carefully administered and thus involve high administrative costs. Last but not least, the outcome of encouraging innovations is not ideal. Since there is no guarantee that these methods reward innovators according to their contribution to the economy, innovation funding may not be used efficiently to achieve the best outcome for society.

Compared with the above methods, patents are an ideal way to stimulate innovation. The method requires no extra funding because the rewards come from the sales of the innovated products or from the patent transactions; it requires minimum administration; it can reward many inventors and can reward them appropriately according to the contribution of their inventions. However, patents also have their downsides. They can cause social welfare losses due to patent monopoly power. Nevertheless, the net social benefit of patents is far greater than the net gains from other methods of stimulating innovation, so patents can and should be used as an ideal and primary tool to stimulate innovation (some might disagree. We will have more discussion on this shortly).

The above is a basic or simplified explanation of the links between patents and economics. Chapters 2–5 are mainly on economics, but they form the fundamental argument that innovations and thus patents are crucial for economic growth. Built on this foundation, Chapters 6 and 7 discuss the patent system in detail, suggest a thorough reform and project the impact of a new patent system. However, the reality is much more complex and this has generated seemingly endless and heated arguments both in economics and in patent studies. To a large extent, this book is a response to these arguments and is an effort to expose the essence of issue and to find a solution.

## 1.2   Cutting Through Economics Jungle

Economists' long-lasting arguments on business cycles and economic growth have generated a thick jungle in which the role of innovation is buried. Classical, neoclassical and new classical economists believe that market mechanism works well to achieve supply/demand equilibrium and that the economy is always in full employment excluding the very short periods of disequilibrium. The repeatedly occurring economic recessions lasting many months or even years and the associated large number of unemployed workers are a powerful rejection to the claims by those economists. Now many economists are converted to econometricians or empiricists. They think economic growth (or everything in the world) follows statistical laws, thus they enjoy playing with statistical models and are uninterested in the nature and the real causes of economic issues. From economic recessions and stagnation, heterodox economists see problems in a market economy, but they attribute these problems to various relevant but non-essential factors. Underconsumptionists believe that the shortage of consumption is the cause of economic recessions, so they advocate luxury spending as the way to promote economic growth. Keynesian economists blame the deficiency of effective demand, which stems from uncertainty and the lack of 'animal spirits' (according to Keynes and post-Keynesians), from nominal wage rigidity, liquidity trap and inelastic investment demand (according to Keynes and orthodox Keynesians), or from nominal and real wage/price rigidities due to imperfection of markets (according to new Keynesians). Monetarists and Austrian School economists attribute the cause of economic recessions to speculative activities and credit cycles. Marxists, socialists and institutionalists see the capitalist institution as the ultimate source of problems in a market economy. Schumpeterians highlight the importance of innovations but deplore the disappearance of entrepreneurship.

Orthodox economists' view that economic recessions are short periods of disequilibrium appeals to no one but themselves. However, it seems that econometrics has gained popularity among orthodox and heterodox economists and among policy makers. Although there are only a limited number of published papers criticizing the trend of empiricism, econometricians' blind faith in statistics was questioned unopenly by many economists through their disbelief of statistical modelling results. In response to orthodox economists' difficulties in explaining the reality and to econometricians' downplaying or even discarding logical power, Chapter 2 of this book emphasizes the importance of logic demonstrated in the history of political economy. Chapter 3 discusses statistical theory and practice. It reveals that probability law is only a law of ignorance and contains neither mechanism nor causality. In the case where the condition for random experiments is not satisfied, probability law is not applicable. In this case, the statistical modelling results are, strictly speaking, not valid. At best, these results can be used only as a rough guide for research and other practice.

Heterodox economists reveal various relevant factors causing economic problems, but they have failed to uncover the key factor underpinning economic growth and business cycles. Uncertainty and the lack of animal spirits can explain economic recessions and stagnation, but they also cause difficulties in explaining the existence of long periods of economic growth in economic history. Rigid wages, sticky prices, and other market imperfections may exist in an economy, so new Keynesian economists have provided microeconomic foundations for Keynesian theory. However, these market imperfections should be a minor aspect of a market economy because the market mechanism has been proven to be an efficient way to allocate resources. Any promotion of non-market approaches (e.g. planned or command economy) is rejected by the failure and replacement of non-market economies in recent history. Starting from Kondratiev (1922) and Schumpeter (1939), more and more economists have realized the contribution of innovations to economic development, but very few of them have understood the mechanism by which innovations affect the economy. Chapter 4 and the first half of Chapter 5 of this book review various economic thoughts and discuss their deficiencies.

In rejecting the existing economic theories on business cycles and economic growth, the second half of Chapter 5 aims to establish a new theory which can reveal the key role of innovations in the economy. The new theory starts with three simple axioms: (1) everyone has a satiety point in consuming any type of good or service; (2) saving acts as both precautionary premium and saved resources, so it can bring satisfaction not only in the future when the saved resources are consumed, but also at the current time through the sense of security obtained. (3) the expected future consumption is the key element of profitability and thus is the base for investment decision. Chapter 5 discusses the implications of these axioms and then explains the vital role of product innovation in sustaining economic growth. The chapter also employs different versions of mathematical models featured by the three axioms. Both the static and dynamic models show that without product innovation, the economy will inevitably go into recession or stagnation, and that adequate and timely product innovations can achieve fast and smooth economic growth.

The above three axioms and the modelling results are easily comprehensible and supported by daily experience, so there is no need to collect trivial evidence to prove this common-sense reasoning. However, the current fashion of empiricism in economics indicates that economists would not accept any common-sense view without an empirical proof. Considering this, at the end of Chapter 5, the author includes some macro data and uses statistical estimations to examine the axioms and the conclusions derived from the theoretical models. The statistical models used are very simple ones considering the inaccurate nature of any statistical models and the large measurement errors in macro data. A simple model can avoid introducing more errors through complex models. Since macroeconomic phenomena are always complex, the empirical results in this book (as in any studies) is only indicative.

## 1.3 Wading Through Patent Turbulence

The arguments on the role of patents in stimulating innovation are passionate ones. The argument that the extremely high prices of patented medicines (e.g. for the treatment of HIV) leave poor patients to die is empathetic. The argument that patents prohibit downstream innovations causes public disquiet about patents. The phenomenon of sleeping patents (some firms purchase patents and shelve them in order to prevent technological progress so as to protect their production using old technology) and the behaviours of patent trolls (people who use patent litigations to threaten many businesses) triggered widespread public condemnation. On the other hand, some professionals argued that without patent protection, many important innovations including medical innovations would not be available or would come out after a long delay, so society would be much worse off.

All arguments on patents stem from the upside and downside of patents. As we showed earlier, patents can be used as an ideal tool to stimulate innovation. However, the negative side of patents may also be significant. As patent holders are awarded monopoly rights in making, selling and using their patent products, they can use their monopoly rights to command very high prices and limit the quantity of patented products or even refuse to produce or license. This will cause consumers to pay high prices for patented goods and limit the benefits that innovations can bring to society. Facing this side effect of patents, some economists (e.g. Granstrand 1999; Encaoua et al. 2006; Winter 1993; Dutton 1984; Mandeville 1996; Jaffe 2000; Lerner 2002; Bonatti and Comino 2011; Moir 2013) argued that the negative impact of patents is so large that it may outweigh their positive effect on stimulating innovation, so the patent system should be abolished. Some economists (e.g. Scherer 1970; Needham 1975; Mansfield et al. 1981; Levin et al. 1987; Baumol 2002; Pretnar 2003; Aghion et al. 2005; Bessen and Maskin 2007) argued that there are other incentives for innovation (such as professional interest of inventors, first-mover advantage, market pressure on oligopolistic firms, imitation cost.), so they argue that it is not necessary to use patents to stimulate innovation.

The voices against the patent system have become louder and louder since the 1990s. Scherer and Weisburst (1995) and Challu (1995) studied the effect of 1978 Italian legislation which allows patenting of medicine. Sakakibara and Branstetter (2001) examined the effect of the 1988 Japanese patent law reform that increased the patent scope in Japan. They all concluded that patent reform did not lead to an increase in research and development (R&D) spending. Lerner (2002) and Lerner (2009) investigated the impact of the changes in the patent systems in 60 countries over 150 years. He claimed that changes enhancing patent protection increased the propensity to file patent applications (indicated by increased patent applications from foreign inventors) but has little or even negative impact on innovation (indicated by decreased patent applications in the UK by inventors in countries of policy change). He also claimed that enhancing patent protection is less effective when patent protection was already strong. Bessen and Meurer (2008) estimated the benefit and cost of patents and claimed that patents brought about net negative effects to firms. Many argued that the oligopolistic market can generate moderate competition which can stimulate innovations to the highest level so that the patent system is not necessary (e.g. Pretnar 2003; Aghion et al. 2005; Bessen and Maskin 2007). Landes and Posner (2004) and Scherer (2009) raised the issue of the political economy of the patent system and cast doubt on patent reform. Boldrin and Levine (2012) even claimed that the patent system should be abolished because the political economy of the system makes it impossible to accomplish the required patent reform.

Chapter 6 discusses both sides of arguments on patents, but with an emphasis on addressing the arguments against patents. The reasoning supporting patents is quite straightforward. That is, although the patent system has both positive and negative effects, the positive effect should be greater; otherwise, there would not be a patent system in history. Some historical events and data are used to support this claim. People on the other side argue that the claim that patents have a net positive effect has oversimplified reality, and then they put forward various reasons and data to show that patents either have a net negative effect or are totally unnecessary. These anti-patent arguments are put into five categories and are commented here.

First, it is argued that patents have no positive effect on stimulating innovations. The reasoning for this type of argument is that inventors are not motivated by financial gains, but by other incentives, such as their interest in creating new things, the desire to satisfy market demand, and the enjoyment of challenges and successes in doing innovations. People also argue that the disclosure of patent information is useless in disseminating information on patent technology and that patentees receive the benefit of patent but produce no further innovation.

The claim that patents have no positive effect is an extreme type of argument which can be defeated easily. Non-financial incentives such as interests and challenges are important factors motivating innovators, but financial incentives are also essential for innovators, especially in modern times. The fundamental reason is that innovations come out of numerous experiments so they are costly and need financial support. For individual professional innovators, financial gain is important for them to make a living. Some innovators are even mainly motivated by financial incentives. For example, the inventors of the steam engine, the Internet search engine, Facebook, Wechat and Paypal. In modern times, corporates are a big player in innovation activities. For them, financial gains from patents are apparently the most important incentive for conducting or supporting innovation activities. This is evident in their activeness in applying for and purchasing patents. The mechanism of patents to stimulate innovation is very clear. Patents give the innovator a power to obtain profit and this benefit is important for inventors, so they should respond positively. If one agrees with this mechanism, one cannot deny the positive effect of patents.

Second, most opponents of patents admit the positive impact of patents on stimulating innovation, but they argue that the negative effect of patents is greater than the positive effect. Many examples are used to back up this claim, e.g. sleeping patents, patent trolls, excessively high prices of patented products (especially patented medicines), obstacles to competition, obstacles to downstream innovations, obstacles to knowledge diffusion, etc.

There is no dispute on the negative effect of patent monopoly, but these negative effects may not lead to a **net** negative effect. It is hard to prove if patents have net positive or negative effects. Reasoning based on theoretical models can arrive at either conclusion and empirical results on this are mixed. However, patents are more likely to generate net positive social and economic results for three reasons. The first reason is that it can be shown that the negative effect of a patent accounts for only part of the positive effect of the patent product, so if the patent system has some positive effect, it should have a **net** positive effect. If the patent system is responsible for some innovations, i.e. some innovations do not come out (or come out much later) without a patent system, these innovations are of extra benefit brought out by the patent system even though people have to pay a higher price for the benefit. In other words, it is better to have innovations at higher prices than no innovation available. If patent laws can only accelerate innovations for a few years, e.g. 5 years, people can enjoy the patent product 5-year earlier than in the case without a patent system (for simplicity we ignore patent approval time), so the duration of patent protection should at least be 5 years so that innovations will come out 5-year earlier and thus bring more benefit. The second reason is that the sheer fact of the existence of the patent system may indicate that the system has a net positive effect. This is a kind of 'existence is reasonable' argument, but this argument has a point here. People have established the patent system because they have realized the unattractive features of innovations as well as the importance of innovations to society. If it is proven that the patent system has no net positive effect, the system would not have been created or, if it had been created by mistakes, it would have been abolished a long time ago. The existence of the patent system at least shows that one cannot reject the possibility that the system has a net positive effect. The final reason is that some of the negative effects of the patent system may stem from the problematic design of the system. It is common wisdom that there are many problems in patent law and its practice. These problems may cause large negative effects, so it is not necessarily the principle or mechanism of the patent system itself has problem. The problems caused by bad design can be overcome by a thorough patent reform.

Third, some opponents admit that patents may have net positive effects, but they argue that the patent system is unnecessary because there are plenty of other incentives to stimulate innovation, for example, market pressure, first-mover advantage, lead time and the protection of high imitation costs.

The obvious response to this argument is that if patents have net positive effects, it would be beneficial (or at least there would be no harm) to use patents as additional incentives to stimulate innovation. More importantly, the patent system is necessary because there are inadequate innovations in the economy. The scarcity of innovation is evident by the repeated occurrence of economic recessions. This forms the fundamental argument in Chapters 4 and 5 that more effort is urgently needed to stimulate innovations. Following this argument, patents have a great role to play and have major responsibility in sustaining economic growth. People saying patents are unnecessary may also have failed to recognize the features of innovation—high possibility of innovation failure and high possibility of being imitated if an innovation is successful after a number of failures. These features make people shy away from innovation activities and thus extra incentives are needed to change this situation.

Fourth, some claim that the patent system is a corrupted system and is exploited by patent lawyers. For example, Boldrin and Levine (2012) argued that the patent system is designed and operated by interest groups while consumers are excluded, and that the patent system is highly technical and thus it is hardly accessible and understandable by the general public. They further argued that this political economy made the patent system beyond repair.

This argument has some elements of truth, but is an exaggerated claim. It is true that interest groups may have influences on the design and administration of the patent system, but this is a problem common to any other laws and regulations. Patent law is enacted by parliament, operated by the government, and arbitrated by the courts. This is the same for other laws. Three agents are involved in the patent system: the patent holder, the producer and the consumer. The patent holder and producer are represented in the courts by their lawyers and the consumer is indeed absent in the court hearing. However, the benefit to consumer in this case is directly related to the producer: if the producer is obliged to pay patent licence fees, the consumer will pay a higher price for the patented products. Hence, the interest of the consumer is indirectly represented by the producers' lawyer. The highly technical nature of a patent case is indeed a specialty of the patent system which imposes some challenges in operating the system, but this is not unique to the patent system. Other specialized courts dealing with intellectual property rights, military and marriage also face this challenge. Since the public or the consumer is indirectly represented by the producers' lawyers, it is not necessary for the consumer to have full understanding of the technical issues. In short, the issue of political economy raised by some economists is not a unique issue in the patent system; the relevant problems can be overcome through a reform of the patent system and through the dynamics of the different players in the system.

Finally, based on some empirical studies, some claim that the performance of the patent system is poor and thus the system should be abolished. Regarding this claim, one must be conscious that empirical studies on patents are not accurate and show mixed results. Although most empiricists do their best to produce reliable results, the empirical studies on patents are only indicative of two reasons. The first is due to the data issue. Data on patents are not always available so sometimes the researcher has to rely on estimated data to do the estimation. Even if the data are available, the measurement errors in patent data (especially data at the macro-level) may have a significant impact on results. The other reason for inaccurate empirical results is related to modelling methods. Statistical modelling is now popularly used in empirical studies, but the probability law on which statistical models rely requires that the data must satisfy the condition of random experiments. This condition is generally not satisfied in empirical studies, thus the statistical models are, strictly speaking, invalid, or at least have large margins for error.

When there is a controversy or argument, the issue involved is possibly complex and often each side finds it difficult to prove its point of view. Patent controversy is of no exception. On theoretical grounds, patents have both positive and negative effects on the economy and society; no one can prove convincingly if the positive effects outweigh the negative effects or vice versa. In terms of empirical evidence, the effects of patents are hard to be measured accurately, especially at the macroeconomic level. Hence, it is of no surprise that the empirical results are mixed and criticized by each side. Some so-claimed state-of-the-art and scientific empirical studies on patents also have a number of errors or margins for error. To get out of the endless arguments on patents, in Chapter 6 the author suggests a systematical historical approach—to study systematically the impact of patents on one or more industries. The author uses the brilliant work of Bright (1949) on the electrical lighting industry to illustrate this approach.

In considering the theoretical framework for the patent system design, the appendix of Chapter 6 discusses and rejects the balanced approach. The negative effect of a patent monopoly has been recognized by the designers of the patent system. To mitigate the negative effect, a patent law generally awards the patent holder a monopoly right for a limited period of time, i.e. limited patent duration. This design is a compromise between stimulating innovations and disseminating innovation. The longer (shorter) the patent duration, the more (less) reward for the patent holder and thus higher (lower) encouragement to innovate, while on the other hand, there is more (less) restriction on the diffusion of the patent technology. However, starting from Nordhaus (1967), some studies have argued that there is an optimal patent duration which maximizes the **net** social welfare (i.e. the positive effect on stimulating innovation minus the negative effect on restricting patent technology diffusion). Following this 'finding', the balanced approach in designing a patent system is suggested in order to find the optimal patent duration. Chapter 6 reviews these studies and identifies the flaws in them. Through graphic demonstration and mathematical proof, the chapter shows that it is very unlikely that there exists an optimal duration or breadth for patent protection.

After considering existing patent arguments and theories on the patent system design, the author draws the following conclusions in Chapter 6: it is likely that patents have overall positive effects and that some negative impacts of patents result from the issues in the patent system design. The inefficiency in the patent system are largely due to the shortcomings in patent laws resulting from the compromising design. Although there are other ways of stimulating innovation, patents are not only a necessary addition to other stimuli but also should play a major role.

## 1.4   Necessary Patent Reforms

On rejecting the balanced approach, Chapter 6 also proposes necessary reforms to establish a new patent system. These reforms include:

1. prohibiting both exclusive patent licences and patent assignments.
   Both practices give monopoly power to people or entities other than the patent holder, so the practices magnify the negative effect of the patent monopoly while failing to produce significant positive impact on stimulating innovation. These practices also greatly restrict patent demand and cause a thin patent market. Since the negative effect of these practices is substantial while the positive effect is negligible, they should be banned in the new patent system.
2. simplifying and standardizing patent transactions.
   This reform can prevent abuse of patent power such as tie-ins, i.e. the attempt to require the licensee to purchase inputs from the patent holder. It can also prevent the patent holder extending patent monopoly power by his/her adding requirements other than a licence fee or royalty (e.g. restriction on the quantity the licensee can produced, etc.) into the patent licence agreements. With simplified and standardized patent agreements, the current high cost of patent transactions can be reduced remarkably. The reduced transaction cost can also help to increase demand for patents and thus lead to a thicker patent market.

3.  redefining patent rights and confining patent rights to distribution or commercial activities.

The practice of exclusive licences and unstandardized patent agreements directly stems from the definition of patent rights. To ensure these practices are banned, patent rights must be redefined. Considering the negative effect of patent rights on research and on downstream innovation, the patent right should be limited to distribution or commercial activity only. Hence, in the new patent system, patent rights should be defined as the exclusive right of the patent holder to grant a standardized, non-exclusive and non-restrictive licence for distribution and commercial use of the patented technology.

4.  improving the patent quality standard and widening the scope of patent protection.

The necessity to improve the patent quality standard can be illustrated by the one-click patent, which was granted to Amazon by the US patent office and subsequently caused a public outcry. This kind of innovation does not satisfy the 'non-obviousness' condition and thus is not a true innovation. Granting patents on this kind of pseudo 'innovation' fails stimulating true innovation but results in a large negative impact on normal business operations. On the other hand, if an innovation can satisfy patent requirements such as usefulness, novelty and non-obviousness, the innovation should be granted a patent regardless of the area of innovation. The widened protection scope can stimulate innovations in more areas and thus bring more benefits to society.

5.  prolonging patent duration infinitely.

There should be no restriction on patent duration considering that it is unlikely to have a trade-off in patent duration and that patents have overall positive effects. This reform is also consistent with property right law if a patent is viewed as the intellectual property of the patent holder. In considering other reforms proposed, prolonging patent duration is feasible and appropriate in the new patent system because the negative effect of patents is greatly reduced due to a ban on exclusive licences and on patent assignments and also due to the simplified and standardized patent transactions. Hence, removing restriction on patent duration can encourage innovation to the greatest degree while having very limited negative effects.

6. abolishing the compulsory licence rule.
   The compulsory licence rule is adopted by many countries to address the problem of the abuse of patent right. The rule allows anyone to apply to the patent authority for a patent licence if the patent has not been used in the 3 years after it was granted. Although this rule is seldom used, it exerts great pressure on patentees in their negotiation of patent licence prices. If the patentee fails to agree on a licence agreement, the rule effectively reduced the patent duration to 3 years and thus acts as an invisible sword hanging over patent protection. In short, the compulsory licence rule greatly restricts the returns to patentees and reduces the power of the patent system in stimulating innovation.
7. enhancing international coordination in patent protection.
   Generally speaking, the country implementing the patent reform first will enjoy an enriched patent technology and thus will have more benefit on its economy. However, without an international implementation of patent reform, the returns to the patent holder will be reduced substantially. Moreover, the producer in the country implementing patent reform will be disadvantaged compared with producers in other countries. To protect the benefit of the patent holder and thus encourage innovation to the greatest degree, it is desirable to enhance international coordination in patent protection.

## 1.5    The Post-capitalist and Post-patentist Eras

If the patent system is reformed according to the above suggestions, it is expected that the new patent system will bring dramatic changes to the economy as well as to society. Chapter 7 provides a projection on the post-capitalist economy and society (after a new patent system is established) and further on the post-patentist economy and society (after a discoverer's right law is established). Projection into far future is always a risky business, especially in social science. However, scientific projection is valuable. It can also be used as a test to prove or disprove a theory. Sometimes a prediction is proven correct so the theory is accepted. For example, the confirmation of the prediction of the returning time of Halley's Comet led to the final acceptance of Newton's Gravitation laws;

the confirmation of the prediction of gravitational waves by Einstein put his relativity theory more firmly on the ground. If a theory includes the key factors and reveals the correct mechanisms, the projection from the theory should have some elements of truth and thus have the potential to be scientific. It is hoped that the projection provided in Chapter 7, an outcome of the author's confidence, will be examined by economists and others, and will stand the test of time.

# References

Aghion, P., Bloom, N., Blundell, R., Griffith, R., & Howitt, P. (2005). Competition and Innovation: An Inverted-U Relationship. *Quarterly Journal of Economics, 120*(2), 701–728.

Baumol, W. J. (2002). *The Free Market Innovation Machine.* Princeton: Princeton University Press.

Bessen, J., & Meurer, M. (2007). *The Private Costs of Patent Litigation* (Boston University School of Law Working Paper No. 07-08).

Bessen, J., & Meurer, M. (2008). *Patent Failure: How Judges, Bureaucrats, and Lawyers Put Innovations at Risk.* Princeton, NJ: Princeton University Press.

Boldrin, M., and Levine, D. (2012). *The Case Against Patents, Federal Reserve Bank of St. Louis* (Working Paper Series 2012-035A).

Bonatti, L., & Comino, S. (2011). The Inefficiency of Patents when R&D Projects Are Imperfectly Correlated and Imitation Takes Time. *Journal of Institutional and Theoretical Economics, 167*(2), 327–342.

Bright, A. (1949). *The Electric-Lamp Industry: Technological Change and Economic Development from 1800 to 1947.* New York: Macmillan.

Challu, P. (1995). Effects of the Monopolistic Patenting of Medicine in Italy Since 1978. *International Journal of Technology Management, 10*(2), 237–251.

Dutton, H. (1984). *The Patent System and Inventive Activity During the Industrial Revolution 1750–1852.* Dover, NH: Manchester University Press.

Encaoua, D., Guellec, D., & Martinez, C. (2006). Patent Systems for Encouraging Innovation: Lessons from Economic Analysis. *Research Policy, 35*(9), 1423–1440.

Granstrand, O. (1999). *The Economics and Management of Intellectual Property: Towards Intellectual Capitalism.* Northampton: Edward Elgar.

Jaffe, A. (2000). The US Patent System in Transition: Policy Innovation and the Innovation Process. *Research Policy, 29*(4–5), 531–557.

Kondratiev, N. D. (1922). *The World Economy and Its Conjunctures During and After the War (in Russian)*. Moscow: International Kondratieff Foundation.

Landes, W., & Posner, R. (2004). *The Political Economy of Intellectual Property Law*. AEI-Brookings Joint Center for Regulatory Studies, Washington D.C.

Lerner, J. (2002). 150 Years of Patent Protection. *American Economic Association Papers and Proceedings, 92*(2), 221–225.

Lerner, J. (2009). The Empirical Impact of Intellectual Property Rights on Innovation: Puzzles and Clues. *American Economic Review Papers & Proceedings, 99*(2), 343–348.

Levin, R., Klevorick, A., Nelson, R., & Winter, S. (1987). Appropriating the Returns from Industrial Research and Development. *Brookings Papers on Economic Activity, Special Issue on Microeconomics, 18*(3), 783–831.

Mandeville, T. (1996). *Understanding Novelty: Information, Technological Change, and the Patent System*. Norwood, NJ: Ablex.

Mansfield, E., Schwartz, M., & Wagner, S. (1981). Imitation Costs and Patents: An Empirical Study. *The Economic Journal, 91*(364), 907–918.

Moir, H. (2013). *Patent Policy and Innovation: Do Legal Rules Deliver Effective Economic Outcomes*. Cheltenham: Edward Elgar.

Needham, D. (1975). Market Structure and Firms' R&D Behaviour, XXIII. *Journal of Industrial Economics, 23*(4), 241–255.

Nordhaus, W. D. (1967). The Optimal Life of a Patent. Cowles Foundation Discussion paper no. 241, New Haven.

Pretnar, B. (2003). The Economic Impact of Patents in a Knowledge-Based Market Economy. *International Review of Intellectual Property and Competition Law, 34*(3), 887–906.

Sakakibara, M., & Branstetter, L. (2001). Do Stronger Patents Induce More Innovation? Evidence from the 1988 Japanese Patent Law Reforms. *RAND Journal of Economics, 32*(1), 77–100.

Scherer, F. M. (1970). *Industrial Market Structure and Economic Performance*. Chicago: Rand McNally.

Scherer, F. M. (2009). The Political Economy of Patent Policy Reform. *Journal of Telecommunication and High Technology, 7*, 167–216.

Scherer, F. M., & Weisburst, S. (1995). Economic Effects of Strengthening Pharmaceutical Patent Protection in Italy. *International Review of Industrial Property and Copyright Law, 26*(6), 1009–1024.

Schumpeter, J. (1939). *Business Cycles: A Theoretical, Historical, and Statistical Analysis of the Capitalist Process*. New York and London: McGraw-Hill Book Company Inc.

Winter, S. G. (1993). Patents and Welfare in an Evolutionary Model. *Industrial and Corporate Change, 2*(2), 211–231.

# 2

# Logic, Politics and Economics—A Brief History of Political Economy

Through reviewing the development of our knowledge in politics and economics, this chapter aims to draw a few conclusions from history and to shed light on the future of economics. Another purpose of the chapter is to demonstrate the importance of logic, which has been highlighted by human history but is partially denied or downplayed by the statistical or econometric approach.

The long history of human civilization has generated a rich knowledge base regarding political economy. Here the author uses the term political economy in a broad sense—it includes philosophical, political and economic thought relevant to a society. Since the purpose of the chapter is to draw lessons from history, only the most important and relevant schools of thought are discussed here. For readers interested in the history of economic thought but with only limited time, the author suggests the book by Landreth and Colander (2002). If you have enough time to go deeply into the development of economic thought, the History of Economic Analysis by Schumpeter (1986) can give much insight into various schools of economic thought as well as relevant historical economic environment.

© The Author(s) 2019
S. Meng, *Patentism Replacing Capitalism*,
https://doi.org/10.1007/978-3-030-12247-8_2

## 2.1   Formulation of Ideas from Practice

Political and economic ideas came from social economic requirement and practice. Ancient civilizations generated invaluable ideas which formed the foundation for the development of economic and political thoughts. Of these ideas, Chinese thought is the oldest recorded thought, while Greek thought had a substantial influence on modern economics in Western countries.

### 2.1.1  Chinese Thought

Human civilization has a long history. For example, the Egyptian civilization started from about thirty-first century BC, Chinese twenty-ninth century BC, Assyria twenty-fifth century BC, Babylonian twenty-third century BC, Greek twelfth century BC and the Israel/Hebrew state about tenth century BC. These civilizations around the world must have generated many political and economic thoughts from a very early stage. However, the written evidence of these thoughts is hard to find. Two exceptions are the Chinese and the Greek. The Greek political and economic thought as the foundation of Western thought will be discussed in the next section. The Chinese thought is discussed here as representative of non-Western political and economic thought.

During the Eastern Zhou Dynasty (772–256 BC) in China, the central power was very weak so the country was ruled by numerous states. During the Spring and Autumn period (770–476 BC), these states fought with each other and reduced the number of states to seven. These seven states were at war with each other during the period of Warring States (475–221 BC) until the establishment of the Qin Dynasty. The conflict between states generated high demand for ideas which could increase the power and wealth of states, so the period of Spring and Autumn and Warring States saw the flourishing of political and economic ideas.

The book *Guan Zi* by Guan Zhong (725–645 BC) recognized the desire for happiness and avoidance of pain in human nature and emphasized the importance of economic policies in managing a state and in increasing the wealth of a nation. The thought of Daoism initiated by Lao Zi (fourth century BC) suggested human activity should follow natural laws. Confucianism put forward by Confucius (551–478 BC) and further developed by Mencius (372–288 BC) heightened the importance of ethics in ruling a state. It admitted that pursuing wealth is a natural motive, but any profits must be obtained in an ethical way. The thought of legalism started by Shang Yang (390–338 BC) and formalized by Han Fei (280–233 BC) promoted the use of standards and punishment for ruling a state. It also regarded agriculture as being the only source of power of a state. The book *Xun Zi* by Xun Kuang (third century BC) emphasized the importance of thrift and of increasing production. The book *Mo Zi* by Mo Di (470–391 BC) advocated thrift and love. These different schools of thought were generally in a form of blended philosophy, ethics, politics, economics and management. However, the book *Guan Zi* illustrated a significant amount of economic thought, so we discuss it in more detail.

Guan suggested that the government managed its people by working with human nature. He said: whenever men see profit, they cannot help but chase after it and whenever they see harm, they cannot help but run away. If the government controls the presence of wealth wisely, people will be naturally content. Without pushing them, they go; without pulling them, they come (Guan 1998). Here Guan actually suggested a way of managing the economy through market forces.

The centre of the economic thought of *Guan Zi* was its light/heavy theory: when a good is abundant, it becomes light (cheaper); when it is locked away, it becomes heavy (dearer). This was essentially a primitive form of supply/demand theory. Using modern economics jargon, the light/heavy theory can be explained by supply and demand curves: with a downward-sloping demand curve, a right shift of the supply curve (an increase in supply or more abundance of the good) would cause a fall in price, and vice versa. Guan used his theory to solve the problem of high prices of grain, animal skins and horns.

Guan was the chief minister of the state Qi. The king of Qi asked Guan: 'the prices of grain are very high. I want to provide some relief to the families of sacrificed soldiers, but we do not have much money. What can we do?' Guan suggested that the King should have a meeting with the rich and powerful to solve this problem. The King followed Guan's suggestion and said to the rich: 'it is urgent to have enough grain reserve for the safety of the city in case the city is besieged. If you have grain stored at home, do not dispose of it by yourself. I will purchase it from you at a fair price'. In this way, the King increased the grain reserve dramatically and so he gave some grain to the poor families and the families of the dead soldiers.

On another occasion, the King asked Guan: 'the prices of animal skins and horns are very high probably due to taxes. How can we reduce the price of these items without affecting tax revenue?' Guan suggested that the King employ people to dig deep channels and ponds on the main roads in the country and to increase the height of the bridges. The King was puzzled. Guan explained: 'when the roads are flat, the carts and carriages can travel easily. High bridges and deep pools make travel difficult, so the animals used to pull the carts and carriages such as horses, cattle and donkeys, become exhausted and are likely to die early or even die on the road. As a result, the price of these animals will increase substantially and this will attract supplies from other states. The dead animals on the roads will decrease the price of animal skins and horns'. One year after the King adopted Guan' suggestion, the price of the animal skins and horns halved. The King also reduced the tax rate on these items but the tax revenue increased because of a larger tax base thanks to increased supply.

Guan even used the light/heavy theory to conquer two neighbouring states Lu and Liang without a military action. Based on the fact that Lu and Liang were very good at producing silk clothing, Guan first forbade all silk clothing production in his state of Qi. He then suggested to the King to wear silk clothes and he also ordered all government officials to wear clothes made of silk. Silk clothes became a fashion in the state of Qi—even ordinary people want to have them. The price of silk clothing increased dramatically. Guan also encouraged businessmen to import silk clothing and this caused a silk boom in Lu and Liang. These two states shifted to producing silk and abandoned their grain production. After three years, Guan asked the King to wear clothes made of cotton and also ordered government officials to do so, he also forbade grain trading with Lu and Liang. As a result, the price of silk tumbled and the price of grain skyrocketed in these two states. People could not bear the high prices of grain and fled to Qi. With a much-reduced population and thus reduced tax income, the Kings of Lu and Liang could not finance and control their states and had to surrender to Qi.

## 2.1.2  Greek Thought

For any community or society, behavioural standards—morals and laws—are necessary binding elements. Initially, these standards were held up by conventions, which may have been deemed the result of divine sanctions of religions. However, during the fifth century BC, religions in ancient Greece were undermined by atheists and Sophists. Atheists claimed that gods did not exist. Sophists believed that it was impossible to obtain absolute knowledge so it was unknown to the human being whether gods existed or not. For Sophists, truth was anything one believed or could be persuaded to believe, so everything, including justice, was relative, individual and temporary.

Primitive democracy was a feature of ancient Greek society: everyone had a chance to have a political position because of the rotating system that run the state. Consequently, there was a great need for speech training. Sophists found a great market. They practised rhetoric and taught the art of persuasion. Sophists proposed that laws were social contracts between the state and individuals; however, the conditions of the contract were subject to interpretation so the attitudes towards laws were quite different among Sophists. Protagoras (c. 481–411 BC) believed that laws were necessary for the state to protect individuals so he urged submission to the laws. Thrasymachus (c. 459–400 BC) thought that the ruling powers made laws for their own benefit so justice was nothing but the interest of the stronger. In other words, law and justice actually meant injustice. Lycophron (c. 320–280 BC) regarded laws as a means of guaranteeing an individual's rights against his fellow citizens but the laws had no concern with morality. Glaucon (c. 445–after 339 BC) thought justice was never practised from choice but from the fear of suffering from punishment, so he argued that what matters was to appear just and lawful rather than in fact to be just. In the view of

Sophists, it was fine to pay lip-service to justice or to disobey the law as long as one was not caught.

Socrates (470–399 BC) disagreed with Sophists. He thought virtue was knowledge, arguing that, if a person truly had good knowledge, he/she would act morally. Socrates also regarded law as a social contract based on the belief that a contract is necessary to hold the society together. So he advocated obeying the law even if it were wrong. When Socrates was on death row for political reasons, his friends tried to rescue him from prison. However, he chose to be executed, saying: 'Do you think a city can go on existing, and avoid being turned upside-down, if its judgements are to have no force but are to be made null and void by private individuals?' Socrates' thought on the city state was further developed by Plato and Aristotle.

Plato (427–347 BC) believed that the soul belonged to the ideal eternal world outside space and time and that the best and healthiest condition of the soul depended on the presence of order. Plato argued that, since the performance of a function depended on structure, it was necessary for parts to be subordinated to the whole. These beliefs formed the foundation of his design of the perfect stationary city state. In the perfect state depicted in his book *The Republic*, activities were strictly regulated. The governing class used their intellectual power to plan and direct the policy of the city, the warriors or soldier class used their courage to defend the city, and the rest (farmers, artisans, etc.) provided for the economic needs of the city. Plato had a view that material greed of politicians was the worst evil of political life, so economic activities and private properties were restricted from the governing class. Class division in Plato's book might have reflected the concept of division of labour—allowing everyone to specialize in whatever he/she did best. Considering that a domestic currency would be useless abroad, Plato also regarded money as a symbol or token devised for the purpose of facilitating transactions.

Aristotle (384–322 BC) shared Plato's view that reality lay in form or structure, but the ideal spiritual world of Plato was replaced with the perfection of god in Aristotle's book *Politics*. Aristotle also introduced the concepts of 'immanent form' and 'potentiality', which allowed the material world to make ordered progress towards perfection without changing certain basic principles or elements. When it came to the everyday life of human beings, Aristotle allowed a departure from his philosophic view by claiming that the aim of ethical study was practical. In his view, man was by nature a political animal and the goal of human life was to seek happiness. The main purpose demonstrated in his two books *Politics* and *Ethics* was to look for the best State which would realize a good life and justice. Different from the perfect city of Plato, Aristotle argued for private property and family over community life. To support the idea of the good and virtuous life, Aristotle was judgemental on Virtue and Vice, rejecting the hedonist idea of pleasure and pain.

A small part of *Politics* (I, 8–11) and *Ethics* (V, 5) concerned economic problems. In these chapters, Aristotle distinguished value in use and value in exchange. He had a concept of just exchange value for every commodity based on the community's evaluation. He condemned monopoly prices as unjust. On the theory of money, Aristotle agreed with Plato that the fundamental function of money is to serve as a medium of exchange in the markets of commodities, i.e. the Cartel theory of money; however, Aristotle further claimed that, to serve as a medium of exchange, money itself must be one of the commodities so money itself had a value. This implied a metallism or metallist theory of money. Aristotle condemned interest or usury as unjust: because money was merely a medium of exchange, there was no justification for money to increase in value when money changed hands.

It is worth mentioning that ancient Greek thoughts on politics and economics were greatly influenced by philosophic development— Plato and Aristotle were the greatest philosophers of their times. It is said that 'many problems in Greek philosophy resulted from a confusion of grammar, logic and metaphysics' (Guthrie 1967, p. 47). The logic issues have been clarified during philosophical debate. One example is the debate of Parmenides' logic net, which still puzzles many today. Aristotle is regarded as the first person to establish a logic system—hence the term logic. Later a Greek Stoic philosopher Chrysippus (c. 280–206 BC) further developed a system of propositional logic. These systems of logic have formed the foundation of valid inference in philosophy, politics, economics and other areas.

### 2.1.3  Arab-Islamic Thought

From the eleventh to fifteenth century, Arab-Islamic scholars translated Greek thought into the Arabic language and examined the economic issues in the context of their religious life. They also translated Greek thought, particularly Aristotle, from the Arabic language into Latin, and this was later used by Scholastics.

Aristotle's economic thought was adapted and developed by Arab-Islamic scholars to fit the Islamic religious world which was governed by the Divine Law. The most important Arab-Islamic scholars were Abu Hamid Al-Ghazali (1058–1111) and Ibn Khaldun (1332–1406). Al-Ghazali described the role of voluntary exchanges and how markets coordinated economic activities. He realized that the advent of currency resulted from economic exchange activities, which in turn stemmed from specialization of production and division of labour. He also examined other economic topics such as interest and usury, public spending and taxation. He designed the optimal ways to levy taxes so as to increase tax base and spread tax burden appropriately. Khaldun noticed the societal change when people moved from a low-income, nomadic desert life to an agricultural-dominated life with higher labour productivity and income. He also examined a wide range of economic topics, including population, profit, price, luxuries and capital formation.

### 2.1.4  Scholasticism

During the thirteenth century, medieval society was dominated by feudalism, i.e. an agricultural society managed through tradition, custom and authority, rather than by markets. However, manufacturing and changing technology led to a slow increase in economic activity and the decline of feudalism. Churches were concerned that increasing economic activity would swing people's minds away from their religious and ethical values, towards materialism. Consequently, the efforts of scholastics were similar to that of Arab-Islamic scholars: adapting Aristotle's economic thought to their religious beliefs.

St. Thomas Aquinas (1225–1274) was an important person to reconcile economic activity with Christian belief.

Since the early Christians believed that communal property was in accord with natural law, it was a struggle to make private property ownership compatible with religious teaching. Aquinas followed Aristotle's idea in approving the regulation of private property by the State. In response to the concerns about Christian values, Aquinas claimed that private property ownership was not a contradiction, but an addition, to the Christian natural law, just like clothing is an addition to a naked body. The ethical aspect of commodity prices was also a major concern to Scholastics. Aquinas regarded a just price as a price to meet the needs of the trading parties, not a price to generate a profit. If an individual anticipated a profit but the motive was for charity, self-support or to contribute to public well-being, then the price was also just and no ethical issues were involved. Aquinas adopted Aristotle's view that interest and usury were unjust, but this view was moderated by later scholastics who accepted the taking of interest in the case of business lending.

## 2.1.5  Mercantilism

Mercantilist economic thought occurred between 1500 and 1750. During this period, self-sufficient Feudalism gradually gave way to manufacturing and trade. The substantially increased international trade generated a large number of merchants. Based on their trade practices, they formed their opinions and published pamphlets advocating the best policy for promoting the power and wealth of the nations. Most important mercantilists were Thomas Mun (1571–1641), William Petty (1623–1687) and Bernard Mandeville (1670–1733).

The opinions of mercantilists were quite diverse, but they shared some common ground. For most of them, the goal of economic activity lay in production rather than in consumption and the power and wealth of the nation was measured in precious metals like gold and silver. They shared the view of scholastics that the total world wealth was fixed so international trade was viewed as a zero-sum game: the gain of one nation was the loss of another. Their policies were generally aimed at increasing the inflow of precious metals or maintaining a favourable trade balance by increasing exports, encouraging production and restraining consumption. Keeping wages as low as possible was also a key policy of mercantilists, which aimed at reducing the cost of production and thus obtained a price advantage over other trading nations.

## 2.1.6 Physiocracy

Although the self-sufficient manor economy was replaced by manufacturing in other European countries around 1750, agriculture was still a key part of the French economy. This formed the background of the short-lived but important intellectual movement of the physiocracy theory in France, led by Francios Quesnay (1694–1774). Physiocrats believed that the operation of the economy was governed by natural laws. By studying the market price, they concluded that allowing individuals to follow their self-interest could generate free competition, which could produce the best prices for society. As a result, the economic policy of Physiocrats is a laissez-faire approach—to leave things alone.

Another important contribution of physiocrats was their insight into the interdependence of different economic activities. At that time northern France introduced advanced technology in agriculture but southern France was relatively underdeveloped. The resulting different levels of wealth prompted Physiocrats to uncover the nature and cause of wealth by analysing the value of inputs and outputs. In 1758, Quesnay produced the Tableau Economique and a circular flow diagram, which showed the transactions in a three-sector society: farmers, landowners, and artisans and servants. Landlords received income (land rent) from farmers and paid farmers and artisans for goods. Farmers sold agricultural goods, received money from landlords and artisans, paid land rent to the landlord and paid artisans for non-agricultural goods. Artisans received money from both landlords and farmers and paid farmers for agricultural goods. Since Quesnay considered manufacturing and other economic activities as 'sterile', he attributed all net products of the society (equivalent to GDP values in modern terms) to land rent.

## 2.2   Contest of Ideas Based on Logical Reasoning

To highlight the importance of logic in economics, this section is organized by contrasting different economic ideas. Since economists may have many different ideas and even one school of thought may have different aspects, an economist or a school of thought may appear in different contests of ideas. For example, mercantilists held the idea that the source of wealth was precious metals while advocating government intervention consequently, international trade; some mercantilists were underconsumptionists; so mercantilists would be involved in defending three types of ideas. For readers who wish to have the whole picture on different schools of economic thought, it is suggested they consult some traditional texts on economic history.

### 2.2.1   Moralists v.s. Underconsumptionists

We are often told that thrift is a virtue and that extravagance is vice. This is the view of moralists. In contrast, underconsumptionists viewed spending as the source of economic growth, so thriftiness is not advisable for an economy. Who is right? For the sake of the economy, what should we do? To shed light on this issue, a detailed investigation into the argument between moralists and underconsumptionists is worthwhile.

Starting from Plato through to Aristotle and Scholastics, moral values had an important influence on political and economic thinking. At the time of mercantilism, moralists had an enormous influence. One of the most important moralists was the third Earl of Shaftesbury, Anthony Ashley Cooper (1671–1713), who had a significant influence on Adam Smith through his teacher, Francis Hutcheson. Moralists believed that the innate goodness of human beings could guide people to distinguish right from wrong and to choose right action. Moralists believed that saving was a virtue and spending a vice. However, this belief was ridiculed by a mercantilist Bernard Mandeville (c. 1670–1733).

In his poem *Fable of the Bees*, Mandeville (1723) teased the moralists by hypothesizing a beehive economy which is driven by vice instead of virtue. In the poem, the moralists persuaded the bees to replace the private vices of prodigality, price and vanity with usual virtues such as modesty and thrift. The result of the virtuous practice was the collapse of the economy because of the shrinking of consumption spending. Through this poem, Mandeville illustrated the importance of consumption for an economy: it is meaningless to produce more products than needed for consumption. Here Mandeville was a mercantilist but also an underconsumptionist.

Mandeville's view was later refuted by moralists. Hutcheson (1750) admitted that a small part of consumption was owing to Vices, but he thought an equal consumption of manufactures, and encouragement of trade may exist without these Vices. Later, Hutcheson and Leechman (1755) continued to refute the importance of luxury consumption:

> By abating of his own expensive splendor could by generous offices to his friends, and by some wise methods of charity to the poor, enable others to live so much better, and make greater consumption than was made formerly by the luxury of the one. (Hutcheson and Leechman 1755, Vol. 2, p. 320)

While Hutcheson refuted the ideas of underconsumptionists based on different types of consumption, the argument by Adam Smith was much deeper because he linked savings to capital accumulation, which he thought was the source of the Wealth of Nations. The central idea of capital accumulation spreads throughout *the Wealth of Nations*, e.g. Adam Smith (1776) said:

> In all countries where there is tolerable security, every man of common understanding will endeavor to employ whatever stock he can command, **in producing either present enjoyment or future profit**...A man must be perfectly crazy who, where there is tolerable security, does not employ all the stock which he commands, whether it be his own or borrowed of other people, in some one or other of those...ways. (Smith 1776, p. 268, bold type added)

Later, Smith claimed that savings (unconsumed goods) are employed as capital immediately:

> That portion of his revenue which a rich man annually spends, is in most cases consumed by idle guests, and menial servants, who leave nothing behind them in return for their consumption. That portion which he annually saves, **as for the sake of profit it is immediately employed as a capital**, is consumed in the same manner, and nearly in the same time too, but by a different set of people. (Smith 1776, p. 321, bold type added)

Smith's expression here can be summarized as the famous theorem—savings are spent as quickly as consumption because no one will hoard money for its own sake. This implies that products are used for either consumption or investment (capital accumulation) so underconsumption is not possible. Smith's argument temporarily silenced underconsumptionists. However, Smith overlooked the link between investment and consumption: the purpose of investment is to make profit through catering for future consumption. If consumption capacity is in question, investment is apparently not a wise decision and thus savings may not be spent immediately as Smith claimed. This unsettled issue sowed the seed for later debate. This thread will be picked up later.

## 2.2.2 Nature of International Trade and Causes of Wealth of Nations

At the time of Mercantilists, the manor economy declined and the nation state was on the rise, so it was important to determine the best policies to consolidate and increase the power and wealth of the nation. Since precious metals like gold and silver were commonly used as money by each nation, mercantilists thought that wealth was measured by precious metals so that the way to increase wealth was to obtain more precious metals. Given the fixed amount of precious metals in the world, the purpose of international trade for a nation was to obtain as much precious metals as possible by encouraging exports and discouraging imports through government policy such as tariffs, quotas, subsidies and taxes. To stimulate exports, the nation needed to encourage production and reduce domestic consumption. Mercantilists also suggested the government cut wages so as to lower the price of exports. As a result, the wealth of nations was based on the poverty of many ordinary people.

The idea of maintaining a favourite or positive trade balance was proven impossible by David Hume (1711–1776), a close friend of Adam Smith. Hume's argument was known as the price-specie flow mechanism: a favourite trade balance meant exports greater than imports and thus a net inflow of precious metals (species) from other countries to the home country. Since the species were used as money, the change in the amount of species implied an increase in the money supply of the home country and a decrease in money supply in other countries. As a result, the prices in the home country rose while the prices in other countries dropped. The price difference would cause both an increase in imports to the home country and a decrease in exports from the home country, so the favourite trade balance could not be maintained.

In fact, Mercantilists' logic in claiming that precious metals were the most valuable and thus were the key source of the wealth of a nation had some elements of truth. In the third century BC, Aristotle had pointed out that money itself had value. Adam Smith also agreed at this point. However, Smith disagreed that money was identical to the wealth of a nation or was the source of wealth. For Smith, the wealth of nations was the goods and services the nation produced and consumed, so he viewed the source of wealth of a nation as the factors determining the capacity of the nation to produce goods and services. One such factor was the productivity of labour and another was the proportion of labour employed productively. He also thought the productivity of labour depended on the specialization or division of labour. Interestingly, Smith viewed the labour employed in producing a commodity as productive while labour employed in producing services as unproductive. He said: 'a man grows rich by employing a multitude of manufacturers; he grows poor by maintaining a multitude of menial servants' (Smith 1776, Book II, p. 112).

Being consistent with his view on the wealth of nations, Smith viewed the role of exports as the means of obtaining imports and, more generally, the role of international trade as a way of utilizing the division of labour. He put forward a trade theory of 'absolute advantage'. If both trading countries specialized in producing different products, e.g. France specialized in wine and England specialized in wool, the cost and thus the price of a specialized product would be lower. By engaging in trade, both countries would benefit. Due to the mutually beneficial nature of international trade, Smith advocated for unregulated international trade. Smith's absolute advantage theory was advanced by David Ricardo (1772–1823) to a 'comparative advantage' theory, which was further advanced by Eli Heckscher (1879–1952) and Bertil Ohlin (1899–1979) to a Heckscher-Ohlin model to reveal the sources of comparative advantage.

## 2.2.3  Government Intervention v.s. Laissez-Faire

Mercantilists advocated government intervention through export subsidies, import tariffs or quotas, and through regulations to keep wages low. Adam Smith reviewed many regulations proposed by mercantilists and found that they had led to less desirable resource allocation than those produced by competitive markets. Based on his reasoning, international trade was mutually beneficial, so he supported free international trade without any government interference.

Adam Smith also extended his argument to domestic trade. According to him, the voluntary and mutually beneficial nature of trade rendered a natural solution, while any government intervention would damage the natural rights and liberties of the individual. He also successfully related market price formation to resource allocation. Here is his reasoning. If the market price of one type of commodity was significantly higher than the cost of production, the producer of this commodity would earn significantly more profit than other producers. Under a competitive market, the higher profit would attract more producers to produce this commodity. As a result, the price of the commodity would drop until it equalled the production cost. This reasoning indicated that, under the competitive market, the self-interest of traders could achieve desirable social outcomes—optimal resource allocations and natural prices in the long run.

Even as an advocate of free markets, Smith did notice the areas where the market might fail to deliver a desirable social outcome. He listed these areas as national defence, building and maintaining roads and school, and administering justice. The services in these areas tended to be under-supplied by private markets because of insufficient profit, so he suggested government provision for these services. He also argued that, for national defence and for protecting infant industries, trade regulation was necessary.

Smith's 'laissez-faire' creed has been adopted by most economists since, but the creed has been challenged by some from time to time. One major successful challenge was done by Keynes (1883–1946). At the tail of the Great Depression, Keynes (1936) published his famous book *General Theory of Employment, Interest, and Money*, in which he argued that savings might not be fully invested because people hoard money or prefer liquid assets (liquidity preference), or because investors are reluctant to take risks due to lack of 'animal spirit'. When savings are not fully invested, there will be oversupply of goods, or deficiency of effective demand, in the economy, so it is necessary for the government to intervene.

The foundation of Smith's laissez-faire approach—the assumption of competitive markets—was often criticized by heterodox economists, especially Marxists, socialists, and institutional and historical critics. With the development of capitalism, many industries became highly concentrated. This changed industry structure certainly offset the market competition and thus might warrant government intervention. Marxists, socialists and institutional critics claimed that market concentration signalled the end of both capitalism and private property right so an institutional change was necessary. The proposed communism or socialism is featured by a central planning government, so it is the highest degree of government intervention.

Although the heterodoxy's rejection of Adam Smith's 'invisible hand' is somewhat radical, their criticism has some elements of truth. Self-interest and social benefit are two conflicting goals, so it is not easy to reconcile them. Market mechanism may establish some links between them but the transformation of self-interest to social benefit cannot be achieved purely through a free market. For example, without a government regulation of minimum age to work in a firm, the use of child labour may be still in practice. Without a workplace safety standard, workers may still suffer from unsafe practice. Without standards on food and drugs, consumers may still be affected by unsafe or unhealthy produce. The self-interest of capitalists means that they only care about their profits so they are indifferent to workers' and consumers' health.

To help the market mechanism to transform self-interest into social benefit, the government has a great role to play. First, the government can lay down laws and rules to foster a free market, e.g. the property right laws, antitrust laws, forbidding violent and menace behaviour. Second, the government can improve the equality of income distribution. In an capitalist society, capital is in the hands of relatively fewer capitalists, which have monopoly power over a large number of workers, so capitalists tend to have more income share. Although the market mechanism can channel the resources efficiently to the right sectors, it does not address the fairness issue. Hence, the government needs to adjust income distribution through taxation policies. Third, the government needs to establish necessary standards and requirements. Without these standards, capitalists' self-interest will cause great harm to workers and lead to a miserable society. Last, the government should have tight regulation on monopolistic sectors. History shows that monopoly always presents in an economy. People with monopoly power definitely will abuse the monopoly power. The recent examples include the bad behaviours in the US financial sectors exposed during the global financial crisis, the bad behaviour in the Australian financial sectors and in retirement villages exposed through the investigations by Royal Commissions. To prevent the abuse of monopoly power, the government must keep a close eye on the behaviours of monopolies.

## 2.2.4 The Puzzle of Value Determination

Why do some goods have very high prices (e.g. diamonds and gold) while other goods are very cheap even though they are very important (e.g. water and salt)? Why are the prices of some goods quite stable while those of other goods change frequently and substantially?

Who plays a key role in determining the value of goods—producers or consumers? Uncovering the value of goods has been a long journey and it has become a key element in economics.

As early as the third century BC, Aristotle differentiated between 'value in use' and 'value in exchange' and had a concept of just value. He claimed the monopoly price was unjust but he was unable to give a definition of 'just value'. This vagueness on value was carried on to Scholastics who regarded a 'just price' as a price meeting the needs of the trading parties. This definition is apparently subject to moral judgement.

Adam Smith further clarified the difference between value in use and value in exchange by employing the diamond-water paradox: diamond has little use (low value in use) but is very expensive (high value in exchange) while water is very useful (high value in use) but is very cheap (low value in exchange). To determine the value in exchange, Adam Smith developed a labour theory of value and a production cost theory of relative prices. His production cost theory stated that the relative prices of commodities were dependent on the payment to production factors such as land, labour and capital. His labour theory of value indicated that the value of a commodity was determined by labour hours (to be precise, this is the theory of labour command in a primitive society. Smith had three labour theories of value: labour-cost theory in a primitive society, labour-command theory in a primitive society, and labour theory in an advanced economy). Smith explained that prices are determined only by the supply or production side because perfect competition in the market will eventually equate the commodity price with the cost of production. He called the short-run prices 'market prices', which could differ from the production cost, and the long-run prices 'natural prices', which would be equal to the production cost.

In the early 1810s, Smith's 'value theory of production cost' was used by protectionist to argue for tariff protection. During the period of the Napoleonic wars, British agriculture was protected from competition from continental Europe by trade barriers such as the wars and tariffs. For example, in 1791, the British Parliament passed the Corn Laws which restricted importation of grains so as to maintain a grain floor price of 50 shillings per quarter hundredweight (English weight unit, 1 quarter hundredweight is about 13 kg). In 1803, the floor price was raised to 63 shillings per quarter hundredweight. After Napoleon was captured in 1813, the landlords and farmers worried about the increased crop imports during peacetime so they went to parliament to ask for an increase of the floor price to 80 shillings per quarter hundredweight. This caused a great public debate, which generated new economic ideas.

Protectionists argued that, based on Smith's theory of production cost, the increase in wages would not necessitate a decrease in the profit rate because the increase in production cost would be offset by the increase in price of grain. They further argued that lowering the tariff on grain would cause a fall in food prices and wages. This would cause a deflation and thus a depression, because grain was a necessity and was widely used in the economy. On the other side, Ricardo argued that high tariffs on grain would lead to, on one hand, the use of low-quality land and, on the other hand, the increase in wages in the agricultural sector. The former would decrease the return of per unit of land and the latter would increase the production cost. As such, the profit rates and thus capital accumulation would decrease. To win this argument, Ricardo discarded Smith's value theory of production cost and tried to find an invariable measure of value or absolute value of a commodity. This led him to adopt Smith's labour theory of value and try to further develop it into a labour theory of the modern economy. He did overcome many problems in measuring the different skills of labour and linking capital goods, land rent and profit to labour, but he was battled by the fact that labour value is not invariable, so he failed to establish a labour theory of value for a modern economy.

Both production cost theory and labour value theory have limitations because they attribute value only to the supply side. In the 1830s, Richard Whately criticized Ricardo's labour theory of value by saying: pearls are valuable not because men have dived for them; on the contrary, men dive for them because they are valuable. This criticism implied that the usefulness of a commodity (value in use) or demand for commodity had something to do with the price of the commodity (value in exchange). By the same reasoning, the production cost theory of value was limited because it could not explain the diamond-water paradox: given the fact that diamonds and water are both produced by nature and involve little production cost, then, why is diamond very expensive while water is very cheap?

Unsatisfied with the production cost theory of value, Menger (1840–1921), Jevons (1835–1882) and Walras (1834–1910) developed a marginal utility theory of value. They claimed that goods have value because they are useful to us, i.e. because they can provide us with satisfaction or happiness or utility. As we consume more of a type of good, the utility of an additional good (marginal utility) decreases. For example, when you eat the first apple, you may feel it is so sweet and fresh (especially if you are hungry and thirsty); as you eat more apples, you will feel the subsequent apples become less sweet and less fresh to you. The marginal utility theory claimed that it is the marginal utility that determines the price of goods. In the case of diamonds, we can have only a very small quantity due to their scarcity. This leads to high marginal utility and thus high prices. On the other hand, water is plentiful. This necessitates a low marginal utility and thus a low price. This marginal utility theory of value marked the beginning of marginal analysis.

Marginal analysis was extended to the supply side to establish the marginal productivity theory by Gossen (1810–1858), Longfied (1802–1884), von Thunen (1783–1850) and Alfred Marshall (1842–1924). Actually, David Ricardo had argued that more use of less-fertile land would lead to a decrease in profit rate, so he had actually implied a decreasing marginal productivity of the land. But this line of research was not pursued until during the marginal revolution. The marginal productivity theory was based on the concept of 'marginal product': the quantity of products produced by an additional factor, for example, the marginal product of labour is the output produced by additional labour input. Similar to marginal utility, marginal product decreases as the amount of the factor used increases. The marginal productivity theory claimed that the price of a good is determined by its marginal cost. This is in conflict with the marginal utility theory.

It was Alfred Marshall who unified both marginal utility and marginal productivity theories and established a framework of supply and demand which can explain the value of a commodity in any situation. In this framework, the value of a commodity is determined by both marginal productivity on the supply side and marginal utility on the demand side. In a supply-demand framework, the production cost theory of value is correct when the demand is assumed to be fixed. In this case, the demand curve is vertical so the price of a commodity is totally determined by the supply curve. On the other hand, when the supply is assumed to be fixed, the marginal utility theory of value is correct because the supply curve in this case is vertical and the price of the commodity is determined solely by the demand curve. The equilibrium price in a supply-demand system determines the value of goods. Since the equilibrium price equals both marginal cost and marginal utility, both marginal utility theory and marginal productivity theory hold. To measure the relative importance of supply and demand on the value of a commodity, Marshall also suggested the concept of price elasticity.

## 2.2.5  Income Distribution and the Fate of Capitalism

One of the important purposes of value determination is to measure the welfare of economic activities because value determination has significant implications for income distribution. For example, a high price of grain may bring more income to landlords and farmers. In reality, income distribution is also affected by social structure or institutions. In a capitalist society, capitalists or producers have a bigger say on how much to pay for workers' wages, while in a socialist country the government plays a greater role in determining the workers' wages. In an economic theory, income distribution also reflects the belief or the preference of the author or advocator. The notion of just value or just price by Aristotle or scholastics is based on the fair benefit to each person involved in trade. For mercantilists, the price of goods can be manipulated by government policies to obtain benefit from international trade at the expense of other countries. The Physiocrat Quesnay (1694–1774) viewed the net product (i.e. added value) as land rental because it came from the productivity of nature. However, Marx believed that only labour can increase the value of products so all added value should belong to workers. Since production activity in reality involves many inputs, the distribution of added value is more complicated than Physiocrats and some Marxists have claimed.

Adam Smith proposed three parts of net value: land rent, wages and profits. The land rent was determined by the monopoly demand of the landlord and by the different levels of productivity of land. Wages were determined by the inventory of goods or capital used by labourers, such as food, clothing and housing. The cost of reproducing labour is assumed as fixed, so the capitalist will reserve a fixed amount of funds to pay wages (wage–fund doctrine). As a result, the wage rate is the wages fund divided by labour force. After deducting land rent and wages from the net product, one can obtain the profits—the payment to the capitalist as an interest return on capital and as a return for risk.

In extending Smith's approach, Ricardo developed a model to determine land rent based on intensive and extensive margins. In the case of agricultural production (e.g. grains), as demand for grains increased, people had to utilize less-fertile land, which caused a decrease in land productivity. The term 'extensive margin' was used to describe the effect of land of different fertilities. On the other hand, the productivity of the land may change as more labour and capital is applied to the same land. The effect of successive doses of capital and labour on a given plot of land is called 'intensive margin'. Ricardo assumed no rent to the last plot of land used and no rent to the last dose of capital and labour involved in production, so the land rental is the sum of all extensive and intensive margins.

The consequence of the income distribution theory of Smith and Ricardo on the fate of capitalism is quite pessimistic because of the decreasing rate of profit over time. Smith reasoned that: (1) accumulation of capital will lead to more competition for labour (i.e. more capital competes for the same amount of labour). This would lead to a rise in wages and a fall in profit; (2) the competition in the commodity market would lead to a fall in commodity prices and thus a fall in profit; and (3) the competition in the investment market for a limited number of investment opportunities would lead directly to a fall in profit. Ricardo disagreed with Smith's reasoning. By combining the wage–fund doctrine with a conclusion from Malthus' population theory that an increase in income will lead to an increase in population, Ricardo argued that, if the competition for labour bid up real wages, according to Malthus's population theory, the rise in labour income would lead to an increase in population. Given the fixed wage fund, the wage would start to fall. In considering an increase both in demand for labour and in population, the wage rate would remain unchanged, so Smith's first concern was dismissed by Recardo. Using the law of markets proposed Say, Ricardo argued that the products resulting from the increased investment can be sold at previous prices, so he concluded that the second and third reasons provided by Smith were not true either. However, Ricardo thought that the land rent would rise because extensive and intensive margins were pushed down as production increased. The increase in land rent would drive down profit and this would also lead to a pessimistic stationary state of zero profit. Economics is a dismal science!

This pessimistic view was considerably improved during the time of James Mill (1773–1836). By that time, the population thesis of Malthus (1766–1834) was rejected, and the prediction of diminishing returns was proven wrong, thanks to the technological improvements in agriculture. James Mill eventually rejected the wage–fund doctrine by saying that labour force and wage rate might not exhaust the fixed amount of the wage fund. He argued that the laws on production are fixed but the laws of distribution are determined by institutions and cultures and thus can be moderated through societal reform. He still held the view of the stationary state, but he thought this might not necessarily be a bad thing because the criteria for a good society was the happiness and well-being of individuals rather than the amount of material goods they possessed.

Mill's view of flexible income distribution theory was challenged by the conclusion of marginal productivity analysis. According to this analysis, the prices of factors (e.g. wages, land rent and capital rent) are equal to their marginal product, so everything is objective and fair. Marshall ended this controversy by pointing out that the conclusion that prices of factors equal to their marginal products do not mean that prices of factors are determined by their marginal products. Rather, factor prices determine the amount of factors to be employed and thus determine marginal products. The prices of factors are ultimately determined by the supply and demand curve, so they are subject to the influence of market forces as well as institutions.

In short, income distribution is necessarily connected to different value judgements which are consistent with different predictions of the fate of capitalism. Physiocrats thought all added values were from the land so a feudalist economy would be the ultimate solution. Classical and neoclassical economists like Smith, Ricardo, Mill and Marshall believed the source of wealth of a nation was capital accumulation, so a stationary state of capitalism was the future of society. Marxists believed that the diminishing return and other internal conflicts in capitalism indicated that capitalism would be overthrown by proletarians and would be replaced by communism, so they adopted the labour theory of value and attributed all added values to labour inputs. Schumpeter placed much value on the entrepreneurs' ability to take risks and introduce innovative products and thought the success of capitalism in capital accumulation would cause the demise of entrepreneurship and thus the end of capitalism itself. Although he was very critical of Marx (1818–1883), regarding Marx as a prophet and regarding Marxist socialism as religion, Schumpeter (1883–1950) speculated that capitalism would be replaced by democratic socialism.

## 2.2.6　Say's Law of Market v.s. Keynes' Excess of Saving

As discussed previously, the issue of underconsumption was popularized by Mandeville in 1723, but this concern was dismissed by Smith's claim that the portion which a person annually saved was immediately employed as a capital so as to obtain profit. With the industrial revolution occurring from about 1760, however, the English production capacity soared and the difficulty in selling products concerned a number of people. Reflecting this concern, Lord James Maitland Lauderdale (1759–1839), Jean Charles Sismondi (1773–1842) and Malthus (1766–1834) claimed that an economy might produce more commodities than can be disposed of (Lauderdale 1804; Sismondi 1819; Malthus 1820). In the second edition of Say's treaties, Say acknowledged the concerns of underconsumption: 'It is common to hear adventurers in the different channels of industry assert, that their difficulty lies not in the production, but in the disposal of commodities' (Say 1821, I XV, 1). However, Say dismissed the concern by proposing a 'Say's law of market'.

Say's argument was centred on purchasing power. The implicit expression of the law of market appeared in the first edition of Say's *Treatise*, Chapter 22, of markets.

> Imagine a very industrious individual having everything he needs to produce things: both ability and capital; if he were the only industrious person in a population which, aside from a few coarse foods, does not know how to make anything; what could he do with his products? He will purchase the quantity of rough food necessary to satisfy his needs. But what can he do with the residue? Nothing. **But if the outputs of the country begin to multiply and grow more varied, then all of his produce can find a use, that is to say, it can be exchanged for things which he needs or for additional luxuries he can enjoy, or for the accumulation of the stocks that he considers appropriate.** (Say 1803, Vol. I, Book, 1, pp. 152–153, bold type added)

From this extract, it is clear that Say's idea was that, if the other people do not have enough ability to purchase, the individual will have unsold products (residue), but as an economy grows, the output (and thus purchasing power) of the other people increases, so the individual can sell his/her products. This idea is repeated in other chapters, for example, in Chapter 5, Say (1803, Vol. II, Book 4, p. 175) said:

> In order to consume it is necessary to purchase; now, one can make purchases only with what one has produced. **Is the quantity of outputs demanded consequently determined by the quantity of products created? Without any doubt.** Everyone can, at his pleasure, consume what he has produced; or else he can buy another product with his own. **The demand for products in general is therefore always equal to the sum of the products available.** (bold type added)

The definition of demand is crucial for Say's proposal that demand is determined by supply. In the first sentence, Say simply equalled these terms with 'the power to demand for outputs or products' or 'potential demand capacity'. This logic subtlety caused the long controversy on Say's law. If one defines demand as the 'will or desire to demand', Say's law is an absurd assertion, but if one defines demand as the 'power to demand', Say's law holds. Since the generally accepted meaning of 'demand' is the 'will or desire to demand', without a special definition for 'demand', Say's law is a false statement (the supporters of Say's law often change the meaning of 'demand'. This results in inconsistency and thus controversies). Facing attacks concerning his Law of Market, in a letter to Malthus, Say reformulated the concept of production as 'the production of things, the price of which will cover cost'. This statement implies that all products can be sold at production cost, so Say effectively excluded overproduction by his definition of production and made his Law of Market inapplicable to any economic activities.

Say's law triggered a long debate between Ricardo, Malthus and James Mill. When defining the meaning of market, Mill (1808, p. 81) treated 'the national market, the power of purchasing and the actual purchases of the nation' as the same thing. Ricardo noticed the different meanings of the word 'demand' at the very beginning. In a letter as early as 1811, Ricardo (1952, Vol. 6, p. 56) replied to Mill:

> You observe that the demand for corn is unlimited. It is clear that you attach a different meaning to the word demand to what I do. **I should not call the mere desire of possessing a thing a demand for it**… By demand I should understand a desire to possess with the power of purchasing. (bold type added)

Mill replied that 'I follow Dr. Smith's rule, which is to call it **effectual** demand, as often as it means the will to purchase combined with the power' (Ricardo 1952, Vol. 6, p. 58, bold type added).

When Malthus declared to Ricardo (Ricardo 1952, Vol. 6, pp. 132, 133) that 'I by no means think that the power to purchase necessarily involves a proportionate will to purchase', Ricardo replied that 'We agree too that effectual demand consists of two elements, the power and the will to purchase, but I think **the will is very seldom wanting** where the power exists' (bold type added). From this conversation, it is clear that the heart of the Say's law controversy is whether or not the will to purchase is always abundant.

The argument continued. McCulloch (1864, p. 146) joined the argument against Malthus by claiming the will to purchase is never in shortage:

Malthus has justly stated that the demand for a commodity depends 'on the will combined with the power to purchase it', that is, on the power to furnish an equivalent for it. But **who ever heard of a want of will to purchase**? If it alone could procure necessaries and luxuries, every beggar would be rich as Croesus, and the market would constantly be understocked. The power to purchase is the real desideratum. It is the inability to furnish equivalents for the products necessary to supply our wants that 'makes calamity of so long life.

It is apparent that, in the argument against the underconsumptionist, Ricard, Mill and McCulloch simply assumed away underconsumption by claiming that the will to demand is abundant. Keynes totally disagreed with this assumption. In the second last chapter of his popular book *The General Theory*, Keynes reviewed the thought of mercantilism and the theory of underconsumption and regarded underconsumptionists, including Mandeville, Malthus, Gesell and Hobson, as 'who, following their intuitions, have preferred to see the truth obscurely and imperfectly rather than to maintain error, reached indeed with clearness and consistency and by easy logic but on hypotheses inappropriate to the facts'. From his preference, it is clear that Keynes regarded demand as the will to purchase.

Given this definition in the mind of Keynes, it not surprising that, in VI of Chapter 2 of his book, Keynes interpreted Say's law as 'supply creates its own demand'. This interpretation was criticized as the well-known naive rendition of Say's law (e.g. Jonsson 1995). Although Keynes' interpretation was not exactly the same as what Say's law means—supply created a demand for other goods, Keynes' expression was a logical deduction from Say's interpretation: 'the mere circumstance of the creation of one product immediately opens a vent for other products' (Say 1821, p. 167). Following Say's reasoning, the creation of other products should also open a vent for the product of concern. Consequently, the creation of one product itself opens its own vent, or, supply creates its own demand. With such an implicit deduction, Keynes's definition highlights the absurdity of Say's law.

Keynes put much effort into rejecting Say's law. In VII of Chapter 2, Keynes listed Say's law as one of the three key assumptions underpinning classical economics. His rejection of Say's law was made based on his claim that it is possible to have an excess of saving over investment.

It seems quite easy to establish a claim of excess saving. Investment must be financed by savings, so the level of investment in a closed economy must not be greater than the level of saving. If some savings are not utilized in investment, then the amount of savings will be greater than the amount of investment and thus there is an excess of saving. However, to establish the possibility of excess saving, one must overcome the issue resulting from saving and investment identity, or the income–expenditure identity.

The income–expenditure identity necessitates that the value of the total output created in a closed economy must equal the total income of that economy. The total output consists of two parts—consumption and investment; similarly, the total income is used for two purposes—consumption and saving. From the income–expenditure identity, one can easily arrive at the saving-investment identity:

value of output = income;
value of output = consumption + investment;
income = consumption + saving;
consumption + investment = consumption + saving;
investment = saving.

The above equation means the amount of investment must always be equal to the amount of saving. If this is true, Keynes's claim of excessive saving above investment is invalid. Keynes had a significant difficulty in breaking this saving-investment identity—he spent two chapters wrestling with this issue. In Chapter 6, he said, in order to reach clearness of mind we need to consider the decisions to consume rather than the decisions to save, then he said an individual's decision to consume and to invest is related to the decisions of other individuals. The interaction may present a difficulty in obtaining simultaneous decisions of consumption and investment and thus may cause the saving-investment identity to break down. In Chapter 7, he relied on the role of a banking system to create credit without saving or to absorb saving without making an investment. To justify this ability of the banking system, he eventually went back to the interdependence of saving and consumption between different individuals.

For although the amount of his own saving is unlikely to have any significant influence on his own income, the reactions of the amount of his consumption on the incomes of others makes it impossible for all individuals simultaneously to save any given sums. (*The General Theory*, p. 84)

However, Keynes had a hard time to explain his reasoning regarding simultaneous decision. In Chapter 7, he admitted that the old-fashion view that saving always involves investment is formally sounder than his concept of excessive saving. In the preface to the French edition, he stated that his claim of divergence of saving and investment 'has been considered a paradox and has been the occasion of widespread controversy'. Then, he spent a long paragraph explaining this. Essentially, he said, for individuals, the amount of saving and investment can differ, but the aggregate saving and aggregate investment for an economy must be one of exact and necessary equality.

Does the saving-investment identity always hold for an economy? The answer lies in careful logical thinking. When we talk about 'income = value of output', i.e. the income–expenditure identity for an economy, we mean the money value of income the economy received equals the value of the real output. The former is the value in form of money while the latter is the value in the form of goods and services. These are the same value in different forms, so they must be equal. When we say the value of output encompasses two parts—consumption and investment, we simply regard the amount of unconsumed goods and services as an investment, which actually means the amount of savings in the form of goods and services. So, the saving-investment identity actually means that the savings in the form of money must equal the savings in the form of goods and services. This is of course true because they are also the same value in different forms. As a result, the saving-investment identity did not claim that the amount of unconsumed goods and services (the investment in the saving-investment identity) and the amount of invested saving (the true meaning of investment) must be equal.

In viewing the world as a closed economy (and we also ignore assets or view assets as past savings for simplicity), investment is always the whole or a part of savings (the amount of invested savings), namely, savings $\geq$ investment.

What is the likely case in reality? Generally, people like to obtain wealth and prefer more wealth, so the supply of goods and services has no constraint. On the other hand, a firm's demand for goods and services to invest depends on the prospect or expectation of making a profit (the author prefers to use the word 'prospect' which conveys an objective meaning. An incorrect expectation may also lead to an investment demand, but it inevitably is corrected by the market), so the demand for goods and services is contingent on the prospect of profit earning or investment opportunity. When investment opportunities are abundant, the demand for investment is abundant, but the total investment available is determined by savings. In this case, we have an equality: saving = investment. If the investment opportunities are very limited, the investment demand will be very small and savings are relatively abundant, so we have in this case saving > investment. As a result, Keynes's claim of excess saving holds.

## 2.2.7 Profit and Interest Explained

Profit and interest are important concepts in the discipline of economics as well as in our daily lives. Although the investigation and debate about the basic concepts are less conspicuous than those on major theories like marginal analysis and the theory of income distribution, the investigation of these two concepts by previous economists showed how logical thinking on basic concepts can deepen our understanding and thus make a subject more rigorous.

In a stylized analysis, these two concepts can be explained easily. For example, in the case of production with three inputs, labour, land and capital, profit is the revenue residual after deducting the payment related to labour and land (i.e. wages and land rent), so profit is the return on capital, which can be called capital rent or interest. In this sense, profit and interest are the same thing—the return to capital. However, in the business world, the calculation of profit is much more complicated than this due to various factors involved, such as the compensations for the risk of business and the efforts of the entrepreneur. Should these elements be a part of profit?

J.B. Clark (1847–1938) argued that pure or economic profit should be defined as the revenue residue of a firm after all inputs are paid at a price equal to their opportunity cost (the price for alternative use of these inputs) (Clark 1908). By this definition, both payments to the entrepreneur and to capital should be excluded, so pure economic profit is not a payment to any inputs but a payment compensating for the risks of business activity. F.H. Knight (1885–1972) further argued that the business risks (e.g. the impact of weather on agricultural output) can be covered by insurance. Since the insurance premium belongs to business cost, it should also be deducted to obtain pure profit (Knight 1921). After this is deducted, the pure profit should be zero because the total cost must equal the output value in a competitive market (alternatively, we say that, in the long run, the competitive market will drive down the price of output until it equals the production cost). However, empirical calculations show that pure profit tended to be nonzero (positive or negative).

Where does this pure profit or pure loss come from? One obvious answer is market power. In reality, markets are not perfectly competitive, so monopoly power can generate a margin or positive pure profit for a firm. However, this explanation cannot explain the nonzero profit in some seemingly very competitive industries such as agriculture and electronics. Moreover, this explanation cannot explain the pure negative profit (pure loss) for some firms because the market power can generate only a positive pure profit. A new explanation was proposed by J.B. Clark, Alfred Marshall and J.A. Schumpeter, who argued that profit is temporary income resulting from economic dynamics, i.e. profit is the windfall or loss associated with economic dynamics.

Next, we turn to the concept of interest. We say capital rental and interest are the same thing in a stylized case, but this statement is conditional on the situation where the producer owns the capital. If he/she obtains the capital from other sources, an interest has to be paid. Where does this interest come from? If capital deserves an interest payment, this seems quite different from other inputs like labour and land. We have already seen that pure profit would become zero in the competitive long-run equilibrium, but the interest payment to capital seems to persist. How is this persistence of interest to be explained? This question generated controversies and stimulated development of the theories of interest. Bohm-Bawerk, Knight and Fisher are the most important contributors to this development.

In his book *Capital and Interest, a Critical History of Economic Theory*, Bohm-Bawerk (1884) criticized Marxist-socialist condemnations of interests as forms of capitalistic exploitation. Bohm-Bawerk (1888) further established the cornerstone of interest theory: 'present goods are, as a rule, worth more than future goods of a like kind and number. This proposition is the kernel and center of the interest theory which I have to present' (Bohm-Bawerk 1888, p. 237). Fisher (1907, 1930) further developed Bohm-Bawerk's theory. Fisher reasoned that all productive agents (e.g. capital, land and labour) yield streams of income, which can be viewed as either interest or rent. Knight (1949) elaborated Fisher's point: 'only historical accident or psychology can explain the fact that interest and rent have been viewed as coming from different sources'. In other words, the return to land is historically called rent, the return to labour is historically called wage and the return to capital is historically called interest. On returning to our case of two payments to capital, the interest paid to investors is the economy-wide capital rent, while the capital rent paid to the producer is the differential capital rent paid to different firms and industries based on their performance, just like the differential wages earned by workers in different industries.

## 2.2.8 Nominal Value, Real Value, Relative Value and Money Illusion

Nominal value and real value are related to money and inflation. The use of money has a long history. It is believed that primitive money like shells was used in 3000 BC. The use of metal money appeared in 800–300 BC. The concern about inflation came much later. The most significant inflation in ancient times is the inflation that occurred in Roman Empire due to excessive debasement (e.g. adding more proportion of cheap base metal copper to a silver coin) and multiplication of the standard coins in order to finance government spending. As money supply increases, the value or price of goods in terms of money increases while the true value of goods is unchanged. The value in terms of money is loosely defined as the nominal value while the true value of the goods is called the real value. Since it is hard to measure the true value of goods when the price level has changed, the real value of goods is generally obtained by using an aggregate price index. For example, if the nominal value of a house is $1.5 million currently, but the overall price for all houses in a country increase from 100 in the base year to 150 in the current year, the real value of a house in the current year is: 1.5/150*100 = $1 million.

Distinguishing nominal and real values is important in measuring the cost of borrowing. The nominal cost of a loan is measured by nominal interest rate, which is determined by the equilibrium in money market. The real cost of a loan is measured by real interest rate, which is determined by the equilibrium in capital market. Fisher (1930) linked the nominal interest rate i and real interest rate r through the expected inflation rate $\pi$: $1 + i = (1 + r)(1 + \pi)$, or approximately, $i = r + \pi$. This is called the Fisher effect. Similarly, differentiating nominal and real values is also important for wage or salary payment. If a worker's nominal wage is unchanged at $2400 per month but the consumer price index increased from 100 in the base year to 120 in the current year, the purchasing power of the worker's wage decreases to $2400/120*100 = $2000. This purchasing power of wage is called real wage. In a monetarist's adaptive expectations model, workers are satisfied with the unchanged nominal wage and thus are fooled by the presence of inflation. In other words, the workers suffer from money illusion.

Although money illusion has been seen through by economists, it is still baffling professionals in finance. The exchange centres in New York have a tradition to celebrate when the stock price index reaches certain points (e.g. the Dow Jones Industry Average index reached 20,000 points or the Nasdaq index reached 4000 points). When the stock price index dropped heavily during a recession, it is often reported that many billion dollars of wealth were wiped out. Personal and corporates' wealth is tied to the prices of stock held and the rewards for corporate executives are also tied to stock price. Stock prices are just nominal value and the stock price indexes are only an indication of how the nominal value departs from the real value, just like the CPI is an indication of inflation, so we should not put much emphasis on them. It is point-less to celebrate a high inflation figure; similarly, there is no need to cel-ebrate a record-high stock price index. On the contrary, the nominal value of a stock should be discounted by the stock price index to obtain a real value of the stock. This is the same way as we discount nominal GDP by GDP deflator to obtain real GDP.

However, if we study nominal and real values further, we may find that their definitions are too loose and sometimes cause confusion. Since nominal value is measured in terms of money and is calculated based on money price (i.e. the price in terms of money), any factors causing a change in money price will cause a change in nominal value. If the money price change is caused by a change in money supply, the resulting value change is indeed a nominal change. However, if the money supply is fixed, a change in the demand for goods (e.g. consumer become extremely active in purchasing because of improved consumer confidence in future) may also cause a change in money price. This change in money price and thus value of goods should be the change in real price and real value because they result from the real change in underlying demand or supply. By this reasoning, nominal value should be related directly to the money supply. If the money supply is fixed, any changes in value should be viewed as a real change. If so, the differ-ence between nominal and real value/price hinges on the growth rate of the money supply, which can be used to obtain the real value from the nominal value of all goods and services.

The concept of real value/price is also often confused with the relative or physical value/price. Relative or physical value/price means the value of one good in terms of the value of another good (one may argue that relative price can be nominal price when it is relative to a special good 'money'. Considering the paper money in a modern economy, here we do not treat money as a good), e.g. 1 kg of lamb = 6 kg of rice. This price is real because it avoids the inflation issue caused by a change in the money supply. However, relative price is suitable only for a primi-tive economy. For an economy with numerous goods and services, there will be numerous relative prices and this will cause confusion and inef-ficiency in market transactions. Real price avoids this by measuring the value of all goods and services in terms of money. Real price also avoids the influence of inflation by fixing the money supply. The confusion of real value/price and relative or physical value/price often occurs when the concept of 'real income' is considered.

When we say income is a determinant of demand, we refer to real income which means two things. One is that the income is expressed as the amount of money, rather than the amount of goods. The other is that the income is measured under the assumption of a fixed money supply; otherwise, a change in money supply will cause changes in prices of goods and this necessitates that the same amount of money income may obtain different quantities of goods demanded by the consumer. When macroeconomists talk about real income, however, they often mean physical income—the amount of goods and services obtained. This changed definition of real income causes a lot of confusion and argument in macroeconomics. This can be shown by an aggregate supply and aggregate demand (AS/AD) model in Fig. 2.1.

The horizontal axis in Fig. 2.1 is labelled 'output/income', which means the physical output/income of the economy (the quantity of output equals the quantity of income because the amount of goods and services produced equals the amount received in the economy). The vertical axis is labelled 'price level', meaning the aggregate price in terms of money. We can assume, initially, the aggregate demand $AD_0$ and aggregate supply $AS_0$ determine the equilibrium output/income level $Y_0$ and price level $P_0$. Macroeconomists generally interpret the output/income $Y_0$ as real output/income. Using the notion of 'income determines demand', we can conclude that the real income $Y_0$ determines

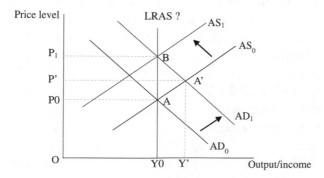

Fig. 2.1   Confusion caused by nominal, real and physical values

the aggregate demand curve $AD_0$. This is apparently incorrect because income $Y_0$ is the equilibrium quantity of the economy, which is determined jointly by two curves $AD_0$ and $AS_0$. Here the incorrect conclusion results from the confusion about real income and physical income. When we say 'income determines demand', we refer to the real income measured by $P_0{}^*Y_0$ (the area $OP_0AY_0$), rather than the physical income $Y_0$ in Fig. 2.1.

Figure 2.1 also shows the confusion caused by nominal and real price level. To be able to draw aggregate supply and demand curves $AS_0$ and $AD_0$, one must assume that the money supply is fixed, so the price level in Fig. 2.1 should mean real price level. If money supply changes, there are a number of price levels associated with each level of physical output and thus it is impossible to draw a demand or supply curve. For example, at money supply $M_0$, we may draw $AS_0$ and $AD_0$. If money supply increases to $M_1$, the price level will increase at each physical output level, so aggregate supply will shift to $AS_1$ and the aggregate demand curve will shift to $AD_1$. Hence, the economy will be settled as point B with the same output level but a higher price level. The shift of equilibrium point from point A to point B and thus the change of price level from $P_0$ to $P_1$ indicate a change of base for real price and thus for our AS/AD analysis.

However, macroeconomists regard the price level in Fig. 2.1 as nominal price level. Their interpretation of the effect of an increase in money supply is as follows. First, with the increased amount of money the aggregate demand increases so $AD_0$ shifts to $AD_1$. Second, firms will continue to supply according to supply curve ($AS_0$) because their production cost has not yet been affected by the increased money supply, so the new equilibrium is achieved at point A' with an output level $Y'$ and price level $P'$. Third, the increased price level feeds into the production cost of firms, so the supply curve shifts up to $AS_1$ and the economy settles at point B with an output level of $Y_0$ and price level $P_1$.

This explanation of the inflation process by macroeconomists is correct, but the problem is that they regard the aggregate supply curve $AS_0$ and $AS_1$ as short-run supply ($SRAS_0$, $SRAS_1$), and the vertical AB curve as the long-run supply (LRAS). With a vertical long-run aggregate supply curve LRAS, any shifts of aggregate demand curve will not affect the long-run output level. Hence, macroeconomists claim that aggregate demand has no influence on the output of the economy in the long run because it is fixed by the long-run aggregate supply curve. They term this long-run output level as the potential output level, full-employment output level, or the output level at the natural rate of unemployment.

The mistake of macroeconomists is that they fail to realize that Fig. 2.1 describes nothing else but the process of inflation when there is an increase in money supply. It has nothing to do with economic behaviours in the short run or long run. The vertical AB line shows that money is neutral once the inflation process completes, so it is not a long-run supply curve. The vertical long-run supply curve LRAS claimed by macroeconomists leaves them no way to explain the apparent economic growth in the long run. The only explanation they can put forward is that the vertical long-run aggregate supply curve will jump to the right so as to allow economic growth. This is totally against common sense. Economic growth in the short run can experience jumps (e.g. due to a scientific breakthrough and a sudden increase in productivity), but how can the economy performance jump in the long run (e.g. from an output level of $20 billion in 1970 jump to $20 trillion in 2018)? Some macroeconomists explain the vertical long-run aggregate supply curve using the constraint of technology and resources. It is true that one needs to consider the constraint of the economy, but the constraint of resources and technology is more relevant in the short run than in the long run. For example, in microeconomics, it is in the very short run that the supply curve is vertical. The truth is that, in the long run, technology is more likely to make large progress and thus will relieve the resource constraint. Hence, the long-run aggregate supply curve should be flatter than the short-run ones. These implausible explanations in macroeconomics all result from confusion about nominal value, real value and physical value, and they can be fixed by a careful logical thinking.

## 2.2.9 Contextual v.s. Abstract Approaches

Economics is intimately related to politics and human history, so it is not surprising that the subject was much contextualized at the early stage. Adam Smith's wealth of nations reflected not only his economic ideas but also his view on philosophy, social science and ethics.

Even after he extracted the theory of competitive markets, his economic policy was still contextualized. He acknowledged the limitations of competitive markets in some situations and suggested government provision or government interference in defence, education and community infrastructure. His contextualized approach was adopted by many economists including Malthus, Say, J.S. Mill, Marx and Menger.

However, Smith's contextualized approach was not taken up by Ricardo. To defeat the arguments of protectionists during the Corn Law debate, Ricardo developed abstract models which employed a number of assumptions, concepts and analytical tools. This abstract approach was developed further by Cournot, von Thunen and Walras through introducing general mathematical expositions of economic models. With the aid of mathematics, Walras developed the general equilibrium framework which provides much insight into how an economic system works. This abstract approach is still followed by neoclassical and new classical economists today.

Contextual or abstract? Which approach is better? Both approaches have their advantages and drawbacks. The contextual approach often provides interesting stories and concrete and useful background information. We can easily see the logical links based on detailed information, so research using this approach is understood by a wide audience. However, since there are too many factors involved in any real cases, this approach is not particularly good at identifying the essential factors and the main mechanisms behind the reality. On the other hand, the abstract approach is concise, rigorous and capable of finding the mechanisms or principles behind each case. For example, F.Y. Edgeworth (1845–1926) examined the microeconomic theories on input demand, production, exchange and consumption, and he pointed out that basic microeconomic theory is simply a repeated use of the same mathematical tool—constrained maximization. He argued that, by extracting the mathematical core from its institutional context, one can reflect the essence of the issue and apply it to similar problems. By relying on mathematics, however, this approach requires more assumptions, tends to overlook non-economic factors and is less accessible to a wide audience.

Realizing the benefit of both approaches, some economists such as J.S. Mill and Alfred Marshall, adopted an eclectic approach. Given the broad knowledge of different disciplines he had acquired, it is not surprising that Mill regarded economics as only a part of knowledge related to mankind. He admitted that the abstract analysis yielded some useful results, but he emphasized that these results must be integrated into a more complex aspects of human society. This indicates that Mill was more supportive of the contextual approach.

Marshall's background was different. He was trained in mathematics for his undergraduate degree and translated Ricardo's and J.S. Mill's economics into mathematics. However, his wide reading of history made him believe that classical economists often failed to realize that society changes. He proposed a method of blending the theoretical, mathematical and historical approaches, but he leaned towards a pro-contextual approach. In this principle of economics, he said:

> But I know I had a growing feeling in the later years of my work at the subject that a good mathematical theorem dealing with economic hypotheses was very unlikely to be good economics: and I went more and more on the rules – (1) Use mathematics as a shorthand language, rather then as an engine of inquiry. (2) Keep to them until you have done. (3) Translate into English. (4) Then illustrate by examples that are important in real life. (5) Burn the mathematics. (6) If you can't succeed in (4), burn (3). This last I did often. (Marshall 1920, p. 339)

Marshall's eclectic approach angered economists who used either contextual or abstract approaches. The advocates of the abstract mathematical approach were definitely irritated by the above comments about the limitation of mathematics. On the other hand, the historical and institutional schools were critical that the mathematical method in Marshall's approach was abstract, rigid and meaningless because it took information out of historical and institutional contexts, and that the classical and neoclassical approaches were unscientific because the basic assumptions were untrue.

Although Marshall's eclectic approach was followed by a number of economists, notably, Pigou, Keynes, and Friedman, the battle was won by economists using the abstract approaches, including Edgeworth, Pareto, Abraham Wald, Kenneth Arrow, Gerard Debreu, Robert Solow, John Hicks and Paul Samuelson. Nowadays, the papers in economic journals are highly mathematical. Not only do people outside of economics discipline have difficulty in understanding them, but also people from different schools of economic thought, or even people using different models, have a difficulty in understanding each other's work. The direct impact of the abstract approach is that economics has become less practical and less useful, and that fewer people can understand or show an interest in economics. Although the abstract approach won popularity in the economics discipline, the discipline itself became less popular because people shy away from economics. This is evident by the worldwide decrease in the number of students studying for an economics degree.

So, what approach should we use? Continue with the abstract approach or restore the contextual approach or become an eclectic like Marshall? The author would say none of these approaches would work given the problems of each approach shown by history. Here the author suggests a diversified approach. Both the contextual and abstract approaches have their advantages and disadvantages, so we should use the merits of each approach. However, this does not mean that we should combine both approaches to form an eclectic approach. Essentially, the eclectic approach is generally impractical. Some researchers have specialized in mathematics while others are good at historical narrative and institutional analysis. Although training may improve the ability of both types of thinkers, generally the two abilities do not come together except in a few very unusual and talented individuals. This diversified approach is an application of the theory of labour division to the economics discipline itself.

While the diversified approach allows people to use different methods to address economic problems, this approach also encourages researchers to be aware of the limitations of the adopted method, to take into account the advantages of other methods, and to utilize the research results obtained from the other methods. The diversified approach necessarily involves a contest between different economic ideas. This contest is necessary in order to develop a sound economics discipline.

However, the contest should not be the same type as the current arguments between orthodox and heterodox economics, and between different schools within heterodox economics, with the debaters tending to ignore the limitations of their own methods while exaggerating the disadvantages, and losing sight of the advantages, of the approaches of others. For example, orthodox economists tend to criticize the heterodox for lacking in mathematical training and for focusing on non-economic factors, but they fail to realize that some non-economic factors have vital impact on economic models. Heterodox economists, on the other hand, criticize that the orthodox's models work in historical vacuum or under assumptions that are not realistic or even totally wrong, but they fail to admit that it is necessary to filter out numerous non-essential information and to use assumptions in order to grasp the essence of the problem and find the main mechanism. This kind of one-side argument might help to produce heated and endless debates, but it is not helpful in either producing the right solution to economic problems or developing each school of economic thought and the economics discipline as a whole. If each school of economic thought can see merits and limitations of each side when contesting economic ideas, different schools of economic thought will complement and strengthen each other, rather than compromising or destroying each other. As a result, the economics discipline will evolve into a new stage.

## 2.2.10 Decline of Macroeconomic Theories

It is fair to say that modern macroeconomics started from Keynes' effort to address the issue of business cycles, on which classical economists provided little insight. In adopting both Smith's doctrine that savings are immediately invested and Say's law of market, classical economists believe that the economy will reach equilibrium in which the price will equalize supply and demand, so an overall excess supply or a general glut is ruled out in the classic theory. In response to the Great Depression, Keynes put forward a theory that equilibrium and unemployment can coexist. This opened a new chapter in macroeconomics.

Keynes rejected Say's law and claimed that investment depends on the entrepreneurs' decision to take investment risks and on their liquidity preferences. If entrepreneurs lack the 'animal spirit' and people tend to hoard money, savings will not be fully invested and thus there will be a deficiency of effective demand. Keynes also demonstrated how a change in investment can lead to a change in output level. This link between investment and output is later called the multiplier model or AE/AP model.

However, Keynes' theory has some drawbacks. One is that, although Keynes assumed non-neutrality of money (i.e. people's tendency to hoard money can cause an economic recession), in his theory money plays no role in determining price and output. For example, he believed that inflation depends on the excess of aggregate expenditure over real output so inflation is totally unrelated to money supply. This view is totally rejected by monetarists. Rising inflation in the 1960s cast doubt on Keynes' belief and the stagflation (stagnation of production coupled with high inflation) in the early 1970s led to marginalization of Keynesian economics. The other drawback of Keynesian economics is its rejection of the classical framework and replacement of many rather ad hoc assumptions in Keynesian models. For example, the importance of animal spirits and the liquidity preference in Keynes's theory reject the assumptions of rationality and neutrality of money in classical economics. As a result, Keynesian economics cannot fit into the general equilibrium framework and lack a microeconomic foundation.

In the 1950s and 1960s, economists (notably Hicks, Modigliani, Solow, Tobin, Samuelson, and Patinkin) tried to absorb Keynesian economics into a classical framework to form a neoclassical synthesis. In this synthesis, the IS/LM model to some degree can reconcile the argument between Keynesians and monetarists while the AS/AD model and the Philips curve were used to reconcile Keynesian economics and neoclassical economics. The aggregate supply (AS) curve and the Phillips curve were assumed as a vertical line in the long run, so the aggregate demand and government policies (either fiscal or monetary) should have no impact on either output level or on employment level—this was an outcome predicted by classical economists. However, in the short run, the aggregate supply curve was positively sloped and the Phillips curve was negatively sloped, so aggregate demand and government policy could alter the output or employment level—an outcome from Keynesian economics. In the early 1970s, the general attitude in economics was that classical economics was correct in the long run while Keynesian economics worked well in the short run. This dichotomy in the long run and the short run was later reconciled by adaptive expectations: people form their expectations based on past experience. In the short run, people can be fooled (make mistakes in expectation) so government policy works; in the long run, people can see through the consequences of government policy, so policy cannot affect the output level.

However, this reconciliation was ended by rational expectations theory in the mid-1970s. The rational expectations theory was first put forward by Muth (1961), who applied rational expectations to model the price fluctuations in an isolated market. However, it is Lucas' paper 'Expectations and the Neutrality of Money' (Lucas 1972) that initiated the revolution of rational expectations and transformed neoclassical economics to the new classical economics. According to the rational expectations hypothesis, if people can have all available information, they will make rational expectations about equilibrium, instead of adjusting their expectations in stages. As a result, if a policy is ineffective in the long run it is also ineffective in the short run. In other words, if the Keynesian economics does not work in the long run, it does not work in the short run either.

Facing the challenge from the new classical economics, the new Keynesian economists argued that one cannot consider the rational expectations of one individual in isolation. The reasoning is that the rational expectations of each individual are made contingent on others' expected decisions. This points to expectation conundrums or economic coordination failure. To highlight the importance of economic coordination, the new Keynesian economists maintained that, while Keynesian economics needed a microeconomic foundation, microeconomics needed a macroeconomic foundation. The new classic economists could not fight on this ground, so the debate came to a dead end without a clear winner.

Since the argument arising from business cycles could not be continued, economists put their effort into studying economic growth.

The exogenous growth model initiated by Solow (1956) was further developed, and the endogenous growth model appeared in the 1990s. Since economic growth is generally regarded as a long-run view, all growth models are supply-side models based on production input such as technology, labour and capital. The endogenous model expresses technological change as a function of human capital which is determined by education and research. However, these growth models show only an incomplete picture because economic activities are based on interactions from both the supply and demand sides. By excluding the demand side from the economic growth theory, economists lack a deep understanding of both economic growth and business cycles. As a result, there are few new ideas on economic growth, and theoretical research in this area has also stagnated. This has led to a sharp decline of macroeconomic theories and the only outlet left for economists' energy is empirical research.

## 2.3   Theories and Empirical Evidence

Theories are obtained from both our observation and the analysis of reality. Once a theory is formed, it is also subject to examination by further empirical evidence. The flat earth theory was based on our observation of a wide horizon, but this theory was proven wrong as our knowledge grew and more observations came to light. Many other theories were proven incorrect too, for example, the theory that earth is the centre of the universe, Aristotle's claim that the heavier object falls faster, and the hypothesis that the long necks and legs of giraffes have evolved because of generations of repeated stretching to reach for leaves at the top of trees. Parallel examples in economics include Smith's wage–fund doctrine, Smith and Ricardo's diminishing rate of profit, Malthus' population thesis, and Keynes' theory and policy on economic recessions.

Compared with natural sciences, it is much harder to examine the theories in economics, or social science in general. Scientific experiments are a powerful tool for natural scientists. Aristotle's claim that a heavier object falls faster was defeated by Galileo's experiment at the leaning tower of Pisa. The impact of a drug can be examined by experiments on animals and on humans. However, scientific experiments play a very limited role in social science because it is either unethical, impractical or too expensive to do a scientific experiment to repeat a social event. As a result, the data in the social sciences do not satisfy the conditions of scientific experiments—'other thing being equal'. This causes great difficulty in verifying theory with empirical data. For example, the supply/demand theory in economics claimed that the supply curve is positively sloping while the demand curve is negatively sloping, which basically means that a higher price of a good will induce more supply of and less demand for this good. This theory requires that other things such as income and substitutes are unchanged. Empirical data cannot satisfy this condition and thus cannot prove or disprove the supply/demand theory. The inability of doing scientific experiment in social science not only makes it difficult to examine a theory but also makes it hard to judge an empirical study.

Given that it is impossible to replicate the events studied in reality, it is impossible to verify the empirical studies in economics and in social sciences, so the results from this studies are called 'data-dependent'. This casts doubt on the rigorousness and usefulness of empirical studies. This section briefly reviews the history of empirical research in economics, its fallacies and limitations, and its relation to theories.

## 2.3.1 Common-Sense Empirical Approach

This approach is to conduct research by collecting and using data as regarded relevant by common sense. The common-sense empirical approach has a long history. As early as 3050 BC, ancient Egyptians collected data concerning population and wealth; in 3000 BC, the Chinese conducted accounts of statistical work while the Ancient Greek conducted censuses for the purpose of levying taxes; in 435 BC Romans had extensive surveys of population and also kept records of births and deaths.

William Petty (1623–1687) was a champion in the common-sense empirical approach. He wrote 'Political Arithmetic' in 1676, but it was posthumously published in 1690. In this book, Petty forged analytic tools dealing with data. He explicitly advocated the use of statistics to measure social phenomena and attempted to measure social variables such as national income, exports, imports and population. With his friend John Graunt (1620–1674), he studied the weekly British mortality data—the bill of mortality. The studies on British population were carried on by James Steuart (1713–1780), Anne Turgot (1727–1781), and later by Malthus. However, Petty's methods were so crude that Adam Smith suggested that Petty himself had little use of political arithmetic.

The quantitative study advocated by Petty was exampled by the estimation of demand for wheat by Gregory King (1648–1712) and Charles Davenant (1656–1714). They observed the influence of the wheat harvest and its price. Assuming a normal level of harvest, they discovered that, if the harvest is below this normal level by 10, 20, 30, 40 and 50%, the price will increase by 30, 80, 160, 280 and 450%. This work anticipated the concept of demand elasticity.

Petty's political arithmetic approach was followed by Richard Cantillon (c. 1680–1734). He attempted to establish the basic principles of economics through reasoning and tried to collect data to verify these principles. He viewed elements of the economy as integrated parts of the whole. He also studied the economy by sectors and analysed income flow between them. This thought had much influence on Quesnay, who became the forerunner of the general equilibrium approach.

The empirical application of the partial equilibrium analysis initiated by Marshall generally uses the common-sense approach. Although partial equilibrium analysis considers only one market at a time and involves the assumption of ceteris paribus—other things being equal, the analysis can capture the main factors or main mechanism and requires much fewer data, so it is popularly used by various agents and for various purposes. Moreover, although a partial equilibrium generally involves only the price, supply and demand of a good or a service concerned, the partial equilibrium analysis can be extended to include, in the supply/demand function, the impact of other factors such as income and the prices of other goods.

## 2.3.2 Systematic Use of Data

This approach is largely related to general equilibrium analysis, of which Francois Quesnay (1694–1774) was the forerunner. He realized the interrelatedness of different economic agents and economic activities.

He also tabulated the transactions between farmers, landowners, and artisans and servants in a simple economy. This approach was formalized mathematically as a general equilibrium model by Walrus, whose theoretic model was used by Wassily Leontief (1906–1999) to design an empirical input-output analysis. Leontief also actively advocated for and was involved in data collection, and he also constructed the first input-output table (Leontief 1936, 1941). Later he applied input-output analysis to address various topics, including environmental pollution (Leontief 1970), international trade (Leontief 1974), and the choice of technology (Leontief 1985). Due to its ability to include very detailed sectoral information at a relatively low cost, input-output analysis is still used today by various institutions, especially government and private firms.

Walrus' theoretical framework was further improved by Gerard Debreu and Kenneth Arrow. Arrow and Debreu (1954) laid down the conditions under which the equations of competitive equilibrium in the Walrasian framework have a solution. This paved the way for empirical applications of a general equilibrium model. The first applied general equilibrium model was built by Shoven and Whalley (1972). Later, a large number of applications appeared. With the aid of general equilibrium solutions solving software, general equilibrium models are widely used, with very detailed industrial information. However, building an applied general equilibrium model needs skills and takes time, so it is mainly done by universities and research institutes. The ready-to-use off-shelf commercial models are used by governments and large firms.

The other thread of using systematic data is the study of business cycles. Lucas and Prescott (1971) and Kydland and Prescott (1982) created the real business cycle (RBC) model which utilizes macroeconomic time series data such as GDP, consumption, investment and employment (labour hours). The model uses a production function as well as a consumption function to represent an economy. The technology level in the production function is assumed random but with a persistent feature (the technology level in the previous year affects that in the next year). With the dynamics of capital and technology, a random technological shock can generate fluctuations in output, investment and employment. The approach is to use historical data to calibrate the parameters in the model so as to simulate rather than predict the performance of the economy.

The further development of the RBC model saw many models with an increased size and complexity. The models can include a few industries, can have different production and consumption functions and can have Keynesian assumptions such as monopolistic competition, price and wage rigidities, and non-neutrality of money. The model also acquired a new name—the dynamic stochastic general equilibrium (DSGE) model. The drawback of a DSGE model is that it lacks industrial details and the poor performance of the DSGE models has been highlighted by their failure in forecasting the GFC. Even if a DSGE model could mimic economic performance well, this does not mean the mechanism and parameter values in the model are true in reality, so the model may not be correct in essence.

### 2.3.3 Probability Approach

The probability approach has by and large become a synonym for econometrics. This approach is based on the probability theory developed in statistics. The birth of the probability theory started with the study of gambling activities. Gerolamo Cardano (1501–1576), a renowned Italian physician and mathematician as well as an inveterate gambler, wrote a book *Liber de Ludo Aleae* (*The Book on Games of Chance*) in 1564. Pascal (1623–1662), Fermat (1601–1665) and Huygens (1629–1695) contribute significantly to the calculus of probability. James Bernoulli (1654–1705) continued the contribution to probability for a repeatable trial by establishing the 'limit theorem'.

In the latter part of the sixteenth century, the Danish astronomer Tycho Brahe introduced the probability approach into astronomy in order to find accurate determinations based on repeated observations. In the seventeenth century, Galileo (1564–1642) studied the impact of errors on measurements of stellar distances. Statistical work in astronomy was continued by Thomas Simpson (1710–1761), Daniel Bernoulli (1700–1782), Pierre-Simon Laplace (1749–1827) and Carl Friedrich Gauss (1777–1855).

The probability method in astronomy was brought to the social science domain by Adolphe Quetelet (1796–1874) and Simeon

Poisson (1781–1840). Using the theory developed by Laplace, Poisson estimated the sex ratio at birth in France from 1817 to 1826. Using a concept of 'average man', Quetelet estimated the relative distribution of physical and social attributes of different groups such as height, chest measurement, birth, marriage, drunkenness, suicide and crime. The statistical work in social science in the later 1800s was progressed by Francis Galton (1822–1911). The bivariate normal model was extended to the multivariate normal model by Francis Edgeworth (1845–1926) and Karl Pearson (1857–1936). Another important person who contributed significantly to modern statistics was Ronald A. Fisher (1890–1962). Two of Fisher's major contributions were the development of the theory of estimation for the sampling theory and the development of the design of factorial experiments.

Statistical approaches were used in economics in the early 1900s. Henry Moore (1869–1958) is considered the pioneer in applying statistical methods to economic research thanks to his work in estimating demand curves. Moore (1908) argued that it is necessary to bring together the theory of economics and the science of statistics and regard the theory of probability as a machinery of general application in the study of the mass-phenomena. In an effort to test the marginal productivity theory of wages, Moore (1911) estimated the relationship between wages and marginal productivity of labour, industrial concentration, personal ability and strikes, but his estimations were not rigorous. Moore (1914) estimated a negative-sloped supply curve for pig iron, which attracted wide criticism. Later Henry Schultz (1893–1938) estimated a demand curve and found that different values of elasticity of demand can be obtained depending on the choice of price or the quantity demanded to be used as the dependent variable. E.J. Working (1900–1968) raised the identification problem in estimating the supply or demand curve and provided a general rule to solve this problem.

Moore (1914) also investigated the relationships between rainfall cycles, agricultural cycles and economic cycles. Similar to W.S. Jevons (1835–1982) who attributed business cycles to sunspot activity, Moore attributed economic cycles to the movement between the Sun, Venus and Earth.

The task of investigating business cycles was carried on by Wesley Clair Mitchell (1874–1948), who belonged to the institutionalist school. In 1920, Mitchell founded the National Bureau of Economic Research, which played an important role in economic data collections and analysis. In the book 'Measuring Business Cycles' (Burns and Mitchell 1946), Mitchell tested Schumpeter's theory on different types of business cycles and rejected it. The topic of business cycles was also of interest to Ragnar Frisch (1895–1973) and Jan Tinbergen (1903–1994). They both developed large dynamic macroeconometric models and used data to test various business cycle theories.

The application of probability theory to economics was formally justified by Trygve Haavelmo (1911–1999), who argued that, to use statistical methods, one must accept the probabilistic nature of economic laws. Haavelmo (1944) did realize that random experiments are the foundation for probability theory, but he argued that economic data are the results of 'natural experiments'. Although so-called natural experiments are totally different from random experiments which are conducted under approximately the same conditions, economists seemed to accept the natural experiment argument. As a result, Haavelmo (1944) formally established the probability theory as the foundation for econometrics. Afterwards, the status of econometrics was formally established and large Keynesian-type macroeconometric models were developed at Cowles' Commission. However, these structural macroeconometric models lost support in the mid-1970s because of the prediction failure resulting from changing economic conditions. The failure of the structural models led to the popularity of the vector autoregression (VAR) model developed by Sims (1980). The fact that no econometric model predicted the global financial crisis led to the popularity of the Bayesian approach.

### 2.3.4 The 'Great' Empirical Period

Although the purpose of the three approaches in empirical economics is to test economic theories, from the late 1990s, very few theories have been tested by these approaches. Moreover, papers using the probability approach have been dominant in economic journals with other empirical approaches occasionally making appearance, mostly in heterodox economic journals. Economics has entered a new period of empiricism. A number of factors contributed to this change.

First, the stagnation of economic theoretical thinking. With the establishment of marginalism theory and game theory, microeconomics has approached maturity. At the macro-level, the debate on business cycles came to a deadlock and the research on economic growth ran out of steam. As a result, there are simply not many new theories for testing.

Second, the drawbacks of common-sense and systematic approaches. These two approaches are closely related to economic theories. The stagnation of economic theories has a direct flow-on effect on these two approaches. Moreover, the common-sense approach appears relatively simple and less attractive while the systematic approach is comprehensive but requires a great deal of time to deal with data and model construction.

Third, the attractiveness of the probability approach. The foundation of this approach is probability theory, so this approach can coexist with economic theories or can be totally divorced from any economic theory. In fact, some econometricians argue against all economic theory under the name 'letting the data speak for themselves'. The probability approach also employs advanced mathematics, which may be more appealing to some journals. Most importantly, the application of probability theory to economics encounters many issues such as spurious estimation, autocorrelation, heteroscedasticity and unit root. These issues provide a large playground for the econometrician to practise innovations such as designing new tests and new estimation methods. These innovations are well received by economic journals.

Fourth, the development of information technology. Computers have become more powerful and cheaper, so the cost of doing econometric modelling has reduced substantially. The information technology revolution has also brought about abundant data. Commercial econometric software like STATA, CAT, E-VIEW, MATLAB and R makes econometric estimation an easy task. With all these elements, the provision of econometric work is abundant.

Given the above factors, econometric studies are in high demand and are abundant in supply, so it is no surprise that these types of studies will dominate and create a 'great' empirical period. However, this 'great' empirical period is deeply flawed because 'natural experiments' are not 'random experiments'. Although econometricians are enjoying their time, their econometric exercises including their newly invented tests and models are of little use to business activities. Economics has become more probabilistic, more complex, less accessible and less useful. As people shun away from economics, the base of economics dwindles. Although this 'great' empirical period was dubbed as the 'credibility revolution', i.e. a change giving more credibility to economics, the credibility of the econometric methods is very much in question. Given the problematic theoretical foundation for econometrics and many issues in practice (see Chapter 3 in this book), the validity of econometrics is in doubt, so the 'great' empirical period cannot last long.

### 2.3.5 Fallacies and Limitations in Empirical Studies

Both the common-sense approach and the systematic approach are valid, but both have upsides and downsides. The common-sense empirical approach considers the data and variables in a small area and ignores the impact of the broader environment. However, since the data and variables are contextualized, the approach is more meaningful and gives insight to the main mechanism underpinning the results.

On the other hand, the systematic approach considers all variables in the system, but the approach uses either a highly simplified model like RBC or involves a large number of assumptions which may not hold (e.g. assumptions in a general equilibrium model). Moreover, the mechanism proposed (i.e. the functions used and the parameters calibrated) in the model may not be true. This has significant implications for the results. Even if the model mimicked the historical data very well, the whole modelling results may be in jeopardy because the functions and parameters in the model may not reflect reality. This is demonstrated clearly by the failure of the RBC models in the 1960s.

The probability approach takes into account uncertainty and can utilize various types of data, but the validity and usefulness of this approach are in question because it has a number of theoretical and practical problems. We will discuss them in detail in Chapter 3; here we introduce them only very briefly.

Probability law holds only in the case of random experiments, which are conducted under approximately the same conditions. For example, throwing a die or tossing a coin. Astronomic observations and random-sampling surveys can also be viewed as random experiments. However, economic data often involves observations over time with vastly changed conditions, so the condition of random experiment does not hold. As such, strictly speaking, it is invalid to apply the probability law to economic data.

Theoretically, one can satisfy the condition of random experiments by including in the model all variables involved. However, since there are so many variables involved over time, and some variables are unknown to the modeller, it is impossible for anyone to include all variables involved. Even if one could, he/she also faces the insurmountable (if not impossible) task of figuring out a correct function for a large number of variables.

The theoretical problem embedded in the probability approach has many practical implications. Since probability law reveals no mechanism, one is unable to find the truth by employing an econometric model, so it is not surprising to see that econometricians tried but failed to solve causality issues and spurious regression within a statistical framework. Since the condition for probability law does not hold for economic data involving a time frame, the approach encounters a lot of problems related to data such as autocorrelation, stationarity, heteroscedasticity and inaccuracy. Econometricians developed a number of tools to fix these data problem artificially, but did not fix the source of these problems. Just like the solution of cutting toes short to fit into shoes and covering a wolf with sheepskin, the practical tools of econometricians do not overcome the theoretic issues of the probability approach and are unable to make this approach valid.

Even if the probability approach is valid, the usefulness of this approach is limited because of its inability to reveal the mechanism behind the facts. The popular thought is that probability law is the mechanism behind all econometric models, but natural law and mechanisms are quite different concepts. There are two types of natural laws: primitive and advanced natural laws. Primitive laws reveal no mechanism and can be applied to only a very limited number of situations.

One example is Kepler's laws in astronomy, which describe planetary motion in a solar system. On the other hand, an advanced natural law reveals the mechanism behind the facts and has wide applications. For example, Newton's Law of Gravitation reveals how gravitational force works and can be used to explain the orbits of planets and comets, the movement of objects as big as stars and as small as bullets. Probability law is only a primitive law. Although it describes the uncertainty by the probability of different outcomes, it does not say anything about what causes the probability. In the absence of forces or mechanisms behind probability, our understanding on the research question investigated by the econometric model is very limited, so the approach contributes little to the aim of doing research—to improve our understanding and to find the truth.

One may argue that: 'I am not interested in understanding the mechanism but interested in prediction'. The response to this argument is that understanding the true mechanism is the key for correct prediction. This can be illustrated by weather forecasting. One may agree that the short-term weather forecast is pretty accurate but the long-term weather forecast is just a conjecture. What causes this difference? The short-term forecast is based on data (e.g. humidity and wind speed) and Newton's inertial law while the long-term forecast contains no mechanism because it is purely based on historical data and probability law.

Without a mechanism, a valid prediction from an econometric model can be made only when all the conditions for the sample data period hold as the same. Since it is impossible to maintain similar conditions when one is predicting the future, the prediction from any econometric model is thus invalid. This is proven by the poor performance of econometric models. Even on the eve of the 2008 global financial crisis, no econometric model suggested that the crisis was coming. Ironically, Alfred Cowles, the founder of the Cowles Foundation and the journal *Econometrica*, concluded that stock forecasting is doubtful and abandoned his forecasting business after his failure to foresee the Great Crash. Cootner (1964) found the forecasting of stock prices was at best doubtful. Ormerod (1994) listed the number of forecasting failures by econometricians, including the failure to predict the Japanese recession, the collapse in the German economy, and the turmoil in Europe due to the crisis of the exchange rate mechanism (ERM) system. Generally speaking, econometric predictions were seldom correct. If econometric models could predict the future reasonably well, we would not discuss this issue here—we all would be doing various econometric modelling works right now.

Given the pros and cons of each approach shown above, what is the best way to do empirical research in economics? Two suggestions seem appropriate. One is to use the advantages of each approach and to avoid the drawbacks by the complementary use of other approaches. The other suggestion is that we must focus on the mechanism or the truth behind the facts or data. Since a research question is most likely to be affected by local variables suggested by common sense and theories, one can use, in most cases, the common-sense approach, supplemented by other approaches. On the other hand, the systematic approach is very useful for investigating a system, for example, an economy-wide effect.

Strictly speaking, the probability approach is invalid, but it can still be useful because it can provide information on correlations among different variables. Since the validity of this approach is critically hedged on the condition or assumption that the variables outside the model are unimportant, one must be very careful on three grounds when applying this approach. One is that the modelling results are only indicative because the condition for probability theory only loosely holds even if one has included all important variables. The possibility of omitting important variables always exists. The second aspect is that all attempts at addressing the data problems through econometric methods cannot fix the problem because the source or essence of the problem is not addressed. Moreover, these attempts of data fitting have the potential to do more harm than good because they might cause a loss of important information and might introduce more errors to the data. The third aspect is that the econometric model may reveal the correlation rather than the causality or mechanism between variables, so it is necessary to combine the probability approach with other approaches or with economic theories so as to uncover the mechanism behind the data and so to make a valid prediction.

## 2.4    Progression of Scientific Research and the Future of Economics

What is the future of economics? To answer this question requires a good understanding of the development and nature of the economics discipline. Generally speaking, economics belongs to social science, so the future of economics is related to the pattern of science progression and to the purpose of scientific research. After investigating the nature of scientific research and of the economics discipline, this section forecasts the direction for the development of the economics discipline and suggests the areas of improvement in the future.

### 2.4.1    Nature of Scientific Research and Hypotheses on Progression of Science

The purpose of scientific research is to obtain knowledge by using valid methods. Superficial knowledge can be obtained by using our observation along with our other senses, but deep knowledge regarding the truth or mechanism behind the phenomenon, is more difficult to obtain. While observations, instruments, experiments and other tools are useful in obtaining truth, logical thinking—using both induction and deduction—is an essential tool. However, not everyone agrees with this view. In ancient Greece, sophists denied the existence of truth and they claimed that, if truth did exist, mankind is unable to find it. For sophists, truth is anything which one can be persuaded to believe. Although the view of sophists was defeated by Plato and Aristotle through logical thinking, denying of both truth and our ability to obtain it is still consciously or unconsciously in the minds of some people. The probabilists today are essentially statistical sophists. For them, everything is governed by probability, so there is no mechanism, no truth, or there are different versions of truth dependent on how the probability is played out. As a result, scientific research for them is just to do probabilistic modelling and the results from the modelling are the truth.

Here the author must declare that he is not against probability law—probability is part of our daily life so probability law is useful.

However, probability law is a primitive law because it reveals no mechanism—it tells us the probability under various situations but tells us nothing about what causes the probability. Probability law can give us some low level of knowledge, but much more beyond the statistical work needs to be done for the purpose of scientific research. Given the long history and the customized view of treating probability law as a mechanism, it is an arduous battle to overturn this view. Nevertheless, the battle must be won because this view is the biggest obstacle in our search for truth.

Different views on the nature of scientific research generated different hypotheses on how science progresses. Logical positivism originated with a group called the Vienna circle and gained popularity from the 1920s to 1930s. This group regarded the task of scientists as to develop a logical theory and to form empirically testable propositions. If the propositions are empirically tested and verified, the theory is accepted as true. This view was adopted by many economists who argued that economics is a fact-based study and thus is objective, so economics should be considered a science or positive subject.

Karl Popper (1935) argued that one cannot perform all the tests related to a theory, so the theory cannot be verified as true, but the theory can be disproven by one negative test. This approach is called falsificationism. Thomas Kuhn (1962) proposed a hypothesis of paradigms. A paradigm is regarded as a widely accepted approach or knowledge used by researchers. All researchers solve a problem within an existing paradigm. Existence of anomalies unable to be explained by the existing paradigm may not cause a paradigm change, but accumulation of anomalies will eventually lead to a new paradigm which can better explain the anomalies. So a popular existing paradigm may not embody the truth, and thus may not necessarily be a better paradigm.

Imre Lakatos (1968, 1970) advanced the paradigm hypothesis to the research programme hypothesis. According to him, a research programme involves some hard-core assumptions from which a set of peripheral implications can be developed. Falsifying a few peripheral implications will lead to an adjustment in the research programme but does not lead to a change regarding the hard-core assumptions. Only sufficient falsification will lead to a change of hard-core assumptions and thus a change of research programme. In this view, research is less about the truth than about the programme.

Paul Feyerabend (1975) developed an anarchistic theory of knowledge. He argued that acceptance of any method will limit the creativity of problem-solving so the better science should not be confined to a particular method. He argued that any methods, including rhetoric, can be used by scientists as long as the methods convince people. This essentially rejected the existence of both methods and truth because truth is viewed as anything that people can be convinced to believe. This is essentially a renewed version of the Sophists' view.

These hypotheses each have some valid points, but they are basically the reflection of the proponents' views on the nature of scientific research. The author of the present book views scientific research as a journey to find the truth or mechanism. This view is an evidence-based conviction. The long journey of mankind has found that the earth is a sphere rather than flat, that the sun, instead of the earth, is at the centre of the solar system, and that mankind has been evolved from animals and further from single-celled microbes. This knowledge may not be the ultimate or absolute truth (e.g. we know that the sun is not exactly the centre of the solar system because the sun wobbles or rotates around a point close to the centre of the sun), but at each step we are getting more understanding of the situation and thus getting much closer to the truth. We will never regress to believe that the earth is at the centre of the solar system. However, the search for truth is not a smooth one. It involves the contesting of different ideas; the correct idea may not be established or even be recognized in a short period of time; and overcorrections—throwing out the baby with the bathwater—happen all the time. Based on this view, the author here proposes a new hypothesis: science progresses through digression.

Some analogize knowledge as a snowball that gets bigger and bigger in a rolling process. This snowball analogy captures the feature of knowledge accumulation. Our knowledge base does increase over time and new evidence coming to light can reveal more truth and causes us to reject false theories which may have dominated for centuries. This is just like a snowball rolling down a hill: it picks up more snow and throws out rubbish because of weakness or low cohesiveness.

However, when the rubbish is thrown out, along with it may be a chunk of pure snow. In terms of our knowledge, when an old idea is rejected and is replaced by a new idea, the correct elements in the old idea may also be rejected. For example, when Aristotle's free fall theory was rejected by Galileo, we might have overlooked the impact of air in the case of the free fall of a feather and a stone; when the flat earth theory was rejected, we might have failed to recognize the reasonable elements: due to the enormous size of the earth compared to a person, there is no harm in regarding as flat the patch of earth in our daily life (e.g. a football field). In economics, overcorrection happens too. When Adam Smith and Ricardo rejected the claim of underconsumptionists, they also rejected the link between investment and consumption which had been clearly implied in the claim of early underconsumptionists and explicitly proposed by Malthus. When Keynes rejected classical economics and established the theory of deficiency of effective demand, he rejected the long-established and verified general equilibrium framework. When neoclassical and new classical economists rejected Keynesian and new Keynesian economics, the influence of the demand side in the long run was wholly disregarded.

Overcorrection caused a temporary decrease in the size of the snowball or the knowledge base. However, this result is not a regression but a digression. The rejection of Keynesian economics did not lead to a return to classical economics, but to neoclassical economics; the rejection of neoclassical economics led to new Keynesian economics; and the rejection of new Keynesian thought led to the new classical economics. This type of progression by digression also appears true for our society. The contest between slaves and masters did not lead to a return to a primitive society but to feudalism; the contest between farmers and landlords led to capitalism. As will be shown in this book, the contest between capitalists and workers (or proletarians in Marx's language) will lead to a new invention-based society—patentism.

Understanding the way science progresses has important implications for improving our journey towards truth. Since overcorrection tends to occur, we need to try to absorb the valid elements from a theory while rejecting most of it. In this way, we can avoid the loss of valuable knowledge while discarding incorrect components. In other words, we can reduce the degree of digression and increase the amount of progression. As a result, our journey towards truth becomes smoother and faster.

## 2.4.2   Economics as an Art and as a Science

Is economics an art or a science? This question was in the minds of economists from the very beginning. From Aristotle to scholastics, economists realized the function of the market (voluntary exchange) but in the meantime, condemned the exchange for economic gains. Economics is an area of study involving both objective laws about the economic system and judgement on human life, so it is understandable that early economists tended to mix the function of the market with value judgements. David Hume realized in 1724 that previous scholars tended to make a mistake: they tried to justify a value judgement based on what they discovered about how the economic system worked. Hume proposed a dictum: what ought to be cannot be derived from what is.

Since 'what is' indicates objective or positive laws while 'what ought to be' is a subjective or normative judgement, Hume's dictum can be restated as 'normative statements cannot be derived from positive statements'. As a result, Hume's dictum became the source of division between positive and normative economics. The former indicates that economics is a science and the purpose of studying economics is to uncover the objective laws governing economic activity, just like natural laws in natural science. The latter denies the existence of objective economic laws and regards economics as a judgement of social justice and as a tool to improve human life. Putting it differently, the former regards economics as a science while the latter regards it as an art.

Although having value judgements in their books, Smith and Ricardo generally regarded their production theory and income distribution theory as a science. The distinction between positive and normative economics was formally made by Nassau Senior (1836). This distinction was largely followed by economists later. James Mill saw the role of institutions in income distribution, so he separated his income distribution theory from production theory and allowed social reform to create more equitable income distribution. In this way, James Mill regarded his income distribution theory as normative and his production theory as positive. Marshall adopted a similar eclectic attitude. Keynes obviously believed that economics was an art to save capitalism. Nowadays, most classical (neo- or new) economists regard economics as a science while most heterodox economists view it as an art.

However, positive statements and normative statements are not totally separate ones. Just like the case of 'value in use' and 'value in exchange', there are some internal links between positive and normative statements. One example was the value judgement of Smith and Ricardo on landlords and capitalists. Both economists highly praised the capitalists while regarding the landlords as parasites on the economy. The reason for this value judgement was that they both perceived that capital accumulation by the capitalist was the source of economic growth. By contrast, Marx viewed capitalists as parasites who exploited proletarians because he reckoned that only labour could create added value.

The connectedness of positive and normative statements was also revealed by marginal productivity analysis from the 1870s to 1920s. According to marginal principles, the price of a production factors (e.g. wages or capital rental) equals the marginal product of the factor (labour or capital) under the competitive market. Based on this, John Bates Clark (1847–1938) concluded that the distribution of income induced by competitive markets is an ethical distribution because it rewards the factors of production according to their economic contribution to the social product. To avoid the violation of Hume's Dictum, Marshall argued that marginal productivity determined only the demand for factors while the price of factors was determined by market supply and demand. Even considering Marshall's argument, Clark's conclusion still had some elements of truth as long as the price of a factor equals its marginal productivity.

Can we view economics as a whole as both an art and a science? The answer seems in the negative. Any discipline as a science must be objective in its approach. For example, in a natural science such as physics or chemistry, the purpose of research is to uncover the truth or mechanism, so the researcher should always have an objective mind.

To treat economics as a science, we must admit that objective laws govern the economic system. On the other hand, economics is quite different compared with natural science. The discipline of economics is closely related to politics and the ultimate purpose of economic research is to improve our living conditions, so a subjective value judgement is an inevitable element. Treating economics as an art we emphasize the flexibility and the power of mankind to determine the economic system. These two views seem to contradict each other.

This seeming contradiction can be reconciled by the peculiarity and subtlety of economic laws. Like natural laws, economic laws are objective, so the economic system can operate by itself eventually no matter whether we like it or not; however, unlike natural laws, economic laws are laws about human society so the laws work through human behaviour. Take the market mechanism as an example. Without government intervention, the market can work out equilibrium prices and allocate resources efficiently and effectively. However, to have a functioning market, we must establish a legal framework for private property and for antitrust behaviour. Society may have the free will and power not to establish this legal framework, but this will lead to inefficient resource allocation and undermine the development of society and the welfare of everyone in the society. This has been proven by the attempts to establish a Utopian society, socialist society and communist society. In the end, people will learn the lesson through interactions between social classes and through different attempts, and thus will lay the foundation for a functioning market eventually. In this sense, the market mechanism is objective but is also realized through human actions.

Economics as both a science and an art can shed light on how to develop and benefit from this subject. On the one hand, we need to respect economic laws and do our best to uncover them. Because economic laws are objective, it is no use to act against them. Otherwise, one will learn the lesson in a hard way. On the other hand, we should advocate views and policies in accordance with economic laws. As these laws work through our actions, active actions will shorten the trial-and-error process and lead to early adoption of economic laws. In this way, the development of the human civilization will speed up and will enter a new phase.

## 2.4.3  Future of Economics

In uncovering and advocating economic laws, economists have an important role to play. However, economists must change the current state of economics. The necessary changes include emphasizing logical thinking, focusing on the essence of the theory rather than the appearance of a theory, allowing for contests between different economic ideas, uncovering and improving valid empirical methods and bridging the gap between theory and data.

### 2.4.3.1  Emphasizing Logical Thinking

As a result of the dominance of the probability approach, the importance of logical thinking is significantly downplayed. According to the probability approach, everything is determined by probability law, so there is no need to use logical thinking to uncover the causes or mechanism behind the statistics; rather, what we should do is to accept the results and the predictions from the econometric models. This attitude downgrades human intelligence and has hijacked the purpose of scientific research—to uncover the truth. For the development of our society as well as the economics discipline, this attitude must be rejected and logical thinking must be highlighted.

The importance of logical thinking cannot be emphasized enough because it is an essential tool to uncover the truth. One may argue that a 'logic jump' is necessary for a scientific breakthrough because an invention, innovation or new idea is most likely the result of a logic jump rather than logical thinking. Moreover, it can be argued that many theories in physics (e.g. quantum theory and Einstein's relativity theory) are contradictory to logical thinking and to our common sense. These arguments confuse the use of brainstorming process to generate new ideas with the logical thinking in order to uncover the truth.

Logical thinking plays a very minor role in generating new ideas because everything based on logical thinking is foreseeable outcome of premises and thus by nature is not new. However, the new ideas obtained through brainstorming or logic jumps must be tested by evidence or experiments. Not all logic jumps are valid. We can accept the results of logic jumps only if they are supported by evidence, for example, the idea of a 'perpetual machine' was not supported by evidence and thus was rejected; on the other hand, Einstein's claim of constant speed of light in vacuum is accepted based on observed evidence, although based on our common sense this claim is illogical. Even in the case of the valid logic jump like Einstein's claim, the illogical nature of the claim indicates that there must be something deeper which can explain the apparent contradiction with common sense. In other words, Einstein's theory may be not deep enough and may need to be developed further. By this reasoning, once a new idea is formed and is proven to be valid, logical thinking plays a vital role in developing a theory and in uncovering the truth.

### 2.4.3.2 Focusing on the Essence of a Theory and the Contest of Economic Ideas

With the abstract or formative approach having defeated the contextual approach, the economics discipline is focused more on appearance and fashion rather than on the essence and usefulness of a theory.

One example is the use of mathematics. It is undeniable that mathematics plays an important role in economic thinking. It expresses an economic problem concisely and can capture the essence of a problem. It allows a problem to be addressed quantitatively and rigorously. However, the use of mathematics has its drawbacks. It requires more advanced mathematical training and thus is less accessible to the general public. It is a powerful but complex instrument. In the wrong hands, it expresses a research question incorrectly, it can conceal a logical mistake in a way hard to be detected (e.g. an implicit change of condition for an equation), and thus it may confuse both the researcher himself and the audience. In other words, mathematics can be an error-prone blunt instrument in an inexperienced hand and a magic tool in the hands of a skilful mathematic magician. The excessive use of mathematics in economics contributed to both the decline of readership of economic journals and the decline of people interested in studying economics. It also contributed to the pursuit of the cutting-edge appearance of journal articles and books while neglecting the purpose and usefulness of a theory.

The other example of focusing on appearance and fashion is the economists' obsession with 'innovations'. There is no doubt that innovations are important and should be encouraged for the purpose of our society and for the development of the economics discipline, but the purpose of innovation is to solve a problem and to benefit society. Most importantly, the innovation must be valid. Currently, the economists' pursuit of innovation is driven by fashion and by the desire to publish a paper or to make a household name, rather than by the usefulness to the discipline and to society. Numerous estimation methods and testing tools are invented by econometricians. These innovations do make popular some journals and some econometricians, but how much impact do they generate on our economic system and to our daily life? Almost none. Most importantly, these estimation or testing methods are not valid because the condition for the probability theory does not hold in the case of time series economic data. Any innovation attempting to address artificially the problems in economic data are the same as the plans to renovate a mansion without fixing its weak foundation. These types of innovations are invalid and useless, and thus should be abandoned.

The pursuit of fashion and appearance and neglecting the essence of a theory has led to a reduced interest in developing and contesting economic ideas. Essentially, the great empirical period produced very few economic ideas so there are very few new ideas to contest. Moreover, the pursuit of fashion and appearance led mainstream economists to focus on specific types of models and thus they have difficulty communicating effectively with people outside of their specialized area, let alone to contest ideas. It is heterodox economists who are interested in and have made a significant contribution to the contest of economic ideas. While the immediate purpose of contesting ideas is to popularize the proposed ideas, the ultimate goal should be to uncover truth. A false idea may gain popularity for a while or even for an extended period of time, but the truth will eventually defeat the false idea, so only the truth will become popular in the end.

### 2.4.3.3 Uncovering Valid Empirical Methods and Bridging the Theory-Data Gap

Along with the development of information technology, the amount of data available today grows exponentially. In order to utilize these data, it is necessary to develop empirical methods. However, these empirical methods must be valid ones. As we have discussed briefly in this chapter and will discuss further in Chapter 3, many methods developed in econometrics are invalid, because they require the application of probability theory while the data used do not satisfy the condition for probability theory. A careful examination is required to verify which econometric methods are valid and whether the conditions for these methods are valid. Using invalid methods not only causes unnecessary waste of resources and time, but also produces misleading results and causes detrimental impacts on research and on economic activities.

In developing empirical methods, it is important to bear in mind that the purpose of empirical research is to form or test economic theories so as to uncover the mechanism governing economic activities. The nature of social science dictates that we have difficulties (either moral or economic) in obtaining experimental data, so many factors have influence on the data. However, an economic theory normally involves a number of assumptions in order to exclude the influence of a wide range of factors. This causes the gap between economic theory and economic data. This gap is hard to bridge because economic data generally do not satisfy the condition of economic theory. So, how do we bridge this gap?

Various approximations and empirical methods must be used to close the gap between the conditions of the data and the conditions of the theory. Moreover, since the conditions for the data and for the theories are not the same, the empirical results might be unable to prove or disprove a theory due to the approximate nature of the methods used and the disparity between empirical and theoretical conditions. When the difference between theory predictions and empirical results is found, we need to examine which factors on both the data and theory sides may cause the difference. Based on this, we can improve both the theory and empirical research. This process can be continued again and again. If the difference between theoretical and empirical research becomes reasonably small, we may conclude that the theory is supported by empirical data. If the difference is still large, we may conclude that the empirical data do not support a theory and we can pinpoint where the theory is wrong. Compared with the econometric methods used in current empirical research—simply to put some data into a model and generate results, the way suggested here is an arduous one, but it is a necessary and valid one to bridge the gap between economic theory and economic data.

# References

Arrow, K. J., & Debreu, G. (1954). Existence of an Equilibrium for a Competitive Economy. *Econometrica, 22,* 265–290.

Bohm-Bawerk, E. (1891 [1888]). *The Positive Theory of Capital* (W. Smart, Trans.). London: Macmillan.

Bohm-Bawerk, E. (1922 [1884]). *Capital and Interest: A Critical History of Economic Theory* (W. Smart, Trans.). New York: Brentano's.

Burns, A., & Mitchell, W. (1946). *Measuring Business Cycles.* New York: Columbia University Press.

Clark, J. (1908). *The Distribution of Wealth: A Theory of Wages, Interest and Profits.* New York: Macmillan.

Cootner, P. (1964). *The Random Character of Stock Prices.* London: Risk.

Feyerabend, P. (1975). *Against Method: Outline of an Anarchist Theory of Knowledge.* London: Verso Books.

Fisher, I. (1907). *The Rate of Interest: Its Nature, Determination, and Relation to Economic Phenomenon.* New York: Macmillan.

Fisher, I. (1930). *The Theory of Interest.* New York: Kelley and Millman.

Guan, Z. (1998). *Guan Zi* (W. Allyn Rickett, Trans.). Princeton, NJ: Princeton University Press.

Guthrie, W. (1967). *The Greak Philosophers—From Thales to Aristotle.* Norwich, UK: Fletcher & Son Ltd.

Haavelmo, T. (1944). The Probability Approach in Econometrics. *Econometrica, 12*(Supplement), 1–115.

Hutcheson, F. (1750). *Reflections upon Laughter, and Remarks upon the Fable of the Bees.* Printed by R. Urie for D. Baxter.

Hutcheson, F., & Leechman, W. (1755). *A System of Moral Philosophy.* Sold by A. Millar.

Jonsson, P. O. (1995). On the Economics of Say and Keynes' Interpretation of Say's Law. *Eastern Economic Journal, 21*(Spring), 147–155.

Keynes, J. M. (1936). *The General Theory of Employment, Interest, and Money*. London: Macmillan.

Knight, F. (1921). *Risk, Uncertainty, and Profit*. Boston: Houghton Mifflin.

Knight, F. (1949). *Capital and Interest, in Readings in the Theory of Income Distribution*. Philadelphia: Blakiston.

Kuhn, T. (1962). *The Structure of Scientific Revolutions*. Chicago: University of Chicago Press.

Kydland, F., & Prescott, E. (1982). Time to Build and Aggregate Fluctuations. *Econometrica, 50*(6), 1345–1370.

Lakatos, I. (1968). Criticism and the Methodology of Scientific Research Programs. *Proceedings of the Aristotelian Society, 69*(1), 149–186.

Lakatos, I. (1970). Falsification and the Methodology of Scientific Research Programs. In M. Lakatos (Ed.), *Criticism and the Growth of Knowledge* (pp. 91–195). New York: Cambridge University Press.

Landreth, H., & Colander, D. (2002). *History of Economic Thought* (4th ed.). Boston: Houghton Mifflin.

Lauderdale, J. (1804). *An Inquiry into the Nature and Origin of Public Wealth*. Edinburgh; London: Constable & Co.; T.N. Longman & O. Rees.

Leontief, W. (1936). Quantitative Input and Output Relations in the Economic Systems of the United States. *The Review of Economics and Statistics, 18,* 105–125.

Leontief, W. (1941). *Structure of the American Economy, 1919–1929: An Empirical Application of Equilibrium Analysis*. Cambridge: Harvard University Press.

Leontief, W. (1970). Environmental Repercussions and the Economic Structure: An Input-Output Approach. *The Review of Economics and Statistics, 52*(3), 262–271.

Leontief, W. (1974). Structure of the World Economy. *The American Economic Review, 66,* 823–834.

Leontief, W. (1985). The Choice of Technology. *Scientific American, 252*(6), 25–33.

Lucas, R. (1972). Expectations and the Neutrality of Money. *Journal of Economic Theory, 4,* 103–124.

Lucas, R., & Prescott, E. (1971). Investment Under Uncertainty. *Econometrica, 39*(5), 659–681.

Malthus, T. (1820). *Principles of Political Economy*. London: W. Pickering.

Mandeville, B. (1723). *The Fable of the Bees, or, Private Vices, Public Benefits* (2nd ed.). London: Printed for Edmund Parker.

Marshall, A. (1948 [1920]). *Principles of Economics*. London: Macmillan.

McCulloch, J. R. (1864 [1965]). *The Principles of Political Economy, with Some Inquiries Respecting Their Application* (5th ed.). New York: Hugustus M. Kelley.

Mill, J. (1808). *Commerce Defended*. London: C. and R. Baldwin.

Moore, H. L. (1908). The Statistical Complement of Pure Economics. *Quarterly Journal of Economics, 23,* 1–33.

Moore, H. L. (1911). *Laws of Wages: An Essay in Statistical Economics*. New York: Macmillan.

Moore, H. L. (1914). *Economic Cycles: Their Law and Causes*. New York, NY: Macmillan.

Muth, J. (1961). Rational Expectations and the Theory of Price Movements. *Econometrica, 29*(3), 315–335.

Ormerod, P. (1994). *The Death of Economics*. London, UK: St. Martin's Press.

Popper, K. (1935 [1992]). *The Logic of Scientific Discovery*. London: Routledge.

Ricardo, D. (1952). *The Works and Correspondence of David Ricardo* (P. Sraffa & M. Dobb, Eds., Vol. 6). Cambridge: Cambridge University Press.

Say, J.-B. (1803). *Traite d'Economic Pol itique* (1st ed.). Paris: Deterville.

Say, J.-B. (1821). *A Treatise on Political Economy*, translated from the fourth edition of the French. London: Paternoster-Row.

Schumpeter, J. (1986). *History of Economic Analysis*. London: Routledge.

Senior, N. (1836 [1951]). *An Outline of the Science of Political Economy*. New York: Kelley.

Shoven, J. B., & Whalley, J. (1972). A General Equilibrium Calculation of the Effects of Differential Taxation of Income from Capital in the U.S. *Journal of Public Economics, 1*(3–4), 281–321.

Sims, C. (1980). Macroeconomics and Reality. *Econometrica, 48,* 1–48.

Sismondi, J. C. L. (1819). *Nouveaux principes of d' economie politique* [New Principles of Political Economy]. Paris: Delaunay.

Smith, Adam. (1776 [1904]). *An Inquiry into the Nature and Causes of the Wealth of Nations* (E. Cannan, Ed.). London: Methuen.

Solow, R. (1956). A Contribution to the Theory of Economic Growth. *Quarterly Journal of Economics, 70*(1), 65–94.

# 3

# Statistical Sophistry

The application of probability theory to macroeconomics has created macroeconometrics and changed the course of development in macroeconomics. The studies on macroeconomic theories have been marginalized and macroeconomics has almost become an empirical subject. In this chapter, the author argues that the application of probability theory to macroeconomics is theoretically flawed and, in practice, is plagued by insolvable problems and logical issues. The inappropriate use of the probability theory in macroeconomics has dire consequence in economic research, so the trend of econometricalization must stop.

Since macroeconometrics is essentially an application of statistical theory, it is necessary to review the development of statistical thought in order to have a concrete grasp of the foundation of macroeconometrics. Section 3.1 serves this purpose. Since the purpose of our historical review is to provide the reader with sufficient background information, Sect. 3.1 focuses only on the main storyline of the development of statistics. To make the review accessible to a general audience, the author has used very little mathematics and jargon. Some readers may be interested in a detailed review of the history of statistical thought. The author recommends the book by

© The Author(s) 2019
S. Meng, *Patentism Replacing Capitalism*,
https://doi.org/10.1007/978-3-030-12247-8_3

Chatterjee (2002) for those with an intermediate level of knowledge in mathematics and statistics and the book by Hald (1998) for those with an advanced level of knowledge in these areas. Based on the historical review in the previous section, Sect. 3.2 discusses the premise and limits of probability law, which form the crucial foundation of this chapter.

Section 3.3 provides a historical account of the development of macroeconometrics, including criticism and debates of different stages. Through examining the theoretical foundation of macroeconometrics and illustrating the problems in the practice of macroeconometricians, Sect. 3.4 rejects the model fitting approach. Sometimes it is argued that the performance of a theory serves as the criterion for the validity of a theory, so Sect. 3.5 assesses the performance and impact of macroeconometrics. Section 3.6 concludes the chapter by summarizing the lessons to be learned from the long detour in the development of macroeconomics and by providing some suggestions for dealing with macroeconomic data.

## 3.1    A Brief Review of History of Statistical Thought

Statistics has a long history. The ancient Egyptians collected data on population and wealth as early as 3050 BC. China also had accounts of statistical work from 2300 BC. A census was conducted in Greece in 594 BC and another in Athens in 309 BC. Later, Romans conducted extensive detailed country-wide surveys. Statistical work also was conducted by almost all countries in history, notably, the census known as the 'Doomsday book' prepared for William the Conqueror in 1088, the deaths registration commenced in London in 1532, and the decennial census of population in the USA commenced in 1790. The census taking activity spread to other countries such as France, England and Belgium. In 1902, the US census bureau became a permanent institute. These statistical activities demonstrate the importance of statistics in our society.

Although statistics in daily parlance has a broad meaning, statistical thought is largely related to probability and belongs to the domain of epistemology. Chatterjee (2002) regard the core of statistics as 'a prolongation of inductive reasoning' even though deductive reasoning is also necessary in statistical work. There are some disputes among philosophers as to whether inductive reasoning can reveal true knowledge or not. Hume (1739) puts up the problem of the derivation of natural laws of the causal type, i.e. the potential problem in deriving a natural law by generalizing the results from particular observations. There are several solutions to Hume's problem, of which the solution by Popper (1963) can be accepted by most people: induction can lead to a hypothesis (potential truth) which is true until it is falsified by further observations. From this point of view, the progress of science is the process of successively setting up and falsifying hypotheses in the light of new evidence available.

The key feature of statistical induction is related to uncertainty or, more precisely, probability. Uncertainty is very common in our daily life. We may spot a lot of wild mushrooms after a rain, but we may not always find mushrooms after a rain. After a couple of years of experience, we can conclude that we will have a high chance (probability) of finding wild mushrooms after a rain. Human beings have used the concept of probability for a very long time but in an implicit way.

The birth of probability theory started with the study of gambling activities. According to Hacking (1975) and Hald (1990), Gerolamo Cardano (1501–1576), a renowned Italian physician and mathematician as well as an inveterate gambler, wrote a book 'Liber de Ludo Aleae' (the Book on Games of Chance) in 1564. Galileo Galilei (1564–1642) wrote a small piece around 1620 to answer a puzzle of dice-players: in throwing three 6-facet dice, the chance of getting a total of number 10 is higher than 9 (Hald 1990). Later, Pascal, Fermat and Huygens contribute significantly to the calculus of probability.

A French nobleman and gambler, Chevalier de Mere, asked Blaise Pascal (1623–1662) to solve the problem of a minimum number of trials: throwing a single 6-facet die 4 times is enough to have a better-than-even chance to get at least one outcome of 6. Pascal solved the puzzle by calculating probability: the chance of getting a number 6 in each throw is 1/6, so the change of getting a number other than 6 is 5/6. In throwing the die four times, the chance of getting a number other than 6 is $(5/6)^4 = 625/1296$, and the change of getting at least one 6 is $1 - (5/6)^4 = 671/1296$. So the chance of getting one 6 to the chance of getting none of number 6 is 671/625, which is better than even. Pascal, Fermat (1601–1665) and Huygens (1629–1695) also solved the division problem in die throwing games. The book 'De Ratiociniis in Ludo Aleae' (Computations in Games of Chance) in 1657 by Huygens remained the standard textbook for about half a century.

After Pascal, Fermat and Huygens, James Bernoulli (1654–1705) continued the contribution to probability for a repeatable trial by establishing the 'limit theorem'. More importantly, Bernoulli contributed to the development of the probability concept. Before Bernoulli, probability was regarded as an objective concept. The book of 'Ars Cogitandi' (the Art of Thinking) by Arnauld and Nicole has four chapters devoted to probabilistic reasoning. The book suggests probability should objectively be judged on the basis of past events. According to Chatterjee (2002), this objective approach to probability reflected the view of Pascal because he was associated with the book's authors at that time. In his book 'Ars Conjectandi' (the Art of Conjecture), James Bernoulli expressed an impersonal subjective view on probability by arguing that things are uncertain to us only because of the limitedness of our information. Later, Bayes proposed a personal subjective view of probability based on the concept of expectation. With the development of the 'sampling theory' in the 1900s, the objective view of probability again becomes a dominant view.

The application of probability to early social work is limited due to the limited data available. The most notable work is Gaunt's work on the Bills of Mortality and on plague epidemics. On the other hand, the application of the probability theory to astronomy and geodesy was flourishing and fruitful thanks to a large amount of observation data. In the latter part of the sixteenth century, the Danish astronomer Tycho Brahe introduced the probability approach into astronomy in order to find accurate determination based on repeated observations. In the seventeenth century, Galileo (1564–1642) studied the impact of errors on measurements of stellar distances and made three observations: (a) errors are unavoidable in making instrumental observations; (b) small errors occur more frequently than large errors; and (c) errors tend to be equally distributed in two directions. These features are the early statements of normal distribution—a symmetric bell-shaped probability distribution centred at 0. Along this line, Thomas Simpson (1710–1761) and Daniel Bernoulli (1700–1782) made significant contribution. Statistical work in astronomy was continued by Pierre-Simon Laplace (1749–1827) and Carl Friedrich Gauss (1777–1855). As a result, the theory of errors was developed.

The probability method in astronomy was brought to the social science domain by Adolphe Quetelet (1796–1874) and Simeon Poisson (1781–1840). Using the theory developed by Laplace, Poisson estimated the sex ratio at birth in France from 1817 to 1826. He showed that the probability of male birth for the whole country was stable but not for each administrative district. Based on this result, Poisson put forward the law of large numbers—as numbers become larger (e.g. the population for the whole country), the ratio is stable. Using the same method, Poisson also estimated the conviction rate (the proportion of the convicted among the total accused) based on the French judicial statistics from 1825 to 1839. He also found that the overall conviction rate was stable for the whole country but not for each district, nor for the particular type of crime (e.g. crimes against person and crimes against property). This further supported his claim of the law of large numbers.

Using a concept of 'average man', Quetelet estimated the relative distribution of physical and social attributes of different groups such as height, chest measurement, birth, marriage, drunkenness, suicide and crime. One of his studies used the data on the chest measurements of about 6000 soldiers to fit a curve of normal distribution. He also examined the distribution of heights of 100,000 French conscripts. He found that, as long as the group was homogeneous, the normal curve fitted the empirical distribution well.

Interestingly, Quetelet's research inspired Maxwell to develop the kinetic theory of gases. Using a probability approach, Maxwell found that, although individual gas molecules behave erratically when temperature changes, the ensemble of gases demonstrate regularities in behaviour. Boltzmann went further with this idea and successfully explained the second law of thermodynamics.

The research of Quetelet in social science and Maxwell and Boltzmann in physics caused considerable disquiet in social science, physics and philosophy. Quetelet's study demonstrated that similar to the situation in natural science, there are regular laws governing social behaviour. This means that there are some restrictions on human freedom. On the other hand, the application of the probability theory by Maxwell and Boltzmann to the behaviour of gases indicated that the physic world may be governed by probability law, instead of by deterministic law. Heisenberg's uncertainty principle is a further development along this line. The debate about the nature of the social world and the physic world continues to the present day.

The statistic work in social science in the later 1800s was progressed by Francis Galton (1822–1911). As a cousin of Charles Darwin, Galton was inspired by the Origin of Species and held an idea that the quality of the human population may be improved by the promotion of scientific artificial selection (i.e. judicious marriages). As a result, Galton was interested in the deviation from the average (i.e. how much the quality of a group is improved by judicious marriages compared with the average quality of the whole population). Along this line, Galton's work later moved to genealogical studies and developed the theory of correlation and the bivariate normal model.

The bivariate normal model was extended to the multivariate normal model by Francis Edgeworth (1845–1926) and Karl Pearson (1857–1936). Since the model residuals (i.e. the difference between the historical data and the predictions from the model) in social studies are often found non-normal, Pearson and other statisticians developed different types of distributions (notably $\chi^2$ distribution and student's-$t$ distribution) through the methods of mixture, compounding and formal embedding. Based on these distributions, a model can be specified and fit with empirical data. The model-fitting approach developed by Pearson was spread to diverse fields and continues to the present day, so Pearson is widely regarded as the founder of modern statistics.

Another important person contributing significantly to modern statistics was Ronald A. Fisher (1890–1962). Two of Fisher's major contributions are the development of the theory of estimation for the sampling theory and the development of the design of factorial experiments. The randomization in his factorial design tends to balance out the conditions of the experiments so that the heterogeneity decreases and the experiments become relatively closer to random experiments.

Further development in modern statistics is largely along the line of model fitting, with improvements including more types of models, tests and estimation methods. With more data of large sample sizes becoming available, the model-free or non-parametrical approach also appeared. The application of statistical method on economic data created a new subject—econometrics. The application on economic time series data created macroeconometrics, which is used in macroeconomics so widely that it has essentially replaced macroeconomics. We will discuss the application of the probability theory in macroeconomics in Sect. 3.3.

## 3.2 Premise and Limits of Probability Law

Based on the historical review in Sect. 3.1, we can discuss the premise and limits of probability law, which are crucial to a valid application of the probability theory. In practice, the premise and limits of probability law may not be satisfied due to limitations in the data. In this case, how far the model departs from the premise and its limits can be used to assess the reliability of results from a probability model. This section forms the foundation for our criticism of macroeconometrics in this chapter.

### 3.2.1 Premise of Probability Law

From Sect. 3.1, we know that formal studies on probability started with gambling events such as throwing dice or tossing a coin. This kind of event can be called random experiments, i.e. experiments conducted under approximately the same condition. For random experiments, probability law is an indisputable scientific theorem because the law was proved in the past and can be examined by random experiments anytime. We propose random experiments as the condition or premise of probability law here and we will examine the consequences of moving away from this premise.

In some cases, it is impossible or unethical to perform a random experiment. One of these cases is in astronomy—we can observe the movement of planets and stars but, so far, we have no ability to do experiments with them. Even so astronomic observations can be viewed as the results of random experiments. Compared with the life of planets and stars, the repeated observation period is very short, so the orbit of planets and the position of stars are almost unchanged during the repeated observation periods. As such, observations are repeated under approximately the same condition and the observation activity can be regarded as a random experiment. In dealing with the observation data, Laplace and Gauss used a probability model and developed the theory of error. Their model assumed the measurement error as random and this assumption was proven plausible by the normal distribution of residuals. This method used in astronomy was also proved successful.

The method used in astronomy was used by Quetelet for his research on attributes of the human population. As shown in Sect. 3.1, Quetelet studied the data on the chest measurements of about 6000 Scottish soldiers and on the heights of 100,000 French soldiers. In both cases, the data fit a curve of normal distribution. He concluded that homogeneity of the group was the condition to fit a curve of normal distribution. Here, a homogeneous group means that the conditions other than the variable of concern are approximately the same, so the data can be viewed as an outcome of a random experiment. It is of no surprise that the normal distribution can fit the data well.

However, when the probability model was generalized in social science, the model residuals are often found non-normal. This is not a surprise given the heterogeneous nature of social data. For example, the distribution of income of a country depends on many factors such as education, race, age and government policy. For a country with wide income gap, the income distribution is most likely not normal. However, the income distribution for a relatively homogeneous group (e.g. of the same age, same race, same level of education, etc.) would be expected to be normally distributed because, in this case, the survey data can be viewed as an outcome of random experiments. The homogeneity requirement is emphasized by Pearson et al. (1899) and is implicitly addressed by Fisher (1947) in his emphasis on the design of experiments.

The importance of homogeneous groups is often understated. Using our previous example, Poisson proposed the law of large numbers based on his study on the male-birth rate (the proportion of the number of males at birth in the total number of births) and the conviction rate (the proportion of the number of convicted to the total number of the accused) in France. He found that, for the studied period, the overall rates are stable for the whole country but not for each administrative district. This gives us the impression that when the sample numbers get large enough, the heterogeneous issue is not important. This is a misunderstanding.

Armed with modern genetic knowledge, we can explain why the seemingly heterogeneous issue is not relevant in the case of male-birth rate. The sex at birth is determined by the sex gene from the male parent. There are a pair of different sex chromosomes X and Y in the male human genome but a pair of same-sex chromosomes X in the female human genome. As the male human genome splits up to form sperm cells, chromosomes X and Y are separated and go to different sperm cells, so the number of sperm cells containing X sister-chromatid, which will produce female babies, equals the number of sperm cells containing Y sister-chromatid, which will produce male babies. As a result, the chance of producing a male and female child is 1:1. This explains why the ratio of the male-birth rate is close to ½. The perceived heterogeneity in a population such as income difference and education difference does not affect, and thus is irrelevant to, male-birth rate. In other words, in terms of the probability of sex at birth, there is no heterogeneity in a population. The law of large numbers does not contribute to solving the heterogeneity issue, but it does contribute to a stable male-birth rate when a population is large. The role of the law of large numbers here is the same as the role of large repeated trials in random experiments.

The same reasoning can be applied to the case of conviction rate. The stable overall conviction rate for the country as a whole is due to the same judicial system in a country: the judges or juries judge the case based on the same law, so the conviction rate is largely determined by the standard of the law and will become more stable as the number of cases becomes larger. If we consider the conviction rate for a number of countries, we will find that the conviction rate for each country tends to be stable but the overall conviction rate for all countries may not be stable because of the different law system and standard in each country and the variation of the number of cases in each country over time. The reasoning regarding heterogeneity here is the same as that in the example given by Pearson et al. (1899). This indicates that, not only is the law of large numbers unable to solve the problem of the heterogeneity issue but that the presence of heterogeneity can invalidate the law of large numbers—if the population is heterogeneous in terms of research purposes, the large numbers will not produce a stable expected probability.

Although the non-normal distribution in social studies is an indication that the data cannot be viewed as an outcome of random experiments due to heterogeneity problems, statisticians (including Pearson and Fisher) at the turning of 1900 started to derive other distributions to be used to fit a model. This type of model fitting marks the beginning of 'modern' statistics. The application of model-fitting techniques to macroeconomics created macroeconometrics, which necessitates a more severe heterogeneity problem because the condition of generating economic time series data changes vastly over time. The trend of model fitting regardless of the heterogeneity issue drove statistic study further away from the requirement of the random experiment and made statistical studies less and less scientific. We will discuss the implication of this trend in Sect. 3.4.

## 3.2.2   Limits of Probability Law

Every scientific law has its limits, probability law is of no exception. Too often the limits of probability law are ignored so the scientific statistics are pushed into the space of pseudoscience. In this section, we will examine the limits of probability law.

1. Probability law is only applicable to limited uncertainty cases

Probability is related to uncertainty, so probability law has been widely regarded as a law to deal with any uncertainty case. It is widely accepted that uncertainty is a common phenomenon in our life and that certainty is a special case of uncertainty, i.e. a certainty event is an uncertain one with a probability of 100%. Because of this, probability theory is viewed by some statisticians and economists as a super theory to deal with everything in the world. This view vastly exaggerates the function of probability theory. For example, can the probability theory predict how long someone will live or can it at least provide a probability of the lifespan of an individual? Econometricians may claim that they are able to do just that. How reliable is their prediction? How can we benefit from these predictions? Even if the statistician can have some clues from the lifespan of his/her parents, from his/her current health situation and lifestyle, any prediction will fail because there are so many uncertainties in the future, e.g. there may be an earthquake or a cyclone in the area, he/she might be caught in a car accident or an airplane crash, there may be a war in 10 years' time. So many known and unknown uncertainties make a reliable prediction impossible. In short, the probability law is not a magic tool to deal with all uncertainties.

However, probability law does have the power to deal with uncertainty proved by random experiment. Although the law cannot predict the outcome of the next throw of a die, the law can predict that if one repeatedly throws a die of 6 facets, the chance of winning is 1/6. This prediction can be proven by random experiments. In the case where it is infeasible to do random experiment, the closer the case is to the condition of random experiments, the more reliable the prediction that the probability law can make.

## 2. Probability law is only a rough guide for our ignorance

What is uncertainty or probability exactly? There are both objective and subjective views. The outcome of a coin toss game or the movements of glow-worms in a cave is not affected by anybody's mind, so the outcome is definitely objective. But this does not mean that the uncertainty or probability is objective. If one assumes that there is no way to obtain more information on the uncertainty and probability, then the uncertainty and probability are independent of the observer's mind and thus can be viewed as objective. However, this is not always the case. For example, the movements of glow-worms are not uncertain to the worms themselves. In the case of coin tossing, if a very small and sensitive creature can detect and calculate accurately the force of tossing and the movement of air, it can tell the outcome of each toss once the coin is tossed. In this case, the uncertainty and probability to the small creature are totally different from those to human beings. As such, uncertainty and probability can be viewed as an 'impersonal subjective' view: uncertainty or probability of an event is the same to any individual human being due to our shared limitation in perception or in knowledge. If the person who tosses the coin can, like the tiny creature mentioned above, detect and calculate the force of toss and the movement of air, he/she will know the outcome before the coin is tossed, so the uncertainty and probability to him/her are dramatically different from those of other human beings. In this case, uncertainty and probability can be viewed as 'personal subjective'. The common example of 'personal subjective' probability is car insurance. Due to the information asymmetry, the probability to the insurance policy issuer and the policyholder is quite different.

From the above reasoning, it is clear that different views to uncertainty and probability are all valid. The key here is the different assumptions about the limitations in our perception and knowledge. It is common sense that human beings have limitations in perception and knowledge, but history also tells us that our limitation in perception and knowledge can improve over time. Hundreds of years ago, we could not find out what is in the air. Nowadays, with the help of instruments, we can detect electrons and even smaller particles like quarks. A more convincing case is our understanding of cholera. This was a terrifying disease and people had no idea about the chance of catching the disease. John Snow did a lot of data collection and research work on the disease and found the epidemic tended to strike certain areas of London. Finally, it was found that the disease was spread through contaminated water. Modern microbiology has identified the germs that cause cholera and thus, this disease is no more uncertain to us.

Natural disasters caused by earthquakes, eruption of volcanos, hurricanes, etc., are still uncertainties to us because of our limited knowledge. If we have more knowledge and more and relevant data on these events, we may foretell when and where these events will happen. As such, probability law can be viewed as a law of ignorance. With limited knowledge, we have to use probability law as a guide to cope with these uncertainties. However, this guide is a rough one, because it cannot predict events with certainty but only gives a chance of events occurring.

3. Probability law contains no causality and no mechanism

Probability law can tell the probability of the outcome in random experiments, but the law tells nothing about what causes the outcome. The results from a probability model indicate only the correlations rather than causality. So probability law is quite different from other natural laws such as Newton's three laws on force and Darwin's evolution theory, which manifest the causes or mechanisms for the cases of concern. From this point of view, probability law represents a lower level of understanding than the other natural laws. Ignoring the differences between probability law and other natural laws, Quenetlet created the conundrum of the restriction on the mental world (social physics) and the indeterminacy of the physical world.

Many statisticians and econometricians tried but failed to include causality into a statistic framework. This is not surprising because probability law does not reveal causality. By the same reasoning, the predictions from a probability model can only be a rough guide because there is no causality in the model. Sometime, the prediction from a probability model may even be misleading. For example, some people regressed the data of ice cream consumption and crime rate and found significant positive correlation. It would be ridiculous if someone predicts that people eating more ice cream are more likely to commit a crime. This type of regression (or correlation) is called spurious regression (or spurious correlation). To avoid spurious correlation and improve the reliability of prediction, the low level of understanding presented by probability law needs to be upgraded by stepping outside of the statistical domain and using logical thinking to find out the causality or mechanism in each case.

## 3.3   Rise of Macroeconometrics Despite Numerous Grave Criticisms

In the late 1800s and early 1900s, statistical methods and probability theory experienced significant development, hence it was not surprising that statistical approaches were introduced into economics in the early 1900s. Henry Moore (1869–1958) is considered the pioneer in applying statistical method to economic research thanks to his work in estimating demand curves (Moore 1911, 1917). Ragnar Frisch (1895–1973) and Jan Tinbergen (1903–1994) are regarded as the pioneers in formalizing the statistical approach in economics. Frisch (1934) argued that most economic variables were simultaneously interconnected in 'confluent systems', which needed to be analysed by means of multiple regression systems. Tinbergen's pioneering work was contained in his report for the League of Nations, Statistical Testing of Business-Cycles Theories (Tinbergen 1939), in which he introduced multiple correlation analysis, built a parametrized mathematical model and applied statistical testing procedures.

The introduction of statistical methods into economics was highly suspected of being inappropriate by the then influential economists such as Alfred Marshall (1842–1924), Francis Edgeworth, and John Keynes (1883–1946), who argued that the 'ceteris paribus' (i.e. other things being equal) assumptions in economics made it difficult to apply statistical methods. In fact, Tinbergen's work was reviewed and heavily criticized by Keynes. His criticism was published in the Economic Journal (Keynes 1939). Keynes first pointed out that Tinbergen did not explain clearly the conditions that the economic data must satisfy in order to apply the statistical method. Then, he pointed out the specific flaws in Tinbergen's work: (1) statistical tests cannot disprove a theory because multiple correlation analysis relied on economic theory to provide a complete list of significant factors; (2) not all significant factors are measurable so the model may not include all relevant variables; (3) different factors may not be independent of each other; (4) the implausible assumption that the correlations under investigation are linear; (5) the treatment of lag and trends is arbitrary; and (6) Tinbergen's work is a piece of historical curve-fitting and description and thus is incapable of making an inductive claim.

Tinbergen's reply to Keynes was published in the Economic Journal the following year (Tinbergen 1940). He avoided the issue of logical conditions for applying multiple correlations but answered the other questions. Regarding the complete list of relevant factors, Tinbergen claimed that it is not clear beforehand what factors are relevant, so they are determined by statistical testing; for independence of variables, he argued that the statistical requirement for explanatory factors was uncorrelated rather than independent. For the induction issue, he maintained that, if no structural changes take place, it is possible to reach conclusions for the near future by measuring the past. Tinbergen's reply did not answer Keynes' crucial questions such as the conditions for applying statistical method and the non-homogenous economic environment over time. These questions were formally answered by Haavelmo (1943, 1944).

Haavelmo (1943) spent a large part of his short paper in demonstrating that the time series can be viewed as random variables and that random variables have to be introduced in order to confront theory with actual data. Once the random variables are introduced, the non-logical jump from data to theory is justified and thus a complete list of causes is not necessary for an econometric model. He considered the data were produced by probability laws. He claimed that if these laws were to persist, the model could predict the future. He also demonstrated that statistical tests can prove or disprove a theory subject to type I and/or type II error. These ideas were formalized in Haavelmo (1944), in which he introduced the natural experiment concept to justify the foundation for applying probability theory in macroeconomics. Haavelmo (1944) also distinguished observational, true, and theoretical variables, and demonstrated at length how to use a joint probability function to estimate simultaneous equations and to do projections.

Haavelmo's arguments silenced Keynes. After Haavelmo's work, the probability approach was firmly established in macroeconomics despite continued criticisms from different schools of economic thought. The representative criticisms and defences are briefly reviewed here.

Milton Friedman strongly opposed the econometric approach in macroeconomics. Friedman (1948) also reviewed the work of Tinbergen (1939) and concluded that Tinbergen's results could not be judged by the test of statistical significance because the variables were selected after an extensive process of trial and error. Friedman (1948) pointed out that Cowles Commission work (econometrics) were built on two articles of faith, one of which is the possibility to construct a comprehensive quantitative model from which one can predict the future course of economic activity. He believed this kind of comprehensive model is possible only when decades of careful monographic work in constructing foundations had been done. Friedman (1951) proposed a naïve model to examine the predictive performance of econometric models. Ironically, this naïve model was later adopted by the Cowles Commission and, following the naïve-model approach, Sims (1980) developed the vector autoregression (VAR) model.

The poor forecasting performance of large structural macroeconometric models in the early 1970s discredited macroeconometrics and sparked a battery of criticisms. Leontief (1971) regarded econometrics as an attempt to use more and more sophisticated statistical techniques to compensate for the glaring weakness of the database available. Worswick (1972) claimed that econometricians did not forge tools to arrange and measure actual facts but only to make a marvellous array of pretended-tools. Brown (1972) concluded that running regressions between time series is only likely to deceive. Against this backdrop, Lucas (1976) published his famous paper 'econometric policy evaluation: a critique'.

By introducing rational expectation into policy evaluation, Lucas (1976) demonstrated that any change in policy will systematically alter the structure of econometric models. Thus, he claimed that there is a theoretical problem in structural macroeconometric models. However, the issue raised by Lucas (1976) is not fatal. An expectation variable was added to the structural macroeconometric model by econometricians (e.g. Wallis 1977; Wickens 1982; Pesaran 1987) and thus the dynamic stochastic general equilibrium (DSGE) model was developed. The rational expectation was able to be modelled within a structural macroeconometric framework. The other consequence of Lucas' critique is that some econometricians (e.g. Sims 1980) moved away from a structural model and proposed univariate time series naïve models and subsequently the VAR model. Although this approach has no economic theory backing and thus incurred heavy criticisms (e.g. Cooley and Leroy 1985), it produced much better forecasts than the structural models and thus was popularly used by banks. The criticism of the VAR model being theory free in turn led to the popularity of the Bayesian method.

The poor performance of macroeconometric models also led to blaming the poor techniques used by the modellers. Hendry (1980) demonstrated how an inappropriate model can generate spurious regression and suggested the use of statistical tests to increase the robustness of econometric models. Leamer (1978) commented: 'the econometric modeling was done in the basement of the building and the econometric theory courses were taught on the top floor'. Leamer (1983) proposed the use of sensitivity analysis in order to take the 'con' out of econometrics and thus restore the credibility of econometrics. Spanos (2011) blamed the impropriate statistical model specification and validation such as the theory-driven approach and the Akaike-type model selection procedures, and advocated the error statistical approach and the general to specific procedures.

The claim that statistic testing can give econometrics a scientific status was not supported by other economists, notably, Summers (1991) and Keuzenkamp (1995). Summers (1991) argued that formal econometric work had made little contribution to macroeconomics while informal pragmatic empirical work has a profound impact. He pointed out that the role of a decisive econometric test in falsifying an economic theory looked similar to, but was actually totally different from, the role of empirical observations in natural science. By examining the statistical tests in Hansen and Singleton (1982, 1983), Summers revealed that the test is just a confirmation of common sense because consumption by different consumers is not perfectly correlated with their wealth. Keuzenkamp (1995) provided a review of the book by Hendry (1993). In his review, Keuzenkamp rejected Hendry's use of the 'data generation process', the general-to-specific approach, and the 'three golden rules' of econometrics—'test, test and test'. Hendry's 'three golden rules' was rejected because (1) the test is not genuine (the macroeconometric test is different from the test used on experimental data), (2) the selection of the appropriate significance level is arbitrary, and (3) a statistical rejection may not be an economically meaningful rejection.

Freedman (1995) highlighted the crucial role of the assumption of an independent and identically distributed (IID) disturbance in computing statistical significance and in legitimate inferences. Freedman (1999) pointed out that econometricians tend to neglect the difficulties in establishing causal relations and that the mathematical complexities tend to obscure rather than clarify the assumptions for a statistical model. However, Freedman regarded these problems as the limits of econometrics. Freedman (2005) further identified many problems as limits of econometrics, including the faith in IID disturbance, the use of linear functions, model selection problems, asymptotic assumption and the inconsistency in Bayesian methods.

The outbreak of the global financial crisis (GFC) stunned macroeconometricians as no econometric model had forecast this event. This triggered widely spread criticisms on econometrics both within and outside of the economics profession. In the New York Times, Krugman (2009) claimed that the macroeconomics of the last 30 years was spectacularly useless at best and harmful at worst; in the Financial Times, Skidelski (2009) blamed banks for their blind faith in forecasting from mathematical models. Kling (2011) questioned the integrity of the macroeconometric models and claimed that macroeconometric models are unscientific because statistical models are based on repeatable events and thus are not suitable for accurate prediction or historical explanation. Heterodox economists (e.g. Nell and Errouaki 2013) blamed the assumptions used in econometric models. Interestingly, it were the rational expectation and optimization assumptions in the DSGE model, rather than the statistic assumptions, that were criticized by Nell and Errouaki (2013). In defense of the stochastic approach, Hendry and Mizon (2014) attribute the failure of the model to the probability distributions shift during economic recessions. An interesting book 'Economics as a Con Art' by Moosa (2017) showed a number of econometric modelling cons and suggested that empirical research should rely on clear thinking, intuition and commonsense.

## 3.4 Flawed Theoretical Foundation for Macroeconometrics

Why have numerous criticisms not had any impact on the popularity of macroeconometrics? The author's view is that these criticisms failed because they either tried to attack econometrics from the outside without the use of a statistical framework or focused too much on methodological details. The former approach (e.g. Keynes 1940; Friedman 1948; Leontief 1971; Solow 2010) attacked econometrics as a whole but failed to pinpoint where and why econometrics is wrong. The latter approach (e.g. Liu 1960; Lucas 1976; Freedman 2005) pointed out some technical flaws in econometrics but missed the main target—the framework of econometrics. In other words, the existing criticisms did not attack the foundation of macroeconometrics successfully. This section first focuses on the theoretic foundation of macroeconometrics and then on logical problems and issues in the practice of macroeconometricians.

Studies on the theory of econometrics tend to be cluttered with equations and econometric techniques. Here, we try a different approach. The strategy of this section is to attack the framework of macroeconometrics. Although some econometric techniques are mentioned, we will not go into details. Once macroeconometrics is proved baseless, the entire approach of macroeconometric modelling is invalid and thus, any econometric techniques are incapable of establishing the legitimacy of macroeconometrics.

## 3.4.1 Violating the Condition for Probability Theory

It is widely accepted that the probability theory is the foundation of macroeconometrics. However, macroeconometrics is built on very shaky foundations and has a logical problem that is insolvable. From the very beginning of the econometric approach, Keynes (1939) pointed out that the econometricians must address the conditions for applying statistical method not just applying it. Keynes' injunction has been addressed by several econometricians over time but in an inadequate way.

In Chapter 3, 'Stochastical schemes as basis for econometrics', Haavelmo (1944) successfully argued that the probability theory should be applied to economics. He made a valid point by saying that: even in the case of an exact economic relationship, or when we say something is certain, we mean the probability nearly equals one. The author agrees that statistic theories should be able to be used in any discipline but only when the conditions for probability theory hold.

The vital condition for probability theory is that of random experiment. Probability theory is valid because it has been proven by experiments. For example, if you toss a coin for a large number of times, you cannot predict the outcome (heads-up or tails-up) of each toss. However, a large number of experiments reveal that there is a 50% chance of heads or tails occurring. This type of experiments is called 'random experiment'. Haavelmo (1944,

p. 49) correctly stated that 'the notion of random experiments implies, usually, some hypothetical or actual possibility of repeating the experiment under approximately the same conditions'. Note that 'under approximately the same conditions' is actually very similar to the term 'other things being equal' used in economics and similar to the 'same condition' used in scientific experiment. All sciences have something in common. It is necessary to discuss 'approximately' here. Experiments can be repeated either at different times and/or at different places, so one cannot repeat an experiment under exactly the same conditions. Nevertheless, 'approximately the same conditions' for random experiments is a crucial element for probability theory. If the experiment is conducted under significantly different conditions, the law of large numbers and the central limit theorem do not hold, so the probability of an event may not be obtained, or the previously obtained probability is invalid. In the case of coin-tossing experiments, if different types of coins are used in a different environment, e.g. metal coin, wood coin, metal-wood coin and iron-lead coin, under calm or windy weather, or in a magnetic field, the probability of 'heads/tails' up will not be 50/50 due to the influence of various factors in the different experiments.

Based on the definition of 'random experiment', correctly implemented surveys and scientific experiments can be viewed as random experiments. However, it is very obvious that the macroeconomic time series data do not fit with the concept of 'experiments', let alone 'random experiments'. To justify the use of the probability theory in time series data, Haavelmo (1944) introduced a concept of 'natural experiments', namely 'the experiments which, so to speak, are products of nature and by which the facts come into existence' (Haavelmo 1944, p. 50). Even if we view macro-time series data as the result of such 'natural experiments', these experiments are not repeated under approximately the same conditions because conditions change remarkably over time. Since 'natural experiments' do not satisfy the conditions for 'random experiments', the former cannot be put under the umbrella of 'random experiments' and thus, the time series data do not satisfy the condition for applying probability theory. Consequently, it is invalid to apply probability theory to macroeconomic data.

Confronting this logical problem, Haavelmo's successors do not bother to reconcile the difference between random experiments and natural experiments. Rather, they just try to downplay the importance of this difference. Based on the finding by Mason (1962) that 'the history of natural science provides many instances of ideas derided at conception which are taken as axiomatic later' and the finding by Kuhn (1962) that 'science actually progresses through revolutionary changes in basic theoretical frameworks', Hendry (1980) stated that 'this characterization experimentation may be useful, but is not an essential attribute' (Hendry 1980, p. 388). Leamer (1983) argued that no scientific experiment is designed free of bias, so 'the difference between scientific experiments and natural experiments is a difference in degree, but not in kind' (Leamer 1983, p. 33).

Discounting the condition for applying probability theory is unscientific or even irresponsible because, as shown earlier, this condition is vital for probability theory to be valid. Leamer's argument is simply a case of changing the wording. The errors and bias in any scientific experiments are extremely tiny when compared with the vastly changed conditions in natural experiments which generate time series data. A difference in degree can lead to a difference in kind. It is reported that the difference in gene sequence between a human and a chimp is less than 1%. This small difference distinguishes humans from chimps. Using such reasoning, the difference between scientific experiments and natural experiments is both in degree and in kind. In terms of Hendry's argument, it is true that, in the course of the development of science, conditions or theoretical frameworks can be formulized later or can change over time. However, for a theory or study to be rigorous or just valid, its conditions must be held. If there are any changes in conditions, they must be proved to be valid in order to upgrate the theory.

The importance of valid conditions for a theory can be illustrated by examples in physics, the model natural science that inspired Haavelmo and other econometric theorists. Newton's inertial theorem necessitates that the speed of a bullet fired from a moving train must be different from the speed of a bullet fired from a stationary platform even if the gun and the bullet used are exactly the same. However, Einstein proposed that light does not follow this rule, i.e. the speed of light is the same no matter if it is emitted from a moving or stationary object. Based on this proposition Einstein developed relativity theories, which are proved to be correct by some astronomic observations, despite that some predictions from his theories (e.g. time and mass dilations, and length contraction) are counter-intuitive. Einstein's theories are correct (or have a chance to be correct if you do not believe the evidence proving time dilation) because his assumption that the speed of light is the same is based on (or supported by) the evidence provided by the 1887 Michelson–Morley experiments. The experiment was meant to prove the existence of ether; instead, it demonstrated that the measured light speed was the same regardless of the direction of the instrument placed in relation to the earth's motion. Similarly, there is also evidence to support the validity of assumptions for the quantum theory and the big bang theory (although the big bang theory has not yet been proved to be correct). The problem for macroeconometrics is that no evidence shows that probability theory can be applied to non-experiment time series data. On the contrary, the long history of statistics development, or even the simple experiment of tossing a coin, shows that, when the experimental condition changes, probability theory will be rendered invalid. In the case for natural experiments or for time series data, the condition changes considerably and this invalidates probability theory.

In short, applying probability theory to time series data violates the vital condition for random experiments and thus invalidates the probability theory. This makes the entire macroeconometric approach baseless. To make macroeconometrics scientific or just to make it valid, econometricians must either prove that natural experiments somehow can satisfy the conditions for random experiment or prove that the conditions for random experiments can be removed safely. For more than half a century, nobody has accomplished this task and the author believes that in the future nobody will be able to succeed in this task. Given this, macroeconometrics is fundamentally unscientific.

### 3.4.2 Inability to Include All Factors in a Model

There is a way to circumvent the condition for random experiments. That is, if all variables are included in the model, natural experiments can be viewed as under approximately the same condition and thus can be treated as a random experiment or a controlled experiment (no econometrician has formally argued in this way, but this is the underlying reasoning for multiple variable regression, especially the 'from general to specific' approach). This argument is the underlying reason for Keynes's request that Tinbergen should provide a complete list of significant factors. However, the unfortunate thing is that this approach is pragmatically impossible to implement because, given the complexity of natural experiment, no one can claim that he/she has included all significant variables in a model.

There are an unknown number of factors which might affect time series data, and some factors themselves may be unknown to the modeller, so it is impossible for the modeller to make a valid claim that he/she has included all important variables in the model. Because of this, random experiment in statistics and a controlled experiment in natural science are necessary. If the attempt by macroeconometricians to take into account all factors is valid, then it is not necessary for scientists and statisticians to keep the same condition for their experiments. This is apparently rejected by the practice of scientists and statisticians.

Facing countless numbers of factors involved, macroeconometricians have a 'magic' tool—testing—to uncover important variables. According to macroeconometricians (e.g. Tinbergen 1940; Hendry 1980; Leamer 1983), testing can discover important variables, transforming econometrics from alchemy to science and taking the 'con' out of econometrics. It is worth mentioning that here econometricians actually change the task of including all important variable to the task of detecting important variables. Even if econometric tests are able to verify the importance of a variable, econometricians cannot claim that they include all important variables because there may be important variables which have not been tested.

Without going into the details about the subjective and unscientific nature of econometric testing (e.g. the selection of the number of lags in the testing), the author claims that all tests will fail if the assumption of random variable breaks down. The random variable assumption ensures that the explanatory variables are stochastically independent, that is, the probability density function of variable $X_1$ does not affect the probability density function of variable $X_2$. However, for macroeconomic variables, this assumption often does not hold—it is well known that multicollinearity is very common in macroeconomic data. As a result, some variables may be insignificant if tested alone but can be significant if jointly tested with some combination of other variables. This means the tests have a low power to rule out insignificant variables. On the other hand, a variable that the test shows is highly significant may be an irrelevant variable due to spurious correlation. Econometricians claim that tests can detect and prevent spurious correlation, but it will be shown later that this claim is not true. In short, the test itself cannot determine whether a variable is important or not. It is the econometrician who finally has to make subjective decisions. This is part of the reason that it is better to classify macroeconometrics as an art rather than as a science.

In summary, since an unknown number of known and unknown variables can be involved in time series data, it is impossible to obtain a complete list of significant variables and thus, the omission of significant variables in a macroeconometric model is a most likely outcome. In practice, neither econometric testing, or economic theory, or common sense is able to select all significant variables for a macroeconometric model. The intention to include all significant variables is valid in theory but is impossible to put into practice.

### 3.4.3 Least Likelihood of Selecting a Correct Function for a Macroeconometric Model

Even if the econometrician can overcome the difficulty of including all important variables in a model, he/she faces another difficulty: selecting a correct function for a macroeconometric model because of the multiplicity of variables in the model. A simple example can demonstrate this difficulty. Suppose there are 10 variables (this could be a very small number compared with the complete list of all significant variables) in a macroeconometric model and the influence of each variable may be expressed by 5 different functions, e.g. linear, quadruple, logarithm, exponential and logarithm-quadruple. The number of possible functions for the model will be $5^{10} = 9,765,625$. Since there is only one truth in the model, only one function form is correct in the almost 10 million possible function forms in our simplified case. It is obvious that the likelihood of the econometricians' obtaining a correct function is extremely small. In other words, function misspecification is very likely in a macroeconometric model.

Most econometricians are not patient enough to try out the 10 million possibilities and they should not be blamed for this. Although some econometricians are interested in new functions (e.g. wave function), linear functions are used in the vast majority of macroeconomic models because of the convenience of linear regression. The underlying reasoning is that there are many mechanisms (i.e. probability laws according to Haavelmo 1944, or data generation processes according to Hendry 1993) which govern the economic data and that, if the linear model can fit the data well, the model uncovers one of the mechanisms.

The above reasoning is not only flawed and deceptive but it is also very dangerous because it hijacks the purpose of doing research (i.e. finding the truth or the true mechanism). The deceptive use of words helps the macroeconometricians to 'justify' their goal for research—mimicking reality and predicting the future. Haavelmo (1944) regarded probability laws as mechanisms of macroeconometrics, but his use of 'probability law' is very deceptive. The truth is a probability law is anything but a mechanism because mechanisms are not the concern of probability theory. For example, probability theory can tell you, in the experiment of tossing a coin, the chance of heads-up and tails-up is 50/50, but it tells you nothing about what mechanism causes this probability. Regarding probability law as a mechanism is simply saying that there is no mechanism and thus there is no truth. As a result, what Haavelmo said is that there is no need to uncover the truth because there is no truth (or we never know the truth), or simply that fitting the data is the goal of research because the data are the truth.

Although different data generation processes do exist in a computer-aided numerical exercise such as the Monte Carlo studies, the author agrees with Keuzenkamp (1995) that there is no such thing as different data generation processes for real economic data. If the data generation process is referred to the true mechanism (i.e. the truth), then there is only one data generation process because there is only one truth. This claim can be easily illustrated by a mental exercise. Suppose two models on real GDP are estimated and both fit the given data very well.

The results of one model show that the marginal effect of consumption on GDP is 1 (i.e. a $1 million increase in consumption will lead to $1 million increase in GDP), but the results of the other model indicate that the marginal effect of consumption on GDP is 0.5. Hendry would say that both modelling results are valid because they may come from different data generation processes. However, there is only one truth. If the marginal effect is 1 then it cannot be 0.5 or vice versa. So, it is impossible for both models to be correct. In this reasoning, the term data generation process used for econometric modelling is also deceptive and it tends to turn people's intention and attention away from finding the truth.

The importance of using a correct function which reflects the true mechanism is highlighted by the GFC. If a theory or model reveals the true mechanism, it should be able to predict the events in its domain. For example, gravitation law can predict or explain the movement of objects in the cosmos. By this reasoning, the failure of econometrics in predicting the GFC (or 'the profession's blindness to the very possibility of catastrophic failure' according to Krugman [2009]) indicates that the econometric models did not reveal the truth. However, the macroeconometricians still have not understood that there is only one true mechanism for one reality which their models have failed to uncover. Instead, they continue to harbour their statistical illusions and explain vaguely that the shift of underlying distribution of shocks causes the breakdown of their model during a recession.

The other deceptive reasoning related to the function specification is that macroeconometricians claim that they can use controlling variables to 'control' the model estimation. The reasoning is, if all possible variables can be included in the model but only a few variables are of interest to the modeller, the modeller can focus on the variables of their interest but use other variables as controlling variables to filter out the impact of these variables. By this reasoning, the modeller can turn the uncontrolled non-experimental data into controlled experimental data, so the empirical econometric modelling is similar to the empirical study based on controlled experiments in natural science. This reasoning is seriously flawed and deceptive because the modeller is not able to filter out correctly the impact of controlling variables. The controlling variables are controlled

by mathematical functions selected by the modeller. There is no indication and thus no reason to believe that these functions are correct, so the
controlling variables in a macroeconometric model are unable to exclude
the impact of the controlling variables in the same way as do the controlled conditions in scientific experiments. Otherwise, scientists working
in their laboratories should give up scientific experiments and join the
econometricians to conduct controlled modelling at a much lower cost.

## 3.5   Problems in Macroeconometric Practice

The previous section shows that the theoretical foundation for macroeconometrics is flawed due to its inability to satisfy the condition for random
experiment, so macroeconometrics should be dead at its birth. However,
it has thrived and dominated macroeconomic studies. Why has the validity issue been ignored? One reason is that, apparently, no one is able to
make time series data satisfy the condition for random experiment.
However, the inability to do random experiments is no excuse for macroeconometricians to assume that data are from random experiments and
thus for macroeconometricians to generate invalid results. The other reason is that macroeconometricians have successfully convinced most economists that econometric methods are able to fix the problems arising from
data and thus make econometrics a rigorous science. This section will
show that what macroeconometricians have produced is simply a mirage.

The macroeconometricians' approach is first to make an assumption
of strict conditions, then to relax these conditions and use econometric
tools to bring the conditions of real data close to the strict conditions.
An econometric textbook may start with the five or six conditions for
simple or multiple regression. The important ones are: random error
term assumption, constant covariance assumption, zero multicollinearity (i.e. multiple collinearity) assumption, zero autocorrelation assumption and stationary time series assumption. These assumptions have
never been true in reality, so the econometric estimation suffers from a
number of problems, e.g. the endogeneity problem, the heteroscedasticity problem, the multicollinearity problem and autocorrelation problem. However, econometricians then use a number of tools or models

to address these problems 'thoroughly'. For example, the Hausman test and instrument variable method are used to address the endogeneity problem, the general least square method is used to address the heteroscedasticity problem, the AR model and ARMA model are used to address the autocorrelation problem, the ARCH or GARCH model is used to address both the heteroscedasticity and the autocorrelation problem. A large body of literature is devoted to address the stationarity issue, e.g. the unit root tests, the Granger causality test, the error correction model and the cointegrated VAR model.

This approach indeed looks rigorous, but it is only a game of distorting statistics theory. Using random variable/disturbance as an example, this section starts with the base for the strict conditions proposed by macroeconometricians, then reveals the nature of using econometric method to bring the estimation conditions close to the strict condition required and finally illustrates the problems related to econometric practice.

### 3.5.1 Baseless Assumption About Random Variable/Disturbance

Since it is impossible to justify natural experiments as random experiments, macroeconometricians have built their theory on the assumption of random variables and random disturbance. This assumption is vital for valid model estimation and testing.

Macroeconometricians may argue that any discipline or theory needs to make basic assumptions and that entire econometric theory is valid with the assumption of random disturbance or random variable. It is true that assumptions are used for any disciplines or theories, but these assumptions must be valid or at least plausible, i.e. they must be proven to be correct or be based on solid foundations. One may argue that there is no such thing as a correct/valid/realistic assumption because assumptions are normally based on simplified or idealized reality, and thus are different from reality. This claim confuses correct/valid assumption with reality. An assumption can be a simplified reality or be different from reality, but it must be supported by evidence. For example, the constant-returns-to-scale assumption in economics is not always true in reality, but the case of constant returns to scale does exist in reality. Similarly, the assumption for gravitation law is supported by evidence such as falling apples and the earth's orbiting the sun. A correct/valid assumption is important because, as Friedman (1953) pointed out, an incorrect/invalid assumption leads directly to an incorrect theory. There are many examples of invalid assumption/theory in the development of physics or other disciplines, e.g. the flat earth theory, the earth-central theory, the ether assumption, the assumption of a fire-like element, racism and sexism.

Haavelmo did give a rigorous definition for random variables but did not use the term according to his own definition. Haavelmo (1944) gave two types of systems of random variables. One refers to 'random sampling' (Haavelmo 1944, p. 46). This type is clearly unrelated to macroeconomics because time series are not survey data. The other type refers to 'stochastically independent' variables, which obey the joint elementary probability law:

$$p(x_1, x_2, \ldots, x_r) = p_1(x_1) * p_2(x_2) \ldots * p_r(x_r)$$

To satisfy the above equation, each of the $r$ variables $(x_1, x_2, \ldots, x_r)$ must have an independent probability function, i.e. must have an independent dimension in space $R$ (so space $R$ is of $r$-dimension). In other words, the probability function of each variable does not affect the other. How can one be sure whether this condition holds or not? Without a pre-existing theory, the only way to find the answer is to conduct random experiments. In this sense, the two concepts—random variable and random experiments—are two sides of one coin. However, it is impossible to do random experiments for time series data, so the assumption of a random variable in a macroeconometric model is made based on the need of a macroeconometric model rather than on evidence.

Trying to justify that the probability theory can be applied to time series and that a complete list of significant factors is not necessary, Haavelmo (1943) stated that 'there can be no harm in <u>considering</u> economic variables as stochastic (random) variables having certain distribution properties' and 'only through the introduction of such notions are we able to formulate hypotheses that have a meaning in relation to facts' (Haavelmo 1943, p. 13, underline is added by the author). Apparently,

Haavelmo was simply 'considering' the economic variables are random variables because he <u>needed</u> this assumption. When illustrating the way to estimate a linear consumption-income function, Haavelmo (1943) added two random disturbances after stating 'to make it a real hypothesis to be tested against facts, we <u>have to</u> introduce some random variables' (Haavelmo 1943, p. 17, underline is added by author). Nowadays, the assumption of random disturbance is simply used by theoretic or applied econometricians even if they know this is not true in reality. The reason is that they <u>need</u> this assumption so that estimation and testing of the macroeconometric model are valid. Once the economic variables are considered random, it is fairly easy to dismiss Keynes' request of a complete list of variables because the probability of one random variable does not affect that of another. The reality is that most economic variables cannot satisfy the definition of random variable because of multicollinearity, and thus, Haavelmo's assumption is untenable.

It is arguable that the Bayesian method may be free of the problem of random disturbance because there is no random disturbance in that method. However, the assumption of random disturbance is embedded in the Bayesian assumption of random coefficients. Since the coefficients in the Bayesian method are assumed to be random, one must calculate the mean of the coefficients. Random disturbance is assumed implicitly by Bayesian econometricians when they apply the law of large numbers to calculate the mean and variance of random coefficients. In this sense, the means of random coefficients are equivalent to the point estimates in traditional econometric estimation, so the random disturbance assumption is converted or equivalent to the Bayesian assumption of random coefficients.

When can the assumption of random variable/disturbance be true? We need to trace back to the origin of statistics theory. The random disturbance/variable assumption originally used in statistics is not a mysterious assumption resulting from the imagination of statisticians. The assumption comes directly from and can be proven by random experiments. There are differences between the outcome of each random experiment (e.g. coin tossing) and the statistical mean. These differences form random disturbance, which is caused by random variables. In other words, random variable, random disturbance and random experiment are a trinity, e.g. when we use the concepts random variable/disturbance we imply that random experiments are conducted. If the conditions for random experiments do not hold, the assumption of random disturbance/variable is invalidated, and thus, all econometric estimations based on this assumption are invalid. This is the case for time series modelling. In practice, the assumption of random variable/disturbance is simply assumed by macroeconometricians and is used carelessly. This assumption has become the magic boxes or trash bins in macroeconometrics—any excluded (omitted, unwanted or unexplained) factors are thrown into it.

The function of random variable and random disturbance as a trash bin is also very important for macroeconometricians in order for them to accommodate uncertainty. As we discussed in Sect. 3.2.2, the probability law is related to uncertainty, but the law is unable to solve all uncertainty issues because it is applicable only to uncertainty cases proved by random experiments. However, macroeconometricians are very interested in modelling uncertainty and use it as an excuse for treating time series data as experimental data (e.g. the GDP next year is random because it can be affected by uncertainties). Since an uncertain event (e.g. cyclones) may 'randomly' happen at any time in the future, macroeconometricians think uncertainty is random and hence, it can be accommodated by random disturbance. When macroeconometric models failed to predict the GFC, macroeconometricians blamed the underlying distribution of a shock shift. In fact, the randomness of a shock and its underlying distribution do not exist but are carelessly assumed by macroeconometricians.

To sum up, in order to include the future uncertainty in the disturbance, one must judge if uncertainty is random according to the 'random' concept in the statistic theory. That is, one must do random experiments. There is no way to do random experiments regarding uncertainty, so the claim of random uncertainty is purely based on macroeconometricians' imagination and thus is unscientific.

## 3.5.2 Cutting Short Toes to Fit the Shoes

Although the strict conditions assumed by macroeconometricians are never met, this never bothers macroeconometrians because they claim that they can use a battery of econometric tools to bring the conditions close to the strict conditions. For example, if the error term is not random because it is correlated to the dependent variable, an econometrician can use Hausman test to detect it and use instrument variable (or 2-stage estimation) method to bring the error term close to random. If the error term is not random due to the autocorrelation, an AR term, MA term or ARMA term can be introduced into the model and bring the error term closer to random. If macroeconometricians' claim is true that they can bring the conditions close to the strict conditions required (e.g. bring a non-random error term close to random), does this make macroeconometrics valid? On the surface, it appears this approach is logical and rigorous. Thinking one step further, we can find that this is hardly the case.

Where do the strict conditions for regressions come from? Most people would answer statistical theory. However, this is not the ultimate source of the above-mentioned conditions. Any theory originates from reality and can be examined by reality. Statistical theory is no exception as it comes from and can be examined by random experiments. The strict conditions proposed by econometricians are the features of, and can be verified by, random experiments. If these strict conditions are not satisfied, this indicates that the experiments are not random, in other words, the condition for random experiments is not satisfied. In this case, the statistical theory is invalidated and so is the statistical regression. Using econometric methods to perform a 'cosmetic surgery' to bring the conditions close to the strict condition does not make a historical event a random experiment. using an analogy, covering a wolf with a sheepskin does not make the wolf a sheep! The practice of macroeconometricians reminds the author of a scene in the movie 'Cinderella': Lady Tremaine tries to fit her daughter's foot into the glass slipper by cutting her toes short!

### 3.5.3 Tricky Ways of Using Concepts

Concepts are the building blocks of logical reasoning, so it is important to have clearly defined concepts to ensure correct inference. This is demonstrated by the philosophic debates in ancient Greece and by the long history of logic. However, in order to justify econometric theory and practice, econometricians created and redefined a number of concepts. Similar to the sophists in ancient Greece, this trick on concepts by econometricians conceals their logical mistakes and confused others and sometimes, themselves.

As discussed earlier, since economic data cannot satisfy the condition of random experiment, Haavelmo created a concept of 'natural experiment' as a counterpart of 'random experiment' in economics. The condition of 'natural experiments' is obviously not the same. Because random experiments require to be conducted under approximately the same condition, it is apparent that natural experiments are not a type of random experiments. Facing this logical problem, the econometricians later used the assumption of 'random variable' and 'random disturbance' to justify the application of probability law. If the variables in their model are random as econometricians claimed, the use of probability law is indeed justified. However, this assumption breaches the trinity of random variable, random disturbance and random experiments: random variable and random disturbance come from and can be proved by random experiments. Since economic data are not obtained from random experiments, their assumption of random variable/disturbance is implausible.

Realizing that it is hard to satisfy the condition of random experiments, econometricians have also used words like 'controlled experiments' or 'controlling variables'. These words imply that an experiment is conducted under controlled conditions, just like an experiment in a scientific laboratory. However, econometricians 'conduct' their controlled experiments by including a large number of controlling variables in a model. The controlled condition is 'achieved' through regression based on an assumed function. As we discussed previously, given the large number of variables, the possibility of specifying a correct function is extremely small. As a result, the control experiments performed by econometricians are not only inaccurate but also subject to more distortion because of the inclusion of irrelevant variables and because of the use of incorrect function forms. Nevertheless, econometricians have successfully deceived the outsiders that their estimation is as scientific as the controlled experiments in a laboratory.

The other example of misusing words is related to the concept of mechanism, which is key for us to understand the world and to make a prediction. What is the mechanism of an econometric model? Haavelmo substituted mechanism for probability law or probability schemes. In fact, a probability law contains no mechanism, so the use of the word 'mechanism' in describing probability laws is simply saying that there is no mechanism or no truth. Similarly, Hendry (1980) used the 'data generation process' to describe the mechanism. Since the same results can be generated by different processes, the term 'data generation process' also implies that there are more than one mechanism or truth for a research question. However, the 'data generation process' only exists in the computer experiments played by econometricians. If the 'data generation process' refers to the truth behind the real economic time series, then there is only one truth for a realized time series (e.g. how the GDP is determined) and there is no way that an econometrician can claim that all data generation processes which his/her model finds are the true data generation process (i.e. the truth).

Creating or using words deceptively is not the only trick macroeconometricians use to disguise their logical problems. They also openly redefined the meaning of commonly used concepts to something having similar but different meanings. The notable examples include the definition of 'endogeneity', 'significance' and 'causality'. The usual meaning of 'endogeneity' is the feedback effect between two variables (e.g. between independent and dependent variables in an economic model), but in macroeconometrics this definition is changed into the correlation between the dependent variable and the disturbance. Based on this definition the Hoffman test is designed to 'detect and solve' the problem of endogeneity!

The usefulness of 'statistical significance' was profoundly rejected by Ziliak and McCloskey (2008). To explain this concept we need some statistical knowledge. We explain here in layman's term. For readers who have no economics or statistical background but are interested in a deeper understanding, an elementary statistical or econometric textbook may be useful.

'Statistical significance' is related to a number of concepts, including 'variance', 'standard error', 'probability distribution', 'confidence level' and 'critical t-value'. Variance of a variable is basically the overall difference between the estimated value and the data. The square root of variance is called 'standard error'. Probability distribution describes the probability of different outcomes from random experiments. The commonly used probability distributions are the bell-shaped standardized normal distribution and the transformed students' $t$-distributions with a flatter bell-shape. The area under the bell shape indicates 100% probability. When we cut the edge areas (or two tails), the probability will reduce. If the probability obtained in this way is 95%, the confidence level is 95% or the significance level is 5% or 0.05; if the probability is 90%, the confidence level is 90% or the significance level is 10% or 0.10 and so on. For a $t$-distribution, we can find the $t$-value corresponding to the probability or confidence level. This value is called 'critical t-value' at a confidence level. One can compare the absolute value of estimated coefficient with the absolute value of (standard error) * (critical $t$-value). If the former is greater than the latter, the coefficient is significant at certain confidence levels; otherwise, it is insignificant. If one varies the estimated coefficient value by the range of (standard error) * (critical $t$-value), one can obtain the confidence interval estimate. For example, the 95% confidence interval estimate of estimated coefficient $b$ is: [$b$ − (standard error) * (critical $t$-value at 95% confidence), $b$ + (standard error) * (critical $t$-value at 95% confidence)].

From the above explanation, it is clear that the meaning of statistical significance depends on three items: the value of estimated coefficient, the value of standard error and the critical value. If the estimated coefficient is very small (e.g. 0.0001) but the value of the (standard error) * (critical $t$-value) is even smaller (e.g. 0.00001), we still say that the estimated coefficient is 'statistically significant'. It would be fine if the concept of 'statistically significant' has nothing to do with a judgment whether or not the coefficient is 'significant'. However, in the applied econometric practice, 'statistical significance' is often interpreted as 'significance' (otherwise the term 'statistical significance' would be useless in interpreting econometric results) in explaining research questions.

In connection with 'statistical significance' is the concept of 'confidence level' or 'significance level'. These concepts give the reader the impression of how accurate the estimation is, but the two concepts critically rest on the assumed and untested probability distribution. For example, the normal distribution is commonly used ($t$-distribution and other distribution are transformed based on normal distribution), but only random experiments exhibit normal distribution. Since economic data are not generated by random experiments, assumed distribution based on normal distribution is implausible and thus the confidence level in econometrics is invalid but acts as a smoke screen.

Finally, we consider the word 'Granger-causality' created by econometricians. Causality commonly indicates a cause-consequence relationship. The strict definition can be found in 'An Enquiry Concerning Human Understanding' (Hume 1748): 'we may define a cause to be an object, followed by another, and where all the objects similar to the first are followed by objects similar to the second. Or in other words where, if the first object had not been, the second never had existed.' By this definition, A causes B means (1) A occurs before B; and (2) if A does not occur, B will not occur. Hence, the causality can only be examined by experiments for both instances (when A presents and when A does not present).

A similar definition is used by the forerunner of mathematical statistics, Karl Pearson. He stated that 'whenever a sequence of perceptions D, E, F, G is invariably preceded by the perception C..., C is said to be a cause of D, E, F, G' (Pearson 1892, p. 155) and that 'if the unit A be always preceded, accompanied or followed by B, and without A, B does not take place, then we are accustomed to speak of a causal relationship between A and B' (Pearson and Lee 1897, p. 459). Although Pearson was fully aware of the difference between causality and correlation, from the point of view of statistics he put more emphasis on correlation by saying 'it is this conception of correlation between two occurrences embracing all relationships from absolute independence to complete dependence, which is the wider category by which we have to replace the old idea of causation' (Pearson 1910, p. 157).

Despite the cliché in statistics that 'Correlation does not imply causation', modern econometricians have tried to force the causality concept into the probability theory by associating causality with correlation in different ways. Suppes (1970) defined: A causes B if the conditional probability of B given A is greater than B alone, and A occurs before B.

Granger (1980) further defined: a variable A causes B, if the probability of B is conditioned on its own past history and the past history of A does not equal the probability of B conditional on its own past history alone. The statistical language tends to prevent the general readers from understanding these definitions. These statistical definitions can be translated into common language: A causes B if A occurs before B and if the chance of the occurrence of B is positively or negatively correlated with the occurrence of A. This statistic definition does not address Hume's second condition for causality 'if A does not occur, B will not occur'. Yet, based on his statistical definition, Granger was able to develop a Granger causality test. Although Granger pointed out the Granger causality test is not a test of causality, i.e. X Granger-causes Y does not mean X causes Y, the introduction of new but very similar concepts 'Granger-causality' just causes more confusion. The inability of the Granger causality test to detect causality and to solve the problem of nonsense correlation was demonstrated in an example in Sect. 3.5.7 in this chapter, which proved that Granger's definition of causality is indeed not causality.

By redefining common concepts using statistic language, it seems that macroeconometricians have transformed a logical problem like 'endogeneity' and 'causality' into a statistic problem and then they have designed statistical tests to detect and solve the problem. As shown in Sect. 3.5.6, this is simply an illusion because the statistic problem which the tests are trying to prove or disprove is different from the logic problem the researcher is trying to solve. The statistical redefinitions of concepts change the meanings of commonly used concepts. This becomes a logical deficiency imbedded in macroeconometrics and this deficiency confuses people and downgrades people's logical reasoning power.

### 3.5.4  Inability to Model Uncertainties Based on Time Series Data

If all the problems discussed so far can be ignored, i.e. if it can be assumed that both economic variables and uncertainties are random and that the econometrician is able to include all possible variables in a model and to find out the correct function, will macroeconometric modelling be scientific? It depends on how 'uncertainty is random' is defined. Generally, we cannot say uncertainty is random over time, e.g. a drought can randomly occur last year, or this year, or some years into the future. The reason is obvious: the future is unknown to us so we cannot claim that uncertainty is random along a timeline. However, it is valid to view uncertainty as random at any given time, e.g. the GDP next year may be affected by some random factors. Even in this case, the results from a statistical model are still invalid because the econometrician is estimating a model based on a sample size of one. This can be shown by an example of estimating a demand curve—the relationship between the quantity of a good demanded and the price of the good.

The true demand curve (e.g. $D_1$, $D_2$ and $D_3$ in Fig. 3.1, which was named 'reversible demand schedule' in Haavelmo [1944]) is hard to estimate because the requirement of the 'other things being equal' for the demand curve does not hold for time series data. This is the well-known identification problem in estimating a market demand curve. However, macroeconometricians claim that they can solve this problem. Haavelmo (1944) suggested that the estimated demand curve using time series data will be the long-run or dynamic demand function (shown as $D_L$ in Fig. 3.1, which Haavelmo [1944] called the irreversible demand curve).

However, the estimated long-run demand curve may not be a downward-sloping curve, as we expected. It may be upward sloping as shown in Fig. 3.2. This is not merely a hypothetic case. Moore (1914) estimated a positive demand for pig iron. Does this prove that the demand theory is wrong? It does not, because this demand curve does not satisfy the 'ceteris paribus' or 'other things being equal' condition. For example, there may be changes in household income, the price of other goods, inflation, etc. Suppose the econometrician can take into account these variables and use a multiple variable regression to control the variables other than the quantity demanded and the price of the concerned good. If the impact of other factors is thus excluded (for a moment assuming it can be excluded by the econometricians), the correct points will be the red points in Fig. 3.2. As a result, the new demand curve $D_{L1}$ is estimated. Is this demand curve a valid one? The answer is still 'no'. The data for each year are supposed to be random so the red points are only one sample of many random possibilities for each year. As a result,

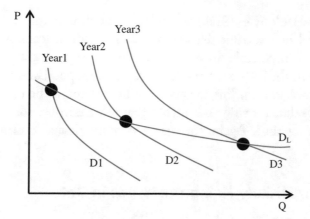

**Fig. 3.1**  Estimating the demand curve in principle

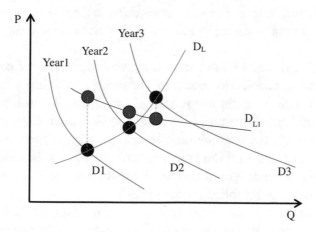

**Fig. 3.2**  Estimating the demand curve in practice

the estimation is based on a sample size of 1 for each year. This is equivalent to the case where an agent conducts a yearly survey by collecting only 1 sample each year.

Macroeconometricians might change the definition of uncertainty and argue that the uncertainty of the GDP in each year is because the GDP is random over time; in other words, the GDP of each year is a random pickup (realization) of GDP values of a certain distribution. To claim a variable (e.g. GDP) is random over time is to play the role of either God or a fortune-teller: if an econometrician cannot foresee the future, it is impossible for him/her to know (or creditably believe) that the GDP in the future is a random draw from a population. Besides the logical problem regarding the guess (or belief) about the future, the past time series data provide no evidence that any macroeconomic variable is random over time. Again, the macroeconometricians' claim or belief is baseless.

### 3.5.5 Artificially Fixing the Issues in Data

Macroeconomic data are featured by high heterogeneity due to vastly different situations over time. The implication of heterogeneity of data on statistical modelling has been demonstrated in Sect. 3.2.1 and an excellent example can be found in Pearson et al. (1899), so we will not discuss it here.

Regarding the data accuracy, we may recall the theory of errors developed from the probability model in astronomy. This theory has successfully dealt with errors in astronomic observations. Measurement errors are impossible to eliminate and can be modelled when the errors are not big. However, if the measurement error is very large, there is no way to sort them out and find the true measurement. For this reason, we are often told in science lessons that accurate observation and measurement are the starting point for scientific research.

It is common wisdom that macroeconomic data are highly inaccurate due to the different sources and methods used in generating such data. Macroeconomic data are generally obtained through aggregation. During the process of aggregation, the measurement errors at micro-level will be aggregated and thus accumulated (one may hope the measurement errors will be evened out during the aggregation, just like diversified investment will reduce risk, but these two processes are quite different. Diversification of investment works because the average return is much less volatile than the return on one stock. There is no mechanism to ensure that the positive measurement error will be offset by a negative measurement error. On the contrary, the different standards and measurements used at the micro-level will increase the measurement error at the macro-level). To make the situation worse, some macro-data are themselves the modelling results from another model. Long time series are even more unreliable because of the changes in the standards for data collection and aggregation. Examples of unreliable macroeconomic data can be found in many studies such as Morgenstern (1950), Reid (1977), and Hendry and von Ungern-Sternberg (1980). Given unreliable macroeconomic data, how can a macroeconometric model produce reliable scientific results? Experience from natural science indicates it is impossible. The author once asked a full professor in the USA who specialized in econometric modelling on China: 'Given that the economic data in China is censored by the government, how reliable do you think your modelling results are?' His answer was: 'It does not matter. Many of my papers have been published in top journals'.

Given the dubious status of economic data, it is necessary to improve the data quality in order to make any economic study reliable. This is not an easy task because the errors in measurement have to be addressed from the very beginning and the data aggregation process must be tracked thoroughly. However, macroeconometricians create some procedures to overcome the shortcomings in the data. Although the intention of macroeconometricians is good, their effort is in vain since these procedures do not address the sources of errors in measurement. Contrary to the desire of macroeconometricians to improve the quality of data, it is most likely that macroeconometric techniques will introduce additional measurement errors and/or magnify the existing errors and thus worsen the quality of the data. The resampling technique used in bootstrap has been heavily criticized by many economists. Here we only considers the consequence of de-trend or differencing, which are commonly used in time series modelling. De-trend and differencing are valid methods to treat data if there is no measurement error in the data. However, if measurement errors present, they can be magnified by de-trend or differencing.

Take the US GDP time series as an example. The annual GDP from 1926 to 2013 is in the magnitude of $US1019.9–16768.1 billion (see

Table 3.2 in appendix at the end of this chapter). A conservative view is taken to assume a relatively small measurement error, which is 1% of measured GDP, so the measurement error for the US GDP is ranging from $US10.2 to $US167.7 billion. The first differences of the GDP are in the range of $US56.0–604.9 billion. Since the measurement error may increase in the operation of first differencing (e.g. there is a downward measurement error in the previous year and an upward measurement error in the current year), a conservative view is again taken to assume that measurement errors are unchanged. Even so, the measurement errors could account for as high as 8.7–91.9% (average 20.2%) of the first-differenced value (see the last column in Table 3.2). Keeping in mind the magnitude of the measurement error, how can anyone rely on the modelling results based on the first-differenced or de-trended data? How can the tests based on the variation of data (e.g. the unit root tests) be reliable?

### 3.5.6   Scientific Illusion of Statistic Tests

In the effort to recover confidence in macroeconometrics after the prediction failure of many macroeconometric models, rigorous statistical testing is highly recommended. As previously noted, Hendry (1993) suggested: 'The three golden rules of econometrics are test, test and test; that all three rules are broken regularly in empirical applications is fortunately easily remedied. Rigorously tested models, which adequately described the available data, encompassed previous findings and were derived from well-based theories would greatly enhance any claim to be scientific.' Are the statistic tests scientific enough to save macroeconometrics?

One big problem about statistic tests is that the theory on test design and the test criteria are established based on artificially generated random series. Since these data satisfy the definition of random experiment and random variable, the tests are proved valid. From this point of view, macroeconometrics can be said to be scientific on computer-generated data. However, because the real macroeconomic time series do not satisfy the definition of random variable, the tests and criteria are invalid when they are applied to macroeconomic data. This is why Keuzenkamp (1995) claimed that statistical tests are not genuine. If someone does strictly follow the statistic test, he/she would be like a pre-programmed robot acting mechanically to unknown conditions.

In practice, the statistic tests in macroeconomics appear to be objective but are often quite subjective. What will an econometrician do if the indication from the test contradicts common-sense or economic theories? Most people will choose common sense over tests. It is an open secret that the results of macroeconometric modelling are very sensitive to the number of lags chosen for the model. It is common that, in a regression model, a macroeconometrician includes both contemporary variable (e.g. current-year GDP) and lagged variable (e.g. GDP in last year, in the year before, etc.). The number of lags to be included in the model has significant implication on modelling results. The subjective (and thus unscientific) choice of time lags in econometric modelling was pointed out by Keynes (1939) and many other economists. Now econometric theory provides tests and selection criteria to determine the number of lags used. This sounds scientific in theory but, as explained previously, it is invalid when applied to real data. In practice, the modeller does not and cannot always follow the selection criteria. If all selection criteria (e.g. the Schwarz criterion and Akaike information criterion) suggest that the modeller should choose 4 lags but this would produce unexplainable results (e.g. the consumption contributes negatively and significantly to GDP), the modeller would most likely ignore the test results and select the number of lags to produce sensible results. Moreover, there are a number of tests for the same problem, e.g. many kinds of tests for unit root, for autocorrelation, for heteroscedasticity, for endogeneity and for cointegration. It is not surprising that the different tests do not agree with each other. What would an econometrician do when faced with this situation? He/she has to make a decision and is most likely to select the one that can produce the most desirable results.

Many tests themselves involve the selection of the number of lags and the test results depend on the lags, for example, the Granger causality test. As will be discussed in more detail later, the rationale of this test it that, if the variation of a variable A in the past (e.g. a change in temperature last week) is correlated to the variation of another variable

B (e.g. the change of the height of a plant this week), then A Granger-causes B. However, should we choose one lag (e.g. the temperature last week), two lags (e.g. the temperature 2-week ago) or 3 lags (e.g. the temperature 3-week ago)? Theoretically, any number of lags is possible and the number of lags may change over time. It is difficult to know the actual number of lags in reality and thus one has no way to select the correct number of lags for the Granger test. So, the test relies on the arbitrary selection of lag numbers. Even with the help of some statistics criteria, e.g. the Schwarz criterion and Akaike information criterion, the selection of lag numbers is more likely to be arbitrary and thus makes the testing subjective.

The GFC destroyed the macroeconometricians' illusion of the power of statistical tests and the ability of their models to predict the future. Those models, built by econometric leaders who advocated statistical tests, were rejected as a consequence. These econometric leaders were unwilling to admit the failure of their statistic tests as well as their econometric models; instead, they come up with an amazing explanation: the underlying distribution of the shock changed during a recession. Based on this explanation, the econometric leaders were effectively stating that even a correct econometric model with robust statistical tests is unable to predict the future.

### 3.5.7  Impossibility of Overcoming Spurious Regression Within a Statistic Framework

In a broad background, the econometricians' untenable confidence in statistic tests results from their conviction that they can use the statistic framework to solve reasoning problems in economics. In the statistic framework, the name for a time series (e.g. real GDP data) means nothing, so there is no such thing as spurious correlation in this framework: a statistic theory says nothing about whether a correlation makes sense or not; therefore, there is no need, and no means, to test a spurious correlation in the statistic framework. When it is said that there is a spurious correlation, it means that the correlation does not indicate any causality. This statement is a logical judgement which cannot be solved within a statistic framework but needs to be solved by experiments and logical reasoning. Thus, any attempt to overcome the problem of a spurious correlation by developing statistic tests is misguided and futile. This is demonstrated in the macroeconometricians' long journey of fighting against spurious correlation.

The problem of nonsense correlation or spurious regression has a long history. As early as 1926, Yule (1926) found high nonsense correlations between the proportion of Church of England marriages to all marriages with the standardized mortality during 1866–1911. Hendry (1980) demonstrated an almost perfectly fitted curve between UK inflation and the cumulated rainfall. Macroeconometricians always try to solve this nonsense correlation problem within a statistic framework. Yule (1926) found that, in the case of nonsense correlation, there are strong serial correlations in the estimated residual, so he thought the autocorrelation in the estimated residual must be the reason for a nonsense correlation. This reasoning is fatally flawed because a feature associated with something is not necessarily the cause. Using an analogy, a person may develop an ulcer on one leg which attracts a number of flies. A fake doctor sees this and thinks that the flies must be the cause of the ulcer and his prescription to the patient is to kill the flies in order to cure the ulcer.

Yule's wrong logical reasoning was followed by his successors and a large body of literature has been devoted to testing and modelling non-stationary time series. Is this practice able to overcome the problem of spurious regression? As will be discussed in Sect. 3.5.10, the univariate time series (stationary or non-stationary) study is based on an imaginary assumption and thus is flawed. Even if it is true that the correlation between two non-stationary time series is temporary and thus false and that a non-stationary model (e.g. cointegration or first-differencing model) can overcome the non-stationary issue, the modeller cannot guarantee that the revealed correlation indicates causality and thus the modeller cannot rule out that the correlation is not spurious. The flaw is that the econometrician actually changed the task of testing and avoiding nonsense correlation to the task of detecting and avoiding autocorrelations. They accurately hit the wrong target.

To solve the problem of spurious regression, Granger developed an econometric method to test causality,[1] i.e. the Granger causality test. Granger correctly pointed out the difference between correlation and causality; however, in designing a test, he assumed that a change which appeared earlier and correlated to a change coming later is the cause of the second change. So the logic behind the Granger causality test is that, if an earlier change in time series A correlates (positively or negatively) to a later change in time series B, A must cause B. While it is true that a cause is always occurring earlier than the consequence, the reverse is not necessarily true. For example, lightning is always seen first and thunder is heard later, but the lightning is not the cause of thunder. They all are the results of electrical discharge. We see lightning first because light travels much faster than sound. In short, a logical mistake is embedded in the Granger causality test. This mistake can be easily illustrated by a hypothetical experiment.

Assume that a scientist is conducting an experiment on the impact of temperature on the growth of plants A and B and that, in the selected range of temperature, both plants respond positively to the increase in temperature but plant A is more sensitive to the temperature change (i.e. plant A responds faster). As the temperature increases, both plants A and B grow faster, so there is causality between temperature and the growth of both plants. Meanwhile, since plant A responds faster to the temperature change, the recorded change of growth of plant A will be always earlier than that for plant B. Therefore, the Granger causality test in this case would conclude that growth of plant A causes the growth of plant B. It is apparent that the Granger causality test is flawed and has failed in this case.

Heckman (2000) reviewed the causality and policy analysis in economics and concluded that 'the information in any body of data is usually too weak to eliminate competing causal explanations of the same phenomenon. There is no mechanical algorithm for producing a set of 'assumption free' facts or causal estimates based on those facts' (p. 91). He correctly pointed out the fact that econometric models cannot solve

---

[1]Granger differentiated Granger causality from causality, i.e. A Granger-causes B does not mean A causes B. This just causes more confusion. If Granger causality is not causality, the Granger causality test has no ability to solve the spurious regression issue, thus the test itself is pointless.

the causality issue but he incorrectly attributed this to the 'too weak' data. Regarding this conclusion, the author of the present book would say that Nature will never specially design experiments in order to produce suitable data which can reveal the causality to mankind. The essential of this issue is that causality is a logical issue which cannot be solved within a statistical framework. The only way to find out causality is to conduct more experiments and to exercise our logical reasoning power.

### 3.5.8 Fallacies of Univariate Model, Instrument-variable Method and Bootstrap

Econometricians have developed some methods or models which sound reasonable and thus are still popularly used; however, once they are scrutinized, it is apparent that these models involve absurd propositions.

The use of the univariate model is an interesting case. The model was first proposed by Friedman as a naïve model to ridicule the performance of large macroeconometric models. The naïve model only consists one variable—a regression of current variables on its past values (since the model used variables in different time periods, the model is also called a dynamic model). Friedman successfully showed that the prediction from large macroeconometric models is not better than his naïve model and ridiculed the modelling results as GIGO (garbage in garbage out). However, this naïve model became the standard testing tools for econometricians. Moreover, some econometricians added more variables into the model (while keeping the past value of the dependent variables), and the resulting model has a new name VAR model. It is claimed that the VAR model has better forecast performance and this claim led to a group of econometricians advocating 'let the data speak for itself'.

Does regressing a variable on its past values make sense? On the surface the answer is positive: just like the inertial law in physics, the past value may have an influence on current value. The obvious example is capital formation or the household debt level—the current value is an accumulation based on the past value with some depreciation or interest payment. Another example may be the consumption level for a household: if a household has high consumption in the past, it would tend to have high consumption in the future. In short, studying history can inform our future, so it is arguable that the dynamic nature of the univariate model may be valid to predict the future.

However, even in the above cases of inertial nature, it is pointless to build a dynamic econometric model to explore this nature. In the case of capital formation or household debt, we can calculate the influence of past value accurately based on depreciation and interest rate. In the cause of consumption, the trend or even average growth rate can be used predict the future consumption no worse than a dynamic univariate model does: the only difference is that the econometric model can produce a smoke screen—the confidence interval.

In practice, if we regress the current value with 1-lagged past value, we have a positive coefficient—this is easily explained by the inertial effect. However, if we regress the current value with multiple past values, we may find some past values have a positive effect while some have a negative effect, e.g. consumption of 1-lag has a positive effect on current consumption, consumption of 2 lags has a negative effect and so on. Does anyone really believe this negative effect and explain it satisfactorily? The author doubts any econometrician can win this challenge.

Just like the inertial effect in physics, the inertial effect in social science is easy to understand and to estimate, so it is not the main task for a researcher. The important task in both physics and social science is to uncover the force causing the change in variable value. In the case of consumption, if a household has suffered a significant decrease in income and/or in asset value, the consumption level has to decrease. It is the mechanism or interaction between variables that is the emphasis of research. Building a dynamic univariate econometric model excludes any interaction between variables and reveals no mechanism of a system, so it is simply a waste of time and valuable resources.

The second case to be discussed is a method called 'bootstrap'. This method is devised to overcome the issue of a small sample. For example, an econometrician only has data (e.g. output, profit, employment, etc.) of 50 firms. This small sample may affect the credibility of modelling results. In this case, he/she can apply the bootstrap procedure: put 50 slips with the name of 50 firms in a black box and randomly draw one each time, record the data of the firm, put the slip back into the black box, draw one slip again (a computer is commonly used to mimic this procedure). If you repeat the random draw 1000 times, you have a sample size of 1000. This looks a convenient way to increase sample size. Moreover, the econometricians claim that the bootstrap utilized random experiments, so the data are from random experiments and thus qualified for the application of probability theory.

This absurd practice overlooked one important fact, that bootstrap is a sampling from samples, instead of sampling from a statistical population (i.e. all firms in a nation). There is no new information added to this process, so the results from bootstrap are not better than the results from the original sample. It is true that the modelling results based on the data from bootstrapping are different from the modelling results based on the original sample and, if one repeats the bootstrap again (e.g. randomly drawing another 1000 samples), the estimation result will change again. However, if one increases the bootstrap sample to an extremely large number (e.g. repeatedly draw 1 billion samples from the 50 samples), based on the theorem of large numbers, one would find the distribution and the modelling results are very close to those from the original sample. This is apparently only statistical magic.

The last case considered is the instrument variable (IV) or 2-stage Least Square (2SLS) estimation. This method is designed to overcome the endogeneity issue in estimation. For a valid estimation, we require that the dependent variable is endogenous while the independent variables are exogenous. By exogeneity we generally mean that the value of the variable is determined outside of the system, similarly endogeneity means the value of the variable is determined within the system. In a demand and supply system, the price and the quantity demanded and supplied are all endogenous—any two can determine the third, so it is pointless to regress either demand or supply on price. The idea of instrument variable estimation is that, in the case that endogenous explanatory variables are in the model, the modeller can find some instrument variables which are correlated to the endogenous explanatory variables but uncorrelated to the error term in the model. The modeller first regresses the endogenous explanatory variable on the instrument variables to obtain the estimates of endogenous explanatory variables, then replaces the original data of the endogenous explanatory variable with estimates, which are assumed to be exogenous, and obtain modelling results.

It seems that the instrument variable method solves the problem of endogeneity, but there are many problems with both its theory and practice.

First, there are problems with the definition of endogeneity. The normal definition of endogenous variables—variables determined by other variables in the system—relies on the causality relationship. However, econometricians change the definition of the endogenous variable as an explanatory variable correlated to the error term. As emphasized previously, correlation does not mean causality, so a correlation between an explanatory variable and the error term does not mean the variable is determined by other variables in the system and thus correlation is not an indicator of endogeneity. As such, any econometric work based on the altered definition does not address the endogeneity issue at all.

The new definition of endogeneity by econometricians does highlight a severe problem in econometric estimation—the correlation between the error term and explanatory variables. This correlation is a clear indication that the estimation is invalid. However, the effort of econometricians to address the correlation issue is just window-dressing. If econometricians have managed to avoid correlations between the error term and explanatory variables, or the error term is somehow made random artificially by econometricians, this still does not make econometric estimation valid. A random error term is only an indicator of random experiments and a valid estimation must be based on random experiment. Econometric manipulations are unable to address the essence of the issue. They only change the appearance or perception.

Secondly, there are problems with the selection of instrument variable. By generating an estimate of endogenous explanatory variables through instrument variables, the approach may circumvent the problem of correlation between endogenous explanatory variables and the error term if the instrument variables chosen are indeed strongly correlated to endogenous variables but uncorrelated to the error term. However, this approach is problematic when the relationship between the endogenous variables and the instrument variables is considered.

There are two possibilities. One possibility is that the endogenous variables are the agents or media of the instrument variables, i.e. the instrument variables generate effects through the endogenous variables. A good example of this is the impact on human health of poisonous elements accumulated in food. Some herbicides (e.g. now banned herbicide DDT) are cumulative and the farmers' use of these herbicides may accumulate in various food sources (e.g. wheat or meat) for humans. Here, the yearly consumption of DDT-affected food by humans is an endogenous explanatory variable for human health and the yearly use of DDT by farmers acts as the instrument variable. Based on this instrument variable, we can estimate the yearly amount of DDT in food consumption and thus provide a more accurate estimation of the impact on human health. In this case, the instrument variable estimation is a reasonable but unnecessary approach. The yearly use of DDT by farmers is a better explanatory variable than the consumption of DDT-affected food because the concentration of DDT in food may vary each year, i.e., the change in DDT-affected food is not proportional to the amount of DDT used. As a result, replacing the consumption of DDT-affected food by the amount of DDT used will produce the same results as those by the IV approach, so the IV approach is redundant.

The unnecessity to use IV or 2SLS estimation for the case where endogenous variables act as media can also be shown by a structural model or a reduced form model. For example, a demand for a commodity may be expressed as a function of both the price of the commodity and the income of consumers, while the supply of a commodity may be expressed as a function of the price of the commodity and of the cost of production factors (e.g. wages and capital rentals). Since supply must equal demand at equilibrium, we can solve two equations (the supply function and the demand function) to obtain the reduced form equation, i.e. supply and demand are both functions of consumer income and of the cost of production factors. In this reduced form, the endogenous explanatory variable—commodity price—is eliminated. As such, there is no need to go through a 2-stage estimation (i.e. estimate the price first and then use it to estimate supply or demand).

The other possible relationship between the endogenous variable and the instrument variable is shown in a partial endogeneity case, where the instrument variable partially affects the endogenous explanatory variable. An example of this is the relationship between the use of fertilizer, food output and human health. In this case, human health is a dependent variable, fertilizer is an instrument variable, and food output is a partial endogenous explanatory variable because it is only partially determined by the amount of fertilizer used. Applying instrument estimation to this case, one would, in the first step, regress food output on the amount of fertilizer to obtain the estimated food output due to the use of fertilizer and then, in the second step, use this estimated food output to estimate the impact on human health. The flaw of the IV estimation in this example is obvious: the original food output data are much more important than the estimated food output which mainly concerns the contribution by the use of fertilizer. Econometricians are clearly aware of this flaw and thus have stressed the importance of strong instrument variables (e.g. high correlation between endogenous variable and instrument variables). No matter how strong the instrument variables are, the estimates for endogenous explanatory variable are not as good as the original data because the estimates disregard the impact of factors other than instrumental variables. As such, the instrumental variable estimation in this case is always worse than the one-step estimation using original data.

There are also general problems associated with the IV approach. One is that, by including instrument variables, the approach simply enlarges the system to be considered. By enlarging the system, the IV approach increases the complexity of estimation and reduces its ability to solve the problems. The other general problem associated with the IV approach is related to measurement errors and estimation errors. These errors present in any estimations. By involving 2-stage estimations, the IV approach is essentially to estimate a model based on other estimations. This will enlarge measurement error and cause more estimation errors.

### 3.5.9  Issues with the VAR Model, the DSGE Model and the Bayesian Method

The failure of forecasts from the structural macroeconometric models in the 1960s and early 1970s led to the popularity of the VAR model. By incorporating multiple variables, the VAR model generalizes the univariate autoregression (AR) model, which examines the interdependence between the current observation and the past observations in time series. If some or all variables in the VAR are non-stationary but are cointegrated, an error correction term is included in the model and the VAR model becomes a Vector Error Correction Model (VECM) or a restricted VAR. If the non-stationary variables are not cointegrated, these variables are differenced and the model becomes a VAR in difference.

Since this approach disregards any theory and lets the data speak for themselves, heavy criticism from economists is focused on its naive theory-free state. However, VAR modellers claim that their models forecast reasonably well. Indeed, the projection from VAR models is much better than other models, e.g. the DSGE model. This is of no surprise. Because a VAR model focuses on producing the best fit for the historical time series data, so the model will capture the trend of the time series well. As long as nothing unusual happens (i.e. no new factors come into action), the captured trend will continue and the projection appears quite accurate. If the situation changes, the projection from a VAR will be far away from reality. The inability of the VAR models to predict the GFC is a good example. From this point of view, a truly reliable prediction must come from a correct understanding of the true mechanism behind the data, i.e. a proven theory.

The DSGE model has been developed along the line of structural macroeconomic models. The utility maximization and cost minimization are well defined in a general equilibrium framework of a DSGE model. The dynamics of the economy are normally realized in capital accumulation and the stochastic feature is introduced into the model by random technological shocks. The time series data are used to calibrate the parameters and perform simulations. The forecast performance of a DSGE model is generally poor and this is heavily criticized by VAR modellers (e.g. Spanos 2011).

The unsatisfactory performance of the DSGE model is also not surprising because in reality the outcome is affected by so many factors which may not be included in the model. Even if the time series data can calibrate the parameters in the model to produce a close fit for historical data, this does not mean that the theory or function in the model reflects the true mechanism of the economy. In this sense, the poor performance of the DSGE model does not necessarily reject the theoretical approach but it may indicate that the theory or the functions in the model are incorrect or that they do not capture the main drivers in the economic system.

The Bayesian method provides a good way to blend theory with data. This method treats the modeller's knowledge (e.g. common sense, beliefs or theories) as prior information, which can be put into the model through the prior density function (or posterior) to restrict the estimation. This way of utilizing theory or common sense is similar to but much more flexible than the restricted VAR model (e.g. the sign restriction in the VAR model). The other profound difference from the traditional estimation method is that the Bayesian method assumes that the parameters to be estimated are random and there is no random disturbance in Bayesian simulation. Bayesian claim that their models forecast well.

The better performance of the Bayesian method compared with the theoretic econometric model approach is explained by its data-driven nature. Except that the Bayesian approach provides a more flexible way to incorporate prior knowledge, this approach is similar to the VAR approach. Both approaches aim at producing the best fit for historical data, so they are data-driven and thus can capture the trend in time series better than a theory-driven structural model. This leads to the better forecasting performance of the data-driven approach. As stated earlier, this good performance in forecasting is not reliable if the circumstance changes. The main criticism of the Bayesian method is on its reliance on prior information, namely, including in a model untested common sense, belief or theory is unscientific. If the prior information is wrong, the estimates from the Bayesian method will be contaminated. It appears that the assumption of random parameters might protect the Bayesian method from the critiques on random disturbance but, as discussed earlier, the random disturbance assumption is embedded in its assumption of random parameters. As a result, all criticisms on the assumption of random disturbance are equally applicable to the assumption of random parameters.

### 3.5.10 Fortune-Teller Style of Forecasting

One assumption for macroeconometric modelling and forecasting is that the existing (or realized) time series data are a sample randomly drawn from a hypothetical large-size time series population from the past into the future. Does this fictitious time series population exist? Because this involves a future, nobody can confirm this except perhaps God or a fortune-teller who 'know' the future. However, macroeconometricians have made this claim and this in turn suggests macroeconometric modelling and forecasting is similar to the work of a fortune-teller.

The key for macroeconometrical forecasting is the statistical relationship within time series, e.g. how the GDP last year affects the GDP this year. Does a statistical relationship within a time series exist, or does the seemingly existing statistical relationship within a time series imply any causality? If the answer is yes, it is valid to use macroeconomical models to project the future and vice versa. The author uses two obvious examples to illustrate the statistical relationship within a time series. If one does a time series analysis using the match records of Roger Federer in the period from 2003 to 2011, he or she must forecast that Roger has a high chance of winning a Grand Slam from 2013 to 2016. This projection turns out to be deadly wrong. A macroeconometric modeller may argue that there is a structural change in 2012. If a modeller is unable to predict the structural change, his/her prediction about the future is unreliable. A better explanation of this prediction failure is that the 'statistical relationship' reveals or indicates no causality so it is invalid to claim that the existing statistical relationship will definitely continue into the future. As a result, 'proven' statistical relationship cannot be used for forecasting. Another example is about the length of days and nights. Based on the existing records at a location, a time series modeller may proudly predict that roughly 12 hours of daytime will be followed by 12 hours of night-time and that the daytime will be longer in summer seasons and be shorter in winter seasons. This type of project looks well performed but it is actually useless. Firstly, with common sense based on a few years of observation, anyone can predict as well as any econometrician does, so the econometric work is redundant in this case. Secondly and more importantly, the econometric projection fails to reveal the cause of the daytime night-time pattern and contributes nothing to our understanding of this pattern. If, for some reason, the earth's rotation rate changes significantly, the econometric projection will be in limbo and the econometrician has no clue as to why the prediction has gone wrong. In short, a statistical relationship does not necessarily indicate a causality relationship and only the latter can form a foundation for valid forecasting.

The study of statistical relationships within a time series leads to a differentiation of stationary and non-stationary time series. If a time series is stationary, the correlation between this time series and other stationary time series is regarded as a meaningful correlation. On the other hand, if a time series is non-stationary, with the exception of cointegration, the correlation between this time series and other time series is regarded as not true, because the time series explodes or drifts over time. Based on this, a large body of literature has been developed in order to model non-stationary time series. Are the studies on stationary and non-stationary time series valid? The answer is no. Like other macroeconometric studies, these studies make baseless assumptions that there are some mysterious statistical relationships within any time series. For accumulative economic variables such as asset value and debt value, one can appreciate some links between current and past data. For non-accumulative variables such as GDP, unemployment rate, inflation rate, interest rate, no one can figure out a link between current and past data because there is none. Moreover, the tests for non-stationary time series (i.e. unit root tests) have found that most macroeconomic time series, including inflation rate, unemployment rate and interest rate, are non-stationary, so the time series on above rates should either explode or drift over time, but history shows otherwise—these rates have been fluctuated within certain ranges.

If there is no relationship within a time series, the assumption for time-series studies is false and thus invalidates the studies. It is worth pointing out that, even though some macroeconomic variables appear to have a relationship within a time series, e.g. GDP is highly correlated with lagged GDP, this correlation does not indicate causality between past data and current data; rather, the correlation is caused by a similar environment (i.e. the determinants of GDP). In this case, whether or not a time series is stationary is determined by the environment: it may be stationary in some periods but becomes non-stationary when the environment changes. As such, the seeming relationship within the time series is a spurious correlation so it is pointless to study it; instead, it is more important to study the environment (or determinants) affecting the time series.

For some macroeconomic variables, there indeed is a logical link within their time series, e.g. due to the inertial effect described in Sect. 3.5.8, the accumulated debt level or capital stock can be expressed as follows:

$$\text{debt}_t = (1 + r) * \text{debt}_{t-1} + \text{consumption} - \text{income},$$
$$\text{capital}_t = (1 - \phi) * \text{capital}_{t-1} + \text{investment}$$

In this case, the logical link between past and current data is basically an indication of the cumulative nature of these variables. The correlation between past and current data is straightforwardly determined and thus not worth studying; rather, it is more useful to study what <u>causes</u> the change in this kind of time series.

It has been demonstrated that studies on stationary/non-stationary time series are either pointless or not worthwhile. What is the implication of these studies for macroeconometric forecasting? For accumulative variables, it is of no surprise that the forecast may not be too far away from reality because of the obvious accumulative trend. For non-accumulative variables, the forecast from a macroeconomic model depends on luck because the seeming relationship (stationary or non-stationary) within a time series is determined by other variables in the environment. To forecast under different conditions, one must find out the driving force or the mechanism behind the empirical data.

One may disagree with the above statement, saying that the VAR models developed by Sims (1980) and the Bayesian models recommended by Leamer (1983) have been able to forecast very well. One reason for the 'effective' performance of these models is that they include more information than other models. In particular, the Bayesian approach includes some prior information (economic knowledge, common sense or beliefs of the modeller) as a vital element for estimation. Based on this, it can be argued that, if Bayesian models do forecast more effectively, it is the triumph of prior knowledge or common sense. The other reason is that the economic environment did not change much when the forecasting of these works appeared to be effective.

One may defend the failure of econometric models in predicting the GFC by saying that it is hard to forecast a rare event like the GFC because these rare events are not in the data often. One response is that rare events or normal events should all be governed by the same law. A model failing to predict the rare events must fail in finding this law. More importantly, the purpose of forecasting is to prevent or to be prepared for a disastrous rare event. Using an analogy, if a seismologist has predicted correctly that there is no earthquake for 364 days of a year but failed to predict an earthquake of 9.0 on the Richter scale in one day, does this seismologist forecast well? In this sense, the inability of macroeconometric models to predict a severe economic recession like the GFC shows that macroeconometric forecasting is no better than the above-illustrated seismological forecasting.

## 3.6     Performance of Econometrics

We have seen in the previous section that macroeconometrics is weak in foundation and problematic in practice. However, some econometricians argue that the validity of an approach is not dependent on how realistic the assumption and logic of the approach are but is dependent on how well the prediction is in line with reality. In response to this argument, we examine the performance of econometrics in this section.

### 3.6.1  The Saga of Econometric Modelling

One purpose of econometric modelling is to predict the future. If the variable of concern follows a trend, e.g. the annual GDP growth in recent years is about 3%, the prediction can generally be made according to the trend. If it turns out to be true, an econometrician may claim triumph. However, this is a triumph of experience. When the condition changes, the prediction from econometric models fails spectacularly. This was strikingly demonstrated by the projection failures of macroeconometric models to predict the stagflation of the early 1970s and by the failure to predict the GFC of 2008. The golden growth period of the 1950s to 1960s gave people confidence and, understandably, the economic data of this period helped to produce optimistic modelling results. Then, economists were dismayed by the phenomenon of stagflation in 1970s. Similarly, the long economic growth from the mid-1980s to the early 2000s led to light-hearted modelling results but econometricians were hammered by the outbreak of the GFC. The failure of macroeconometric models indicates that they have not captured the key mechanism of the economy. The failure of econometricians and economists indicates that their understanding of the economy is not good enough.

In fact, there are plenty of signs of an instability in the economy prior to economic recessions and stagnation. The increased volatility and disruption of housing markets and the stock market started in 2006 were two of these signs. It is reported that Prof. Nouriel Roubini from New York University presented all the signs of economic recession to a seminar at IMF in December 2006 and he predicted an economic recession in the USA. However, this prediction was rejected by the Chief Research Officer of the Economic Cycle Research Institute, Anirvan Banerji, on the basis that Roubini's analytical framework was overly subjective and did not give the actual likelihood of a recession. In other words, Banerji thought Roubini's work was not credible because it was not based on econometric modelling.

The poor performance of econometric modelling is also shown in the debate about abortion and crime. In 1999, John Donohue and Steven Levitt wrote a paper entitled 'The Impact of Legalized Abortion on Crime'. Using the abortion and crime data for different states of USA, they argued that the legalization of abortion in the 1970s explained as much as 50% decline in crime in the 1990s. Their reasoning is that abortion reduces the number of unwanted children who are at the risk of committing crimes when they grow up. The paper was published as a lead article in the Quarterly Journal of Economics in 2001 and attracted a large amount of attention from the media.

Criminologists dismissed the claim by Donohue and Levitt (2001) because the crime rates of people born before and after the legalization of abortion showed a similar pattern: if the explanation of Donohue and Levitt is correct, the aggregated crime rates for people born after the legalization of abortion should be much lower than those born before the legalization of abortion. However, economists focused on criticizing the econometric techniques of Donohue and Levitt. Joyce (2004, 2009) argued that the relationship between crime and 'wantedness' of a child are affected by a complex structural model involving sexual activity, contraception, marriage and birthing. Foote and Goetz (2008) and Joyce (2009) claimed that it is inappropriate that Donohue and Levitt regressed the log of crime counts on the log of abortion rates. When regressing the log of crime rate on the log of abortion rate, they found that the results changed greatly. They also criticized Donohue and Levitt for not adjusting the standard errors for serial correlation. Foote and Goetz (2008) also uncovered that Donohue and Levitt did not include the fixed effects in their computer codes although their paper claimed that the fixed effects are included. Lott and Whitley (2007) modelled age-specific homicide rates and claimed that legalized abortion raised murder rates because forbidding abortion leads to more unmarried births—the impoverishment of women reduces investment of children and thus leads to more crime when their children grow up. Facing these criticisms, Donohue and Levitt collected more data, re-specified their model and provided responses. No one won the debate. Most people now agree that the epidemic of crack cocaine in the late 1980s and early 1990s had much more impact on the decline in crime in the 1990s.

This type of debate on econometric techniques is basically a waste of time and resources. There are so many issues involved in data. For example, Steve Sailer (2005) pointed out issues of the occurrence of abortions before the legalization, cross-state abortions and demographic composition. There are also so many variables involved during the period of 20 years. Using both time and state fixed effect variables may reflect the issue but is not able to solve the issue satisfactorily. Even if one can fix all the problems regarding data, variables and the functions forms, the estimation results simply show a correlation between abortion and crime, not necessarily a causality.

In short, econometrics itself is a gross approach so it is incapable of obtaining accurate results. Debating econometric techniques does not change the nature of econometric modelling. Nevertheless, the debate benefited Levitt significantly. Extending the data mining and economic modelling on other topics such as the cheating of teachers and Japanese sumo wrestlers, information control by real-estate agents, and Stetson Kennedy's investigation of the Ku Klux Klan, the effect of parenting on education, and the patterns of naming children, Steven Levitt and Stephen Dubner in 2005 published a book, 'Freakonomics', which is a runaway success thanks to the debate.

Controversial modelling results also appear in other social sciences. Petersen et al. (2013) conducted a survey in Argentina, the USA and Denmark on upper-body strength (indicated by the strength of dominant arm), social economic status, and the attitude to economic redistribution. Based on their survey data, they made a number of claims. For men of high social economic status, arm strength was negatively and significantly related to supporting redistribution of income and wealth; for men from low social economic status, arm strength was positively and significantly related to supporting redistribution of income and wealth; for women, the correlation between arm strength and the attitude to redistribution of income and wealth was statistically insignificant. Their explanation was based on previous studies. For example, Campbell et al. (1999) and Sell et al. (2009) who claimed that greater up-body strength indicated greater fighting ability and thus more ability to assert self-interest during a conflict. They claimed that this worked for men but not for women, because direct physical aggression was a less rewarding strategy for women over human evolutionary history.

Durante et al. (2013) studied the relationship between women's ovulatory cycle and the female vote based on a survey in the USA. They claimed that, for unmarried women, ovulation made them become more liberal, less religious and more likely to vote for Barack Obama while, for married women, ovulation have an opposite effect. Beall and Tracy (2013) studied female ovulation and the colours of clothes they wear and claimed that women are more likely to wear red or pink at peak fertility because men are more sexually attracted to women wearing red.

Gelman and Loken (2014) criticize the above controversial research as 'overstate the significance level by ignoring multiple-comparison problems', e.g. women's vote may correlate with ovulation, but may also correlate with age, social economic status and political preferences.

The overstated statistical significance caused by the multiple comparisons issue was termed as p-hacking by Simmons et al. (2011) or p-fishing by others. The concern about the reliance on $p$-value as criteria of refutation of the null hypothesis was regarded as the statistical crisis in science (Gelman and Loken 2014). This concern led to a 2016 announcement by the American Statistical Association on statistical significance and $p$-values, and a 2018 announcement by a prestigious journal 'Political Analysis' that the journal will not report $p$-values.

Although the criticism on relying of $p$-value to reject null hypothesis has some valid points, it did not go far enough to reveal the grossness nature of the statistical approach. The multiple comparisons issue can be taken care of by a multivariate model with a number of controlling variables, but even this approach is not the same as doing controlled or random experiments. Since the condition for controlled or random experiments—other things being equal—does not hold in applied research, the $p$-values based on random experiment is misused and thus is simply not valid in social science.

More importantly, even if the significance of a correlation is established unequivocally, it is worth emphasizing that correlation does not mean causality. Extra experiments and reasoning are necessary to establish a causality. This is what is lacking in the above controversial research. For example, the explanation of Petersen et al. (2013) seems reasonable, but there are holes in their reasoning. Although up-body strength is important for animals or for primitive humans in obtaining resources, it is less important for modern humans: the competition in our society is largely based on intelligence rather than on body strength. Physical fighting in our society is rare and is not the normal situation of determining resource distribution, so applying up-body strength to social affairs is questionable. Moreover, the ability to obtain resources and the attitude to income/wealth redistribution are totally different things. Although both are related to self-interest, the former is a measurable objective ability while the latter is a subjective intention. Physical strength may lead to stronger ability to obtain resources, but this does not mean a stronger person is more selfish while a weaker person tends to be more altruistic. With these logical holes, the causality cannot be established even if the survey results are unquestionably significant.

## 3.6.2   Why Econometric Models Fail: An Illustration*

Macroeconometricians attributed these failures of econometric models to the incorrect implementation of econometric estimations by amateur modellers (Leamer 1978), lack of rigorous testing (Hendry 1980), or an unexpected shift of underlying distribution (Hendry and Mizon 2014). While bad modelling techniques do cause failure, using a demonstrative estimation of the GDP identity, in this section the author will show that the failure of macroeconometric models is fundamentally due to the flaws exposed in the previous section. For the details of estimations, please see appendix at the end of this chapter.

# 3.7   Detrimental Impact of Econometrics

Facing embarrassing performance, econometricians have to admit the limitations of econometric modelling. Nowadays, the popular expression regarding the performance of econometrics is: 'all models are wrong but some are useful' or simply 'harmless econometrics'. Is econometrics harmless? It appears so—no matter whether the modelling results/projections are right or wrong, there is little impact on the economy. However, econometrics can cause harm in a number of ways. This section discusses the impact of econometrics on society and on research in social science.

## 3.7.1   Deadweight Loss of Econometrics

Econometric modelling can have a significant impact on our society. First, econometric modelling is dominantly used by governments to project future economic performance and to formulate or support economic policies. At the tail of the outbreak of the GFC, many people blamed then Chairman of US Federal Reserve Greenspan for lax monetary policy. If Greenspan was responsible for causing the GFC, the econometric modelling that Greenspan relied on to set his policy must also be blamed. Despite a notorious reputation regarding accuracy, econometric modelling is still widely used by governments, banks and research institutes, so the modelling results are still influencing people's decision making.

Secondly, the projection from econometric models about the economy can be misleading and result in complacency. This is harmful because people are vulnerable when they are unprepared for disasters. When the GFC occurred, Queen Elizabeth II famously asked why nobody had seen it was happening. No economist can answer this question and econometricians provide no way to react to the economic shock. It is Keynesian economics and the old IS/LM model that came to the rescue of the governments around the world.

Thirdly, econometrics results in a great waste of resources. Many institutions are engaged in econometric modelling, and time series analysis is taught in most universities. Meanwhile, econometric models proliferate in academic journals. All these activities are a waste of time and money because the outcome of econometric modelling is of little use due to the serious flaws in this approach. Econometric modelling was compared to a sausage factory. For the author, a 'garbage generating machine' or GIGO (garbage in garbage out) termed by Milton Friedman is a more accurate analogy.

Last but most importantly, econometrics delays or even suppresses truly useful research in economics. One piece of evidence is the dwindling of economic theories and the empiricalization or econometricalization in economic research. This trend is very obvious since the 1970s. Nowadays, not only the vast majority of economic journals but also the journals on economic theory are flooded with empirical papers. If econometrics had not dominated macroeconomics for over 60 years, the cause of and remedy for economic recessions might have been found and the GFC might have been avoided.

### 3.7.2  Impact of Econometric on Research Philosophy

Compared with its harm on society, the impact of macroeconometrics on scientific research work is more devastating because macroeconometricians have hijacked the well-established scientific research goal and method.

The goal of scientific research is to find causality, truth or the true mechanism, which can be proven by experiments or events, and can be used to explain phenomena, predict future events and provide solutions to existing problems. The established method to achieve the goal of scientific research is an ongoing process of 'data-hypothesis-theory'. From experimental or empirical data, we can derive a hypothesis which can be further developed into a theory. The theory will be examined by newly available data and this will lead to a new hypothesis and an improved theory. This procedure goes on continuously and leads us closer and closer to truth. However, macroeconometrics suggested a totally different approach.

Using correlation or significance level, macroeconometricians obtain the important variables for a research problem. These variables are put into a function (usual a linear function or a transformed linear function) to form a model. The model is then estimated by historical data using various econometric methods. Because the econometric methods can produce a best fit for historical data, the econometricians claim that the model can explain the research question well and can be used to predict the future. Thus, the task of research has been achieved and there is no need to go through the 'data-hypothesis-theory' process.

Can an econometric model explain the research question? To some degree the answer is positive: the model can show the correlation between the dependent variable and the explanatory variables, so there may be some causality relationship between dependent and explanatory variables. However, the explanation from an econometric model cannot be taken seriously because correlation does not necessarily mean causality. In other words, an econometrician is unable to claim that the econometric model reveals the true mechanism.

Since an econometric model does not reveal the true mechanism behind the research problem, the prediction from an econometric model is unreliable. With no knowledge of the true mechanism, the prediction from an econometric model is no different from the prediction based on common sense or based on ordinary observations. This kind of prediction is correct if the situation remains unchanged. However, if the situation changes, the prediction will be false.

Moreover, macroeconometrics diverges from the 'data-hypothesis-theory' process, and requires little logical reasoning. This degrades human intelligence to the computer or robotic level and causes a waste of research resources. Being an obstacle in scientific research, macroeconometrics causes slow progress in scientific research and has a negative impact on our knowledge base and our living standard. For this reason, it is necessary to reject the research method used by the macroeconometrician.

## 3.8    Lessons to Be Learned

If you have followed the author's reasoning so far, you would agree that applying probability theory to economic time series data has severe defects and has dire consequences. Keynes regarded econometrics as 'statistical alchemy', not ripe enough to become science. In commenting on Tinbergen's approach, Keynes (1940, p. 156) stated: 'That there is anyone I would trust with it at the present stage, or that this brand of statistical alchemy is ripe to become a branch of science, I am not yet persuaded. But Newton, Boyle and Locke all played with alchemy. So let him continue'. Alchemy had positive contribution to Chemistry because Alchemists did real experiments in their hope of transforming lead into gold. In comparison, the theoretical econometricians are doing 'experiments' with computer-generated random variables. Since these are not real experiments, the author doubts if econometrics can contribute anything at all to any science in the future. Nevertheless, the great cost due to the development of econometrics can teach us a few lessons so that we can avoid this kind of pseudoscience in the future.

1. Setting a feasible research task and never trading truth for anything else

The world is full of the unknown just waiting for people to explore it. However, we are unable to uncover each mystery when we encounter it, so we need to decide our research task based on available information. For example, the big bang theory appeared only after Hubble discovered that all stars are moving away from us and that the stars at a distance move away faster. Due to lack of information, history witnessed many failed research attempts, such as alchemy, the perpetual motion machine and Leonardo da Vinci's flying machine. Macroeconometricians set out to establish a permanent law in economics but they have so far failed.

Even so, one should not prevent someone from trying to discover the unknown. Curiosity is in the nature of human beings and many successful discoveries and inventions have come after numerous failures. However, the most important lesson from all successful or unsuccessful journeys of discovery is to persevere in the search for the truth. Finding the truth is the northern star for doing research and so it should be a criterion for scientific research. When the alchemists realized that they were not able to transform cheap metals into gold, they avoided the truth by focusing on their secret methods and recipes. Similarly, when macroeconometricians found they could not establish a permanent law in economics and their projections were failing, the appearance of research, e.g. the use of mathematics and so-called techniques in dealing with data has often become the goal of macroeconometric research. From this point of view, the criticism of Krugman (2009) is seen as appropriate that 'the economics profession went astray because economists, as a group, mistook beauty, clad in impressive-looking mathematics, for truth'. In doing research, we should never trade truth for anything else. Otherwise, creditable science will degenerate into alchemy or magic.

## 2. Never discarding or downplaying the power of logical reasoning

Logical reasoning is the foundation for any scientific research. If logical errors exist in a theory, then the theory is flawed. Logical errors in macroeconometrics were pointed out by Keynes and other economists but were downplayed by Haavelmo and other econometricians.

Haavelmo (1943, p. 15) claimed: 'we shall not, by logical operations alone, be able to build a complete bridge between our model and reality. We might be able to push rather close to the other side, that is to reality, but we finally have to make a non-logical "jump"'. To support this claim, he used Keynes' consumption function (consumption is proportional to income) and an investment function as an example.

Haavelmo (1943, p. 17) stated: 'as it stands, this theory is not yet a hypothesis suitable for testing; all we can say is that, if actual observations be tried, the theory is almost certainly wrong'. It seems that with this example he illustrates or justifies the non-logical jump in doing research, but the truth is that here Haavelmo just used the term 'non-logical jump' to confuse the reader's logical thinking.

Since there are measurement errors in the data, even if Keynes' consumption function is correct, the data and theory will not match. The same can be said for theories in natural science: the measurement errors in scientific experimental data necessitate an imperfect match between data and theory. Unlike the data from either scientific or random experiments, macroeconomic data can also be affected by an unknown number of variables, so the impact of variables outside of the model also contributes to the gap between economic data and theory. Just adding a random error term superficially to 'bridge', the gap between data and theory is unscientific and non-logical. This makes macroeconometrics flawed.

During logical thinking, one must be very careful about the accurate meaning of the concepts used. An incorrect use of concepts can disguise the logical problems and thus disguise the flaws in reasoning. There are a handful of examples of deceptive use of concepts in macroeconometrics. As mentioned earlier, the concept of 'mechanism' as misused by Haavelmo (1943), who defined mechanisms as probability laws or probability schemes. In fact, probability laws/schemes provide the probability outcome but provide no mechanism behind this outcome. Haavelmo's use of the word 'mechanism' in describing probability laws gives the reader the impression that a macroeconometric model reveals some sort of truth. However, an econometric model actually reveals nothing except the fitted data because there is no mechanism in the model.

An argument for discounting logical reasoning is that logic alone is unable to lead to scientific progress. It is true that many scientific ideas come from non-logical imaginations or even dreams. For example, it is widely reported that the advent of gravitation law occurred because Newton was accidentally hit by a falling apple and was inspired by Galileo's experiments; Archimedes' principle of buoyant force was discovered after he saw the water overflow when he took a bath; the structure of Benzene (C6H6) was discovered after Kekule had a dream in which a snake ate its own tail. However, logical reasoning is necessary in transforming these insightful ideas into theories or theorems and in testing them. In short, logic alone is not enough but is necessary for any scientific research. One vital logical mistake in a theory is enough to disprove it.

3. Applying a theory with its full conditions or making assumptions based on evidence

Every theory is based on its conditions or assumptions which are essential for the theory to be valid. Thus, when applying a theory, one must be sure that all the conditions for the theory hold so that the conclusions drawn from the theory are correct. For example, if one applies gravitation law to an electrical field or to the particles in the micro-quantum world, the user is making a mistake and will obtain an incorrect answer.

Macroeconometricians do not deny that applying probability theory to macroeconomic data does not satisfy the condition for random experiments (i.e. the condition for the probability theory) and that this practice causes a number of problems such as non-random disturbance (autocorrelation), heteroscedasticity, and endogeneity. The interesting thing is that macroeconometricians first assume an independent and identically distributed population (i.e. there is no problem such as autocorrelation, heteroscedasticity, endogeneity) and then relax their assumptions by fixing the problems through the changing of estimation methods, e.g. adding an AR, MA or ARMA term in the model to fix autocorrelation, using GLS estimation to fix heteroscedasticity and using instrument variable to fix endogeneity. This practice does not fix the cause of the problems—the condition for random experiments does not hold but seeks to fix the problems superficially.

It is claimed that changing the condition of a theory or extending the scope of a theory can lead to an advance in science. This kind of situation did occur in the development of natural science, e.g. extending the wave theory to particles gives the matter wave theory in physics.

The key issue here is the new condition must be supported by evidence. The matter wave theory is proved by the interference of moving particles. The assumptions of any valid theories are all proved by facts. Changing/extending the condition of theory without the support of facts is like a standardized prescription for patients of a new medicine successfully tried on animals but without human trials. Such a medical practice is illegal because of the risk to patients. Similarly, imposing an assumption which is not consistent with facts can lead to a false theory and thus should be opposed in scientific research.

4. Advancing empirical study to theory

Usually, empirical study is a necessary starting point for the development of science. To improve our understanding and thus make a positive contribution to the improvement of people's lives, an empirical study is needed to advance towards the formation of theories. The discovery of the laws of gravitation is a good example of such a development. The empirical studies on the movement of planets led to a remarkable achievement when Kepler's laws of planetary motion could describe fairly accurately the orbits of planets in the solar system; however, people had no idea of the nature of the force or reason behind such planetary movements. Only when Newton provided a theory to explain the planetary movements did people's understanding improve dramatically. As a result, we can easily examine not only the movement of planets but also the movement of stars and many objects in the universe. Armed with Newton's law of gravitation, Halley carefully studied the orbital statistics of comets and found that the comet appeared in 1570, 1607 and 1680, should be the same comet of an elipse orbit. After his prediction that the next return of the Halley comet (the comet named after him) is 1758 was confirmed, the Newton's theory is finally accepted and the earth-centred theory was finally put to an end. Physics made a giant leap forward, and people's understanding improved substantially.

Contrary to what happened in natural science, the dominance of macroeconometrics created a period of little advancement in economic theory: the vast majority of macroeconomists become empiricists. This was no accidental change in macroeconomics because econometricians were under the illusion that their models/data contained the truth and were able to predict the future, so there was no need to develop a theory. As a result, macroeconometricians are busy building models to mimic reality and to 'predict' the future. Such studies provided little economic reasoning and thus shed little light on how to improve the performance of the economy. If we wish macroeconomic study can generate some positive impact on the development of the economy, it is necessary that empirical studies should be advanced towards the formation of economic theories.

In order to establish the importance of economic theories, it is necessary to refute the claim that theories are unnecessary because all theories are wrong. This claim is based on two facts: (1) theories by definition are not reality; (2) all theories are ultimately replaced by new theories. The claim based on the first fact proposed confuses the correctness of a theory with reality. Theory and reality are different concepts highlighting the importance of theories. If these two concepts are equivalent, then there is no need to develop a theory, instead, one should be satisfied with mere observation of 'correct' reality with no understanding. If this were the case, mankind would be still in the pre-civilized stage, and would be no different to primitive living creatures. A correct theory comes from reality but goes above the reality to reveal the mechanism behind (or generating) the reality. It is the understanding of mechanism that advances mankind.

The claim based on the second fact is related to the degree of correctness of a theory. If one requires a theory to be correct on every front and at any level of understanding, he/she must be disappointed with every theory except the theory created by God who never makes a mistake. When we say a theory is correct we mean the main element of the theory is supported by or consistent with currently available evidence. For example, we regard Darwin's theory of natural selection as correct because this theory is supported by evidence from fossil studies, studies on similarities between related living organisms and the modern studies on DNA. We do not require or expect Darwin to be correct on everything proposed in his research. Similarly, because Newton's gravitation law can explain or is supported by numerous facts, we regard it as correct even if Newton believed in the existence of the ether and Einstein demonstrated that gravitation is not an instant force as Newton claimed. Any correct theories must have some elements of truth and the advancing of new theories cannot reject these elements. Science is a process of discarding false claims, keeping the elements of truth, and thus moving closer to truth.

5. Studying opposing viewpoints in the search for truth

From the birth of macroeconometrics, many eminent economists (e.g. Keynes, Leontief, Coase, Friedman and Solow) have given very insightful criticism. However, macroeconometrics grew stronger despite their strong rejection. One reason for this may be that some economists have been trying to criticize macroeconometrics from the outside.

To criticize macroeconometrics validly, one must analyse econometric theory as well as the associated conditions. Once the foundation for macroeconometrics is proven problematic and/or the fatal flaws are identified, the incorrect theory and unscientific practice should cease to exist. For econometricians or students studying econometrics, unless they are able to prove that macroeconometrics is a valid approach (which is highly unlikely), it is unwise to take a position to advocate or support macroeconometrics merely because of its current popularity or because of a conflict of interest (i.e. jeopardizing their careers). If a theory is wrong, it will be discarded eventually.

## 3.9    The Role of Statistics and the Way Forward

Based on the lessons that can be learnt from the development of econometrics, what is the way forward for macroeconomic study? No silver bullet can be offered because the essential problem is that economic data are non-experimental data which cannot satisfy the condition of random experiments or the 'other things being equal' requirement in economic theory. However, relying on non-experiment data is not unique in the economics discipline. No experiment data are available for many macro-level natural science studies, e.g. studies of earthquakes, volcanoes, astronomy, wild animals, epidemics and the environment. The development of natural sciences tells us that it is in vain that one hopes to predict the future by trying to include all possible variables in the model. As a step to uncovering or confirming causality or mechanism, correlation between two variables or simple multiple correlations can be used, but they can only be an auxiliary tool for research. This can be demonstrated clearly by the course of the discovery of the cause of cholera epidemics.

Cholera is an old and dreadful disease caused by a type of bacterium called Vibrio. It attacks its victims suddenly, causing vomiting, profuse diarrhoea and life-threatening dehydration. The world has seen many cholera epidemics which have claimed countless lives. The first known cholera pandemic or global epidemic started in 1817 from India. The 7th cholera pandemic began in Indonesia in 1961 and spread to other countries in Asia, Europe, Africa and Latin America. WHO and its partners have verified 41 cholera outbreaks in 28 countries in 2001 alone. The cause of cholera was unknown to us, until John Snow and other researchers worked it out in the early 1850s.

The first cholera pandemic reached England in 1831. Creighton (1894) reported that the pandemic claimed 1960 English lives. Shapter (1849) gave a detailed description of the outbreak of the 1831 epidemic, including a map of the locality of deaths. These studies were published in the middle of the second cholera epidemic in England from 1848 to 1849, which killed about 15,000 people. At that time, the dominant view on the cause of cholera was miasma, or the pollution or noxious air in affected areas. This view was originally accepted by most people in the medical profession, including William Farr, who worked at the General Register Office, responsible for collection of official medical statistics in England and Wales.

Using the mortality records at the General Register Office, Farr studied the 1848–1849 cholera epidemic in a comprehensive way. He compared this epidemic with the 1831–1832 epidemic and other earlier plague epidemics, analysed a host of possible relevant factors such as age, sex, temperature, rainfall, wind, domestic crowding and property value, and published a report consisting of 300 pages of tables, charts and maps, and 100 pages of introduction analysing the outbreak of the epidemic (Eyler 2001). Farr's prized finding is called Farr's law of elevation, which stated that cholera mortality is inversely related to elevation of the land. Farr expressed his law in terms of a simple formula between cholera mortality and elevations and he showed that this formula fitted the empirical data remarkably well. Farr's law was readily accepted because it was consistent with the miasma theory: the moist soil on the margins of Thames River together with the filth and debris on the river banks caused decay of organic materials. This resulted in a large number of airborne organic particles, i.e. concentration of miasma at the low-elevation level. As elevation increases, the miasma became less concentrated, and thus, fewer cases of cholera were expected. Given the dominance of the miasma theory, it is not surprising the Farr's law was well received by the medical profession at that time.

Based on the initial symptoms of cholera—vomiting and acute diarrhoea, John Snow realized that, while most epidemic diseases began with fever, cholera began with local abdominal symptoms so it was most likely to be related with food or drinking. Also, cholera in the early stages can be treated by opium and chalk, which acted locally, so cholera tended to be a local disease. Based on the pathologic evidence, Snow hypothesized that a living organism got into the patients' digestive system through food or drink, multiplied there and generated some poison causing vomiting and diarrhoea, which in turn caused contamination of the water supply and thus infected new victims. This hypothesis was remarkable given the fact that the germ theory was not established until the 1860s although the theory was proposed by Girolamo Fracastoro as early as in 1546.

Snow faced an uphill battle because his hypothesis was against the dominant miasma theory, which attributed cholera to the low standard of general hygiene or cleanness of the environment. Snow found two pieces of evidence against the miasma theory. Snow quoted the account by Dr. Lichtenstadt in 1831 of the cholera attack in Berditscher in Volhynia. Among 764 victims were 658 Jews and 106 Christians. The cholera attacks on Jews were much higher than the proportion of Jewish population. The mortality rates of Jewish and Christian victims were found to be 90.7 and 61.3%, respectively. These statistics were directly against the miasma theory. Because Jews had the reputation for cleanliness, the miasma theory would necessitate that the cholera attack and mortality rate should be low among Jewish inhabitants. The other piece of evidence was that cholera severely attacked the joint towns of Dumfries and Maxwell both in 1832 and in 1894, but these two towns were usually not unhealthy places. Snow explained that inhabitants in both towns drank the water of Nith, which was contaminated by sewage. Snow also explained the severe cholera attack of Glasgow: despite the high elevation of Glasgow, its water supply source—the Clyde—might have been contaminated because the Clyde is a tidal river.

Snow figured out the above mechanism of cholera transmission in late 1848 and discussed it with several medical professionals, but he felt that the evidence was scattered and general in nature, so he hesitated to publish them. However, in 1849 two local events offered him convincing evidence and urged him to publish his theory in a monograph, 'On the Mode of Communication of Cholera' (Snow 1849).

One event was related to two housing courts at Thomas Street, Horsleydown. The two courts consisted of small houses or cottages, which were occupied by poor families. The houses were located back to back on the adjoining side of two courts, with a small intervening back area where the privies of both courts and the shared drainage situated. At the end of both courts was an open sewer. The difference between the two sets of houses was that, due to the slope, the Surrey buildings on the northern side were lower than the Trusscott's court on the south-side, so the dirty water in the drain could get into the well which was used by the inhabitants of the Surrey buildings. The consequence was distinctive: cholera brought fearful devastation in the Surrey buildings while only one fatal case occurred in the Trusscott's court. Snow gave a detailed account of how the drainage and well water could mix and the bowel evacuations of patients could pass into their beds and be washed into the drain. The other event was about the cholera outbreak in the immediate neighbouring houses in Albion Terrace. Snow explained in detail how the drains of cesspools and surface water might leak and contaminate the spring water supplied to the 17 houses in a row in Albion Terrace.

Through detailed investigation and logical reasoning, Snow established that, compared with other transmissions, water contamination had a greater probability of being responsible for cholera transmission. Snow's proposition raised the awareness but was not accepted by the medical profession. Farr (1852) admitted that river water was filthy but he was not convinced of Snow's conclusion. When the third cholera outbreak erupted in England in 1853–1854, Snow investigated and obtained more evidence, substantially revising his 1849 book and publishing a comprehensive second version of 'On the Mode of Communication of Cholera' in 1855.

In the second version, Snow (1855) employed many more cases along with statistics to prove his waterborne cholera transmission theory. One important case was the terrible outbreak of cholera in Broad Street, Golden Square. In ten days from 31 August to 9 September 1849, there were more than 500 fatal cholera attacks in a small area where Cambridge Street joins Broad Street. On the 1 September alone, there were 143 fatal attacks. The intensity of the attacks was unprecedented in England. Snow inquired into the 83 registered deaths from 31 August to 2 September and found that nearly all fatal attacks had occurred within a short distance of the Broad Street water pump. He also found that 61 out of 83 deceased used to drink from the Broad Street pump. Based on this waterborne cholera theory, Snow realized that the water pump on Broad Street might be contaminated. He made a presentation to the Board of Guardians of St. James's parish on the 7 September and this led to the removal of the handle of the Broad Street pump on 8 of September. The cholera attack subsided in the next three days.

Following his enquiry, Snow (1855) detailed the cholera deaths in the area close to Broad Street. One unusual circumstance was a Workhouse in Poland Street. Although three-quarters of the Workhouse were surrounded by houses of cholera deaths, only 5 out of 535 workers in the Workhouse died of cholera. Snow found that the workers were allowed some malt liquor each day and there was a deep well in the brewery of the Workhouse. The property owner assured Snow that the workers did not use water from the Broad Street pump. Snow also inquired into the 10 deaths in houses closer to a water pump in another street and found that, the families of the 5 deceased preferred the water from the Broad Street pump so they always obtained water there; for 3 cases, their children went to school close to the Broad Street pump, and two drank and the other one probably drank water from the Broad Street pump. Snow (1855) recorded a total of 616 deaths in this outbreak on the map, showing the deaths were closely related to the pump on Broad Street. This is regarded by some people as the beginning of spatial statistical analysis. Snow also demonstrated his honesty by acknowledging the limitation of data. For example, some people were too sick to describe their situation when they were admitted to hospital. Some who contracted cholera in the Broad Street area had moved to and died in other parts of London.

The other important investigation recorded by Snow (1855) is about the relationship between the South London water supply and cholera deaths. He first showed that water suppliers for the south districts of London, the Southward and Vauxhall Company and the Lambeth Water Company, used unfiltered water from the Thames, so the cholera attacks both in 1832 and in 1849 were much worse in southern districts than those in other districts. In 1852, however, the Lambeth Company moved its waterworks to Thames Ditton, a location upstream of the sewage of London, so Lambeth obtained a cleaner water supply free from sewage. Moreover, the fierce competition between the two companies for supplying households in the south London area caused a large mixed supply area. This mixed area provided an excellent natural experiment for Snow to study water supply and cholera attacks: the conditions of customers of both companies could be viewed as the same because both companies supplied to both rich and poor people and to houses of various sizes. Snow investigated the cholera deaths and water supply in the whole south London area (including Lambeth Company area, Southward and Vauxhall area, and mixed supply area). He found that, in the first seven weeks of the epidemic, there were 1263 deaths in 40,046 houses, to which Southward and Vauxhall supplied water, and only 98 deaths in 26,107 houses, to which Lambeth supplied water. Normalizing these numbers to 10,000 houses, Snow showed that there were 315 deaths for Southward and Vauxhall customers and only 37 deaths for Lambeth customers. This indicated that the mortality rate of Southward and Vauxhall customers was eight times higher. In the second seven weeks of the epidemic, the mortality rate of Southward and Vauxhall customers was also about 5 times higher.

Snow (1855) also refuted a number of objections to his theory such as the miasma theory, the geographic theory, Farr's elevation theory, and a report that a person drank by mistake the rice-water evacuation of a cholera patient but he did not attack by cholera. He believed that water was the only mechanism of cholera transmission. Using his theory, Snow was able to explain a number of features of cholera epidemics. For example, cholera was more severe in summer but died down in winter and spring. Snow attributed this to the fact that English people liked drinking tea so they did not drink much un-boiled water, except in summer. Farr's statistics revealed another feature of cholera in England. At the start of the epidemic, the deaths of males were much greater than the deaths of females, but during the later stages of epidemic, the opposite was the case. Snow explained that men tended to move about and took food and drinks at varieties of places due to work so they were at high risk when the epidemic started. On the other hand, most females tended to remain almost constantly at home, so they were safer when the cholera started. However, when the epidemic was more widespread, females were at higher risk because they were less likely to drink beer and thus more likely to drink un-boiled water.

The second edition of Snow's book was widely reviewed and triggered more discussion and investigation, but the medical profession still did not accept Snow's waterborne transmission theory. Edmund Parkes, an established medical professional who studied the early cases of the 1848 epidemic for the General Board of Health, was very critical of Snow's book. Although Parkes (1855) admitted the significance of Snow's opinion and his effort in obtaining facts, he regarded Snow's work as only a hypothesis because it could not explain all the phenomena of the spread of cholera. Parkes discredited Snow's pathological evidence simply by refuting Snow's belief that the blood was not under the influence of a poison in the early stages of cholera. Snow's emphasis on the early symptoms of cholera was regarded by Parkes as doubtful and of little value.

Next, Parkes selected and discussed 11 cases and the water supply statistics in Snow's book. On the case of adjacent courts in Horsleydown (Surrey buildings and Trusscott's court), Parkes revealed that the Surrey buildings had more rooms and thus more people, so he argued that this could contribute to the higher number of fatal cholera attacks in the Surrey buildings. On the Albion Terrace case, Parkes argued that the open sewer may also contaminate the air. On the case of the Broad Street pump, Parkes made the critics in that Snow had neither the proof that the water was contaminated nor the proof that there were no other local causes for the cholera outbreak. He also pointed out that Snow's explanation was inconsistent with the fact that cholera attacks increased quickly and then subsided but the situation of the water supply was unchanged. For other cases, Parkes complained that Snow did not provide necessary information.

Parkes regarded Snow's water supply inquiry as elaborate but he criticized Snow for not considering the other living conditions of the London residents, including the effect of elevation, the density of the population and the impurity of the air. On Snow's inquiry into the south London water supply by two companies, Parkes argued Snow for not focusing on the intermingled supply area. Parkes argued that the Lambeth Company supplied to a good neighbourhood on elevated ground while Southwark and Vauxhall Company supplied a greater part of the poorest, lowest and marshiest district in London, implying that this different conditions could lead to the different mortality rates of customers of the two companies. He also criticized Snow for not obtaining the number of houses in the intermingled area supplied by each company, arguing that the number of cholera deaths could be proportional to the number of houses. Parkes also questioned the reliability of Snow's salt concentration test to determine the source of water supply for the houses in the intermingled supply area, saying that the salt level in the Thames can change considerably from time to time. Nevertheless, Parkes regarded Snow as an honest and conscientious observer, who had pursued the inquiry with great diligence.

Parkes' review of Snow's book appeared overly critical. This might partly result from Parkes' critical personality. According to Eyler (2013), Parkes was very critical in his own study of early cases of the 1848 cholera outbreak. To a greater degree, it was Snow's proposal that water was the only way for cholera transmission that entailed strong criticism from Parkes and others in the medical profession. This proposal was quite understandable from Snow's side. Based on his observation and reasoning, Snow believed that he had uncovered the mechanism of cholera propagation, so any factors other than water were irrelevant ones which disguised the true mechanism. However, this proposal required people to give up the widely accepted miasma theory. This was a mountainous task for Snow because, without hard evidence, it was very unlikely that people would change their lifetime belief. Based on the traditional concept of scientific rigour, Parkes thought Snow was obliged to prove that contaminated water caused cholera attacks and that no other factors caused a cholera outbreak. This requirement was suitable for a laboratory science but not for macro-level science like epidemiology. Since macro-level science involved so many factors and each factor could vary unexpectedly, it was impossible to produce a proof with absolute certainty and with the accuracy of lab science.

The argument between Snow and his opponents also illustrates the role of statistics in scientific discovery. Statistics is an important tool for discovering the pattern of behaviours of variables and for identifying or verifying the significance of relevant variables. The importance of statistics was demonstrated by the Broad Street map made by Snow and his collection of the water supply data and cholera death data for the area of south London, as well as the massive medical data collected and used by Farr in the General Register Office. However, a statistical tool is unable to uncover a cause of anything because what the tool can identify is correlation or association rather than a cause. This manifested itself in Farr's elevation law, which indicated a strong association between cholera attacks and elevation. Farr mistook this association as causality, so he made a logical mistake and drew an incorrect conclusion. Moreover, Farr relied only on statistics and did not investigate the scenarios behind the statistics, so he was unable to discover the causes of cholera but relied on the existing miasma theories. On the other hand, Snow uncovered the cause of cholera through his logical reasoning based on pathological evidence and on the detailed information behind the statistics. The different approaches used by Snow and Farr showed that causality can be uncovered only by logical reasoning based on experiments and investigations.

The lessons from macro-level natural science like epidemiology can shed some light on research into macroeconomics. Correlation and simple multiple regressions can still be useful auxiliary tools to examine a theory, but caution must be taken in interpreting the results and making inferences. When these tools are used, we assume that the effect of other variables is negligible. In fact, many other factors may play significant roles in data, so the empirical results can only be indicative. As a result, the correlations or estimation results cannot prove or disprove a theory or predict the future with confidence, but they can indicate how far away the theory is from the data. This kind of 'informal' approach is not an anti-theory approach suggested by Summers (1991). On the contrary, statistic theory supports this approach. Since the time series data do not satisfy the condition of random experiments, the data are affected by an unknown number of factors which are impossible to take into account in a model. What we can do is to be aware of the impact of other factors in the interpretation of modelling results.

After obtaining empirical results of correlation or simple multi-regression, researchers need to investigate the difference between the theory and empirical results in order to discover the causes. In this way, the difference between theory and data can help the researcher to reform or refine the theory. The refined theory is then subject to data testing again. Through multiple procedures both from theory to data and from data to theory and by using the combined induction and deduction of logical reasoning, the gap between data and theory can be reduced and, more importantly, our understanding will become closer to the truth.

This multi-procedure approach can be illustrated by the evolution of Keynes' consumption function to Friedman's permanent income theory. When Keynes' consumption function was confronted by data, the results were mixed. The aggregated time series estimation showed a marginal propensity to consume (MPC) around 0.90, which implied a unit income elasticity of consumption and a constant saving rate in the long run. However, the studies based on household survey data showed a MPC in the range of 0.60–0.80. This inconsistency in empirical results led to new theories on income and consumption, namely the relative income hypothesis by Duesenberry (1949), the permanent income hypothesis by Friedman (1957) and the life-cycle hypothesis by Modigliani (1986). These theories were supported by cross-sectional studies but could not explain the high MPC in the time series study. Bunting (1989) argued that the comparison between the results from the aggregate time series study and those from the cross-sectional study was not meaningful. By dividing the aggregate data by the number of households each year and using the ungrouped dataset of cross-section data, Bunting substantially reduced the gap of MPCs from time series data and from survey data. This example also showed that, since the condition for random experiment does not hold in macro-level studies (i.e. many factors can affect the results), the results of a macroeconometric model can only be indicative. In other words, macroeconometrics is not an accurate science, so it cannot be used to prove or disprove a theory, explain phenomenon or predict the future.

## Appendix (for Section 3.6.2): Why Econometric Models Fail: An Illustration

To demonstrate the performance of macroeconometric models, the author uses the US macroeconomic time series data 1969–2013 (see Table 3.2 at the end of appendix) to estimate the GDP identity, i.e. the expenditure and income sides of GDP. Since there are some statistical discrepancies in the two sides of the GDP, one must choose one side as the GDP value. The author chooses the income side of the GDP. Based on the 1-lagged GDP and the variables on both sides of the GDP, the author uses the OLS method to estimate 7 models. Here, the author has an upper hand over even the most experienced macroeconometricians and can dismiss any criticism on modelling techniques because we have a complete list of all factors and know the true mechanism (the correct function of the GDP equation). The estimated results for various models are shown in Table 3.1. For the benefit of non-econometricians, the author has not only listed the coefficient and standard error for each variable but also listed the $p$-value.

Model 1 use income-side data to estimate the GDP identity. The estimated results for Model 1 are perfect: the adjusted R-squared is 1, all variables on the income side of the GDP are extremely significant (p=0.000) and the coefficients are extremely close to 1. The coefficient for the constant is close to zero with a very high p-value (p=0.732), indicating there is no constant term in the model. These are exactly what is predicted by the GDP identity:

**Table 3.1** Estimation of GDP determinants based on US time series 1969–2013

| Explanatory variables | Model 1 | Model 2 | Model 3 | Model 4 | Model 5 | Model 6 | Model 7 |
|---|---|---|---|---|---|---|---|
| 1-Lagged GDP | NA | NA | 1.027835 (0.0062792) $p=0.003$ | NA | −0.0033368 (0.0571678) $p=0.954$ | NA | NA |
| Constant | 0.0100158 (0.0290984) $p=0.732$ | 81.31557 (27.25534) $p=0.005$ | 167.7843 (53.17188) $p=0.000$ | −0.0072479 (0.0539686) $p=0.894$ | 82.9972 (29.57229) $p=0.008$ | −44.1798 (21.16042) $p=0.043$ | 51.04608 (13.92943) $p=0.001$ |
| Wage | 0.9998437 (0.0000685) $p=0.000$ | NA | NA | 0.9999421 (0.0003693) $p=0.000$ | NA | 1.311298 (0.0320946) $p=0.000$ | 0.9197758 (0.0628478) $p=0.000$ |
| Tax | 1.00024 (0.0008111) $p=0.000$ | NA | NA | 1.000116 (0.0013469) $p=0.000$ | NA | NA | NA |
| Profit | 1.000112 (0.0000859) $p=0.000$ | NA | NA | 1.000202 (0.0002734) $p=0.000$ | NA | 1.231101 (0.0700907) $p=0.000$ | 0.8337058 (0.0588025) $p=0.000$ |
| Fixed capital | 1.00028 (0.0002497) $p=0.000$ | NA | NA | 0.9998864 (0.0005665) $p=0.000$ | NA | NA | NA |
| Consumption | NA | 1.275733 (0.0452806) $p=0.000$ | NA | −0.0000493 (0.0004116) $p=0.905$ | 1.281519 (0.0851167) $p=0.000$ | NA | 0.4509562 (0.0564406) $p=0.000$ |
| Investment | NA | 0.7793903 (0.0635759) $p=0.000$ | NA | −0.0000775 (0.0003573) $p=0.830$ | 0.7772844 (0.0682467) $p=0.000$ | NA | −0.0225597 (0.0461778) $p=0.628$ |

(continued)

**Table 3.1** (continued)

| Explanatory variables | Model 1 | Model 2 | Model 3 | Model 4 | Model 5 | Model 6 | Model 7 |
|---|---|---|---|---|---|---|---|
| Net exports | NA | 0.6962019 (0.0853013) $p=0.000$ | NA | −0.0001073 (0.0003221) $p=0.741$ | 0.694964 (0.0954259) $p=0.000$ | NA | NA |
| Government spending | NA | 0.161384 (0.1271703) $p=0.212$ | NA | 0.0001931 (0.0003509) $p=0.586$ | 0.1587255 (0.1367049) $p=0.253$ | NA | NA |
| Adjusted R-squared | 1.0000 | 0.9999 | 0.9984 | 1.0000 | 0.9999 | 0.9998 | 1.0000 |

Time series data are from the Bureau of Economic Analysis (http://www.bea.gov/national/nipaweb/DownSS2.asp). The estimation is performed using Stata software. The table lists the coefficients, standard errors (in parenthesis) and the $p$-value for each explanatory variable. The $p$-value indicates the chance of rejecting the hypothesis that the variable is significant

**Table 3.2** The expenditure and income sides of US GDP 1989–2013

| Year | GDP | Wage | Tax | Profit | Fixed capital | Consum-ption | Invest-ment | Net exports | Gov | dGDP | Measurement error (%) {1% of GDP}/dGDP |
|---|---|---|---|---|---|---|---|---|---|---|---|
| 1969 | 1018.3 | 586.0 | 79.4 | 228.1 | 124.9 | 604.5 | 173.6 | 1.4 | 240.4 | | |
| 1970 | 1070.5 | 625.1 | 86.6 | 222.0 | 136.8 | 647.7 | 170.1 | 3.9 | 254.2 | 52.2 | 20.5 |
| 1971 | 1158.3 | 667.0 | 95.8 | 246.6 | 148.9 | 701.0 | 196.8 | 0.7 | 269.3 | 87.8 | 13.2 |
| 1972 | 1275.3 | 733.6 | 101.3 | 279.5 | 160.9 | 769.4 | 228.1 | −3.4 | 288.2 | 117.0 | 10.9 |
| 1973 | 1422.4 | 815.1 | 112.0 | 317.3 | 178.1 | 851.1 | 266.9 | 4.1 | 306.4 | 147.1 | 9.7 |
| 1974 | 1541.4 | 890.3 | 121.6 | 323.3 | 206.2 | 932.0 | 274.5 | −0.8 | 343.1 | 119.0 | 13.0 |
| 1975 | 1675.7 | 950.2 | 130.8 | 357.1 | 237.5 | 1032.8 | 257.3 | 16.0 | 382.9 | 134.3 | 12.5 |
| 1976 | 1857.1 | 1051.3 | 141.3 | 405.4 | 259.2 | 1150.2 | 323.2 | −1.6 | 405.8 | 181.4 | 10.2 |
| 1977 | 2066.7 | 1169.0 | 152.6 | 456.8 | 288.3 | 1276.7 | 396.6 | −23.0 | 435.8 | 209.6 | 9.9 |
| 1978 | 2333.4 | 1320.3 | 162.0 | 526.1 | 325.1 | 1426.2 | 478.4 | −25.4 | 477.4 | 266.7 | 8.7 |
| 1979 | 2587.4 | 1481.1 | 171.6 | 563.6 | 371.1 | 1589.5 | 539.7 | −22.6 | 525.5 | 254.0 | 10.2 |
| 1980 | 2818.6 | 1626.3 | 190.5 | 575.7 | 426.0 | 1754.6 | 530.1 | −13.0 | 590.8 | 231.2 | 12.2 |
| 1981 | 3174.2 | 1795.4 | 224.1 | 669.6 | 485.0 | 1937.5 | 631.2 | −12.6 | 654.7 | 355.6 | 8.9 |
| 1982 | 3338.2 | 1894.5 | 225.9 | 683.5 | 534.3 | 2073.9 | 581.0 | −20.0 | 710.0 | 164.0 | 20.4 |
| 1983 | 3584.0 | 2014.1 | 242.0 | 767.4 | 560.5 | 2286.5 | 637.5 | −51.6 | 765.7 | 245.8 | 14.6 |
| 1984 | 4002.0 | 2217.6 | 268.7 | 921.4 | 594.3 | 2498.2 | 820.1 | −102.7 | 825.2 | 418.0 | 9.6 |
| 1985 | 4295.5 | 2389.2 | 286.7 | 982.9 | 636.7 | 2722.7 | 829.6 | −114.0 | 908.4 | 293.5 | 14.6 |
| 1986 | 4513.4 | 2545.6 | 298.5 | 987.1 | 682.2 | 2898.4 | 849.1 | −131.9 | 974.5 | 217.9 | 20.7 |
| 1987 | 4829.7 | 2725.7 | 317.2 | 1058.8 | 728.0 | 3092.1 | 892.2 | −144.8 | 1030.8 | 316.3 | 15.3 |
| 1988 | 5253.1 | 2950.9 | 345.0 | 1174.8 | 782.4 | 3346.9 | 937.0 | −109.4 | 1078.2 | 423.4 | 12.4 |
| 1989 | 5593.5 | 3143.9 | 371.5 | 1242.1 | 836.1 | 3592.8 | 999.7 | −86.7 | 1151.9 | 340.4 | 16.4 |
| 1990 | 5888.2 | 3345.0 | 398.0 | 1258.4 | 886.8 | 3825.6 | 993.5 | −77.8 | 1238.4 | 294.7 | 20.0 |
| 1991 | 6085.7 | 3454.7 | 429.6 | 1270.2 | 931.1 | 3960.2 | 944.3 | −28.6 | 1298.2 | 197.5 | 30.8 |
| 1992 | 6428.4 | 3674.1 | 453.3 | 1341.3 | 959.7 | 4215.7 | 1013.0 | −34.7 | 1345.4 | 342.7 | 18.8 |

(continued)

Table 3.2  (continued)

| Year | GDP | Wage | Tax | Profit | Fixed capital | Consumption | Investment | Net exports | Gov | dGDP | Measurement error (%) (1% of GDP)/dGDP |
|---|---|---|---|---|---|---|---|---|---|---|---|
| 1993 | 6726.4 | 3824.0 | 466.4 | 1432.4 | 1003.6 | 4471.0 | 1106.8 | -65.2 | 1366.1 | 298.0 | 22.6 |
| 1994 | 7171.9 | 4014.1 | 512.7 | 1589.5 | 1055.6 | 4741.0 | 1256.5 | -92.5 | 1403.7 | 445.5 | 16.1 |
| 1995 | 7573.5 | 4206.7 | 523.1 | 1720.9 | 1122.8 | 4984.2 | 1317.5 | -89.8 | 1452.2 | 401.6 | 18.9 |
| 1996 | 8043.6 | 4426.2 | 545.6 | 1895.9 | 1176.0 | 5268.1 | 1432.1 | -96.4 | 1496.4 | 470.1 | 17.1 |
| 1997 | 8596.2 | 4719.1 | 577.8 | 2059.4 | 1240.0 | 5560.7 | 1595.6 | -102.0 | 1554.2 | 552.6 | 15.6 |
| 1998 | 9149.3 | 5082.4 | 603.1 | 2153.6 | 1310.3 | 5903.0 | 1735.3 | -162.7 | 1613.5 | 553.1 | 16.5 |
| 1999 | 9698.1 | 5417.5 | 628.4 | 2251.4 | 1400.9 | 6307.0 | 1884.2 | -256.6 | 1726.0 | 548.8 | 17.7 |
| 2000 | 10,384.3 | 5863.1 | 662.8 | 2344.2 | 1514.2 | 6792.4 | 2033.8 | -375.8 | 1834.4 | 686.2 | 15.1 |
| 2001 | 10,736.8 | 6053.8 | 669.0 | 2410.1 | 1604.0 | 7103.1 | 1928.6 | -368.7 | 1958.8 | 352.5 | 30.5 |
| 2002 | 11,050.3 | 6149.7 | 721.2 | 2517.3 | 1662.1 | 7384.1 | 1925.0 | -426.5 | 2094.9 | 313.5 | 35.2 |
| 2003 | 11,524.3 | 6372.7 | 758.9 | 2665.4 | 1727.2 | 7765.5 | 2027.9 | -503.6 | 2220.8 | 474.0 | 24.3 |
| 2004 | 12,283.5 | 6748.8 | 817.5 | 2885.5 | 1831.7 | 8260.0 | 2276.7 | -619.2 | 2357.4 | 759.2 | 16.2 |
| 2005 | 13,129.2 | 7097.9 | 873.6 | 3175.7 | 1982.0 | 8794.1 | 2527.1 | -721.2 | 2493.7 | 845.7 | 15.5 |
| 2006 | 14,073.2 | 7513.7 | 940.4 | 3483.0 | 2,136.0 | 9304.0 | 2680.6 | -771.0 | 2642.2 | 944.0 | 14.9 |
| 2007 | 14,460.1 | 7908.8 | 980.0 | 3307.0 | 2,264.4 | 9750.5 | 2643.7 | -718.6 | 2801.9 | 386.9 | 37.4 |
| 2008 | 14,619.2 | 8090.0 | 989.3 | 3176.5 | 2,363.4 | 10,013.6 | 2424.8 | -723.1 | 3003.2 | 159.1 | 91.9 |
| 2009 | 14,343.4 | 7795.7 | 967.8 | 3211.6 | 2368.4 | 9847.0 | 1878.1 | -395.5 | 3089.1 | -275.8 | -52.0 |
| 2010 | 14,915.2 | 7969.5 | 1001.2 | 3562.8 | 2381.6 | 10,202.2 | 2100.8 | -512.7 | 3174.0 | 571.8 | 26.1 |
| 2011 | 15,556.3 | 8277.1 | 1042.5 | 3785.9 | 2450.6 | 10,689.3 | 2239.9 | -580.0 | 3168.7 | 641.1 | 24.3 |
| 2012 | 16,372.3 | 8614.9 | 1074.0 | 4153.2 | 2530.2 | 11,083.1 | 2479.2 | -568.3 | 3169.2 | 816.0 | 20.1 |
| 2013 | 16,980.0 | 8853.6 | 1102.2 | 4396.8 | 2627.2 | 11,484.3 | 2648.0 | -508.2 | 3143.9 | 607.7 | 27.9 |

*Source* Except the last two columns, the data are from Bureau of Economic Analysis (http://www.bea.gov//national/nipaweb/DownSS2.asp)

GDP=Wage +Tax + Profit + Capital Formation.

One may hail that the macroeconometric model works! However, this is not the usual case in macroeconometric modelling and the model has worked because the assumptions for estimating an econometric model held. With perfect theoretic knowledge we know that all variables are included in the model so the conditions for random experiments hold. We also know perfectly well that we have the right function for the model. More importantly, the data perfectly fit in with the GDP identity equation except for very tiny rounding errors (about US$0.1 billion for a magnitude of US$1018–16980 billion GDP) for some years, so the OLS method can find the best fit.

Model 2 uses the income-side GDP data to estimate the GDP identity on the expenditure side:

GDP= Consumption + Investment + Net Export + Government Spending.

With perfect knowledge, this model also includes all variables and uses the right function. However, the data do not fit the equation closely because of the statistic discrepancy (measurement error) on both sides of the GDP. The measurement error causes much damage to the estimation. Although R-squared is still very high (0.9999) and most explanatory variables are significant, the results are quite far from the truth: all coefficients are not close to 1. The marginal contribution of consumption is overestimated while the marginal contribution of investment and net export is underestimated; the marginal contribution of government spending to the GDP is only about 16%. Compared with the true marginal contribution of 100% based on our perfect knowledge, the estimated marginal contribution of government spending discounts the true value by more than 80%. Effectively, the marginal contribution of government spending is insignificant even if one uses the 10% $p$-value as a benchmark of rejection of significance of government spending. Moreover, the constant should be zero but modelling results show it is very significant.

Model 3 estimates the impact of a lagged GDP on current GDP to illustrate the common practice of using lagged variables in macroeconometric modelling. The estimation shows a very high $R$-squared (0.9984) and a very significant impact of past GDP. In fact, the coefficient of

1-lagged GDP is close to (or slightly greater than) 1. This confirms the view that most macroeconomic variables are non-stationary. However, the unit root tests on GDP and other variables are mixed, depending on what type of test is employed. If one believes these variables are non-stationary and thus he/she employs a first-differenced model or a cointegrated VAR model, the results may interest a macroeconometrician but this approach is definitely a step further on the wrong way to finding the truth because there are no dynamics in the GDP identity equation.

Model 4 includes all variables from both sides of the GDP. This exercise assumes that we have no knowledge of what variable is relevant or important so we have to include all possible variables. The estimated results show that the coefficients on the income-side variables are very close to 1 while those on the expenditure-side variables are very close to zero. Since the coefficients on the income-side variables are quite close to the results in Model 1, one may conclude that the irrelevant variables added to the model will not change the modelling results. However, here the expenditure-side variables are not irrelevant variables—they are components of the GDP! Their coefficients are zero simply because the model has already found the best fit, so they become redundant variables. This reasoning is confirmed by the fact that when the expenditure side of the GDP values are used as the values for dependent variables, the expenditure side of the GDP components become very significant (with coefficients close to 1), while the income side of the GDP components is insignificant. Hence, these results demonstrate that an econometric model cannot find which variables are relevant or important but can only suggest which variables can fit the data better.

Models 5–7 show different combinations of variable selections. Model 5 includes the 1-lagged GDP and expenditure-side variables. The results show that the coefficients for the expenditure-side variables are very similar to the results from Model 2, while the lagged GDP becomes insignificant. Again, this result does not indicate that the expenditure-side variables are more important than the lagged GDP, but only shows that the expenditure-side variable can fit the data better than the lagged GDP. Model 6 keeps the relatively more important variables on the expenditure side—wages and profits—but excludes the relatively less important variables—taxes and fixed capital formations. The results show the significant overstatement of the contribution of wages and profits. This is simply the consequence of omitting variables in macroeconometric models, but this model represents a likely case in macroeconometric modelling because in real econometric modelling practice no one has perfect knowledge to include all variables. Model 7 includes the most important variables from both sides of the GDP, namely wages, profits, consumption and investment. The estimation results do not make sense in economics: wages, profits and consumption make a discounted contribution to GDP (the coefficients for these variables are significantly less than 1) while investment contributes negatively (albeit insignificantly) to the GDP.

From this exercise of illustrative estimation, it is seen that, if the conditions for statistical theory hold, a statistical model works well (e.g. Model 1). However, this is an unlikely case in macroeconometrics because we have neither perfect knowledge about the factors involved nor the correct functions to be used, and also because the macroeconomic data are not accurate. From the performance of Models 5–7, we can see how misleading a macroeconometric model can be. Considering the possibility of misspecification of function forms in real econometric modelling practice, the estimation results can be even worse than the example displayed. In short, a macroeconometric model is most likely to be unable to find the truth due to measurement errors in data (e.g. Model 2), the inability to include all possible factors (e.g. Models 5, 6, 7), interference between explanatory variables (Models 5, 7) and misspecification of function form.

# References

Beall, A. T., & Tracy, J. L. (2013). Women Are More Likely to Wear Red or Pink at Peak Fertility. *Psychological Science, 24,* 1837–1841.

Brown, P. (1972). The Underdevelopment of Economics. *Economic Journal, 82,* 1–10.

Bunting, D. (1989). The Consumption Function "Paradox". *Journal of Post Keynesian Economics, 11*(3), 347–359.

Campbell, W. W., Barton, M. L., Jr., Cyr-Campbell, D., Davey, S. L., Beard, J. L., Parise, G., & Evans, W. J. (1999). Effects of an Omnivorous Diet Compared with a Lactoovovegetarian Diet on Resistance-Training-Induced Changes in Body Composition and Skeletal Muscle in Older Men. *American Journal of Clinical Nutrition, 70*(6), 1032–1039.

Chatterjee, S. K. (2002). *Statistical Thought: A Perspective and History*. Oxford: Oxford University Press.

Cooley, T., & Leroy, S. (1985). Atheoretical Macroeconometrics: A Critique. *Journal of Monetary Economics, 16*, 283–368.

Creighton, C. (1894). *A History of Epidemics in Britain*. Cambridge: Cambridge University Press.

Donohue, J. J., III, & Levitt, S. D. (2001). The Impact of Legalized Abortion on Crime. *Quarterly Journal of Economics, 116*(2), 379–420.

Duesenberry, J. (1949). *Income, Saving, and the Theory of Consumer Behaviour*. Cambridge, MA: Harvard University Press.

Durante, K. M., Rae, A., & Griskevicius, V. (2013). The Fluctuating Female Vote: Politics, Religion, and the Ovulatory Cycle. *Psychological Science, 24*(6), 1007–1016.

Eyler, J. (2001). The Changing Assessments of John Snow's and William Farr's Cholera Studies. *History of Epidemiology, 46*(4), 225–232.

Eyler, J. (2013). Commentary: Confronting Unexpected Results: Edmund Parkes Reviews John Snow. *International Journal of Epidemiology, 42*, 1562–1565.

Farr, W. (1852). Influence of Elevation on the Fatality of Cholera. *Journal of the Statistics Society of London, 15*(2), 155–183.

Fisher, R. A. (1947). *Design of Experiments* (4th ed.). London: Oliver and Boyd.

Foote, C., & Goetz, C. (2008). The Impact of Legalized Abortion on Crime: Comment. *The Quarterly Journal of Economics, 123*(1), 407–423.

Freedman, D. (1995). Some Issues in the Foundation of Statistics. *Foundations of Science, 1*, 19–83.

Freedman, D. (1999). From Association to Causation: Some Remarks on the History of Statistics. *Statistical Science, 14*, 243–258.

Freedman, D. (2005). *Statistical Models: Theory and Practice*. Cambridge: Cambridge University Press.

Friedman, M. (1948). Memorandum About the Possible Value of the CC's Approach Toward the Study of Economic Fluctuations, Rockefeller Archive.

Friedman, M. (1951). *Comment, in Conference on Business Cycles* (pp. 107–114). New York: Naitonal Bureau of Economic Research.

Friedman, M. (1953). The Methodology of Positive Economics. In M. Friedman (Ed.), *Essays in Positive Economics* (pp. 3–43). Chicago, IL: University of Chicago Press.

Friedman, M. (1957). *A Theory of Consumption Function*. Princeton, NJ: Princeton University Press.

Frisch, R. (1934). *Statistical Confluence Analysis by Means of Complete Regression Systems*. Oslo: Institute of Economics.

Gelman, A., & Loken, E. (2014). The Statistical Crisis in Science. *American Scientist, 102,* 460–465.

Granger, C. (1980). Testing for Causality: A Personal Viewpoint. *Journal of Economic Dynamic and Control, 2*(4), 329–352.

Haavelmo, T. (1943). Statistical Testing of Business Cycle Theories. *Review of Economics and Statistics, 25,* 13–18.

Haavelmo, T. (1944). The Probability Approach in Econometrics. *Econometrica, 12*(Supplement), iii–115.

Hacking, I. (1975). *The Emergence of Probability*. Cambridge: Cambridge University Press.

Hald, A. (1990). *A History of Probability and Statistics and Their Applications Before 1750*. New York: Wiley.

Hald, A. (1998). *A History of Mathematical Statistics from 1750 to 1930*. New York: Wiley.

Hansen, L. P., & Singleton, K. J. (1982). Generalized Instrumental Variables Estimation of Nonlinear Rational Expectations Models. *Econometrica, 50*(5), 1269–1286.

Hansen, L. P., & Singleton, K. J. (1983). Stochastic Consumption, Risk Aversion, and the Temporal Behavior of Asset Returns. *Journal of Political Economy, 91*(2), 249–265.

Heckman, J. (2000). Causal Parameters and Policy Analysis in Economics: A Twentieth Century Retrospective. *The Quarterly Journal of Economics, 115,* 45–97.

Hendry, D. (1980). Econometrics: Alchemy or Science? *Economica, 47,* 387–406.

Hendry, D. (1993). *Econometrics: Alchemy or Science? Essays in Econometric Methodology.* Oxford: Blackwell.

Hendry, D., & Mizon, G. (2014). Unpredictability in Economic Analysis, Econometric Modelling and Forecasting. *Journal of Econometrics, 182,* 186–195.

Hendry, D. F., & von Ungern-Sternberg, T. (1980). Liquidity and Inflation Effects on Consumer's Expenditure. In A. S. Deaton (Ed.), *Essays in the Theory and Measurement of Demand.* Cambridge: Cambridge University Press.

Hume, D. (1739 [1888]). *Treatise of Human Nature* (L. A. Selby-Bigge, Ed.). Oxford: Clarendon Press.

Hume, D. (1748). *An Enquiry Concerning Human Understanding.* Harvard Classics Volume 37. New York: P. F. Collier & Son.

Joyce, T. (2004). Did Legalized Abortion Lower Crime? *Journal of Human Resources, 39*(1), 1–28.

Joyce, T. (2009). A Simple Test of Abortion and Crime. *The Review of Economics and Statistics, 91,* 112–123.

Keuzenkamp, H. (1995). The Econometrics of the Holy Grail. *Journal of Economic Surveys, 9,* 233–248.

Keynes, J. (1939). Professor Tinbergen's Method. *Economic Journal, 49,* 558–568.

Keynes, J. (1940). Comment. *Economic Journal, 50,* 154–156.

Kling, A. (2011). Macroeconometrics: The Science of Hubris. *Critical Review, 23,* 123–133.

Krugman, P. (2009, September 6). How Did Economists Get It So Wrong? *The New York Times.*

Kuhn, T. (1962). *The Structure of Scientific Revolutions.* Chicago: University of Chicago Press.

Leamer, E. (1978). *Specification Searches: Ad Hoc Inference with Non-experimental Data.* New York: Wiley.

Leamer, E. (1983). Let's Take the Con Out of Econometrics. *American Economic Review, 73,* 31–43.

Leontief, W. (1971). Theoretical Assumptions and Non-observed Facts. *American Economic Review, 61,* 1–7.

Liu, T. (1960). Under-Identification, Structural Estimation, and Forecasting. *Econometrica, 28,* 855–865.

Lott, J., & Whitley, J. (2007). Abortion and Crime: Unwanted Children and Out-of-Wedlock Births. *Economic Inquiry, 45*(2), 304–324.

Lucas, R. (1976). Econometric Policy Evaluation: A Critique. In K. Brunner & A. Meltzer (Eds.), *Carnegie Rochester Conference Series on Public Policy* (Vol. 1, pp. 19–46). Amsterdam: North-Holland.

Mason, S. (1962). *A History of the Sciences.* New York: Collier Books.

Modigliani, F. (1986). Life Cycle, Individual Thrift, and the Wealth of Nations. *American Economic Review, 76,* 297–313.

Moore, H. (1911). *Laws of Wages: An Essay in Statistical Economics.* New York: The Macmillan Company.

Moore, H. (1914). *Economic Cycles: Their Law and Causes.* New York, NY: MacMillan.

Moore, H. (1917). *Forecasting the Yield and Price of Cotton.* New York: The Macmillan Company.

Moosa, I. A. (2017). *Econometrics as a Con Art.* UK: Edward Elgar Publishing.

Morgenstern, O. (1950). *On the Accuracy of Economic Observations.* Princeton: Princeton University Press.

Nell, E., & Errouaki, K. (2013). *Rational Econometric Man, Transforming Structural Econometrics.* Aldershot: Edward Elgar.

Parkes, E. (1855). *Mode of Communication of Cholera,* by John Snow, M.D. (2nd ed.). *British and Foreign Medico-Chirurgical Review, 15,* 456.

Pearson, K. (1892). *The Grammar of Science.* London: Walter Scott.

Pearson, K. (1910). *The Grammar of Science* (3rd ed.). Edinburgh: Black.

Pearson, K., & Lee, A. (1897). On the Distribution of the Frequency (Variation and Correlation) of the Barometric Heights at Divers Stations. *Philosophical Transactions of the Royal Society of London, 190,* 423–469.

Pearson, K., Lee, A., & Bramley-Moore, L. (1899). Genetic (Reproductive) Selection: Inheritance of Fertility in Man, and of Fecundity in Thoroughbred Racehorses. *Philosophical Transactions of the Royal Society of London, Series A, 192,* 257–330.

Pesaran, M. (1987). *The Limits to Rational Expectations.* Oxford: Basil Blackwell.

Petersen, M. B., Sznycer, D., Sell, A., Tooby, J., & Cosmides, L. (2013). The Ancestral Logic of Politics: Upper Body Strength Regulates Men's Assertion of Self-Interest Over Income Redistribution. *Psychological Science, 24*(7), 1098–1103.

Popper, K. (1963). *Conjectures and Refutations.* London: Routledge.

Reid, D. (1977, May). Public Sector Debt, Economic Trends, pp. 100–107.

Sailer, S. (2005). *Abortion and Crime: Sailer Responds to Steven "Freakonomics" Levitt's Response.* http://www.unz.com/isteve/abortion-and-crime-sailer-responds-to/.

182     S. Meng

Sell, A., Cosmides, L., Tooby, J., Sznycer, D., von Rueden, C., & Gurven, M. (2009). Human Adaptations for the Visual Assessment of Strength and Fighting Ability from the Body and Face. *Proceedings of the Royal Society, 276*, 575–584.

Shapter, T. (1849). *The History of the Cholera in Exeter in 1832*. London: John Churchill.

Simmons, J., Nelson, L., & Simonsohn, U. (2011). False-Positive Psychology: Undisclosed Flexibility in Data Collection and Analysis Allow Presenting Anything as Significant. *Psychological Science, 22*, 1359–1366.

Sims, C. (1980). Macroeconomics and Reality. *Econometrica, 48*, 1–48.

Skidelski, R. (2009, August 6). How to Rebuild a Shamed Subject. *Financial Times*.

Snow, J. (1849). *On the Mode of Communication of Cholera*. London: John Churchill.

Snow, J. (1855). *On the Mode of Communication of Cholera* (2nd ed.). London: John Churchill.

Solow, R. (2010). Statement of Robert M. Solow. In *Building a Science of Economics for the Real World* (pp. 12–15). U.S Government Printing Office. http://www.gpo.gov/fdsys/pkg/CHRG-111hhrg57604/pdf/CHRG-111hhrg57604.pdf.

Spanos, A. (2011). Foundational Issues in Statistical Modelling: Statistical Model Specification and Validation. *Rationality, Markets and Morals, 2*, 146–178.

Summers, L. (1991). The Science Illusion in Empirical Macroeconomics. *The Scandinavian Journal of Economics, 93*, 19–148.

Suppes, P. (1970). *A Probabilistic Theory of Causality* (Acta Philosophica Fennica). Amsterdam: North-Holland.

Tinbergen, J. (1939). *Statistical Testing of Business Cycle Theories: Part I: A Method and Its Application to Investment Activity*. New Work: Agaton Press.

Tinbergen, J. (1940). On a Method of Statistical Business Cycle Research, A Reply. *Economic Journal, 50*, 141–154.

Wallis, K. (1977). Multiple Time Series Analysis and the Final Form of Econometric Models. *Econometrica, 45*, 1481–1497.

Wickens, M. (1982). The Efficient Estimation of Econometric Models with Rational Expectations. *Review of Economic Studies, 49*, 55–68.

Worswick, G. (1972). Is Progress in Economic Science Possible? *Economic Journal, 82*, 73–86.

Yule, G. U. (1926). Why Do We Sometimes Get Nonsense Correlations Between Time Series?—A Study in Sampling and the Nature of Time Series. *Journal of the Royal Statistical Society, 89*(1), 1–64.

Ziliak, S. T., & McCloskey, D. N. (2008). *The Cult of Statistical Significance: How the Standard Error Costs Us Jobs, Justice, and Lives.* Ann Arbor, MI: The University of Michigan Press.

# 4

# A Critical Assessment of Different Schools of Economic Thought

History shows that economic growth has been repeatedly disrupted by recessions and stagnations. This phenomenon has triggered a variety of economic theories such as the underconsumption theory; the classical, neoclassical and new classical theories; the Keynesian, new Keynesian and post-Keynesian theories; the monetarist theory; the Austrian School; the institutionalism theory; and the Marxist theory. For hundreds of years, the conflicting ideas behind these theories have caused economists to argue their merits and shortcomings but little consensus has been reached. This situation calls for a new theory which unites the existing theories and is able to explain satisfactorily the nature of cyclic economic growth. This chapter examines existing macroeconomic theories and sheds light on a direction to establishing a new theory.

## 4.1 Underconsumptionists' Theory

Underconsumptionists have stressed the importance of consumption in economic growth. The underlying reasoning for their theory is as follows. Since consumption is the purpose of production, it is the driver

© The Author(s) 2019
S. Meng, *Patentism Replacing Capitalism*,
https://doi.org/10.1007/978-3-030-12247-8_4

of trade and economic activities. As such, insufficiency of consumption will lead to the oversupply of commodities, decreasing economic activities and even depressing economic recessions. The two-volume book 'Mercantilism' by Heckscher (1935) provided an account of early underconsumptionist theories. Keynes (1936) also discussed and supported underconsumptionist theory in his then popular book 'General Theory of Employment, Interest, and Money'.

The earliest expression of the underconsumptionist idea is generally attributed to a pamphlet 'The Treasures and riches to put the State in Splendor' by Barthélemy de Laffemas in 1598. De Laffemas argued that the purchasers of luxury goods such as French silk created a livelihood for the poor while the thrift of the rich will cause the poor to die in distress. Here, de Laffemas demonstrated the paradox of thrift: although thrift is generally regarded as a virtue, it is actually an evil because it is an obstacle to economic growth. This idea was reinforced by a handful of writers, e.g. in 1686, Von Schrotter attacked the regulations that forbade excessive display on clothing, and in 1690, Barbon said 'Prodigality is a Vice that is prejudicial to the Man, but not to Trade'.

It was the 'Fable of the Bees' by Bernard Mandeville (1714) that popularized the idea of underconsumptionists. The text was an allegorical poem, which told the story of a beehive in which the fashion of luxury spending was replaced by the virtue of thrift. When private vices such as vanity and envy governed the behaviours of bees, the beehive was prosperous:

> A Spacious Hive well stock'd with Bees,
> That lived in Luxury and Ease;
> And yet as fam'd for Laws and Arms,
> As yielding large and early Swarms;
> Was counted the great Nursery
> Of Sciences and Industry.

However, when the bees adopted virtues such as honesty, thrift and honour, the beehive shrank and collapsed:

> For many Thousand Bees were lost.
> Hard'ned with Toils, and Exercise

They counted Ease itself a Vice;
Which so improved their Temperance;
That, to avoid Extravagance,
They flew into a hollow Tree,
Blest with Content and Honesty.

So the poem concluded that private virtue led to economic depression but private vices could revive the economy:

Bare Virtue can't make Nations live
In Splendor; they, that would revive
A Golden Age, must be as free,
For Acorns, as for Honesty.

In 1723, the grand jury of Middlesex convicted the poem as a nuisance and thus the poem earned a scandalous reputation. However, the paradox presented by the poem intrigued many people. This paradox was successfully solved by Adam Smith.

Smith (1776) claimed that savings, as a result of thrift, are spent in a productive fashion: 'That portion which he annually saves, as for the sake of profit it is immediately employed as a capital, is consumed in the same manner, and nearly in the same time too, but by a different set of people' (Smith 1776, p. 321). Smith's claim can be summarized as a doctrine of 'saving is invested immediately'. According to this doctrine, luxury spending is not necessary for economic prosperity; on the contrary, savings or capital is the source of economic growth and wealth of the nation.

Adam Smith's doctrine of 'saving is invested immediately' was further reinforced by Say (1803, 1821). In his Traite d'Economie Politique (A Treatise on Political Economy), Say constituted a law of markets, commonly known as Say's law, which claimed that supply creates demand. The reasoning behind Say's law is the income–expenditure identity: supply of goods creates purchasing power for other goods, so supply creates demand. If Say's law is true, the oversupply of one type of goods must be matched by excess demand for other types of goods, so there is no possibility of general oversupply in an economy.

However, the claims by Smith and Say never dismissed the concerns about general oversupply or underconsumption. Lord Lauderdale (1804) stressed the regulation role of demand and claimed that accumulation of capital might have its bounds at all times. Chalmers (1808) argued that the amount of expenditure imposed a limit on the use and extension of capital and that, with continuing capital accumulation, the increasing capital and diminishing expenditure would reduce profit to zero and thus further discourage capital accumulation. He also showed the limitations of capital expansion imposed by the state of agriculture. Spence (1808) maintained that there were definite limits to the accumulation of capital and that the continued and progressively increasing expenditure of landlords was essential to prosperity. Spence's pamphlet agitated the minds of the public and led to a strong response from James Mill.

Mill's (1808) defence to Spence's attack is basically an extended reasoning based on Smith's Doctrine and Say's law. He claimed that the whole annual produce of a country will always be consumed completely because no one would let any part of this produce lie useless. Based on Say's law, Mill further expounded on the complete consumption of annual produce of a nation: 'a nation may easily have more than enough of any one commodity, though she can never have more than enough of commodities in general'.

The controversy on this topic formed a heated argument between a pair of good friends David Ricardo and Thomas Malthus. Malthus (1820) was concerned about the post-war depression following the defeat of Napoleon and suggested that demand was insufficient to create a market for the country's greatly enhanced productivity potential. He further claimed that, if saving was pushed to excess, it would destroy the motive for production. On the other hand, Ricardo (1817) thought that, in an eventual stationary state, wages would be at the natural level and profits would be at the minimum level, but he argued that both the marginal physical output per worker (and thus wages) and the real return to capital were far above the stationary state levels, so he concluded that the post-war depression was a normal fluctuation of an economy. Ricardo (1820) attributed any excess supply to the mismatch between production and consumption.

The controversy spread and continued. Sismondi (1819) blamed capitalist competition for oversupply or underconsumption. He argued that, due to competition, each producer strived to produce more and to sell it at a lower price than his/her competitors so there was a permanent tendency towards overproduction. However, McCulloch (1856) dismissed the importance of the will to purchase emphasized by Malthus simply by asking a question 'who ever heard of a want of will to purchase'. Here, he explicitly equalized demand with the power to purchase and thus unreservedly endorsed Say's law. With super clarity, John Hobson rejected the classical economists' claim that saving enriches while spending impoverishes the community. Hobson and Mummery (1889) pointed out that the objective of production was to provide utilities and conveniences for consumers while the use of capital is to aid this objective. They argued that, while saving increased the amount of existing capital, it simultaneously decreased the amount of utilities and conveniences due to reduced consumption, so undue amount of saving undermined utility and convenience and caused general overproduction.

At Hobson's time, the classical economists seemed to have won the argument and the view of underconsumption was suppressed. This is evident by the account of Hobson, who was invited by two organizations to lecture, but later the invitation was withdrawn due to his underconsumptionist opinion. However, the underconsumptionist thought did not die out; it flared up in Keynes's theory in the 1930s and formed an important element of Keynesian economics. Underconsumptionist thought was also an important element in the different schools of economic thought such as Marxism, institutionalism and the Austrian School. Nevertheless, the underconsumptionist argument has never gained widespread popularity.

In hindsight, it is interesting, and of importance, to ask the question: Why has the underconsumption theory never disappeared but has never gained a long-lasting mainstream status? The counter-intuitive (i.e. why do we need encourage consumption considering unlimited human desire?) and anti-moralism nature of the theory may be a reason, and the historical acceptance and dominance of classical theory since Smith's time may be another. However, there are also other reasons.

First, demand and consumption are subtle and complex concepts. Whether or not general oversupply or underconsumption is possible depends on the subtle difference in the meaning of demand. Classical economists regard 'demand' as purchasing power or define that effective demand is demand supported by purchasing power, so they think Say's law is correct and general oversupply is impossible. On the other hand, underconsumptionists regard 'demand' as the desire to consume which may be greater or smaller than purchasing power, so underconsumption or overproduction can be a reality if the desire to consume falls short. When one applies these different definitions of a person's demand to an economy, the situation becomes more complex and thus the arguments on the possibility of oversupply or underconsumption become interminable.

Second, both sides failed to admit and absorb the valid points of their opponents. This may be due to different ways of thinking. People come from different backgrounds, have different perspectives and thus form different approaches. This is evident in the arguments between Ricardo and Malthus. They were good friends but they came from different backgrounds. Malthus was born into English country gentry and was a clerk in a country parish. Ricardo was a son of stockbroker and had financial experience. As a result, Ricardo's thinking was concise and archetypical, while Malthus was motivated by deep common-sense convictions based on the rich complication of real economic life. Given the different preconceptions and approaches, each found it hard to step into the other's shoes. This may explain why one side could not see the point of view of the other.

Having a confined mentality may also matter. For a yes/no question such as 'is general oversupply possible', people tend to think there is only one answer. If one is confident that his/her answer is correct, he/she generally will not see the problem from the perspective of the other side. This situation is similar to that in the parable of six blind men and an elephant. Once a blind man is confident that an elephant is a tree trunk because he feels a leg of the elephant, he disagrees fiercely with the statement made by another blind man that an elephant is a wall. Similarly, classical economists implicitly (e.g. Say, Ricardo and Mill) or explicitly (e.g. McCulloch) assume that people's desire never falls short of purchasing power, so they are confident that their argument is correct and thus there is no need to comprehend the argument from the other side.

Third, underconsumptionists failed to advance their arguments to an advanced theory level. When underconsumptionists repeatedly used logic and examples to demonstrate that capital accumulation has boundaries and that investment depends on expected future consumption, classical economists rejected these claims simply by applying Smith's doctrine and Say's law. Underconsumptionists have not produced a general theory matching the classical theory; on the contrary, they demonstrated why classical theory does not work for various reasons. For example, Sismondi argued that rigidities, immobilities, time lags and other frictions in an economy prevented equilibrium being reached smoothly and painlessly. Like Marx, Sismondi blamed the separation of property and toil and thus the ever-widening inequality of distribution for oversupply or underconsumption. Malthus regarded Smith's definition of wealth of a nation as a dangerous abstraction from the physical characteristics of the commodities produced. Malthus also rejected Smith's claim that relative price is determined by the relative cost of production by arguing that the prices of monopolized commodities and agricultural products were almost always determined upon a principle distinct from the cost of production. Without a theory, underconsumptionists were disadvantaged because they were fighting on an unequal footing.

Although later heterodox economists produced more advanced theories such as Keynesian economics, monetarist theory, socialist theory, the Austrian school and institutionalism theory, they mainly attacked the classical theory from various perspectives based on evidence from reality—in the same vein as the underconsumptionists' approach. Reality is complex, so it is not surprising that any economic theory will find some supporting evidence. However, these heterodox theories are unable to explain the whole, or overall, reality. To achieve this goal, one needs to build a theory based on the most general and simplest case reflecting the main feature of the reality (e.g. a competitive market for a capitalist society or a command economy for a socialist society) and then to advance the theory by adding special features for complex cases. Classical economics has a general theory based on the simplest case such as a competitive firm, but the theory has various shortcomings. Nevertheless, the general theory of classical economics forms the foundation of its mainstream status.

## 4.2* Classical Economics

Classical economists believe in the efficiency of markets. This belief is rooted in the Physiocrats' idea that natural laws govern the operation of the economy. The idea of natural economic laws was formalized by Adam Smith as the doctrine of the 'invisible hand'. Classical economics has developed over time from traditional or old classical, through neoclassical to new classical economics; however, Smith's 'invisible hand' doctrine is the cornerstone of all types of classical economics.

### 4.2.1 Old Classical Theory

Based on the concept of the competitive market, Smith successfully demonstrated that a market mechanism can channel the self-interest of individuals to achieve an optimal allocation of resources, so an economy can achieve the optimal outcome without the intervention of the government. In response to the underconsumptionists claim that consumption is the driver of the economy, Smith highlighted the importance of investment. Later, Jean-Baptiste Say put an even stronger but more controversial claim that supply creates demand. This claim is called Say's law of markets—a law that has been adopted by most classical economists.

Smith's claim that saving is immediately invested effectively established the link between saving and investment. However, he did not go further to investigate the determinants of investment. The purpose of investment is to make a profit, which is realized only when the products brought about by investment are sold, so the growth of consumption in the future is a key condition for successful investment. If the prospect of future consumption is far less than anticipated, the increased products due to investment cannot be sold and thus the investor cannot make a profit. In this case, the investor would be reluctant to invest and saving would not be invested immediately. As a result, underconsumption or overproduction becomes a reality.

Regarding Say's law, as we have seen earlier, there exists a logical problem because Say failed to distinguish between 'the power to purchase' and 'the will to purchase'. Supply or production does create an equivalent purchasing power, but an individual with this purchasing power does not necessarily exercise it. The believers of Say's law claim that the will to purchase is unlimited and thus the power to purchase equals effectual demand. This claim is apparently disproven by the behaviour of wealthy people.

## 4.2.2 Neoclassical Theory

Traditional classical economics was transformed to neoclassical economics through the marginalism revolution, notably by Jevons, Menger, Walras and Marshall. The neoclassical theory is mainly on microeconomics. The signature of neoclassical economists is the supply and demand curves in a partial equilibrium framework established by Marshall and the mathematic formation of the general equilibrium framework accomplished by Walrus. At the macro-level, Marshallian framework was applied to labour market, goods market, money market and bonds market. According to general equilibrium theory, if three of the above-mentioned markets at macro-level are at equilibrium, the fourth market must also be at equilibrium, so commonly only the equilibria at labour market, goods market and money market are considered.

The neoclassical labour market equilibrium and its impact on the economy are shown in Fig. 4.1. Panel (a) shows that labour supply and labour demand determine the real wage rate (nominal wage divided by price level) $w_0$ and full-employment level $L_0$. The full-employment level $L_0$ will determine the output level $Y^*$ through production function in panel (b). The price level is determined by money market through quantity theory of money. A change in price level will affect real rage rate through the definition of real wage. The changed real wage rate will cause a disequilibrium in labour market, but it is argued that the market mechanism will restore the initial real wage rate. For example, if a

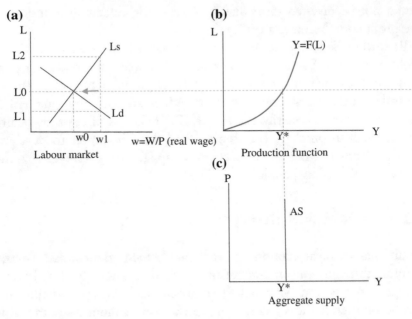

**Fig. 4.1** Neoclassical labour market equilibrium and aggregate supply

drop in price level increases the real wage rate to $w_1$, labour supply $L_2$ is greater than labour demand $L_1$. The excess labour supply will press down real wage until it is back at $w_0$, shown by the arrow in panel (a). Hence, as any price level, the labour market will be at equilibrium with full employment $L_0$ and the output level $Y^*$ will be unchanged. This gives to a vertical aggregate supply curve in panel (c).

On the goods market and money market, neoclassical economists adopted the IS/LM framework developed by orthodox Keynesian economists (for details about the IS/LM framework, see Sect. 4.3.2), but the interpretation of neoclassical is quite different. Figure 4.2 shows neoclassical interpretation of the result of an increase in aggregate demand (e.g. an increase in investment demand). Both Keynesian and neoclassical agreed that IS curve shifts to right (from $IS_0$ to $IS_1$). Keynesian economists conclude that the new equilibrium is achieved at point B

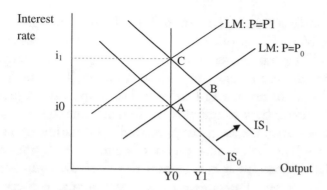

**Fig. 4.2**  Neoclassical interpretation of the IS/LM model

with output level $Y_1$ higher than $Y_0$. However, neoclassical economists argued that the increase in investment demand will push up price level from $P_0$ to $P_1$. This means a decrease in real cash balance or real money supply. Hence, LM curve shifts up and the economy will be in new equilibrium at point C. As a result, an increase in aggregate demand will lead to an increase in price level, leaving the output level $Y_0$ unchanged. This implies a vertical aggregate supply (AS) curve.

The vertical AS curve derived in Fig. 4.1 results from a confusion about real and nominal price. Putting the neoclassical belief that money is neutral into equation in quantity theory of money: $MV = PY$, neoclassical economists assume a fixed money circulation velocity $V$ and a fixed income $Y$, so that any increase in money supply $M$ will lead to only an increase in price level $P$. This varying price levels corresponding to the same income level $Y$ is a pure nominal phenomenon because the varying price is caused by changing money supply. If the money supply is fixed, applying the neoclassical version of quantity theory of money necessitates a fixed price level. As such, the equilibrium point in labour market shown in panel (a) of Fig. 4.1 is corresponding to only one point on the AS curve in panel (c). To derive an AS curve, one needs allow for the shift of labour demand curve to generate different equilibrium points in labour market and thus obtain different points on the AS curve.

The neoclassical interpretation of IS/LM model in Fig. 4.2 has some elements of truth. As price level increases, money demand increases. This would bid up interest rate if money supply is unchanged. Thus, LM curve will shift up and output level decreases. The question remaining is whether or not the LM curve will shift up to the degree that the output level falls back to initial level. To determine the precise position of the LM curve, we need mathematical presentation of IS and LM curves as well as the quantity theory of money. We do not go into too much detail here, but it can be said that the final equilibrium point depends on a set of parameters and assumptions. For example, the money circulation velocity is assumed constant. However, in the case of an increase money demand and a fixed money supply, the money circulation velocity is likely to increase. The violation of this assumption means that the final equilibrium may not be achieved at point C, so the long run AS curve may not be vertical.

Given the apparent economic boom-bust cycles and unemployment in the economy, the claim of neoclassical economists is implausible that, except brief disequilibrium, the economy is always in equilibrium and in full employment. This untrue claim stemmed from a confusion about nominal and real value/price. Neoclassical claim of full unemployment was attacked by both Keynesian economists and monetarists, and this led to the advent of new classical economics.

### 4.2.3 New Classical Theory

New classical theory is formed by introducing rational expectation into the classical equilibrium framework. Although this move is largely a response to monetarist's adaptive expectation or money illusion (to be discussed in Sect. 4.5), the rational expectation revolution extended the static equilibrium framework to a dynamic intertemporal equilibrium framework.

It is generally regarded that Muth (1961) initiated the rational expectation hypothesis in the context of microeconomics, but it was Lucas (1972) who introduced it into macroeconomics, thus marking the beginning of new classical economics. An important implication of applying rational expectation to macroeconomics is that all policies are ineffective unless they are unexpected. With rational expectations, economic agents will react immediately and correctly to any expected policy changes and thus the policy cannot fool people. Only unexpected policy changes can cause misperception and have temporary effects.

This can be illustrated in Fig. 4.3, which shows a typical setting of an AS/AD model, consisting of upward-sloping short-run aggregate supply curves (SRAS), a vertical long-run aggregate supply curve (LRAS) and downward-sloping aggregate demand curves (AD). Suppose the economy is initially at equilibrium point A with output level $Y_0$ and price level $P_0$. An expansionary monetary or fiscal policy will shift the aggregate demand curve from $AD_0$ to $AD_1$.

If this policy is correctly expected, the economic agents with rational expectations can foresee that the policy will increase the price level from $P_0$ to $P_2$. Workers will request higher money wage and the production cost increases. As a result, the short-run aggregate supply curve shifts upwards from $SRAS_0$ to $SRAS_1$. The new equilibrium is achieved at point C where $AD_1$ intersects with $SRAS_1$. The consequence is that the price level increases from $P_0$ to $P_2$ but there is no change in output level, so the policy is ineffective.

If the policy is unexpected, the economic agents are unable to include this information in their rational expectation process. They do business as usual so the short-run aggregate supply curve $SRAS_0$ is unchanged. At point B where the new aggregate demand curve $AD_1$ intersects with $SRAS_0$, the output level increases to $Y_1$ and the price level increases to $P_1$.

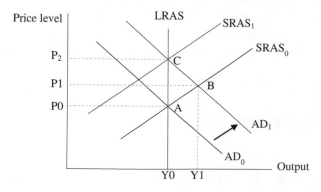

**Fig. 4.3**  Rational expectation and expected/unexpected shocks

However, as the agents realize the inflationary effect of the policy, they require higher wages and higher prices for the commodities supplied, so the short-run aggregate supply curve will shift upwards to $SRAS_1$, the price level increases to $P_2$ and the output level falls back to $Y_0$. As a result, the unexpected policy is effective in the short run but ineffective in the long run.

Hall (1978) applied rational expectation to the life-cycle/permanent income hypothesis and concluded that consumption obeys a random walk. Hall proved that, under certain conditions, e.g. (1) the utility function is quadratic, (2) the utility function is in the form of constant elasticity of substitution, and (3) the stochastic change is small and the interest rate is close to the rate of time preference, consumption itself also obeys a random walk, apart from a trend. As such, Hall claimed that no variable, apart from current consumption, is of any value in predicting future consumption.

Rational expectation overcame the implausibility of adaptive expectation that workers are always fooled by money illusion. It also introduced an uncertainty environment to existing economic theories such as the life-cycle/permanent income hypothesis. The main criticism of rational expectation is that it is an unrealistic assumption that, subject to random errors, all individuals can predict the future economic variable correctly: such a person would need to know all relevant variables and their past values and he/she would also need to know the plausible parameter values of a true economic model. Moreover, the approach puts much effort on expectation and uncertainty but little effort on finding the factors determines the concerned variables (e.g. consumption). As a result, the approach shifts people's attention away from finding the mechanism governing the economic system, which is the essential task of economic research.

## 4.3* Keynesian Economics

Keynesian economics has a historical link to underconsumptionists' theory. Although Adam Smith used the doctrine of 'saving is invested immediately' to defeat the underconsumptionists' view, the concern of underconsumption or oversupply never died out because of the apparent problem of sales stagnation from time to time. The outbreak of the Great Depression entailed massive overproduction leading to unemployment and prompted Keynes to render a revolutionary theory to explain business cycles and to provide policy suggestions for combating economic recessions. Keynes's revolution was successful and became popular in the 1940s. A neoclassical-Keynesian synthesis was formed in the 1950s and 1960s. However, the stagflation (coexistence of economic stagnation and inflation) in the 1970s, largely thanks to the Keynesian-style demand management as well as the negative oil supply shocks, led to the decline of Keynesian economics. Although the new Keynesian economics and post-Keynesian economics survived the counter-revolutions by monetarists and new classical economists, the influence of Keynesian economics has been greatly weakened.

This section reviews and comments on the logic and main proposition of Keynes's General Theory and the different camps of Keynesian economists.

## 4.3.1  Keynes's General Theory

The core element of 'The General Theory of Employment, Interest, and Money' (Keynes 1936) is the deficiency of effective demand. To establish this proposition, Keynes started with a theoretical refutation of classical assumptions, especially Say's law. Although Say's law states that supply creates demand, the logic of Say's law is actually that supply creates an equivalent purchasing power. Since the supporters of Say's law hold a view that human's want is unlimited, in their mind purchasing power can transform the dormant want to achievable demand, so supply creates demand. To highlight the logical problem in Say's law, Keynes interpreted Say's law as supply creates its own demand (Section VII of Chapter 2, Keynes 1936). Based on the experience from the Great Depression, Keynes claimed that the economy has spare production capacity (i.e. a dormant supply). Any increase in demand will activate the dormant supply, so demand creates supply in Keynes's world.

The reasoning behind Keynes's rejection of Say's law and his definition of effective demand is intriguing. He defined an aggregate supply function as the aggregate supply price $Z$ of the output from employing $N$ men, i.e. $Z = \varphi(N)$. Then, he defined an aggregate demand function as the expected proceeds $D$ from the employment of $N$ men, i.e. $D = f(N)$. In modern terminology, $Z$ is the production cost which determines supply price Ps, $D$ is the expected revenue which comes from the consumer's willingness to pay Pd, and $N$ can be viewed as an indicator of quantity of commodities $Q$ as a greater $N$ is directly linked to a greater $Q$. If $D > Z$ (or Pd > Ps), entrepreneurs will employ more people and produce more commodities in order to increase profits. The profit-maximizing point is where $D = Z$ (or Pd = Ps), i.e. where the expected aggregate demand function intersects the aggregate supply function. Keynes defined the value $D$ at this point as the effective demand. The expected aggregate demand with full employment will generate highest amount of effective demand. Any other effective demand will be less and the difference is the deficiency of effective demand. Since the purchasing power supporting the demand is the income from production, the demand at the intersecting point actually means the demand backed up by purchasing power, i.e. the will to purchase accompanied by the power to purchase. As such, Keynes's effective demand is very similar to Malthus's effectual demand.

Armed with the aggregate supply function $Z$ and aggregate demand function $D$, Keynes interpreted Say's law as $Z$ and $D$ are equal for all values of $N$. This necessitates that the aggregate supply function and aggregate demand function must be the same or coincide with each other. This is of course not true. Through these two functions, Keynes also linked Say's law to full employment. If supply can be matched by demand at any employment level ($N$) according to Keynes' interpretation of Say's law, entrepreneurs must employ all labour available in order to maximize profits, so Say's law necessitates that full employment is always achievable.

While Keynes' rebuttal of Say's law shared the view of underconsumptionists that oversupply is possible, he argued this point differently. He differentiated the two components of effective demand—consumption and investment—and argued that investment could be deficient because of uncertainty. Strictly speaking, Keynes was not an underconsumptionist because he did not argue for the possibility of underconsumption but argued for the possibility of underinvestment.

The key assumption (or 'psychology of the community' in Keynes's words) for Keynes's theory is that, as the aggregate real income increases, aggregate consumption increases but by a lesser amount. In other words, the marginal propensity to consume is positive but less than unity. In considering that aggregate real income has two parts, aggregate consumption and aggregate savings, a smaller increase in consumption than income necessitates an increase in savings. If the current investment level can provide enough demand to absorb the savings at the full-employment level, full employment can be achieved. However, investment is often insufficient for a number of reasons. The main factors governing investment level are the marginal efficiency of capital and the liquidity preference—both are affected by uncertainty.

The major elements of Keynes's demand theory are shown in Fig. 4.4. The bold arrow represents a major impact in Keynes' General Theory, the thin arrow a minor impact, the dotted arrow the policy impact, and the dash outlines/arrows the latent elements/effect. Starting from the right-bottom corner, uncertainty affects investment expenditure in two ways: expectation of profit rate and liquidity preference. Facing uncertainty, the investor may be pessimistic about the profitability in the future and thus reluctant to invest; in the words of Keynes, the investor lacks 'animal spirit'. In this situation, investment demand is low regardless of the level of the interest rate. As a result, the interest rate only has a minor impact on investment. On the other hand, uncertainty also affects liquidity preference. If the uncertainty causes a pessimistic view on profitability and a fear of sales stagnation, the firm is likely to hold more cash balance. This means a higher demand for money, a higher interest rate and thus a lower level of investment expenditure. In an extreme case (e.g. in a recession), the firm's desire for money is so high that the money demand curve becomes perfectly elastic (i.e. a horizontal demand curve meaning firms and individuals like to hold as much cash balance as possible regardless of interest rate). As such, any increase in money supply will be readily absorbed by the money demand and thus leave the interest rate unchanged. This case is called the liquidity trap.

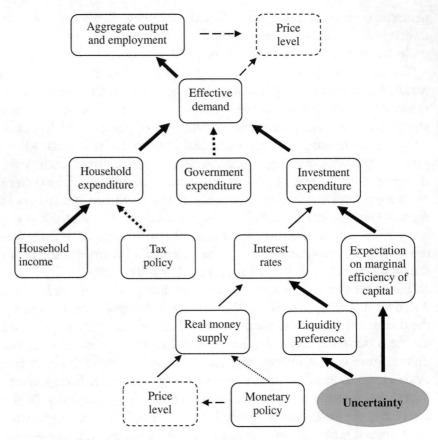

**Fig. 4.4**  Keynes's theory of effective demand (*Source* Adapted from Snowdon and Vane 2005, p. 64)

Keynes regarded monetary policy as inefficient (indicated by a thin dotted arrow) for a number of reasons. First, monetary policy may affect the price level rather than affect real money supply. An expansion of credit tends to increase the amount of money compared with the amount of goods in the economy, so the price of money relative to goods will decrease. As such, there will be little impact on the real money supply and on the output level, so monetary policy is ineffective. Second, in the case of a pessimistic period like an economic recession, extremely high preference for cash (liquidity trap) by firms and individuals can absorb any additional money provided by the central bank, so an increase in real money supply is balanced by the additional real money demand, leaving the interest rates and hence aggregate demand unchanged. As a result, there is no change in investment expenditure and output, so the monetary policy is ineffective. Third, even if a monetary policy can affect the interest rates, the investment demand may not respond to the change in interest rates. For example, an increase in money supply may decrease the interest rate, or the government may stipulate a lower interest rate directly, but firms may not be induced to invest more if the expectation of future profitability is negative. In this case, the investment demand is perfectly inelastic and any impact of monetary policy will be dampened by pessimistic expectation on profitability.

After declaring that monetary policy is ineffective, Keynes relied on fiscal policy such as tax rates and government expenditure to manage effective demand and thus the economy. The impact of fiscal policies can be explained by the multiplier model.

For an economy, total income ($Y$) equals total expenditure ($E$), which comprises two parts: consumption $C$ and investment $I$, i.e.

$$Y = E = C + I \qquad (4.1)$$

Keynes regarded consumption as a linear function of income $Y$:

$$C = a + cY \qquad (4.2)$$

The combination of both Eqs. (4.1) and (4.2) necessitates that an increase in '$I$' or '$a$' will set in place a circular motion: an increase of $\Delta I$ in $I$ causes an increase of $\Delta I$ in $Y$ (according to Eq. 4.1). An increase of $\Delta I$ in $Y$ causes an increase of $c\Delta I$ in $C$ (based on Eq. 4.2) and thus a further increase of $c\Delta I$ in $Y$ (based on Eq. 4.1). This process can continue until reaching an equilibrium. The total effect of a ($\Delta I$) increase in $I$ on $Y$ is:

$$\Delta Y = \Delta I + c(\Delta I) + c^2(\Delta I) + c^3(\Delta I) + \cdots$$
$$= (1 + c + c^2 + c^3 \cdots)(\Delta I) = \Delta I/(1 - c)$$

$1/(1 - c)$ is called the investment multiplier. Since $0 < c < 1$, $1/(1 - c) > 1$. If there is an increase in autonomous consumption a (which is not driven by $Y$), the same multiplier can be derived.

Alternatively, the multiplier can be derived directly by manipulating Eqs. (4.1) and (4.2). Plugging (4.2) into (4.1), we have:

$$Y = a + cY + I, \text{ so } (1 - c)Y = a+I, \text{ or } Y = (a + I)/(1 - c),$$

This equation shows that an increase in either investment '$I$' or autonomous consumption '$a$' can increase the income $Y$ by $1/(1 - c)$ times the amount of increase in either $I$ or $a$.

In the above derivation, the propensity to consume $c$ is a constant coefficient so the consumption function is a linear one. For a nonlinear consumption function, we can use the concept of marginal propensity to consume $c'$ to derive the multiplier.

Define $c' = \Delta C/\Delta Y$, we have:

$$\Delta C = c' * \Delta Y \qquad (4.3)$$

From Eq. (4.1), we can have:

$$\Delta Y = \Delta C + \Delta I \qquad (4.4)$$

Substituting Eq. (4.3) into (4.4), we have:

$$\Delta Y = c' * \Delta Y + \Delta I, or$$

$$\Delta Y = \Delta I/(1 - c')$$

The investment multiplier $1/(1 - c')$ is similar to that for the linear consumption function. For a general case of using a nonlinear consumption function, the marginal propensity to consume $c'$ is used.

The effectiveness of fiscal policy indicated by the bold dotted arrows in Fig. 4.4 can be easily comprehended with the multiplier derived above. An increase in either government consumption expenditure or investment means that the income $Y$ will increase by the amount of $1/(1 - c)$ times the increased government spending. A decrease in the tax rate indicates an increase in household propensity to consume: in considering an income tax rate $t$, the Keynesian consumption function becomes: $C = a + c(1 - t)Y$, where $c(1 - t)$ is the propensity to consume. As $t$ decreases, $c(1 - t)$ increases, $[1 - c(1 - t)]$ decreases, and thus the multiplier $1/[1 - c(1 - t)]$ increases. In considering that $Y = (a + I)/[1 - c(1 - t)]$, the policy will lead to a higher $Y$.

However, Keynes's proposition for effective fiscal policy largely rested on his assumption that a change in effective demand has little impact on the price level, indicated by the thin dotted arrow at the top level in Fig. 4.4. In fact, there is no price variable in Keynes's multiplier model, so his model assumes a horizontal aggregate supply curve (i.e. unlimited supply capacity) and thus he totally ignores the impact of aggregate demand on the price level. In reality, supply capacity is limited, so a change in aggregate demand affects both output level and price level. As such, a substantial or continuous expansion of aggregate demand may lead to a significant increase in price level and little increase in income. This partially explains the stagflation in the 1970s, which happened after the US implementation of Keynesian expansionary fiscal policy in 1960s. The failure of Keynesian economists to foresee and manage the stagflation led to the rejection of Keynesian economics.

Keynes's explanation of persistent involuntary unemployment ultimately rested on the liquidity trap or perfectly inelastic investment. In the case of a liquidity trap, a change in money supply is fully absorbed by changes in liquidity preference, so there is no impact on interest rates and thus no impact on investment. In the case of inelastic investment, investment will not vary due to pessimistic expectations on future profitability, even if interest rates change. Without new investment, the production level, and thus employment level, will not change. If involuntary unemployment existed, it will persist as long as there is no change in the liquidity preference and in the expectation of future profitability (for a more detailed explanation, see Sect. 4.3.2).

A more convenient explanation of involuntary unemployment is based on an assumption of nominal wage rigidity. The explanation can be shown with the aid of Fig. 4.5.

If the labour market is at the equilibrium at point A, the full employment $L_0$ is achieved with nominal wage rate $W_0$. However, if the nominal wage is fixed at an above-equilibrium level, e.g. $W_1$, labour supply $(L_2)$ is greater than labour supply $(L_1)$, so there will be unemployment $(L_2 - L_1)$ in the economy. Classical economists argued that this unemployment is caused by unwillingness of works to take a job with lower nominal wages, so they regarded the unemployment $(L_2 - L_1)$ as voluntary unemployment. Keynes argued that at wage rate $W_1$ the level of

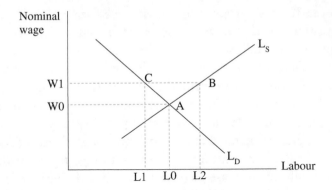

**Fig. 4.5** Involuntary unemployment due to nominal wage rigidity

employment $L_1$ is below the full-employment level $L_0$, so the part of unemployment $(L_0 - L_1)$ is involuntary unemployment.

The determination of interest rates is also a main concern in Keynes's theory because it is a key factor in influencing the level of investment. Keynes explicitly differentiates real and nominal interest rates, but he mainly discusses the rate of interest on money or the money-rate of interest, which can be regarded as a nominal component of interest rates. Keynes listed two determinants of interest rates: the liquidity preference and the quantity of money.

Keynes rejected the classical theory of interest rate determination by using Fig. 4.6. Keynes labelled the horizontal and vertical axes as 'interest rate' and 'investment', respectively. For the convenience of modern readers, the author labels them in reverse order.

In Fig. 4.6, the curves are drawn in a traditional way: the saving curves are upward sloping and investment demand curves are downward sloping. Since income $Y$ has a positive impact on savings, the income levels shown in Fig. 4.6 indicate $Y_3 < Y_2 < Y_1$. The initial equilibrium point is at E with an investment level $Q_1$ and an interest rate $r_1$. Assuming there is a decrease in investment confidence, $I_1$ shifts to $I_2$. Classical theory would tell the reader that the new equilibrium point is at A because the saving curve is unchanged.

However, Keynes argued that, as investment decreases, the income level will decrease based on the investment multiplier. Thus, the saving

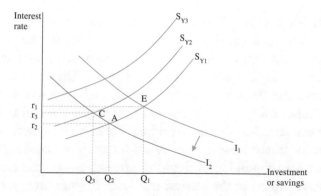

**Fig. 4.6** Keynes's interpretation of interest rate determination

curve would shift up. Since the new income level is unknown, so the new investment level and interest rate are undetermined. Based on Keynes's theory, the state of liquidity preference and the quantity of money can determine the level of interest rate, e.g. $r_3$; then, we can work out the level of investment $Q_3$ and the level of income $Y_2$. Keynes further argued that the classical interest rate theory is mistaken because it regarded interest as the reward for delaying consumption, rather than as the reward for not-hoarding or departing from liquid assets such as cash. Hick pointed out the flaws in Keynes's reasoning: both the demand for money and the interest rate also depended on income level, so one cannot work out $Q_3$ and thus $Y_2$. Besides this, Keynes's argument has following logic problems.

Keynes's arguments against classical theory confused equilibrium with economic dynamics. A change in investment level does cause a change in income, but this will not happen immediately. For example, a firm invests in machinery will not cause income to increase instantly. Income increase occurs only when the firm puts machinery in use and produces more output. In the short run, the income level does not change and thus the saving curve does not shift, so the new interest rate $r_2$ can be determined at point A in Fig. 4.6. In the long run, the increase in investment leads to an increase in income, so the saving curve shifts to $S_{Y3}$ and the new interest rate is determined at point C. In a dynamic system, an equilibrium is a snapshot or a stage in a dynamic process. If one does not allow a moment of relative stability, there will be no equilibrium at all and thus one cannot determine anything.

The argument against the classical definition of interest rates as the reward for waiting has some elements of truth. The classical definition implies that the decision regarding saving and consumption is made assuming other things being equal and thus totally disregards the uncertain environment for decision-making. However, Keynes's alternative definition for interest rate as the reward for the not-hoarding of money contradicts his proposition of liquidity preference—people tend to hold some cash regardless of the interest rate level. Keynes listed a number of reasons for money hoarding or liquidity preference, of which uncertainty plays a vital role. Because of uncertainty, people prefer to keep some cash for safety reason regardless of the level of the interest rate. Also, people need keep some cash for transaction purpose. As a result, people will not be persuaded to give up a certain amount of cash because of a high interest rate. Yet, Keynes's definition of interest rate as the reward for not-hoarding cash implies that people could trade their preference of holding cash for various purposes given the reward from the interest rate.

In short, Keynes's theory rejected Say's Law and proposed a theory of effective demand. Although nominal wage rigidity provided Keynes a useful tool to explain why full employment may not be reached and why involuntary unemployment may be persistent, Keynes's deficiency of effective demand is essentially deficiency in investment which stems from liquidity preference and expectation of future profitability. The ultimate source of underinvestment and economic instability in Keynes's world is uncertainty, which leads to both a liquidity trap and the lack of animal spirit of entrepreneurs due to pessimistic expectations on profitability. Since uncertainty is present all the time, Keynes regarded underinvestment and unemployment as the usual case while economic boom and full employment was a rare case, so he regarded his theory as a general theory and the classical economic theory of full employment as a special case. As a remedy, Keynes advocated fiscal policy which can increase either government investment or household autonomous consumption and affect the economy through the multiplier effect. Keynes regarded monetary policy as ineffective because of the liquidity trap and the perfectly inelastic investment demand.

The limitations of Keynes's theory stem from its assumptions. First, it assumes that uncertainty is the source of inelastic investment demand and liquidity trap. Uncertainty is always present. While the assumption makes it easier to explain deficiency of demand and the involuntary unemployment during a recession, it is an implausible assumption for explaining the adequate demand and full employment during economic expansions and booms. Because nobody can change the uncertain nature of the business world, Keynes's policy recommendation failed to address the problem of private investment deficiency due to uncertainty. As a result, Keynes's policy suggestion relied on government spending, which is effective in the short run but is only a temporary fix because the government cannot increase spending forever.

Second, it assumes that a change in effective demand has little impact on price level. The Keynesian multiplier model has no price variable so the model essentially ignored any role of price change. Although Keynes's theory was later interpreted by a positive sloping aggregate supply curve and a downward sloping aggregate demand curve, which allows for an increase in price level in the face of a right shift of aggregate demand, the price effect is regarded as small because of the assumption of inelastic investment demand and liquidity trap. The ignorance of the impact of demand on price level was proven as a fatal mistake by the 1970s stagflation.

Third, it includes extreme assumptions on the impact of money supply. Keynes assumed that money supply would affect mostly the price level or the velocity of money circulation in the case of liquidity trap, so he regarded monetary policy was ineffective on output level. This assumption was made without any supporting evidence and was proven incorrect later by monetarists.

Fourth, Keynes's involuntary unemployment is based on unusual assumptions such as the liquidity trap, perfectly inelastic investment demand and nominal wage rigidity. Although nominal wage rigidity is a plausible assumption in the short run, it is implausible in the long run—the wage rates have to be determined by markets eventually. These assumptions were based either on special cases or on imperfection of markets, making Keynes's general theory not so general. The limitations of Keynes's theory were attacked by other schools of economic thought such as the monetarists and the new classical economists. Subsequently, these limitations were addressed by new and post-Keynesian economists.

## 4.3.2 Orthodox Keynesian Theory

Keynes's general theory is presented mainly in texts—with a few equations and only one graph. However, orthodox Keynesian economists interpreted and popularized Keynes's theory through graphical demonstration. Hicks (1937) initiated the IS/LM framework, which is further elaborated upon by Modigliani (1944) and popularized by Hansen (1949, 1953). The orthodox Keynesian ideas are still shared by many eminent economists, such as George Akerlof, Paul Krugman and John Quiggin. This section briefly introduces and assesses the IS-LM model and the AD/AS model, and uses them to explain how the liquidity trap and perfectly inelastic investment demand can lead to an underemployment equilibrium.

The derivation of the LM curve starts from the money supply curve Ms and the money (or liquidity) demand curve $L$. Money supply is controlled by central banks so is assumed exogenous, so the Ms curve is a vertical line shown in Fig. 4.7. Money demand is negatively related to interest rate and positively related to income level, so all money demand curves in Fig. 4.7 are negatively sloped and $Y_3 > Y_2 > Y_1$. This necessitates that, at any interest rate (e.g. $r_2$), $L_3 > L_2 > L_1$. Given this setting, a higher interest rate is associated with a higher level of income at the equilibrium points A, B and C. Transferring income and interest rate at all equilibrium points like A, B and C in panel (a) to a graph with income and interest rate as coordinates will produce an upward-sloping LM curve, shown in panel (b). This LM curve represents equilibria in the money market. An increase in money supply necessitates a right shift of the $M_s$ curve, and thus, at the new equilibrium points, a lower interest rate is associated with each income level. As a result, an expansionary monetary policy means a right shift of the LM curve.

The panels (c) and (d) in Fig. 4.7 show the case of a liquidity trap. In this case, interest rates can affect money demand but to a limited degree. When interest rates decrease to a certain level, e.g. $r_2$ shown in

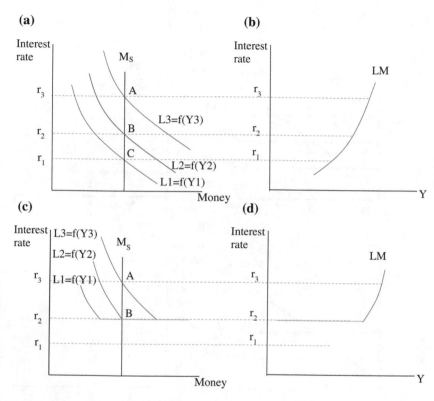

**Fig. 4.7** The LM curve and liquidity trap

panel (c), the demand for money is infinite regardless of income level. As a result, the money demand curves ($L_1$, $L_2$, and $L_3$) at interest rate $r_2$ become horizontal, shown in panel (c). At this interest rate, any level of money supply (i.e. shifts of money supply curve) will be absorbed by money demand, so in panel (d) the LM curve at interest rate $r_2$ will also be horizontal at the low income levels.

The derivation of the IS curve depends on the broadly defined investment (including investment $I$, government spending $G$ and exports $X$, i.e. $I+G+X$) and saving (saving $S$, imports $M$ and taxes $T$, i.e. $S + M + T$) in Fig. 4.8.

In Keynes's model, investment is negatively related to the interest rate but unrelated to income level, so the investment curves $I + G + X$ in panel (a) of Fig. 4.8 are horizontal lines with $r_1 < r_2 < r_3$. Saving,

**(a)**                                    **(b)**

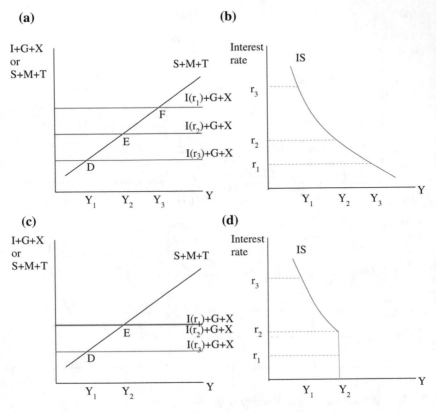

**Fig. 4.8**  The IS curve and perfectly inelastic investment demand

imports and taxes are positively related to income, so the saving curve
$S + M + T$ in panel (a) is a positively sloping line. At the equilibrium
points D, E and F, the income level and interest rate are negatively
related to each other, e.g. low income $Y_1$ is associated with high interest rate $r_3$ at point A while high income level $Y_3$ is associated with low
interest rate $r_1$. Transferring all the equilibrium points like D, E and F
to panel (b), we have a negatively sloping IS curve, which represents
investment-saving equilibrium or equilibrium in goods markets. An
expansionary fiscal policy (e.g. an increase in government spending $G$)
means an increase in autonomous expenditure at each interest rate,
and thus, the horizontal investment curves in panel (a) will shift up.

Hence, at the new equilibrium points the same interest rates will be associated with higher level of income, so an expansionary fiscal policy means a right shift of the IS curve.

The panels (c) and (d) show the case of a perfectly inelastic investment demand at interest rates less than $r_2$. In this case, low interest rates may stimulate investment demand but, when the interest rate is below a certain level, investment becomes insensitive to interest changes [e.g. $I(r_2)$ in panel (c)] because the impact of expectation of profitability overshadows the impact of the interest rate. With a negative expectation on future profitability, a firm would not increase investment even if the interest rate is very low: low interest rates may reduce borrowing costs, but if the firm cannot sell the products in the future it will make a loss. Consequently, the investment demand at the interest rates $r_2$ and $r_1$ in panel (c) is the same so the two investment demand curves coincide. As such, the lower part of the IS curve in panel (d) becomes a vertical line.

Putting the IS and LM curve on the same graph, we can form an IS/LM model. Combining this model with a production function and labour market equilibrium, we can explain the equilibrium adjustment process and the impact of both fiscal and monetary policies on commodity, money and labour markets.

A typical case is shown in Fig. 4.9. Suppose the economy is initially at point $E_0$ where $LM_0$ intersects with IS in panel (a). Both the money markets and goods market are at equilibrium with an output level of $Y_0$, but as will be shown later, the labour market is not in equilibrium. Panel (c) is a 45° line transferring the income from the horizontal axis to the vertical axis. The production function in panel (d) shows that, to produce output $Y_0$, the amount of labour $L_0$ is needed. Panel (b) shows that, with the amount of employment $L_0$, the firm is willing to pay the real wage $V_0$. This real wage fails to clear the labour market: labour supply $L_2$ is greater than labour demand $L_0$. This excess labour supply will push down both nominal wages and real wages. As the real wage decreases to $V_1$, labour supply equals labour demand and thus the labour market achieves equilibrium employment $L_1$. The higher employment $L_1$ than $L_0$ necessitates a higher income $Y_1$ in panel (d). Meanwhile, the reduction in nominal wage leads to lower production costs and this in turn leads to a lower price level, which indicates an

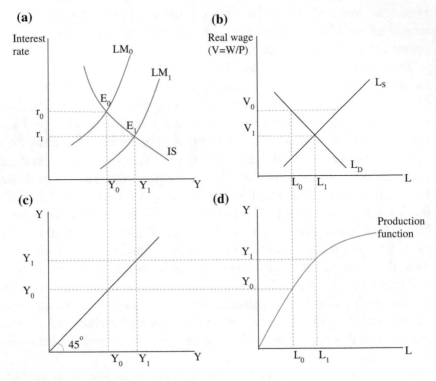

**Fig. 4.9** The classical case of an equilibrium adjustment

increase in real money supply. The consequence of an increase in real money supply is that the money supply curve shifts right to $LM_1$, the interest rate decreases, and thus investment increases at point $E_1$ in panel (a). The effect that falling nominal wages and price levels stimulates investment via reduced interest rates is called the Keynesian effect, which generates a higher output level $Y_1$ and lower interest rate $r_1$. As a result, all markets achieve equilibria.

However, in the cases of both a liquidity trap and the perfectly inelastic investment demand, the Keynesian effect would fail to clear the labour market. The case of a liquidity trap is shown in Fig. 4.10.

Assume that the initial economic condition generates unemployment at a real wage level of $V_0$. An expansionary monetary policy is used to address the unemployment in the labour market, so the LM

**(a)**

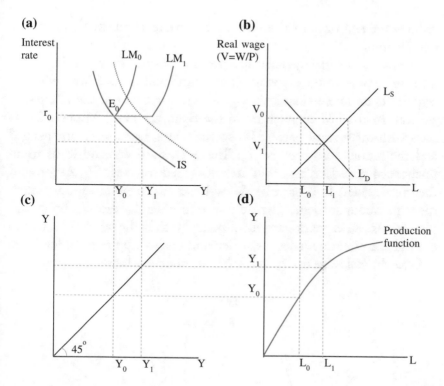

**Fig. 4.10** The impact of liquidity trap

curve shifts to the right from $LM_0$ to $LM_1$. However, since both $LM_0$ and $LM_1$ are horizontal at interest rate $r_0$ in the case of liquidity trap, the shift of LM fails to generate more investment, so the output level stays at $Y_0$ and employment stays at $L_0$. In a liquidity trap, an increase in money supply $M$ is offset by a decrease in circulation velocity $V$, the quantity theory of money ($MV = PY$) necessitates that the nominal income PY is unchanged. As such, the unchanged real income $Y_0$ necessitates an unchanged price level, so the real wage fails to fall and there is unemployment in the labour market. In this case, the only way to achieve equilibrium in the labour market is to shift the IS curve to the right through an expansionary fiscal policy: a right shift of IS increases the output level to $Y_1$ in panel (a). The increased investment demand also pushes up the price level and thus reduces the real wage to $V_1$ in panel (b), so the labour market achieves equilibrium.

The case of a perfectly inelastic investment can be shown in Fig. 4.11. Due to a pessimistic expectation on profitability, the firm will not respond to an interest rate below $r_0$, so the lower part of the IS curve is vertical. As the LM curve shifts to the right, the new LM curve intersects with the vertical part of IS, so the investment level is unchanged and the output level stays at $Y_0$. The unchanged demand leads to an unchanged price level and thus an unchanged real wage $V_0$. As a result, the employment level stays at $L_0$ and there is involuntary unemployment in labour markets. The only way to clear the excess labour supply is to use an expansionary fiscal policy to shift the whole IS curve to the right: an increase investment demand pushes up the price level and reduces the real wage to $V_1$, so the labour market achieves equilibrium.

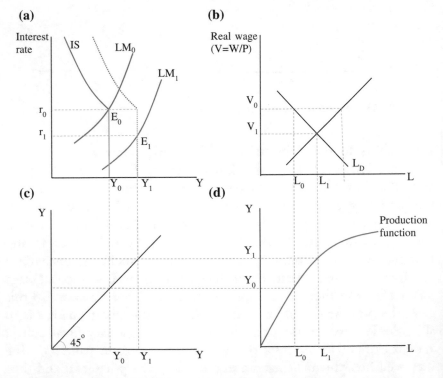

**Fig. 4.11** The impact of perfectly inelastic investment demand

The IS/LM model was further developed by orthodox Keynesian into an AS/AD model, featured an upward-sloping AS curve in the short run. Figure 4.12 shows the AS/AD diagram. The aggregate supply curve (AS) is traditionally assumed upward sloping while the aggregate demand curve (AD) is assumed downward sloping. We also assume that AS has a minimum output level due to the minimum production requirement in an industry or in an economy. In the usual case, AS and AD meet at point E so the goods market achieves an equilibrium output ($Q^*$) and price ($P^*$). However, if the aggregate demand for some reason decreases sharply, the aggregate demand curve will shift dramatically to AD′, and the aggregate supply and demand curves share no common point. The maximum amount of demand is $Q_D$, which is less than the minimum amount of supply $Q_0$. The difference between $Q_0$ and $Q_D$ indicates an overall oversupply in the goods market.

However, Pigou (1941, 1943, 1947) provided a counter-argument regarding the case of liquidity trap and the perfectly inelastic investment. He argued that, as the price level falls, the real wealth value increases and thus autonomous consumption will increase. This is called the Pigou effect. An increase in autonomous consumption means a right shift of the IS curve. As a result, the labour market should be able to achieve equilibrium without any intervention. In terms of the AD/AS diagram, Pigou argued that, as aggregate demand decreases, the price level falls, and the purchasing power of the household increases, so the aggregate demand curve will bend towards the right—it will approach but never reach the horizontal axis (shown as AD″ in Fig. 4.12).

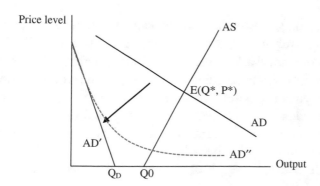

**Fig. 4.12**   AS/AD model and the Pigou effect

As such, the aggregate demand curve AD″ can always meet the aggregate supply curve AS at a positive price level.

Since Pigou's wealth effect largely depended on financial assets, there are arguments about the size of net wealth and its impact on aggregate demand. For example, the outside money—currency plus bank deposit matched by banks' holding of cash reserves—can be viewed as net wealth, but inside money—bank deposits matched by loans to private sectors—is not net wealth. Since deposits make of most of the money supply and since most bank deposits are lent to private sectors, the net wealth is relatively small and thus the Pigou effect is also small. However, generally speaking, Pigou won the academic argument.

Pigou's argument has the same flaw as that in Say's law: the equalization of purchasing power with demand. An increase in purchasing power does not mean an increase in the will to purchase and thus the aggregate demand curve may not change shape when the wealth or purchasing power of households increases. As such, it is not guaranteed that oversupply in a goods market can be eliminated by the price mechanism.

Although the IS/LM model and the AS/AD model visualized and thus popularized Keynes's theory, there is no price level in the model. For this reason, once Phillips (1958) found that there is a negative relationship between unemployment and inflation, the Phillips curve was quickly adopted by orthodox Keynesian economists as a tool complementing the IS/LM model. However, the belief of a stable Phillips curve in the long run was proven wrong by the stagflation in the 1970s.

In summary, orthodox Keynesian economists simplified and popularized Keynes's theory through IS/LM and AD/AS models. However, one drawback of this simplification is that it further weakened the search for the fundamental cause of economic recession and involuntary employment. Keynes attributed the cause to uncertainty, but subsequent orthodox Keynesian economists attributed the cause to the shape of the LM curve or IS curve. Orthodox Keynesian's downplaying the impact of uncertainty has been heavily criticized by post-Keynesian economists. The other drawback of the orthodox approach is that a graphic presentation cannot gauge the impact quantitatively. The inaccuracy of the graphic approach is especially disliked by monetarists.

## 4.3.3 New Keynesian Theory

Pigou's argument discredited Keynes's theory based on the liquidity trap and perfectly inelastic investment demand, so some Keynesian economists think Keynes's theory must be based on wage or price rigidity. They started to seek theoretic support for wage and price rigidity and thus provide microeconomic foundations for Keynes's theory. This group of economists, labelled as new Keynesian economists, includes Gregory Mankiw, Lawrence Summers, Olivier Blanchard, David Romer, Richard Layard and Wendy Carlin.

Rigidity or stickiness of wages and prices means money is not neutral. If a change in the quantity of money cannot be fully transferred to the change in price and wages, it will have an impact on real variables like output and employment levels. The reasons for wage and price rigidity cannot be found in the ideal world of classical economists—perfectly competitive markets require flexible prices and wages. As such, new Keynesian economists focused on market imperfection. The concerned rigidity can be classified into nominal wage rigidity, nominal price rigidity, real wage rigidity and real price rigidity. These rigidities play a vital role in the new Keynesian economists' explanation of business cycles and in their policy implications. This section will review the major points of new Keynesian economics and assess its achievements and shortcomings.

On nominal wage rigidity, Fischer (1977) and Taylor (1980) introduced a theory of long-term wage contracts. The basic reasoning of this theory is that the nominal wages are locked in by the long-term contracts between employers and employees and thus cannot change promptly with changed economic conditions. Phelps (1985, 1990) provided some reasons for the existence of long-term wage contracts, including costly wage negotiations, disruptions by workers' strike actions if wage negotiations break down and the cost of labour turnover when wages are changed frequently. A counter-argument puts that in a long-term contract wages can be indexed to the rate of inflation so as to avoid nominal wage rigidity, but Gordon (2003) argued that it is risky for firms to peg nominal wages to inflation rates. Because the unexpected supply and demand shocks may influence inflation significantly, he argued that wages indexed to inflation rates may cause a significant increase in production costs. Gordon used the oil shock in the 1970s and subsequent inflation as an example for this argument.

Mankiw was not satisfied with the long-term contract theory because it had little microeconomic foundation. He investigated the nominal price rigidity caused by imperfection in goods markets. Mankiw (1985) put forward a menu costs theory. Akerlof and Yellen (1985) and Parkin (1986) also worked on this menu costs theory. This theory claims that, in the case of monopolistic production, if there is a non-trivial cost related to changing prices the firm may opt to keep the price the same in order to avoid menu costs. It was claimed that small menu costs can generate a large change in outputs.

Regarding real rigidity, Ball and Romer (1990) showed that a combination of real rigidities and small frictions to nominal adjustment can lead to substantial nominal rigidities. Mankiw and Romer (1991) demonstrated the interactions between nominal and real rigidities. Real price rigidity can generally be explained by the markup formula for monopolistic production. In the case of demand shock, the marginal cost or marginal revenue may change, but the firm may change its profit margin or markup due to change in demand elasticity, so the relative price does not change. The other sources of real price rigidity include: (1) Thick market externalities (e.g. Diamond 1982): the search cost is low in a thick market, so consumers will be more willing to shop in a thick market. This in turn leads to economies of scale and reduces the marginal cost of firms. The thicker market in an economic boom has lower marginal costs and thus prevents the rise of prices while the thinner market in economic recessions has higher marginal costs and thus prevents the price from falling. (2) Customer markets (e.g. McDonald 1992): when the search cost is not trivial, customers will not put much effort into finding cheaper products of the same quality, so the firms are discouraged from changing prices to attract customers. (3) Complexity of production (e.g. Gordon 1981, 1990): modern production involves a number of inputs and outputs and is connected to a chain of suppliers, so it is hard to gauge marginal cost and marginal revenue in the wake of a negative demand shock. To reduce prices in these circumstances may cause bankruptcy and thus is risky to firms. (4) Capital market imperfection (e.g. Bernanke and Gerlter 1989): borrowers have more information than the lenders about the viability and quality of their projects, so external finance is generally more expensive than internal finance. In a period of economic recession, firms have to rely more on external finance while banks are more sceptical about firms' ability to repay a loan, so the cost of borrowing is much higher. This contributes to a higher marginal cost and thus prevents a fall in the price of the firm's products. (5) Judging quality by price (e.g. Stiglitz 1987): since customers have less information than firms about the quality of the products, they may view price as a signal of quality. In this case, a reduction in price may send a wrong signal about the quality of the products and this makes the firm reluctant to change prices.

There are also a number of theories explaining real wage rigidity, which can be put into three groups: (1) implicit contract (e.g. Bailey 1974), (2) efficiency wage theories (e.g. Yellen 1984), and (3) insider-outsider theories (e.g. Ball 1990). Implicit contract theories seek to uncover what keeps the firm and the workers together in long-term relationships. According to these theories, wage rates represent not only the payment for labour service but also an insurance against the risk of variable income in the face of shocks. As a result, workers prefer stable wages over time to highly varying wage rates.

Efficiency wage theories highlight the interdependence between real wage rates and productivity or efficiency of labour. The consequence of this interdependence is that it is better for the firms to pay a high real wage so as to maintain high efficiency of labour. This general reasoning can be applied to different models. The adverse selection model (e.g. Weiss 1991) emphasizes that firms offering higher wages will attract the best workers. Because of non-trivial hiring and firing costs, firms prefer to pay higher wages to hire capable workers. The labour turnover model (e.g. Salop 1979) concerns costly labour turnover. Because workers' willingness to quit a job is negatively related to pay rates, firms would be willing to pay an efficient wage above the market-clearing level to keep the workers and thus avoid turnover costs. The shirking model (Shapiro and Stiglitz 1984) highlights the inability of a labour contract to specify every aspect of a worker's performance. Since workers have the ability to exercise discretion on their effort levels according to real wage levels, firms are better off to offer workers higher real wages. The fairness model (e.g. Akerlof 1982) states that wage cuts will negatively affect the morale of workers, so the workers would raise their work norms and provide high productivity if firms pay higher wages.

Similar to the efficient theories, the insider-outsider theories are intended to explain the wage rigidity in the face of involuntary unemployment: Why does excess labour supply fail to bid down the wage rates? These theories rely on the advantages of incumbent workers (insiders) over unemployed workers (outsiders) due to labour turnover costs. The theories emphasize the high cost associated with labour turnover such as hiring and firing costs, mandatory severance pay, litigation costs, the insiders' ability and incentive to cooperate or harass the new workers. Due to these costs, the insiders have sufficient bargaining power to keep the real wage higher even if the outsiders are willing to work for less.

With the wage and price rigidity, the explanation of economic recession by new Keynesian economists is straightforward. As shown in Fig. 4.13, with a vertical long-run aggregate supply curve, a left shift of aggregate demand from $AD_1$ to $AD_2$ (a decrease in aggregate demand) should lead to the new equilibrium at $E_3$ where output is unchanged but the price level is decreased. However, due to price rigidity, the economy will move to point $E_2$ with a reduced output level $Y_2$. According to production function, this reduced output level necessitates a reduction of labour demand from $L_1$ to $L_2$. Due to wage rigidity, the labour market will move from $E_1$ to $E_2$. The price level may eventually fall in the long run and the economy will achieve a new equilibrium at point $E_3$, but the new Keynesian economists claim this process may take an unacceptably long period of time due to coordination failure: there is no incentive for any firm to cut prices and increase production.

The other strand of new Keynesian economics explains a business cycle through a dynamic quantity adjustment process. Greenwald and Stiglitz (1993) developed a business cycle model which did not use price and wage rigidity. They argued that, due to the high financing cost

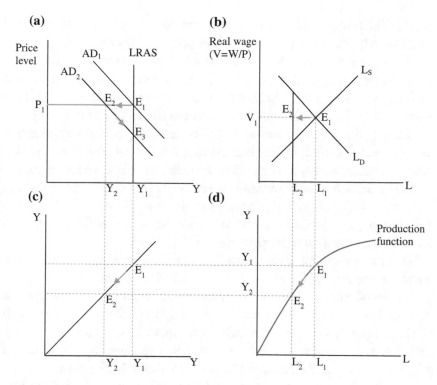

**Fig. 4.13** A demand shock with rigid real wages

and uncertainty about future prices for their products, firms prefer to reduce outputs. As a result, a negative aggregate demand shock induces a negative aggregate supply reduction. As both short-run AS and AD curves shift to the left, the output level decreases and the price level is unchanged.

Since new Keynesian economists believe in price and wage rigidity, money is regarded as not neutral and thus both monetary and fiscal policy would be effective in their theory. Regarding policy choices, new Keynesian economists put less weight on fiscal policy. Taylor (2000a) even argued that fiscal policy should be used in unusual situations, e.g., when nominal interest rates hit a lower bound of zero. The main economic management policy advocated by new Keynesian economists is an interest rate policy of targeting inflation, which is used by most reserve banks. The policy can be illustrated by the following aggregate demand/inflation adjustment (AD/IA) model proposed by Taylor (2000b).

The model has three key assumptions. First, a negative relationship between the real interest rate and GDP, i.e. $y = -ar + \mu$, where $y$ is income or GDP, $r$ is interest rate, the coefficient $a$ is a positive constant, and $\mu$ is a shift variable. This is similar to the IS curve in the IS/LM model.

Second, the targeting inflation policy means a positive relationship between inflation and the real interest rate, $r = b\pi + v$, where $r$ is interest rate, $\pi$ is inflation rate, $b$ is a positive constant, and $v$ is a random variable.

Third, inflation will increase when a lagged GDP increases, i.e. $\pi_t = \pi_{t-1} + c(y_{t-1} - y_f) + w$, where $y_f$ is output or GDP at the full-employment level, $c$ is a positive constant, $w$ is a shift variable, and the subscript $t$ indicates the time period.

The first and second assumptions lead to a negative relationship between inflation and real GDP: $y = -ab\pi + \mu - av$. This is similar to the aggregate demand curve. The third assumption comes from a new Keynesian style of Philips curve, which leads to an inflation adjustment curve (IA): if the output level equals the full-employment output (i.e., $y_{t-1} = y_f$), inflation rate is unchanged (i.e. $\pi_t = \pi_{t-1}$); if $y_{t-1} > y_f$, $\pi_t > \pi_{t-1}$; and if $y_{t-1} < y_f$, $\pi_t < \pi_{t-1}$. The AD and IA curves are shown in Fig. 4.14.

Initially, the economy is at point $E_1$ with output level of $Y_1$ and an inflation level of $\pi_1$. An aggregate expenditure shock causes the aggregate demand curve to shift from $AD_1$ to $AD_2$ (since the vertical axis is

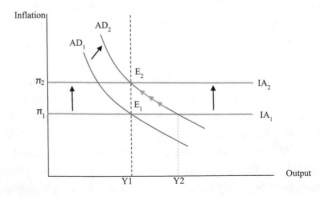

**Fig. 4.14** A demand shock in the AD-IA model

for inflation instead of price level, some economic books name the curve ADI curve). With the initial IA curve, the output level increases from $Y_1$ to $Y_2$. However, as the output increases from $Y_1$ to $Y_2$, the equation in the third assumption necessitates that inflation increases from $\pi_1$ to $\pi_2$, so the IA curve shifts upwards to $IA_2$. Central bank's policy of inflation control (indicated by the equation in the second assumption) leads to a rise in interest rates and thus the economy contracts (indicated by the arrows on the $AD_2$) and finally settles at point $E_2$.

This AD/IA model was further extended by introducing an upward-sloping short-run inflation adjustment curve (e.g. Stiglitz et al. 2015). This type of model is another form of aggregate demand and aggregate supply (AD/AS) model, which will be discussed later in this chapter. Moreover, since the central bank's interest rate policy of targeting-inflation is built in the AD/IA model, the model is critically hinged on the behaviour of the central bank. If the central bank changes its policy, the model becomes invalid.

New Keynesian economics was criticized from different perspectives. Many economists pointed out that new Keynesian economics lacks empirical support. In response, Blinder (1991, 1994) and Ball et al. (1988) provided some empirical evidence based on survey data. Some economists also doubted that small menu costs can generate economic fluctuation at the macroeconomic level, while other economists argued that the use of old IS/LM models or similar models as an analytical tool eliminates key determinants of aggregate demand such as uncertainty. Yet others were critical of new Keynesian economists' acceptance of the rational expectation hypothesis. However, these are not the biggest limitations to the new Keynesian approach.

From the author's point of view, the major shortcoming of new Keynesian economics is its focusing on price and wage rigidity and market imperfection. Due to the complexity of the real world, it is not surprising that new Keynesian economists have uncovered a large number of sources of price and wage rigidity. The reasoning behind most theories is plausible and can be supported by our life experience or empirical evidence; however, the key issue is: Do these theories reflect the general case of an economy? Are the sources revealed by new Keynesian economists the main cause of economic recessions? Market imperfection does exist in any economy, but it is most likely that imperfect markets are not a general feature of a market economy—otherwise, the market economy would be very inefficient and would not function well. If so, market mechanism should give way to other method of economic management, e.g. command or central planning economy. The reality shows that a market economy is generally more efficient than other types of economies. As a result, what new Keynesian economists uncovered were the trivial causes of price and wage rigidity which has little bearing on explaining business cycles and economic growth.

### 4.3.4 Post-Keynesian Theory

Post-Keynesian economists did not accept the interpretation of Keynes's general theory by the orthodox and new Keynesian economists. They thought the orthodox Keynesian's model and neoclassical synthesis were 'hydraulic' and were a retreat back inside the orthodox citadel. They regarded new Keynesian economics as a misinterpretation of Keynes's theory because they thought price and wage rigidity were not necessary conditions for Keynes's general theory. Post-Keynesians viewed Keynes's theory as a radical break with classical thinking, and they sought to represent the true spirit of Keynesian theory.

There are two strands of post-Keynesian economists: the Europe camp and the American camp. The economists in the European camp include Geoff Harcourt, Richard Kahn, Nicholas Kaldor, Michal Kalecki, Joan Robinson and Piero Sraffa. They emphasize the behaviour and functioning of the real economy. Following Kalecki (1943) and Sraffa (1960), the European post-Keynesian economists argued that the division of income between wages and profits is independent of output level, not determined by marginal productivity but by other forces. Kalecki's simple model makes three key assumptions: (1) firms use a cost-plus method of pricing; (2) total output is determined by total demand; and (3) workers spend all their income on consumption while capitalists' investment is unrelated to the level of profit or saving. If capitalists invest all their profit, aggregate demand is able to buy all products so the total output and employment level are high. However, if capitalists are pessimistic and save at least some of their profit, demand could not purchase all products, and thus, the output level will be low and unemployment is inevitable.

Post-Keynesian economists in the American camp include Victoria Chick, Hyman Minsky, Alfred Eichner and Paul Davidson. They have paid more attention to the impact of uncertainty and money. Starting from interpreting Keynes's taxonomic attack on Say's law, they emphasize Keynes's view on saving and liquidity preference. They argued that, since the marginal propensity to consume is always less than unity, a part of income is always saved by purchasing liquid assets, so there is always a tendency that demand for goods is less than supply. Post-Keynesian economists link investment spending directly to the quantity of money, so money is non-neutral in both the short run and the long run. The phenomenon of widely used money contracts is used to justify the non-neutrality of money.

American post-Keynesian economists also differentiated probabilistic risk from uncertainty. The latter is labelled true uncertainty, which is the uncertainty meant by Keynes: 'there is no scientific basis on which to form any calculable probability whatever' (Keynes 1937, p. 113). The former means uncertainty with either objective or subjective probability. With the concept of 'true uncertainty', the American camp rejected the approach of rational expectations and highlighted liquidity preference and 'animal spirits' as the driving forces behind Keynes's analysis of long-period under-employment equilibrium. The policy recommendation stemming from true uncertainty is that the government should play a role in improving the performance of markets, for example, by developing institutions which attempt to reduce uncertainty, provide monetary incentives to encourage individuals to take civilized actions in the interest of society, and by setting up financial safety nets to prevent or offset disastrous consequences.

Although post-Keynesian economics appears to be old-fashioned by sticking to liquidity preference and 'animal spirits' and to offer little new content to Keynes's theory, the interpretation by post-Keynesian economists does capture the essence or major points expressed in Keynes's general theory. By investigating income distribution related to profits and wages, European post-Keynesian economists highlighted the key role of investment in Keynes's general theory—pessimistic expectations of capitalists are the cause of economic recession and involuntary unemployment. However, they go no further than Keynes did in investigating what causes the pessimistic or optimistic expectations of capitalists.

The American post-Keynesian economists highlighted the impact of liquidity preference in rejecting Say's law, the role of money in investment and in the economy, and the influence of true uncertainty on animal spirits. However, the impact of money is greatly exaggerated. Money has a dual function: it is both a type of asset and a medium of exchange. The former function means that money is not neutral—an increase in money supply increases asset value and thus has a real impact on the economy. The function of money as a medium of exchange may affect the real economy in the short run through money illusion but, in the long run, it only affects price levels and thus is neutral. In comparing the two functions, the role of money as a medium of exchange is far greater than its role as an asset, so the assumption of neutrality of money in the long run is not too far from reality. The phenomenon of money contracts emphasized by American post-Keynesian economists may prove the importance of money in a complex economy, but this is not significant enough to prove the non-neutrality of money as money contracts only account for a small part of market transactions.

## 4.4   Marxism

Marxist economic thought is deeply rooted in Marx's theory of history, so it is necessary to review briefly his historical thought. Hegel's theory of history had great influence on Marx. Hegel claimed that history moved forward through the conflict of forces within social systems. According to Hegel, an accepted idea or thesis exists but will soon be challenged by its opposite or antithesis. Out of this conflict is the emergence of synthesis which represents a higher form of truth and becomes a new thesis. Hegel called this process 'dialectic'. Marx (1867, 1885, 1894) extended this idealistic thought into a materialistic theory. Namely, the interaction between the forces of production and the relations of production causes change in social structure. The forces of production, i.e. technology used in production, are dynamic. The relations of production mean the rules in a society, including social relations and property relations, which are inertial in nature.

The contradiction between the forces of production and the relations of production manifests themselves in a class struggle. According to Marx, the struggle between landlords and peasants leads to a capitalist society; the struggle between capitalists and the proletariat will lead to socialist manifestations first and eventually to communism.

In accordance with his theory of history, Marx developed a theory of economic exploitation. He argued that a chief feature of capitalism is the separation of labour from ownership of means of production, i.e. workers no longer own workshops, tools and raw materials in the production process, so capitalism is essentially a society of two classes: capitalists and the proletariats. While investigating the struggle between these two classes, Marx studied commodity prices and wage determination. He thought labour was the common element of all commodities, and based on Ricardo's labour theory of value, Marx developed his labour theory of value.

To avoid the problem of measuring the different skills of labour, Marx created a concept of abstract labour to obtain homogeneous labour quantity (i.e. the amount of labour of the same quality). Practically, abstract labour meant the socially necessary labour time for producing a commodity. To solve the problem of the influence of capital goods on commodity prices, Marx followed Ricardo in regard to capital as stored-up labour. To accommodate the different value caused by different quality of land, he adopted Ricardo's theory of differential rent. Like Ricardo, Marx failed to solve the problem of the influence of profit on commodity prices. On the distribution of profit or surplus value, Marx maintained that the surplus is created by labour but is taken away because labour lacks the ownership of the means of production. In other words, labour is exploited by capitalists.

Using his labour theory of value and a revised version of the wage–fund doctrine, and assuming perfect competition, neutral money, constant return in manufacturing and diminishing return in agriculture, Marx formulated his laws of capitalism. These laws indicate the contradictions in a capitalist economy such as a reserve army of the unemployed, a falling rate of profit, business crises, increasing concentration of industry and capital and increasing misery within the proletariat.

Marx rejected Malthusian population theory but invented a doctrine of the reserve army of labour, which has a similar function in economic theory—to prevent the wage rate from increasing when the capital input increases over time. Marx listed a number of sources of excess labour supply such as the replacement of labour by machines, entry of new members into the labour force and the increase in capital intensity in the economy thanks to technological progress. Marx's claim of a falling rate of profit is essentially based on the same reasoning as classical economists: capital accumulation causes more capital bidding for labour so the wage increases and the rate of profit falls. However, Marx argued that capitalists react to rising wages by substituting labour with machinery which leads to an even lower rate of profit. Marx foresaw the concentration and centralization of capital due to economies of scale, the growth of credit markets and the dominance of the corporation in a business organization. This dominance tends to destroy competition— the cornerstone of a capitalist economy. Marx also predicted increasing misery for the proletariat. This misery can be measured as real income for the mass of society, or the proletariat's share of real income, or the non-economic aspect of life.

Marx claimed that all of these contradictions were made manifest and were exacerbated during business crises. He viewed business cycles as an integral part of the capitalist process. Regarding the causes of business cycles, Marx first rejected Say's law based on the fact that the purpose of capitalist production is profit. He argued that in a barter economy or in a pre-capitalist economy, the purpose of production and exchange was for consumption, so there was no possibility of overproduction. However, overproduction is possible in a capitalist economy because the purpose of capitalism production and exchange is to make profits. Second, Marx conjectured that technological changes could generate business cycles, so he described technological change as 'creative destruction'. According to him, a technological burst generates an increase in capital accumulation and demand for labour, the increase in wages, a decrease in the reserve army of the unemployed and thus an economic boom. The rise in wages will cause a fall in the rate of profit, so capitalists will react by reducing investment and thus the output level falls and the economy goes into recession. Third, Marx also proposed a disproportionality crisis theory:

the unemployment in an oversupplied industry will spread out to the rest of the economy through cross-sector labour movement and thus cause a general decline in economic activity. Finally, Marx claimed that capitalists may periodically react to the falling rate of profit by reducing investment, and this will cause economic fluctuations and business crises.

Marx's theory had a significant impact, both socially and academically. Revolutions influenced by Marxist theory occurred, and socialism is still officially practised in a number of countries. However, the economic performance of these socialist countries is not particularly successful. The breakup of the former Soviet Union marked the failure of socialist practice. Academically, Marx's thought was followed by communists, socialists and other left radicals, including Maurice Dobb, Joan Robinson, Paul Sweezy and Steven Marglin. The influence of Marxism declined after the 1970s.

The great success of Marx's theory is its introduction of dialectical materialism into the evolution of society: the contradictions within a system—the interaction of the forces of production and the relations of production—lead to the evolution of society. This approach can satisfactorily explain the evolution of feudalism to capitalism, so it might also shed some light on the future of society. Marx also highlighted many problems in a capitalist society such as income inequality, unemployment, concentration of capital and industry, and business crises.

The limitation of Marx's economic theory is also related to the grandness of his theory of history. Since Marx developed his economic theory only for the purpose of supporting his vision of social progress, he did not develop a full economic theory and left a number of loose ends in his three-volume work 'Capital'. For example, his labour theory of value did not solve the problem of the influence of profit on commodity prices; without any justification, he regarded the surplus value or profit as solely created by labour; his claim of persistent unemployment is proposed without any convincing justification; his theory of disproportionality crises contradicts market mechanism but was not explained adequately. Because of these loose ends, many regard Marxism theory as unscientific. This might be a reason why support for Marxism has declined in modern times. Nevertheless, Marx's theory provided us with great insight into capitalist society and shed light on some possibilities for the future of society.

## 4.5* Monetarists' Theory

Friedman and Schwartz (1963) and Friedman (1968) form the back-bone of the monetarist school of thought. The term 'monetarists' comes from Brunner (1968), but monetarists' theory—the quantity theory of money—can be traced back to David Hume's influential essay 'Of Money' in 1752. Before the 1930s, there are two versions of the quantity theory of money: the Cambridge cash-balance theory and Irving Fisher's equation of exchange.

The Cambridge cash-balance approach features an exogenous money supply $M$ and a demand to hold nominal money balance Md, which is a fraction of the money value of the national income (PY), i.e. $M = $ Md $ = $ kPY. Any increase in money supply $M$ will result in an increase in price level $P$ if the fraction parameter $k$ and real national income $Y$ are constant.

Fisher defined a concept of circulation velocity of money $V$ as the average number of times a unit of money is used in the course of conducting final transactions for a nominal national income level PY, so the total amount of money required for an economy is given by the following equation: MV $ = $ PY. Apparently, if we let $V = 1/k$, this definition of money supply is equivalent to the cash-balance definition. Since real income $Y$ is exogenous and $V$ is viewed as a constant, this equation also indicates that any change in the quantity of money will lead to a change in price level.

Keynes originally subscribed to the quantity theory of money, but he thought the velocity $V$ would be very unstable and adaptive to whatever changes occurred in money supply. As a result, a change in money supply may be a totally ineffective policy: it affects only velocity $V$ and will have no impact on price level $P$ or national income $Y$. Monetarists totally disagree with the Keynesian view on these issues. Friedman and Schwartz (1963) presented empirical evidence that changes in money supply in the USA played an independent role in cyclical fluctuation. Friedman (1968) employed adaptive expectation to illustrate that there was no trade-off between inflation and unemployment in the long run.

The adaptive expectations hypothesis is easily understood in the case of the labour market, as shown in Fig. 4.15. According to tradition, the

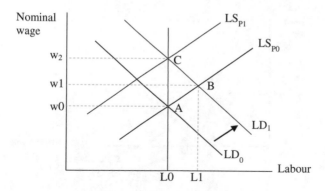

**Fig. 4.15**   Adaptive expectation in the labour market

labour demand curves are downward sloping while the short-run labour supply curves are upward sloping. The initial equilibrium point is at $A$ with a money wage of $w_0$. An expansionary policy (e.g. an increase in money supply) encourages investment and production, so the firm requires more labour, i.e. the labour demand curve shifts from $LD_0$ to $LD_1$. Facing an increase in labour demand, the workers act according to the short-run labour supply curve because they have not thought of the inflation caused by expansionary policy (i.e. workers are fooled by money illusion). Thus, the new equilibrium is achieved at point B with a higher level of employment level and higher money wages. However, expansionary monetary policy will cause inflation. As workers realize that their increased money wages are discounted by inflation to such a degree that they can purchase fewer goods, they start to require higher money wages. As such, the labour supply curve shifts up from $LS_{P0}$ to $LS_{P1}$. The final equilibrium is achieved at point C where the wage level increases to $w_2$ but the employment level remains the same. The policy-induced changes in labour demand or the shift of the labour demand curve has no impact on the employment level in the long run—only the nominal wages have changed.

The same reasoning can be applied to Phillips curves shown in Fig. 4.16.

The short-run Phillips curves $SRPC_0$ and $SRPC_1$ (when the expected inflation rate is 0 and $\pi_1$) are downward sloping, reflecting the

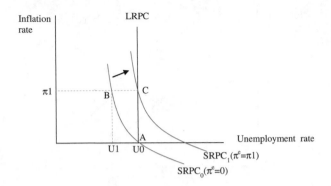

**Fig. 4.16** Long-run and short-run Phillips curves

traditional wisdom that a higher inflation rate is associated with a lower unemployment rate. Initially, the economy is at point A with an unemployment rate of $U_0$ and zero inflation rate. Suppose the government uses an expansionary monetary policy to reduce the unemployment rate to $U_1$. The increase in money supply causes an inflation rate increase to $\pi_1$, so the economy moves to point B with lower unemployment and high inflation. However, this situation cannot last. Once workers detect inflation in the economy, they realize that their real wage actually decreased so they seek a pay rise. As a result, the short-run Phillips curve moves upwards to $SRPC_1$. The long-run equilibrium settles at point C where the unemployment rate comes back to $U_0$, the natural rate of unemployment, or non-accelerating inflation rate of unemployment (NAIRU). Consequently, the monetary policy has no effect on employment level in the long run. In other words, the long-run Phillips curve is a vertical line at the natural rate of unemployment, or there is no trade-off between unemployment and inflation.

Monetarists' theory highlights the role of monetary policy in economic fluctuations. By using adaptive expectation, monetarists successfully explained the short-run effect of monetary policy on output level and the long-run effect on price level. Despite various arguments (e.g. the negative supply shock and the hysteresis of NAIRU), the prediction by monetarists was largely consistent with the stagflation phenomenon in the 1970s and thus led to the success of the monetarists' counter-revolution. However, even though monetarists convincingly demonstrated that monetary policy played an independent role in business cycles, this did not mean that monetary policy is the cause of every economic recession. In other words, monetarists failed to demonstrate that monetary policy is the essential factor behind economic recessions. Moreover, adaptive expectation or money illusion may be able to explain the situation where people have no experience with inflation induced by expansionary monetary policy, but it is implausible given the fact that people have learnt from accumulated experience. Once new classical economists put forward the rational expectation hypothesis, the monetarists counter-revolution swiftly gave way to the new classical counter-revolution.

## 4.6*  The Austrian School

The Austrian school of thought can be traced back to Menger (1871) who developed a production process theory. The theory was further developed by von Bohm-Bawerk (1889) and Von Mises (1912). Friedrich Hayek (1935) combined the production process theory by Menger and the credit theory by Mises to put forward the Austrian theory of business cycles.

The production process or intertemporal structure of production is the cornerstone of Austrian economics. The basic idea of the production process is shown in Fig. 4.17.

To produce a consumable output, multiple stages are required, beginning with the capital as intermediate goods. The value of goods increases as the stages move towards the final goods intended for consumption. Meanwhile, as the final demand passes on to the earlier stages, the level of demand decreases. In this multistage process, entrepreneurs play crucial roles. For firms to make a profit, they must gauge the strength of demand at each stage. Any marginal variation in the early and later stages has important implication for economic growth rates. If the saving rate equals the capital depreciation rate, the production process will repeat and generate the same amount of consumable goods.

To have a growing economy, the saving rate must increase so that it is greater than the capital depreciation rate. An increase in the saving rate will lead to a decrease in consumable goods initially; however, the consumption level will increase later. In Fig. 4.18, periods 1 and 2

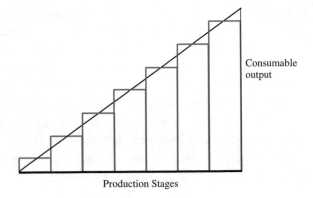

Fig. 4.17   Intertemporal structure of production

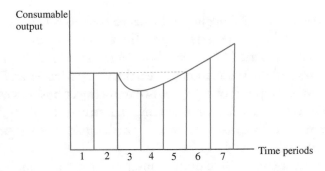

Fig. 4.18   An increase in saving rates in the Austrian model

indicate a zero-growth economy; period 3 indicates a transformation of the economy because an increase in saving rate necessitates a decrease in consumption. Periods 4−7 indicate a growing economy because a saving rate greater than the depreciation rate causes capital accumulation and thus economic growth.

The Austrian theory of economic growth and business cycle is shown in Fig. 4.19. Panel (a) shows the production stages, panel (b) shows the production possibility frontier (PPF) for the economy to produce consumables and capital goods, and panel (c) shows the saving-investment equilibrium. The economy is initially set at point A, where the consumables are produced according to the solid production stages

(a)                                         (b)

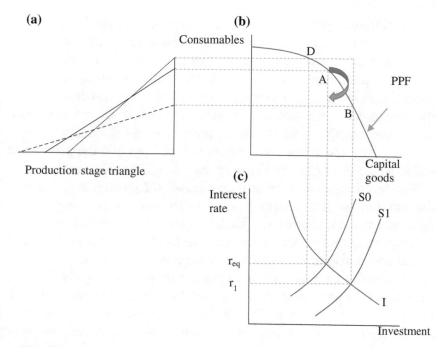

**Fig. 4.19** Credit expansion and economic recession

triangle while the savings and investment are balanced at the equilibrium interest rate $r_{eq}$. If households have a tendency to increase saving rates, the supply of savings will increase and thus $S_0$ shift right to $S_1$ and the new equilibrium in the capital market will generate more investment, so the economy moves to point B where consumption decreases to make up the savings and the production stage triangle is indicated by the dashed line in panel (a). Although the consumption level is lower now, it will increase in the next periods because the dashed line will shift outwards in parallel thanks to the capital accumulation contributed by the increase in saving rates.

However, if the shift of the saving supply curve is not caused by households' willingness to increase the saving rate, rather, it is because of a credit expansion (i.e. an increase in money supply), at the new interest rate $r_1$ firms want to invest more to produce at point B but households want to save less and consume more at point D. The tug of war of households and firms will push the economy to the outside of the PPF to form a credit boom in two ways. If households win the tug of war, the production moves outwards and towards point D. Or, if firms win the tug of war, production moves outwards and towards point B, as shown in panel (b). However, in both cases, the situation is not sustainable: either the investment level is not supported by households' savings at point B or the consumption level is not supported by capital accumulation at point D. As a result, the credit boom will collapse and the production will move inside of PPF and the economy will go into a recession.

The Austrian school uniquely emphasizes the dynamics of the production process and disequilibrium, so its analysis can shed special light on economic growth and business cycles. Since economists of the Austrian school fully adopted the loanable fund doctrine and the role of capital accumulation from classical economics, they are the true believers of capitalism and reject the socialism idea by claiming that there is no rational resource allocation in a socialist system and the coordination of individuals' plans is difficult due to uncertainty. Believing in the role of capital accumulation, the Austrian school shares the same view as classical economists regarding economic growth albeit from a different perspective. Believing the loanable fund doctrine that interest rate can equalize saving and investment, the Austrian school supports Say's law that supply creates demand and thus has to attribute the oversupply during a recession to economic disequilibrium.

The credit bubble explanation of business cycles by the Austrian school is in a way quite similar to the monetarists' explanation based on adaptive expectation: both blamed inappropriate monetary policies for causing boom-bust cycles. However, the explanation of the Austrian school involves no price mechanism. This is starkly different from the monetarists' proposition: an economic boom that leads to a bust must be associated with inflation thanks to an increase in money supply.

Lionel Robbins (1934) and Murray Rothbard (1963) examined the interwar boom and bust. Other Austrian economists (e.g. Littlechild 1990; Horwitz 2000; Garrison 2001) examined the boom in the 1990s, claiming that there is no empirical evidence that inflation is associated with these booms. This seems to add more credit to the Austrian theory of business cycles. However, a better explanation does not mean it reveals the truth. Like monetarists, the Austrian school attributes business cycles to monetary policies. Since money is an important but not the fundamental part of all economies, what the Austrian school reveals may be an important contributing factor but cannot be the key factor underpinning business cycles.

## 4.7   Institutionalism

Institutionalists believe social structure or institutions play a central role in an economy, and thus, they object to the separation of economics from other social sciences such as anthropology, sociology, psychology and history. Thorstein Bunde Veblen (1857–1929) was a forerunner of this school of thought.

Veblen coined the term 'neoclassical' and criticized neoclassical economic theory from three perspectives. First, he discarded Smith's concept of the invisible hand. While Smith saw harmony in the capitalism system such that the competitive markets can channel the self-interest of businessmen into producing social benefit, Veblen claimed that it is obvious to all but economists that making profits and producing goods are totally different things and that businessmen pursuing profits often has deleterious effects on the economy and society. Since Smith's belief of the invisible hand is based on competitive market assumptions, Veblen attacked those assumptions using numerous examples of market imperfections. Second, Veblen attacked the classical assumption that humans are driven by desires to maximize pleasure and minimize pain. He claimed that the behaviour of humans was more generally determined by the institutional environment and culture, rather than being based on hedonistic psychology. Finally, Veblen criticized the failure of classical economists to reconcile their theory with the facts of the real world.

Veblen subscribed to Darwin's evolutionary theory and saw the conflicts in systems. His theory on capitalism, which is manifested in 'The Theory of Business Enterprise' (Veblen 1904), is based on the dichotomy of human behaviours. The instincts of parenthood, workmanship and curiosity lead humans to produce high-quality goods with high efficiency. This behaviour is called industrial or technological employment.

On the other hand, the acquired instincts lead to behaviour that benefits the individual at the expense of the rest of society. The example of this type of behaviour in ancient times included explaining the unknown by appealing to supernatural forces. This non-instrumental, prescientific behaviour is called ceremonial behaviour. In modern culture, the owners of firms are more interested in making money than making goods, so the ceremonial behaviour in modern times is called pecuniary or business employment. In order to make larger profits, firms with monopoly power reduce output and this leads to capitalization of inefficiency, depression and mass unemployment.

This ceremonial behaviour can also be used to describe consumer behaviour. In the theory of the leisure class, Veblen (1899) claimed that ceremonial behaviour leads to wealth-displaying activities, which leads to pecuniary emulation throughout society. Pecuniary emulation leads to conspicuous consumption, conspicuous waste and increased advertising and marketing costs. People are happy only when they consume more than others and this creates tensions in society. As a solution to end this tension, Veblen suggested the end to private property rights.

Compared with Veblen, John Commons (1862–1945) was an institutionalist in action. He rejected classical economists' assumption of hedonistic agents and competitive markets, and thought society and economy were ever evolving and changing. He defined three types of transactions in the economy. Bargaining transactions transfer ownership of wealth by voluntary agreement between legal equals. Managerial transactions involve commands by legal and economic superiors to inferiors. Rationing transactions involve the negotiations of reaching an agreement among several powerful participants. These three types of transactions effectively encompassed all political and economic actions in a society. Commons (1934) put three types of transactions together and defined the result as 'going concern', or 'institution'. Commons recognized that government intervention was necessary to achieve desirable social outcomes, so he was actively involved in and had an impact on a number of social legislations and social reforms.

Following Veblen, Gunnar Myrdal (1898–1987) was also critical of the classical assumption that there is harmony in the capitalist system and that laissez-faire is the best policy for all nations.

Myrdal (1930) envisioned a four-stage development of industrialized nations: mercantilist governmental control, liberalism, the welfare state and the planned economy. Myrdal also criticized neoclassical economics for ignoring the role of normative value judgements and being narrowly focused. He thought it was impossible to completely separate the normative from the positive. Considering classical economists' fixation on equilibrium as inappropriate, Myrdal developed a notion of cumulative causation, which allowed non-economic factors to enter the analysis.

Using his gifted writing skill, John Kenneth Galbraith (1908–2006) criticized classical economics based on the features of American capitalism and society. In his book 'American Capitalism' (Galbraith 1952), he argued that, contrary to the claim by classical economists that monopoly and oligopoly are unimportant divergence from competitive markets, they are the essence of the American economy. Contrary to the common wisdom that concentrated market structure (i.e. monopoly and oligopoly) reduces or even prevents competition and thus causes inefficiency, Galbraith wrote: 'In principle the economy pleases no one; in practice in the last ten years it has satisfied most' (Galbraith 1952, p. 90). He claimed that market concentration generates countervailing power—a self-generating regulatory power. For example, the growth of a large corporation leads to the growth of powerful unions in the industry; large manufacture is counteracted by large retailers. Essentially, Galbraith's countervailing power supersedes competition and Galbraith's visible hand replaces Adam Smith's invisible hand.

In his book 'The Affluent Society', Galbraith (1958) refuted the doctrine of classical economists that firms produce goods to satisfy consumers' needs. He argued that consumers' desires were manipulated by firms so that consumers felt a deep need for the products of an affluent society. As a result, producers created a desire for their products. Galbraith called it a dependence effect that totally reversed the causality chain in classical economics. Moreover, while the dependence effect reminds consumers to buy a new car, an electric toothbrush, etc., there is no such dependence effect to remind consumers of the importance of public goods. As a result, public goods are severely undersupplied. Galbraith conveyed this imbalanced situation in his satirical writing, e.g. luxury cars are driven on badly paved streets, a family picnic with nicely packaged food stored in a portable icebox by a polluted stream; they doze off on an air mattress amid the stench of decaying refuse.

In his book 'The New Industrial State' (Galbraith 1967), the dependence effect was extended to the management of large firms. Modern technology requires large-scale firms. With the separation of ownership and management in large-scale firms, the paid managers are diverted from the goal of maximizing profit—the primary purpose of a firm in an economics textbook. The priority of the managers is to avoid uncertainty and make sure of the continuity of operation or survival of the firms. To achieve this end, they encourage the government to stabilize the economy, cooperate with unions, and manage the preferences of consumers. Once the managers achieve that security, they start to think about sales growth and the price of shares of the firms. In short, the managers' behaviour is dependent on the structure of the firm. The managers themselves become a part of the technostructure of society, and the state supports the technostructure in promoting social attitudes that extol the quantity of goods produced rather than the quality of life in the society. Since the whole state is dependent on the industrial system, liberty may be in jeopardy. In Galbraith's words: 'the danger to liberty lies in the subordination of belief to the needs of the industrial system' (Galbraith 1967, p. 398).

Joseph Schumpeter (1883–1950) can be viewed as a semi-institutionalist. His key contribution was the creative destruction theory in his book 'The Theory of Economic Development' (Schumpeter 1934). Schumpeter thought the principal agents of economic growth were to be found in the institutional structure of society. He identified the activities of entrepreneurs as having a profound impact on an industrialized society. Entrepreneurs take risks in order to introduce new technology and innovative products to the economy, so they are the ultimate source of economic growth. However, as capitalism develops, large firms will become risk-averting and will be run by bureaucratic committees, which will replace the entrepreneurs with prudent managers. This will lead to economic stagnation, the end of the concept of private property and thus the end of capitalism.

Institutionalists emphasize the impact of institutions on economic activity and analyse the economic phenomenon with knowledge from anthropology, sociology and psychology. They like neither mathematics nor statistics, but their analyses are plausible and easily understood by the general public. Institutionalists' criticism that classical economists focus on making economics a positive science and fixate on equilibrium analysis is undeniable. However, while criticizing the classical economists of an abstract approach and implausible assumptions, institutionalists deny any of the valid elements in classical economic theories such as the role of competition and the existence of market equilibrium. The limitations of institutionalism can be summarized as follows:

First, they emphasized the links between economic variables and non-economic factors to such an extent that they opposed an independent economics discipline. While one should acknowledge the influence of non-economic information on economic phenomenon, it is necessary to simplify institutional details in order to study economic activity in depth. By including all types of detailed information in economic analysis, institutionalists may generate a comprehensive result but they also prevent a deeper understanding of the economic problem and thus fail to grab its essence.

Second, Veblen's opposition to hedonistic psychology had some elements of truth, but one cannot totally deny the impact of this aspect of psychology. As social creatures, humans' behaviours may be affected by culture, communities and institutions. Meanwhile, each human as a living being cannot escape from the instinct of maximizing pleasure and avoiding pain. After all, humans are advanced animals. Hedonistic psychology reflects the basic needs of all animals, and thus, it is a fundamental driver of all human behaviours. Besides this driver, the influence of social, cultural and institutional factors should also be considered.

Third, the assumption of a competitive market by classical economists is regarded by institutionalists as being unrealistic, but one cannot deny the existence of competitive markets. As stated earlier, to form a theory, which is an abstraction of reality, one has to simplify reality in order to deal with the essence of the problem. Monopoly behaviour may have existed in the early stage of a market economy but, as long as the economy is dominated by small size of producers, competitive markets should represent the majority of cases of an economy. As time passed, however, the situation has changed so much that monopolies and oligopolies may have dominated the economy. This situation does not prevent the use of competitive markets as a starting point for a study of an economy. The more complicated cases of monopoly and oligopoly can be added to and compared with a fundamental case of competitive markets. The countervailing power and the dependence effect put forward by Galbraith are also of importance for us to understand the consequences of the changed industrial structure. However, more important work should be done on investigating the reason the industrial structure has changed and what are the implications for an economy and for society. Studies on industrial history show that a monopoly will be eventually destroyed by new industries thanks to innovations, so the dominance of monopoly and oligopoly might indicate that the inventions in our society emerge too slowly. Later in this book, we will demonstrate that, with a new patent system, innovations may become abundant and thus have a profound impact on industrial structures.

Finally, to highlight the tension and conflicts in the capitalist system, institutionalists deny any harmony in the market system and essentially reject the market mechanism. Veblen thought businessmen pursued profit at the expense of society rather than with an eye to contributing to social benefits, so he considered pecuniary activity was the source of economic recessions and unemployment. Galbraith thought the development of industrial institutions changed the goal of large firms and this development might have jeopardized the liberty of society. The argument of tension or harmony in the capitalist system critically rested on the assumption of competitive markets: if the markets were competitive, Smith and his followers were right that there is harmony in the market system; otherwise, institutionalists are right. The complexity of markets in reality means both sides may have some elements of truth. Veblen's examples about the detrimental impact of pecuniary activities all came from monopolistic behaviour. By emphasizing on this, he effectively ignored or denied any existence of competitive market. Galbraith's argument that the management of large firms diverted their goal from profit maximization to other things is valid, but he exaggerated the situation and denied that profit making is the main task of the management. After all, profitability is an important index to judge the performance of the management and is a major interest of investors. As a result, although it is not the sole purpose of large firms, profit is a primary concern for both management and the investor.

In short, the capitalist system contains both tension and conflict. Our task is to investigate what causes tension and conflict and to find out how the system can be improved or can be evolved into a better one.

## 4.8* Neoclassical Synthesis

Keynesian-Neoclassical synthesis is a long process, to which orthodox Keynesian, new Keynesian and neoclassical economists have contributed the most. While Marshallian cross (i.e. the equilibrium of supply and demand) is at the heart of neoclassical economics, Keynes's General Theory is a departure from the core of neoclassical economics. The only graph found in Keynes's General Theory was used to reject classical equilibrium theory on the determination of interest rates. The synthesis is an attempt to reconcile these two conflicting schools of economic thought.

In Keynes's mind, uncertainty affected the marginal efficiency of capital and also affected interest rates through liquidity preference (people's tendency to hoard money). He claimed that marginal efficiency of capital and interest rates determined the investment level, which in turn determined income/output levels and employment levels. There was no equilibrium involved in Keynes's theory, but his theory was later incorporated into the equilibrium framework through the IS/LM model, the AE/AP model and the AS/AD model. Although these models also belong to the work of orthodox Keynesian, we discuss them here to show the steps to achieve the Keynesian-Neoclassical synthesis.

### 4.8.1 The IS/LM Model

The first stage of neoclassical synthesis is the IS/LM model developed by Hicks (1937) and Hansen (1949). Hicks rejected Keynes's claim that liquidity preference determined the interest rate and corresponding income level. Hick's argument was that liquidity preference itself

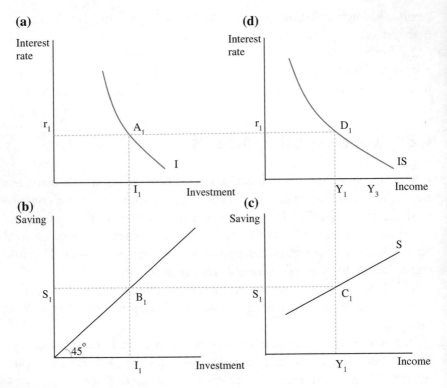

**Fig. 4.20** Derivation of IS curve

(the tendency to hold money) was a function of income: the higher the income, the more cash balance that people will likely hold. Then, Hicks and Hansen developed an IS curve and an LM curve from Keynes's proposition and the quantity theory of money.

The derivation of an IS curve is illustrated in Fig. 4.20. Panel (a) shows the investment demand function. Investment demand is negatively related to interest rate, so the $I$ curve is downward sloping. At interest rate $r_1$, investment level is $I_1$. Panel (b) shows the balance of investment and saving. The 45° line ensures that the investment level is equal to the saving level, e.g. $I_1 = S_1$. Panel (c) shows a saving function. According to Keynes's assumption, the saving level is positively related to the income level with the propensity to save having a value of less than one, so the $S$ curve is a less-than-45° line. At income level $Y_1$, the

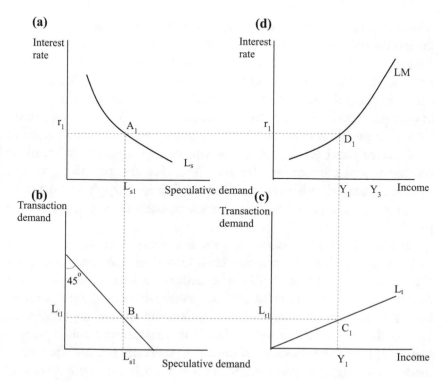

**Fig. 4.21**  Derivation of LM curve

saving function in panel (c) indicates that the saving level is $S_1$. From panels (a), (b) and (c) we can conclude that the interest rate $r_1$ is associated with income $Y_1$. This produces a point $D_1$ in panel (d). In the same way, we can obtain in panel (d) other points when investment is balanced by saving. The collection of all these points forms an IS curve.

The derivation of an LM curve is shown in Fig. 4.21. The total money demand comprises the speculative demand and the transactional demand. The former is negatively related to the interest rate shown in panel (a), while the latter is positively related to the income level shown in panel (c). The total money demand is constrained by money supply, which is assumed fixed so as to simplify the case. This fixed money supply necessitates that the sum of speculative and transactional demand must be unchanged. This produces in panel (b) the downward-sloping-45° line, which ensures that the sum of speculative and transactional money demand equals the fixed money supply. The LM curve can be obtained through the following procedure. The income level $Y_1$ gives a transactional money demand of $L_{t1}$ in panel (c); based on $L_{t1}$, the fixed money supply in panel (b) necessitates a speculative money demand of $L_{s1}$, which in turn produces an interest rate of $r_1$ in panel (a). Therefore, we can conclude that interest rate $r_1$ is positively associated with income $Y_1$ and this produces in panel (d) a point $D_1$ where money supply is balanced by speculative and transactional demand. Repeating this procedure, we can obtain in panel (d) other points of balanced money supply and demand. Connecting all these points gives us an upward-sloping LM curve in panel (d).

Both the IS and LM curves are about combinations of interest rate and income level, so we can put both curves on the same graph, see Fig. 4.22. Any point on an IS curve indicates a balance of investment and savings, i.e. an equilibrium in the goods market. Any point on an LM curve indicates a balance of money demand and money supply, i.e. an equilibrium in the money market. The intersection point of IS and LM indicates the equilibrium of both the goods market and the money market. For example, point A indicates that an interest rate $r_0$ and an income level $Y_0$ can achieve an equilibrium in both markets. The monetary and fiscal policies and other shocks can be expressed as a shift of the IS and/or LM curve. For example, an expansionary monetary policy

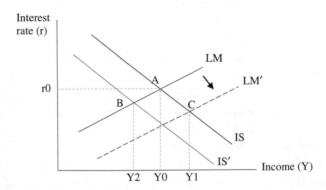

**Fig. 4.22** Monetary and fiscal policies in the IS/LM model

(i.e. an increase in money supply) leads to a right shift of LM to LM' and the new equilibrium at $C$ suggests a decrease in interest rate and an increase in the income level of $Y_1$ (this is a simplified analysis); a contractionary fiscal policy (e.g. an increase in taxation) will lead to a left shift of IS to IS', which leads to a decrease in the income level of $Y_2$ at point B. In this way, the effect of policies proposed by Keynes can be expressed as changes of equilibrium points.

The IS/LM model was very popular in the 1960s, but this model was also criticized for a number of limitations. The most prominent one is that the model assumes fixed prices because it has no price variable, a key element in classical economics. This can be seen in the mathematical form of the IS and LM curve.

The IS curve can be derived from the income expenditure identity:

$$Y = c(Y - T) + I(r) + G$$

Where $Y$ stands for income, $T$ for taxes, $I$ for investment, $G$ for government spending, $r$ for interest rate, $c$ is propensity to consume.

Since investment $I$ is negatively related to the interest $r$, we can write the investment demand function as $I(r) = I_0 - b * r$.

Thus, the IS curve can be expressed as

$$Y = c(Y - T) + G + I_0 - b * r \tag{4.5}$$

The LM curve comes from the real money demand function

$$L(Y, r) = kY - hr \tag{4.6}$$

This money demand must be met by the real money supply $M/P$, so we have the LM curve:

$$M/P = kY - hr,$$

There are three endogenous variables ($Y$, $P$, $r$) in two equations, but one could not solve the equation to find an equilibrium point (i.e. the intersection point of IS and LM). In the IS/LM model, price $P$ is assumed fixed, so we can identify IS and LM and find the equilibrium point for a given price. This fixed price assumption incurred heavy criticism from classical economists and led to the development of the AS/AD model.

## 4.8.2  The AE/AP Model

The second step towards a neoclassical synthesis is the Keynesian cross, invented by Samuelson (1948). The Keynesian cross is also called the aggregate expenditure/aggregate product (AE/AP) model, as shown in Fig. 4.23.

Samuelson conceived a 45° line $Y = AE$, showing all possible points where planned or intended real aggregate expenditure AE equals the total real income or output $Y$. Keynes's consumption function with a less-than-one propensity to consume was presented as a flatter line, AE. The intersection point A determines equilibrium level of total income and aggregate expenditure in the economy. Any change in autonomous investment (i.e. investment not determined by income) causes a shift of the AE curve and thus produced a new equilibrium income level. For example, an exogenous decrease in investment causes a downward shift of AE to AE′ and produces a new income level of $Y_1$. The ratio of income change $(Y_0 - Y_1)$ to investment change (AC) is the investment multiplier.

In this graph, the expenditure function AE is analogous to the demand function. The difference between the two functions is that the aggregate expenditure is a function of income while the demand is expressed as a function of price. The $Y = AE$ line indicates that the source of expenditure is income so it is analogous to the supply function. As a result, the intersection points A and B in Fig. 4.21 are analogous to equilibria in a traditional supply-demand equilibrium model.

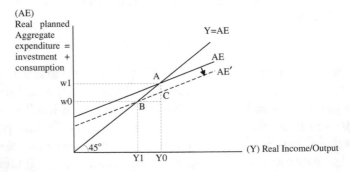

**Fig. 4.23**  A decrease in aggregate expenditure in the AE/AP model

From this point of view, Keynes's idea was thus put into the classical framework. However, this framework is a very simple one—there are no other factors involved except income and expenditure.

### 4.8.3 The AS/AD Model

The advantages of the aggregate supply and aggregate demand (AS/AD) model over both the Keynesian multiplier model and the IS/LM model are that the AS/AD model explicitly includes the price variable and bears a similarity to the Marshallian supply and demand model in microeconomics. It is not surprising that the AS/AD model has replaced the multiplier model and the IS/LM model from the 1980s. Even after 30 years of its dominance, the AS/AD model continues to be an important tool in teaching introductory and intermediate macroeconomics. However, the AS/AD model has also incurred substantial criticism.

Rabin and Birch (1982) pointed out the contradiction between the IS/LM model and the AS/AD model. Their reasoning is as follows. In the IS/LM model, the IS curve indicated the balance between the broadly defined investment and savings, i.e. the equilibrium in the goods market. Since the AD curve was derived or transformed from the IS/LM model, any points on AD should also be the equilibrium points in the goods market. However, in the AD/AS model, only the intersection point of AD and AS indicated an equilibrium in the goods market, so other points on AD should show that the market was in disequilibrium. This argument cast doubt on the validity of the AS/AD model.

Fields and Hart (1990) argued that the derivation of the AD curve from IS/LM implies that firms act implausibly to aggregate demand changes: firms raise output in response to a decrease in the price level and lower output in response to a price increase. They pointed out that this type of response was strikingly inconsistent with modern theories of aggregate supply which predicted that firms would increase output in the face of a price increase. Rao (1991) criticized the inconsistency between the AS and AD curves. His argument is as follows. The AD curve derived from the IS/LM model used a typical Keynesian assumption because the goods market in the IS/LM model assumes a constant price and thus is a market of quantity adjustment. However, the AS curve was derived from marginal productivity, which used profit maximization—a typical neoclassical assumption. These two types of assumptions are totally different, making the resultant AS and AD curves incompatible.

Barro (1994) thought that the AS/AD model was unsatisfactory and thus should be abandoned as a teaching tool. His arguments were: the AD curve derived from the IS/LM model was valid only when there was an excess supply in the goods market; the AS curve derived from the labour market was implausible as it required that the equilibrium in the goods market be accompanied by chronic excess supply in the labour market; the AS curve derived from the price expectation or imperfect information was a special case of rational expectation and thus the AS curve provided no extra value than the rational expectation theory had generated.

Colander (1995) considered the problems in the standard exposition of the AS/AD model and proposed some solutions. One main problem raised by him was the inconsistencies between the standard AD curve and the general definition of the demand curve. He argued that the AD curve derived from the Keynesian multiplier model included the interaction between supply and demand manifested by the multiplier effects while the derivation of a demand curve in microeconomics required the condition of 'other things being equal'. The other problem of the AS/AD model raised by Colander was that the dynamics in the AS/AD model were not supported by empirical data. He provided three possible solutions for the AS/AD model: (1) continuing to act as a rough and dirty policy tool, (2) vanishing due to its serious limitations, or (3) being upgraded to a rigorous model through giving a special definition to the AS and AD curves.

Extending the opinion of Barro (1994), Colander (1995), and Moseley (2010) argued that the AD curve derived from the IS/LM model was based on equilibrium output, so it included information from both the supply side and the demand side. As a result, he regarded the AD curve as both the aggregate demand and aggregate supply curves. The AD curve as an aggregate supply curve conflicts with the AS curve derived from the labour market, production cost or sticky price.

### 4.8.3.1 Difficulties in Constructing Aggregate Supply and Demand Curves

As early as 1936, Keynes in Chapter 3 of the General Theory used the concepts of aggregate supply function and aggregate demand function to refute Say's law and define the concept of effective demand, so Keynes must have had in mind a vision of the aggregate supply and aggregate demand model. However, it is not easy to obtain the aggregate supply and aggregate demand curves for an economy.

One difficulty arises from the aggregation of different commodities. One might think that the aggregate demand/supply can be obtained in the same way as we obtain the market demand/supply from different individuals/firms—that is, to sum all quantities demanded by individuals/firms at each price level to obtain market demand/supply, i.e. a horizontal aggregation. A horizontal aggregation of individual demand curves is shown in Fig. 4.24.

However, we cannot apply this kind of aggregation to obtain the aggregate demand/supply for an economy. One reason is that the commodities are different. In obtaining the market demand curve, horizontal aggregation is valid because the commodity demanded by each individual is the same. When we try to obtain an aggregate demand for an economy, we have to add up all demand curves for different commodities. Since it is pointless to add up different commodities directly (e.g. 1 apple + 1 apple = 2 apples, but 1 apple + 1 pear = ?),

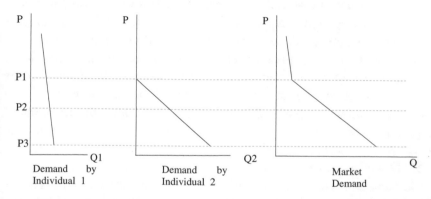

**Fig. 4.24**  Aggregation to obtain market demand

we encounter a difficulty in adding different commodities to obtain an aggregate commodity. The other reason is that, because the prices for different commodities are not comparable (i.e. $1/apple is not comparable to $1/pear), we do not have the same price level for different types of commodities when aggregating the market demand/supply curve. However, as will be shown later, the difficulties of aggregating different types of commodities can be circumvented.

Besides the difficulty in aggregating different types of commodities, there are also difficulties in obtaining and interpreting the aggregate demand/supply curves because of the different meanings of price and income in micro and macro economics. When we talk about the price of a commodity in microeconomics, we imply the relative price (or real price) of this commodity in relation to the price of other commodities. However, when we talk about the price level of an economy, we generally mean a weighted average of nominal price (one may argue that nominal price is relative to money so it is also relative price. This argument treats money as a type of good and ignores the specialty of money that money, especially paper money, cannot be directly consumed to generate utility), which is in relation to money and may be varied by a change in money supply, e.g. an increase in money may increase the price level of the economy. As such, the price used in the graph of the aggregate demand/supply curve is related to the money market and thus is totally different from the price used in the graph of the market demand/supply curve in microeconomics.

Similarly, to derive a demand curve at the micro-level, we assume a fixed or exogenous income. At the macro-level, however, income is endogenized by the aggregate supply and aggregate demand. Real income at the macro-level means either the output or GDP of the economy, which is the counterpart of quantity supplied/demanded at the micro-level. Nominal income at the macro-level is determined by both real income and price level, so it is not only endogenous but also subject to any changes in money supply or changes in any of the other determinants of aggregate demand and supply. The totally different settings about income at the micro- and macro-levels cause insolvable inconsistency between the AS/AD model and the Marshallian supply and demand model.

The use of concepts 'price' and 'income' under the different settings has caused much confusion in interpreting the AS/AD model and has generated a number of arguments. This confusion stems directly from the imprecise nature of defining nominal price/income, real price/income and relative price. According to usual definitions, real price/income is in terms of goods and nominal price/income is measured by money. Relative price is the value of one good relative to the value of another good (or the price ratio of the two goods). To be practical in comparing numerous relative prices in an economy, we need to find a common good (e.g. gold) to be used as a standard. This common good is actually money in the primitive stage of economic development and during the gold standard era. If money is tied to common goods like gold, we can view money as a special good. In this case, relative price, real price and nominal price in this case mean the same thing—the price relative to the value of a special type of good 'money'. However, when paper money is introduced and the gold standard is abolished, the concept of 'nominal price' imply that the price can be affected by a change in money supply, so it should be totally different from real price. As a result, the usual definition of nominal and real prices is inaccurate and can cause contradiction and confusion, so it is necessary to provide a rigorous definition for prices as well as for income.

Since the word 'nominal' implies that the measured value may change if money supply changes, nominal price/income is better to be replaced by a more accurate name: varying-money price/income. The word 'real' implies that the value measured is independent of the change in money supply, so we can replace real price/income by a more accurate name: fixed-money price/income. Since relative price measures the value of one good in terms of the quantity of another goods, we can give a more accurate term: physical price. It is the fixed-money price that is on the graph of both AS/AD model in macroeconomics and the partial equilibrium model in microeconomics. It is the fixed-money income rather than the physical income or varying-money income that determines the demand in both the micro- and the macro-level, so the new concept can overcome the inconsistency between a micro-model and a macro-model.

With these new definitions, we can understand different types of price accurately. In the Marshallian market supply/demand model, the price is measured in terms of money, e.g. the price of apples is $4 per kilogram. This price is a fixed-money price and thus a real price because, in a

Marshallian model, we hold the total money income–expenditure constant. However, the price is a varying-money price or nominal price if the total money income or expenditure changes due to a change in money supply.

The same reasoning can be applied to the price level of an economy. When the money supply is fixed, the price level is a fixed-money price and thus a real price because it indicates the supply/demand condition. However, when money supply changes, the price level is a varying-money price or nominal price because the standard for measuring the value of goods has changed. As such, a change in the price level of an economy can be caused by two factors. One is a change in money supply—we call it a nominal effect. The other is a change in the relative force between supply and demand—we can call it a real effect. The other way to see this consistency between micro- and macro-levels is to consider an economy with only one commodity. In this case, the market demand/supply and the aggregate demand/supply are the same thing, so there is no way one can claim that the price at the macro-level is different from the price at the micro-level.

With the clarified definitions of different types of price and income, we can refute the claim of dichotomy of partial equilibrium and general equilibrium put forward by Owen (1987). Owen illustrated a partial equilibrium model in general form as:

$$q_d = q_d(p, x, Z_1) \qquad (4.7)$$

$$q_s = q_s(p, Z_2) \qquad (4.8)$$

$$q_d = q_s = q \qquad (4.9)$$

where $q_d$ is the planned quantity of demand, $q_s$ is the planned quantity of supply, $q$ is the actual quantity realized. $P$ is the price of goods, $x$ is the exogenous income of the buyer, $Z_1$ stands for other exogenous factors relevant to demand, and $Z_2$ stands for other exogenous factors relevant to supply. These three equations generally can solve for three endogenous variables $q_s$, $q_d$ and $p$, so the equilibrium solution can be obtained.

The general equilibrium in the aggregate goods market was illustrated by Owen as:

$$X_D = X_D(X, r, Z_3) \qquad (4.10)$$

$$X_S = X_S(P, Z_4) \tag{4.11}$$

$$X_D = X \tag{4.12}$$

$$X_S = X \tag{4.13}$$

where $X_D$ is the planned aggregate demand, $X_S$ is the planned aggregate supply, $X$ is the actual real output, $P$ is the general price level, $r$ is the interest rate, and $Z_3$ and $Z_4$ are other relevant exogenous variables. This model contains 4 equations and 5 endogenous variables ($X_S$, $X_D$, $P$, $X$ and $r$), so one cannot find the equilibrium solution. The problem can be solved by adding an extra equation indicating the equilibrium condition in the money market:

$$M_S = M_D(r, X, P, Z_5) \tag{4.14}$$

where $M_S$ indicates the fixed money supply, $M_D$ the money demand, and $Z_5$ the other relevant exogenous variables.

Here, Owen made a mistake regarding the meaning of $x$ and $X$. The $x$ in the partial equilibrium model means the fixed-money income of the consumer, so the $X$ in the general equilibrium model should also mean the fixed-money income of the economy. However, Owen regarded it as the real output of the economy, which is equal to the physical income or aggregate quantity of goods. In Eq. (4.10), Owen showed that real income/output ($X$) determined the quantity demanded ($X_D$); however, he stated in Eq. (4.12) that real income/output ($X$) is equal to the quantity demand ($X_D$). It is apparent that Owen's setting of the aggregate demand function is logically inconsistent.

In both partial equilibrium and general equilibrium models, it is the exogenous money income that affects demand, so the real income/output ($X$) in Eq. (4.4) should be the exogenous money income $M$. Once this mistake is corrected, the general equilibrium model is comparable to the partial equilibrium model, so both the partial equilibrium approach used in microeconomics and the general equilibrium model used in macroeconomics are consistent.

In short, once we have the appropriate definitions of different types of price and income, we can clear up many confusions and improve our understanding. To facilitate our thinking, we initially need to consider the price level in the AS/AD model as a fixed money price (i.e. assuming there is no change in the money supply) and thus a real price and then consider the price level in nominal terms by allowing a change in the money supply. This strategy may help us to clear up many mistakes and confusions regarding the AS/AD model.

### 4.8.3.2 The Ways to Obtain the AD Curve

Due to the difficulties in aggregating market demand curves, methods other than direct aggregation were used to obtain the aggregate demand curve, including methods utilizing the AE/AP model, the IS/LM model and the quantity theory of money. However, these methods of obtaining AD curves resulted in much criticism. We describe and assess the four existing methods first and then introduce four new ways of obtaining AD curves by aggregating market demand curves. For details, please see Appendix 1 at the end of this chapter.

### 4.8.3.3 The Ways to Obtain AS Curve

The arguments regarding the AS curve are less intensive than those on the AD curve, but people still cannot agree upon the different shapes of the AS curve in the long run and also in the short run. Unlike the AD curves, no AS curve is derived from another model. Rather, the derivation of AS curves is based on different assumptions, such as adaptive expectation, wage/price rigidity, imperfect competition and resource constraint. We first examine the existing ways of deriving AS curves and then introduce a new method of deriving an AS curve from the microeconomics foundation.

1. Obtaining AS curves from neutrality of money and money illusion

The dominant explanation of the shape of the AS curve is the neutrality of money in the long run and money illusion or sticky prices or wages in the short run. The AS curve in the short run is generally regarded as upward sloping because of three possibilities when money supply increases. Possibility 1: workers are fooled by an increase in nominal wages resulting from increased money supply (money illusion). They mistake the nominal wage increase as an increase in real wage and thus supply more labour and produce more output, leading to an increase in both price level and output level. Hence, the AS curve is upward sloping. Possibility 2: the price adjustment in the short run is slow for various reasons (e.g. menu costs and long-term contract). This causes partial price adjustment and partial output adjustment and thus an upward-sloping AS curve. Possibility 3: the nominal wage in the short run is fixed or is adjusted more slowly than the price. This causes production costs to increase slower than the price of products. As a result, firms are motivated by profit to produce more output. In the long run, however, the increase in the money supply has no effect on output (neutrality of money) so the AS curve is a vertical line. The shape of the long-run and short-run AS curves can be explained by Fig. 4.25.

In panel (a), labour supply $L_s$ and labour demand $L_d$ achieve initial full-employment equilibrium at point A with labour quantity of $L_0$ and a nominal wage of $W_0$. Panel (b) shows that the amount of labour $L_0$

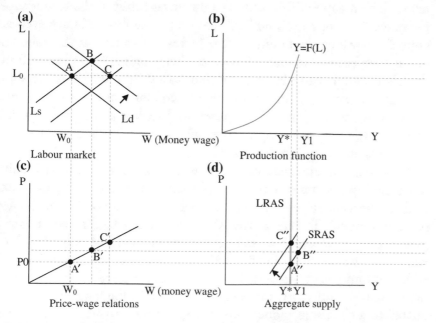

**Fig. 4.25** Deriving the AS curve from adaptive expectation

produces an output level $Y^*$. Panel (c) shows the positive relationship between price level and nominal wage due to the assumptions of neutrality of money and the flexibility of nominal wages. The nominal wage of $w_0$ in panel (c) indicates a price level of $P_0$. Using the output level $Y^*$ and the price level $P_0$, we can obtain a point A′ in panel (d).

If the money supply increases, the demand for goods will increase and the firm will increase its production. As such, the labour demand increases, so the labour demand curve $L_d$ in panel (a) shifts to the right. This will lead to an increase in nominal wages. The workers will be attracted by the higher nominal wage and supply more labour, so the short-run equilibrium is achieved at point B. The increased labour inputs produce higher levels of output $Y_1$ in panel (b) and the increased nominal wage indicates a higher price level in panel (c), so we can obtain the point B″ in panel (d). The line passing through A″ and B″ is the short-run aggregate supply.

Sooner or later, the workers will realize that the commodity-price increase reduces the purchasing power of their wages, so they will press the capitalist to increase their wages to keep up with inflation or simply reduce labour supply. This leads to a downward shift of the labour supply curve $L_s$ in panel (a), and the long-run equilibrium is achieved at point C, at which the nominal wage increases further but the volumes of employment and output fall back to their original level. The shift of the labour supply curve in panel (a) leads to a shift of the short-run supply curve in panel (d), and from the new equilibrium point C in panel (a), we can obtain a point C″ in panel (d). The line passing through A″ and C″ is the long-run supply curve, which is vertical because the output level is unchanged from A″ to C″.

These popular explanations sound reasonable but are actually built upon conceptual confusions. As explained earlier, the aggregate price level can be a varying-money price (i.e. nominal price) or a fixed-money price (real price). To explain the AS curve, we must start with a fixed-money price level which allows no change in money supply. The derivation of the AS curve in Fig. 4.25 omits the fixed-money price and addresses only the non-essential varying-money price level. If this explanation tells the full story, the vertical AS curve in the long run must be related to a change in money supply and the upward-sloping AS curve must be related to a money illusion or sticky prices/wages. These scenarios are possible, but they are not a general case and thus are inconsistent with microeconomics foundations. As a result, the resultant AS curve loses much generality and thus has very little explanatory power.

## 2. Obtaining an AS curve from real and nominal wage rigidity

The approach to explaining or deriving an AS curve from wage rigidity has a strong Keynesian flavour and has some valid points. However, existing derivations suffer from the confusion between equilibrium and supply and between nominal and real prices as well. As a result, the derivations are implausible and inconsistent. We will demonstrate this by using the derivation proposed by Glahe (1977) as an example.

Figure 4.26 shows the way to derive AS based on the labour market equilibrium, production function and nominal wage rigidity.

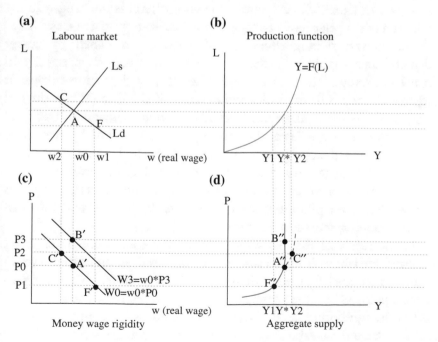

**Fig. 4.26** Deriving the AS curve from nominal wage rigidity

The derivation of the vertical long-run AS curve stems from the equilibrium in the labour market shown in panel (a). The nominal wage rigidity leads to a negatively sloping relationship between real wage and price level shown in panel (c). The equilibrium real wage in panel (a) identifies a price level $P_0$ through the wage-price relations in panel (c), while the equilibrium labour input in panel (a) identifies an output level $Y^*$ in panel (b). This gives a point A on the AS curve in panel (d). If the real wage is too high at $w_1$, the labour supply is high but the labour demand is below equilibrium level, so the output would be below the equilibrium output at $Y_1$. Similarly, if the real wage is too low, labour demand is high but labour supply is below the equilibrium level, so the output level would be less than the equilibrium output too. As such, the equilibrium output is the highest achievable and thus the long-run output level will be optimally fixed at this level. If there is an increase in price level from $P_0$ to $P_3$ due to an increase in money supply, the nominal wage will increase from $W_0$ to $W_3$ and leave the real wage unchanged, as shown in panel (c). As a result, the output level is unchanged in the long run when price level changes, so the long-run AS curve is vertical.

The upward-sloping short-run AS curve is explained by labour demand and supply in panel (a) and the nominal wage rigidity in panel (c). Labour demand is emphasized in this derivation because it represents the profit maximization decision of the firm. When the real wage is above the equilibrium level at $w_1$, labour demand is lower and so the output level is lower than the equilibrium level at $Y_1$. According to the downward-sloping curve in panel (c), the higher real wage under the condition of the fixed nominal wage necessitates a lower price level $P_1$. This means a lower output and lower price when the real wage is higher than the equilibrium level. This gives the point F and forms an upward-sloping short-run AS curve (the A″F″ part).

In the case that the real wage is below the equilibrium level at $w_2$, demand for labour is higher and this would result in an output level $Y_2$ and an price level $P_2$, greater than the equilibrium level. This would produce a point C″ on the aggregate supply curve. However, at a real wage below the equilibrium level, the supply of labour is less than the demand, so the demand for labour cannot be satisfied and the would-be higher level of output is not achievable (indicated by the dashed curve AC″).

This explanation looks plausible, but it does not satisfy the condition of deriving a supply curve. To obtain a supply curve (no matter if it is a market supply or an aggregate supply), one must vary the demand (or shift the demand curve) to reveal the supply under various demand conditions. The demonstrated derivation of the AS curve by Glahe (1977) does not allow the labour demand curve to shift and thus fails to meet the condition of deriving a supply curve. The explanation of the long-run AS curve also relies on changes in money supply, so it addresses only a nominal issue. Thus, the explanation has addressed only one equilibrium outcome or one point on the AS curve. Other points on the AS curve need to be explained by shifting the labour demand curve.

The approach to deriving the short-run AS curve from the labour demand curve is invalid for two reasons. First, although the labour demand curve embodies the firm's profit maximization decision, this decision is related not only to a real wage but also related to the demands in the goods market—it is not profitable to produce more than the market can clear. The approach has totally ignored a key determinant of labour demand and thus is invalid. Second, to derive an AS curve, we should focus on the supply side, i.e. one cannot obtain an aggregate supply curve from market demand curves. We will explain this further shortly.

The invalidity of deriving an AS curve from a labour demand curve underpins the criticism by Barro (1994) that it is implausible that the labour market is in excess supply while the commodity market is in equilibrium. The reasoning behind this criticism can be comprehended by considering the Walrasian general equilibrium framework: a market is truly at equilibrium only when all related markets are at equilibrium.

What is the correct way to derive an AS curve from the labour market? First of all, in order to derive an AS curve, it is the labour supply rather than the labour demand that matters. Labour demand is determined by the firm according to goods demand. For an AS curve, aggregate demand is just a testing tool to reveal aggregate supply. Similarly, labour demand is useful in deriving AS curve from labour market only because it can reveal labour supply curve. Putting it differently, an AS curve represents the response of the aggregate supply to changing demand conditions. In considering an economy with only labour input, this response ultimately comes from the response of labour supply to changing labour demand (i.e. shifts of the labour demand curve). Additionally, labour supply is constrained by the size of the labour force. This is reflected in the horizontal part of the labour supply curve in panel (a) of Fig. 4.27—the labour supply is constrained by the size of the labour force when the real wage is very high.

Second, there is no need to use non-general assumptions such as nominal wage rigidity, which will reduce the scope of validity of the AS curve. For simplification, we assume an economy of one input—labour, so the average production cost, and thus the price of a good, should be proportional to nominal wages (the size of the proportion depends on how much labour input is needed to produce one good). In turn, the nominal wage rate is proportional to the real wage rate and the size of proportion depends on the size of the money supply. In short, nominal wage equals real wage times a price index. Given that the price index must

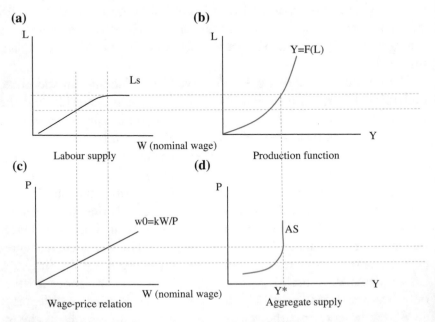

**Fig. 4.27** Deriving the AS curve from labour supply curve

be positive, the nominal wage and price level must be positively related. The exact relationship depends on the behavior of the real wage. For simplicity, we assume a fixed real wage. As such, we have the price level of the economy being proportional to the nominal wage rate, which is shown in panel (c).

Given the labour supply in panel (a), the wage-price relation in panel (c), and the usual upward-sloping production function in panel (b), the AS curve can be derived in panel (d). The horizontal part of the labour supply curve leads to a fixed amount of maximum output, reflecting the limit of the labour force. This leads to the vertical part of the AS curve. Below this maximum output, the AS curve is upward sloping because of the upward sloping of the three curves: the labour supply curve, the production function and the wage-price relation.

### 3. Obtaining AS curves from resource constraints

Considering the resource constraint in an economy, we can build an AS curve as shown in Fig. 4.28. When the output level is low, the production uses relatively few physical resources in the economy, so there will be an excess supply of resources which prevent the resource prices from increasing significantly. As a result, the production cost increases little and this produces an almost flat part of the AS curve. As the output level increases, more resources have to be used and this will push up the prices

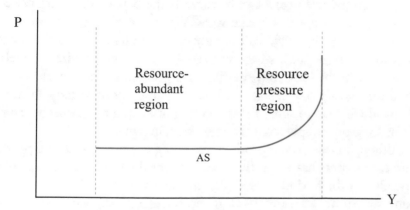

**Fig. 4.28** Deriving the AS curve from resource constraint

of resources, which lead to a significant increase in production costs and thus an upward-sloping AS curve. As the output level reaches the resource limit, the output level is fixed at the maximum level that the resource can support, so any increase in demand will be fully translated into an increase in price level and thus produces the vertical part of the AS curve.

The reasoning behind the resource-constraint approach is consistent with microeconomic foundations. However, it also has a few shortcomings. One is that the approach is ad hoc. The approach is based on an analogy in microeconomics, but it does not show how to obtain an AS curve from principles in microeconomics. Second, there is no differentiation of long run and short run in the derived AS curve. This is totally different from supply curves in microeconomics. The lack of differentiation of long run and short run makes the derived AS curve incapable of explaining the dynamics of the economy. Finally, the explanation of a business cycle based on this AS curve is implausible. When the economy operates at the resource limit, there is no mechanism showing how the economy will fall back into a recession. When the economy is at its trough, there is no explanation why the economy will recover and boom again in the future.

## 4. Obtaining an AS curve from microeconomic theories

We start with the firm's supply curve in microeconomics. In the short run, a competitive firm's supply curve is the part of the marginal cost curve above the lowest average variable cost, i.e. the bold curve shown in panel (a) of Fig. 4.29. This supply curve results from the firm's decision to maximize profit when firm size (or capital in a stylized model) is inflexible. In the long run, the firm will produce at the lowest point of long-run average cost due to the free entry of other firms (any change in demand will lead to some firms entering or exiting the industry), so the firm's long-run supply is shown as point A in panel (b).

Although firms do not produce according to the long-run average cost curve, this curve has a significant bearing for the firm and for the industry. Due to the flexibility of capital and the size of the firm in the long run, the firm's long-run average cost curve is the envelope of the short-run average cost curve, i.e. all short-run average cost curves are

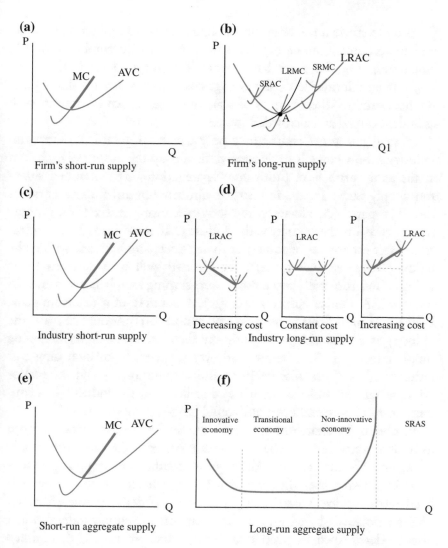

**Fig. 4.29** Deriving AS curves from microeconomic theories

tangent to the long-run average cost curve. When the output level is less than the long-run equilibrium level at point A, the average cost curve is decreasing so the firm enjoys economies of scale. The reason for a decreasing long-run average cost curve may be cost saving from the firm's more efficient use of inputs or shared costs (e.g. only one manager and an accountant are needed for a firm of either 20 employees or 100 employees). On the other hand, when the output level is above the long-run equilibrium level, the average cost curve is upward sloping and the firm exhibits diseconomies of scale due to resource constraints such as limited capital and a crowded space.

To aggregate firms' short-run supply curves to obtain the short-run industry supply curve is straightforward: add up the output of each firm at the same price level (horizontal aggregation). The obtained short-run supply curve for an industry is also an upward-sloping curve as shown in panel (c). However, the long-run supply curve for an industry depends on the nature of the industry. If most firms in an industry exhibit economies of scale (the shared costs among firms within the industry may also play a part), the industry will be a decreasing-cost industry and thus will have a downward-sloping supply curve shown in panel (d). A sunrise industry is a typical example of a decreasing-cost industry. If most firms in an industry exhibit diseconomies of scale, the industry is an increasing-cost industry featured by an upward-sloping supply curve. An old or sunset industry is generally an increasing-cost industry. If the types of firms in an industry are mixed and the economies and diseconomies of scale are cancelled out, the industry is a constant-cost industry and has a horizontal long-run supply curve.

To obtain the short-run aggregate supply for the economy, we need to impose restrictions on the relationship between the short-run supply curves of different markets. Although all methods of imposing restrictions in deriving the AD curve can be used here, the restriction of a fixed relative price is particularly plausible for deriving the AS curve. This restriction is based on the fact that, on the supply side, product price is determined by production cost, which in turn is determined by inputs markets (e.g. intermediate inputs, labour and capital markets) serving all firms. With the assumption of fixed product price ratio, we can obtain an upward-sloping short-run AS for the economy. The details of aggregating industrial supply curves are shown in Appendix 2 at the end of the chapter, but here we can present the resulting short-run AS curve shown as a bold line in panel (e).

For the long-run AS of the economy, the dominant industry type has an important influence. First, if most industries in an economy are new or sunrise industries enjoying decreasing costs, the dominance of decreasing-cost industries necessitates a downward-sloping long-run AS curve in panel (f). Since innovation is the cause and the key feature of a new industry, we call an economy dominated by new industries an innovative economy. Second, if the majority of the industries are old or sunset industries, the dominance of increasing-cost industries results in an upward-sloping long-run AS curve. We call this type of economy a non-innovative economy because it falls short of innovative or new industries. As the output of a non-innovative economy becomes higher, the price level increases faster due to higher pressures on limited resources. This leads to the increasing slope of the AS curve. When the non-innovative economy has used all resources available, the output cannot be increased further. Under this situation, an increase in demand will be transferred fully to an increase in price level, so the AS curve will become vertical. Finally, if an economy has a mixed number of new and old industries and their impacts on the production cost for the economy are cancelled out, the long-run AS will be a horizontal line. This type of economy is a transition between innovative and non-innovative economies, so we call it a transitional economy. For the convenience of presentation, three types of economies are put into one graph to form an integrated long-run AS curve, as shown in panel (f).

One may argue that the short-run AS curve derived from microeconomics is not different from those derived from nominal rigidity of price or wages because the supply curves in microeconomics are derived based on the assumption of constant factor prices and the assumption of given market prices. This argument confuses rigidity with constant prices or wages. Rigidity means fixation when the condition changes, but the assumption of constant prices in firms' plans implies unchanged prices under an unchanged market situation, and thus, it is not the same as the assumption of price/wage rigidity. In deriving a supply curve (or marginal cost curve) from a competitive market, wages and other input prices are assumed as constant parameters. When condition changes, the values of parameters can change, and this will lead to a shift of the supply curve. Similarly, the price-taker assumption in the perfectly competitive market means a given or unchanged price for the competitive firm. It also does not fit into the concept of rigidity either because the market price can change when the market condition changes.

### 4.8.3.4 The AS/AD Analysis

The AS/AD model was criticized for providing implausible explanations for economic events. For example, with a negative demand shock, the AS/AD model based on an AS curve derived from nominal wage rigidity (see Fig. 4.26) would predict a countercyclical movement of real wages. The upward-sloping short-run AS curve used in panel (d) of Fig. 4.26 shows that, as output decreases during a recession, the price level decreases. Meanwhile, the fixed nominal wages assumption used in panel (c) necessitates a rise in the real wage when the price falls. This leads to a strange situation: when the output decreases during a recession, workers are enjoying an increased real wage. The inconsistency of this prediction with empirical results was pointed out as early as in the late 1930s by Dunlop (1938) and Tarshis (1939).

Colander (1995) also pointed out that the explanations from the AS/AD model may be conceptually valid but the effect may be too weak to achieve the results predicted by the model. For example, the traditional explanation of the effect of a negative demand shock is shown in Fig. 4.30. The negative demand shock is indicated by a downward shift of the aggregate demand curve from $AD_0$ to $AD_1$. At point B where $AD_1$ intersects with the short-run AS curve $SRAS_0$, both the output level and the price level were lower. With the lower price, real money supply increases and the interest rate decreases. This leads to an increase in investment and in interest-related consumption, i.e. the Keynes effect. Meanwhile, the lower price indicates an increase in the real value of money balance (i.e. wealth), so household expenditure will increase (the Pigou effect). Both the Keynes effect and the Pigou effect lead to an increase in demand and thus an increase in output. On the supply side,

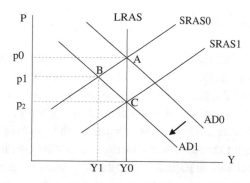

**Fig. 4.30**   Illustration of a standard AS/AD model

the fall in demand for labour leads to lower money wages and thus lower production costs, so the short-run aggregate supply curve shifts down to $SRAS_1$. As a result, the economy moves from B to C—a negative demand shock leads only to a decrease in the price level, leaving the long-run output unchanged.

Colander first pointed out that the prediction of a falling price level was not supported by reality. Second, he argued that the Keynes effect and Pigou effect are simply too weak to drive up the aggregate demand. Evidently, Pigou himself agreed that the Pigou effect was only a debating point and was unimportant to real life. Third, Colander highlighted the detrimental effect of falling prices due to their effect on financial obligations. Finally, Colander gave an intuitively more plausible result: firms will reduce real output supplied in the face of excess supply or insufficient demand.

The implausible explanations in the standard AS/AD model shown by previous studies have exposed the weaknesses of the model. However, the author argues that these unsatisfactory predictions may result from assumptions used for deriving AS and AD curves and the settings of the AS/AD model, rather from the conceptual framework of the AS/AD model. With the AS and AD aggregated from microeconomics foundations, we can demonstrate that the AS/AD model works well in explaining macroeconomic phenomena.

## 1. Economic dynamics

The economic dynamics discussed here include two kinds: the dynamics of different types of economies and the dynamics between short run and long run. We use the traditional graphic presentation to label horizontal axis 'output' or 'income' and vertical axis 'price level'. Using price level rather inflation rate allows us to examine the process of dynamics—from equilibrium through disequilibrium to a new equilibrium.

We start with the dynamics of different types of economies. In the case of an innovative economy, a new product/technology is always expensive when it is first introduced. Plenty of empirical evidence supports this, e.g. the advent of computers, laptops, iPhones and the first commercial trip to space. The reasons for the high price include the high cost of innovation, the high upfront fixed cost of production and the low capacity of production. As consumption increases and the production is standardized, the average cost of production decreases significantly and thus the prices drop. If most industries in the economy are innovative, the effect of the growth of new product/technology will dominate the economy, so the innovative economy would be featured by a growing output accompanied by a decreasing price level, as shown in Fig. 4.31. Although we have not seen this kind of economic growth pattern in history, it is reasonable to foresee this. This type of growth pattern has happened to new industries, why will it not happen to an economy which is comprised of all or most new industries?

As new industries develop over time, they become old industries. If the innovation rate is not high, only a handful of new industries emerge, so the economy will be made up of mixed new and old industries with no dominant side. The opposite growth pattern of two types of industries—the decreasing cost for new industries and the increasing cost for old industries—will lead to a transitional economy, featured by a growth accompanied by no increase in price levels. If innovation still cannot keep up with production, most industries in the economy will become old industries and the economy will become a non-innovative one, which features a rising output and a rising price level. When production reaches the resource limit, the economy will not grow anymore.

Under this circumstance, any attempt to grow a non-innovative economy leads only to price increases.

When it comes to the dynamics of the short run and the long run, one must appreciate that the AD curve is downward sloping while the short-run AS curve is upward sloping. With a positive aggregate demand shock, the AD curve shifts to the right, e.g. from $AD_0$ to $AD_1$ in Fig. 4.31. The new AD curve $AD_1$ intercepts with the short-run AS curve $SRAS_0$ at point A, and this causes an increase in both output and price level. However, point A is not an efficient point because the fixed capital or firm size in the short run can be relaxed in the long run. As time passes, the firm can increase the firm size and produce at point B with the same amount of output but at a lower cost. This causes a drop in price levels. With the increased firm sizes, the short-run supply curve shifts to $SRAS_1$. The new short-run equilibrium is achieved at point C. With the continuous right shifts of the short-run AS curve, the long-run equilibrium will be achieved as point E. The same analysis can be applied to the case when the aggregate demand shifts from $AD_2$ to $AD_3$.

So far, we have assumed that the short-run AS is steeper than the long-run AS. The reasoning behind this assumption is that, due to the fixed capital or firm size in the short run, the cost of production in the short run increases faster than that in the long run. The other factor that needs to be taken into account is the input price. When the output level is high, the pressure on limited resources will drive

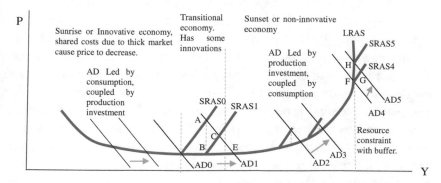

**Fig. 4.31** Economic dynamics between short and long runs

up input prices and increase the long-run production cost sharply. However, firms are unable to perceive resource limits correctly in a timely fashion, so the slope of the short-run AS curve is unlikely to be affected by a future rise in input costs. This will create a situation where the slope of LRAS is greater than that of SRAS when production is approaching resource limits, shown as $SRAS_4$ and $SRAS_5$ in Fig. 4.31.

In this case, when a positive demand shock shifts the aggregate demand from $AD_4$ to $AD_5$, the output level will increase to point G where $AD_5$ intersects with $SRAS_4$. Here, the output can be greater than the resource limit can support, because we assume a soft resource limit, or a limit with a buffer. There is no economy operating at absolute resource limit, but when it is close to this limit (i.e. within the buffer zone), the extremely high price makes it very hard to go further to reach the absolute limit. The buffer zone can be viewed as a soft resource limit. Once the firm realizes the unexpected degree of the input prices hike and revises its production cost by including the factor of input price increases, the short-run supply curve shifts to $SRAS_5$ and the output level falls back to the level supported by soft resource limits. It is worth mentioning that this explanation is similar to but actually different from the money illusion argument because there is no change in the money supply here.

## 2. Business cycles and economic growth

The business cycle is an important phenomenon in macroeconomics, but the explanation provided by existing AS/AD models is unsatisfactory (e.g. Barro 1994; Colander 1995; Mankiw 2003). On another important topic—economic growth, the existing AS/AD model is unable to shed any light at all. A standard AS/AD model assumes a vertical long-run aggregate supply curve. This assumption is problematic in the first place. A general explanation for the vertical long-run AS curve is resource and technological constraint. While an economy is subject to constraints at any time, these constraints are more eminent in the short run. For this reason, the supply curve in microeconomics is vertical in very short run. In the long run, firms can find resources more easily and technology progress more likely occurs and thus overcome resource constraint. Hence, the long-run supply curve is more elastic and thus should be flatter than the short-run supply. The vertical long-run AS curve in a standard AS/AD model contradicts this common sense. Secondly, with a vertical long-run AS curve, the output level in the long run is fixed at the potential or natural output level and thus there is no way to express economic growth in the model. The best a standard AS/AD model can explain about economic growth is that a technological change might cause the jump of the long-run AS from one potential output level to another. Although uneven economic growth is common in short run (e.g. a jump of GDP level due to a burst of innovations or new technologies), this style of growth through jump does not happen in the long run. History shows that the economy shows trended long-run growth rather than jumps.

With the AS and AD curves derived in microeconomics foundations, here we can demonstrate that the new AS/AD model can explain not only both business cycles and economic growth but also the dynamics between them. The explanation depends on different types of economies: innovative or non-innovative.

We start with the analysis of the non-innovative economy near the resource limit buffer zone, as shown in Fig. 4.32. For a non-innovative economy, the demand is driven by consumption and investment. When the output level grows to the level close to the resource and consumption limit buffer zone, the markets are close to saturation

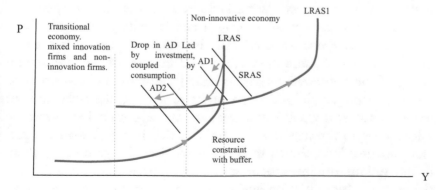

**Fig. 4.32** Cyclical growth of a non-innovative economy

and thus the consumption increase is very limited, so demand growth is largely led by investment in production. However, as investors find that the growth of output is almost zero while the price level increases substantially, they realize that their investment in production is unprofitable in real terms. As a result, they have to pull back their investment in production and this will lead to a substantial left shift of the AD curve, e.g. $AD_1$ shown in Fig. 4.32.

As the AD shifts to the left, the output level and price level will decrease so an economic recession will occur. However, the economy may not go back to the original position according to the original long-run AS curve for various reasons, such as the changed conditions in production (e.g. firm size), the changed households' lifestyle, the change in speculative behaviour of investors (e.g. liquidity preference), and monetary and fiscal policies in place. The decrease in investment and the scaling back of production will cause a decrease in household income, which in turn will lead to a decrease in consumption and a further left shift of the AD curve (e.g. $AD_2$ shown in Fig. 4.32). The weak demand and thus the pressure from sales stagnation force firms to invest in innovation. Eventually, the falling prices will stop when the commodity prices hit production cost limits (e.g. minimum of labour cost); however, the decrease in the output level continues because of the changed consumer expectation—people become more cautious during a recession and cut spending as much as possible. It is at this stage that the Keynesian-style fiscal policy can stop output falling but does not lead to an increase of the price level because of the weak demand as well as idled resources and production capacity.

Now the economy is in its trough and is waiting for innovations. Innovations have not occurred early in adequate quantity because of high possibility of innovation failure and because of the fare of imitation. If some innovations are successful, the resulting new products create demand, so the AD curve starts to shift to the right and the economy starts to recover and expand. As the output level grows to a level close to the new resource limit, which is higher than the previous resource limit thanks to the technological change brought about by innovations, overinvestment in production occurs and this will bring the economy into recession again. As such, a non-innovative economy will experience a cyclic growth.

For an innovative economy, resource limits are not a concern because production innovations can overcome resource limits (i.e. using less resources to produce the same amount of output) and product innovations can create demand. Once the limits on both the production side and the consumption side are overcome, the economy will not experience a persistent output decline or an economic recession. However, the economy may be subject to large cyclical price fluctuations, as shown in Fig. 4.33.

The advent of new products is generally associated with high innovation costs and upfront fixed costs. These high costs necessitate high prices for the new products. Due to the innovation cycle, the advent of successful innovation is not likely to be even. With a burst of innovation and thus new products, the price level of the economy may be pushed up substantially (e.g. point A in Fig. 4.39). With increased demand and the standardization of production procedures, the production cost of new products will reduce significantly and thus bring down the price level (e.g. point B). The next burst of innovation and new products will again push up the price level (e.g. point C). As such, the cyclic innovation burst will cause cyclical price change in an innovative economy.

We have not seen this type of economy in history. The closest evidence would be the dramatic increases in output but very mild increases in the price level during the industrial revolution. The reasons for the absence of an innovative economy or for the scarcity of innovations

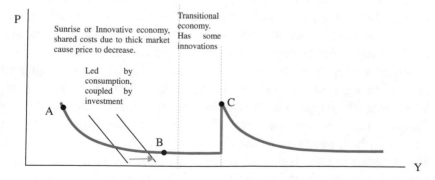

**Fig. 4.33** Continuing growth of an innovative economy

are embedded in the feature of innovation: the high risk of innovation investment and the high externality of successful innovations. The low success rate of innovations makes risk-averse investors shy away from innovation activities. This situation is further exacerbated by the imitation of successful innovations. If an innovation succeeds, it can be used by others and reduces remarkably the return to the innovator or innovation investor. These features make innovation investment unattractive and thus cause innovation scarcity. Some efforts are made to address this issue, notably, the establishment of patent law. However, due to various limitations, these efforts fail to encourage enough innovations and thus fail to create an innovative economy. A thorough revision of relevant laws may change this situation.

3. Fiscal and monetary policies

Fiscal and monetary policies are the main instruments for the government to manage a market economy, so it is crucial that a model can explain and predict the effect of these policies. The difference between these tools is that a monetary policy involves a change of money supply so it may cause a change in nominal price levels (varying-money price). On the other hand, a fiscal policy involves a change in government spending but generally does not involve a change in money supply (unless the government borrows excessively from the central bank and thus forces the central bank to increase money supply), so a fiscal policy may cause a change in fixed-money price, but no change in varying-money price. Since what we have experienced historically is a transitive plus non-innovative economy, we now apply fiscal and monetary policies to this type of economy. Similar analyses can be applied to an innovative economy.

We start with an expansionary fiscal policy. The policy will cause an increase in aggregate demand and thus a right shift of the AD curve. When this shift happens in the transitional region, e.g. a shift of $AD_1$ as shown in Fig. 4.34, the short-run effect is a movement from point O to point A, which causes an increase in both price level and output level. The large rise in price is due to the fixed firm size and thus non-optimal production in the short run. As the firm size increases in the long run, the price level falls back and the output continues to increase to reach

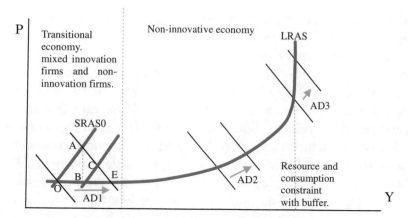

**Fig. 4.34** The effect of an expansionary fiscal policy

a long-run equilibrium at point E. For a transitional economy, the AS curve is horizontal because there is no pressure on resources (or, alternatively, the price effects on innovative and non-innovative firms cancel each other out), so the long-run effect of an expansionary fiscal policy is a large increase in output with little change in price level—a desired effect envisioned by Keynes. However, when the policy is used excessively and thus works on the region of the non-innovative economy, e.g. a shift of $AD_2$, the long-run effect will be an increase in both output level and price level. When the economy reaches the buffer zone of the resource and consumption limit, the long-run effect of an expansionary policy (e.g. the shift of $AD_3$) will lead to a pure price level with no increase in output level.

In our AS/AD analyses, so far we have not mentioned money supply. This omission implies that money supply is unchanged for the foregone analyses and thus the change in price level refers to the change in fixed-money price level, which is in the money price and also in real terms. However, one may still have some doubts on the claim that any change in price level is in real terms as long as money supply is fixed.

One doubt is that, if the money supply is fixed, how can price increases be associated with output increases (e.g. the shift shown as $AD_2$ in Fig. 4.34)? If we treat money as a type of goods to trade for the aggregate good in the economy, given a fixed amount of money, an increase in the quantity of the aggregate good surely will decrease the price of aggregate goods. This puzzle hinges on the circulation velocity of money. Based on the quantity theory of money, $MV = PY$, if $M$ and $V$ are both fixed, an increase in output level $Y$ necessitates a decrease in price $P$. However, the velocity $V$ is not necessarily always fixed. If demand is high, people spend money faster and thus $V$ will increase. If $V$ increases more than $Y$, the price level can increase in the face of an increase in $Y$.

The other doubt is that, when the aggregate demand increases within the buffer zone of the resource constraint (e.g. a shift shown as $AD_3$ in Fig. 4.32), the long-run outcome is an increase in the price level with a fixed output level. One might tend to conclude that this change in the price level surely should be a nominal change because the output is the same. This reasoning confuses the standards for value judgements. When one says that the output is the same so the value is the same, one implies that the standard of value judgement is the output itself. However, the value judgement in the AS/AD model is price, which is based on the value of money (how much money per unit of output). With a fixed amount of money and a fixed amount of output, the price level can increase if the demand is high (as explained, this is due to the increase in circulation velocity). This increase in the price level indicates the force of demand if the money supply is fixed. Because the standard of value judgement has not changed thanks to the fixed money supply, so the increase in price level must be in real terms. A more concrete example may drive the point home. With a fixed number of houses and fixed money supply, housing prices may increase substantially due to a strong demand. This rise in housing prices indicates strong demand and thus indicates an increase in the real value of housing, so the price change is in real terms as long as the money supply is fixed.

Next, we turn to monetary policy shown in Fig. 4.35.

Unless in the case of liquidity trap, an increase in money supply will immediately cause a right shift of the AD curve (e.g. a shift of $AD_1$ shown in Fig. 4.35) because, with more money in hand, people will spend more. However, money supply has no impact on the short-run aggregate supply curve. This is neither because of the money illusion proposed by the monetarists nor because of the sticky price proposed by new Keynesian economists. These two reasons may be valid but are based

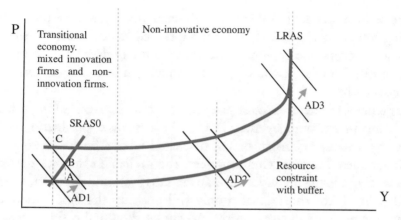

**Fig. 4.35**  The effect of an expansionary monetary policy

on special cases such as information asymmetry and/or imperfect competition. There is a general reason for the unchanged aggregate supply in the short run: the change of price system caused by monetary shock takes time. Like any changes in regulations or standards, the change requires adjustment time. With a change in money supply, the prices of goods will change. However, this change will be completed only after a certain period of adjustment time, so monetary policy will not affect the AS curve in the short run. As a result, an expansionary monetary policy has a short-run effect similar to that of an expansionary fiscal policy: an increase in both price level and output level at point B in Fig. 4.35.

However, the long-run effect is quite different. After the benchmark adjustment is completed, the prices of all goods are inflated roughly by the same percentage, so production costs will increase to the same degree as output prices. This causes an upshift of long-run AS to the same degree as the shift of the AD curve. As a result, the long-run equilibrium is at point C, with an increase in the price level but no change in the output level.

## 4. Supply-side shocks and stagflation

One purpose of moving from the IS/LM model to the AS/AD model is to address supply-side shocks. However, the explanation of supply-side shock in a standard AS/AD model relies on a jump of the vertical long-run AS curve, and this is not supported by the reality of trended long-run economic growth. With an AS and AD derived from microeconomics foundation, the supply-side shock can be explained easily and consistently.

Figure 4.36 demonstrates the effect of a negative supply shock, e.g. an oil embargo by the OPEC, the destroyed production facilities by a natural disaster, or the decreased labour force due to a war. Under these kinds of circumstances, the available resources decrease, so both the long-run and short-run AS curves should shift left. On the other hand, the shortage of supply will push up the input prices and increase production costs, so the AS curves should also shift upwards. As such, both the short-run and long-run AS should shift to the top left, i.e. $LRAS_0$ shifts to $LRAS_1$, $SRAS_0$ shifts to $SRAS_1$. Given a downward-sloping AD curve, both long-run and short-run equilibrium will shift to the top left (e.g. from point A to B, from C to D, or from E to F), and thus, the price level increases while the output level decreases. This result holds for all types of economies (innovative, transitional or non-innovative) and/or for all regions of an AS curve (non-inflationary, inflationary or within the buffer zone of resource or consumption limit).

The result of a negative supply shock—a decrease in output level accompanied by an increase in price level—is named stagflation.

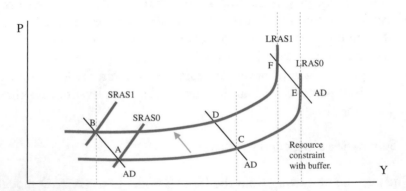

**Fig. 4.36**  The effect of a negative supply-side shock

This happened during the 1970s. There are a lot of explanations for this phenomenon, but the most influential ones are oil price shock (a supply shock) and inflationary monetary policy shock (a demand shock). So far, we see that a negative supply shock can definitely cause stagflation. Can an expansionary monetary policy also cause stagflation? From Fig. 4.35, we see that an expansionary monetary policy will cause an increase in output in the short run because the short-run AS curve is unaffected by the monetary policy while the policy shifts the AD curve to the right. However, under the expansionary monetary policy, the output in the long run is unchanged despite the increase in demand and in price level, because of the long-run AS and AD shift by the same degree. This could be stagflation if the term 'stagnation' is defined as zero economic growth accompanied by high inflation. On the other hand, if the term is defined as a recession (negative growth for a long period of time) accompanied by high inflation, an expansionary monetary policy cannot cause a stagflation (sequential shifts of AD and AS may cause negative growth in a short period, but this negative growth would not last long enough to be called a recession).

Another type of supply shock can be in the form of an exogenous increase in the nominal wage due to an increase in trade union militancy or due to an increase in statutory minimum wages. If an increase in nominal wages is accompanied by an increase in money supply, this leads to a pure nominal wage increase (i.e. real wage is unchanged). In this case, all curves (long-run AS, short-run AS, and AD) will shift upwards and the price level will increase, leaving output unchanged.

If there is no change in money supply, the statutory change leads to an increase in real wages, so production costs increase and both long-run AS and short-run AS shift up. Although the total labour force is unchanged, employable labour decreases due to the increased wage floor, so the actual resource limit decreases and thus the AS curve also shifts to the left. Meanwhile, the AD curve may not shift because the income effect of the increased wage rate may cancel out the effect of increased unemployment. As a result, the situation is similar to the oil price shock.

## 4.9   An Overall Assessment

The foregone review of different schools of economic thought has made it clear that each school reveals some elements of truth but each has failed to provide a complete picture. Just as the parable of the six blind men and an elephant illustrated, each school involves heated arguments with others but fails to reach a consensus.

Classical and neoclassical economists have extracted economic variables from complex social and cultural backgrounds and constructed abstract theories and models. Their efforts have helped us gain a deeper understanding of how an economic system works. However, classical and neoclassical economists have paid relatively limited attention to the conditions and limitations of their theories. When the results from their theories or models have been criticized by others or when the results have contradicted reality, they have tended to downplay the importance of contradiction and have ignored such criticisms completely. On the other hand, while heterodox economists have pointed out the shortcomings of the classical approach, they have often failed to see the valid elements in classical theory and thus have tended to discard classical theories completely. The neoclassical synthesis was an excellent effort to unite neoclassical and Keynesian economics. However, the synthesis failed because of numerous inconsistencies due to the contradictory assumptions used in different theoretical systems and, more importantly, due to misunderstandings and confusions regarding the settings of a macroeconomic model.

A more useful approach is to identify the valid points and limitations of all schools of economic thought and thus to form a new theory by absorbing those valid points and by overcoming the limitations of existing theories. Although it is unlikely for all economists to achieve consensus on what are the valid points and what are the limitations of each school of economic thought, the author provides some judgements to share with the reader.

The invisible hand and the equilibrium theory proposed by classical economists have laid groundwork for our understanding of an economic system. The assumption of competitive markets initiated by Adam Smith has captured the essence of the market mechanism and greatly simplified the complex economy, so it has become the cornerstone of the discipline of economics. However, the invisible hand works only when a necessary legal framework is laid. Classical economists have paid little attention to non-economic factors such as institutions. On this front, the thoughts of institutionalists complement classical economics. Social reforms proposed by Commons, for example, regulation of public utilities, workers' compensation, minimum wage laws and unemployment compensation laws, had a profound impact not only on social life but also on economic analysis. Another important contribution of institutionalists to classical economics is their appreciation of the role of technology, which was not initially considered an economic factor. After Schumpeter demonstrated the importance of innovation to economic growth, the role of technological progress finally was included in the Solow/Swan growth model. Technology is now regarded as a key variable in economics.

Underconsumptionists acutely identified and perseveringly insisted that general gluts or underconsumption caused economic recession. They explained the possibility of general gluts but did not come up with a theory, so they did not convince the classical economists of their point of view. More importantly, underconsumptionists could not solve the contradiction of the phenomenon of underconsumption in an economy with the unlimited desire of human beings for goods and services. Thus, they failed to see that underconsumption is caused by the slow pace of inventing new goods. Pasinetti (1981) and others did see the link between underconsumption and scarcity of invention, but he did not go far enough to uncover the cause of invention scarcity.

Keynesian economists identified the cause of underconsumption as a deficiency of investment due to liquidity preference as well as the lack of animal spirit in an uncertain world. However, they were satisfied with uncertainty as the reason for fluctuation of investment and did not go deeply enough to reveal the dependence of investments on expected future consumption. By combining the multiplier model and investment accelerator model, Hick, Harrod and Samuelson developed a business cycle model, which linked investment to expected future consumption in their rationale, but the future consumption in their model is presented by past consumption growth. In doing so, they assumed away the function of future consumption and failed to uncover the determinants of investment. New Keynesian economists added microeconomics foundations to Keynesian economics, but the foundations were based on market imperfection or even trivial cases such as menu costs, sticky wages and sticky prices. Consequently, their explanation only accounted for a small part of underconsumption problems and failed to reveal the root cause of economic recessions. The solution proposed by Keynesian economists, such as government intervention through fiscal and monetary policies, was only a temporary fix to the market problem and also caused considerable side effects such as stagflation.

Other schools of economics thought also have valid points and shortcomings. The monetarists and new classical economists realized that monetary shocks accompanied by adaptive expectations, or unforeseen shocks under rational expectations, may cause business cycles, but their theories are inadequate in explaining all business cycles. The Austria school emphasized economic dynamics but their theory ignored the role of prices—a key element in an economic system. Marxists argued that the contradiction in the system led to the end of capitalism, while institutionalists validly pointed out that institutions or proper legal framework played a vital role in economic phenomenon. However, both Marxists and institutionalists rejected the role of a market mechanism. They either proposed a planned economy as the replacement for capitalism or shed no light on the type of appropriate institutions for the future.

Based on the above discussion, a new theory must use the classical framework but also be able to incorporate and advance the valid points of other schools of economic thought. For example, the paradox of sale stagnation and unlimited human desire to consume points to the importance of new products. Although our desire to consume is unlimited, the desire to consume any type of goods is limited (e.g. one cannot consume unlimited amount of ice cream in one day). The limited number of types of goods may lead to underconsumption even if our overall desire is unlimited, so creation of new products is the key to avoiding underconsumption. The other example of the need to absorb and advance current schools of economic thought is about the future institutional structure. From different perspectives, Marxists, socialists and institutionalists all have demonstrated that the current form of capitalist institution has a number of problems and thus needs to be replaced or modified. However, most institutionalists have failed to put forward a solution for future institutions. Marxists, socialists and some institutionalists have come up with solutions, but their proposed future institutions all involve a planned economy, which is a rejection of the already proven efficiency of a market mechanism. More work must be done to find a solution which satisfies all proven valid aspects of existing economic thought.

To sum up, the long-standing arguments between different schools of economic thought indicate that all sides must have some elements of truth as well as limitations. The solution to ending these arguments is to build a new theory which includes the valid aspects of all sides. The next chapter is an effort in this direction.

# Appendix 1 (for Section 4.8.3.2): The Ways to Obtain the AD Curve

Due to the difficulties in aggregating market demand curves, methods other than direct aggregation are used to obtain the aggregate demand curve, including methods utilizing the AE/AP model, the IS/LM model and the quantity theory of money. However, these methods of obtaining AD curves resulted in much criticism. We describe and assess the four existing methods first and then introduce four new ways of obtaining AD curves by aggregating market demand curves.

1. Obtaining the AD curve from the AE/AP model

The difficulty in obtaining an AD curve from the AE/AP model is that there is no explicit price variable in the model. A simple AE/AP model can be expressed as:

$Y = I_0 + C_0 + c * Y$, or $(1 - c)Y = I_0 + C_0$, where $Y$ is income, $I_0$ and $C_0$ are autonomous investment and consumption, respectively, and $c$ is a parameter indicating the propensity to consume.

To derive the AD function, we can regard the variables in the model as the function of price because price affects $Y$, $I_0$ and $C_0$. As such, the AD/AE model can be written as:

$$(1 - c)Y(P) = I_0(P) + C_0(P).$$

To gauge the impact of a change in price $P$ on income $Y$, we differentiate the above equation to obtain:

$$(1 - c)dY(P)/dp = dI_0(P)/dP + dC_0(P)/dP.$$

Once we have measured the impact of a change in price on autonomous investment/consumption ($dI_0$ and $dC_0$), we can use the above equation to obtain the impact on total expenditure/income ($dY$). This can also be shown by employing a graph of multiplier models. As the left panel of Fig. 4.37 shows, when there is a price hike from $P_0$ to $P_1$, autonomous consumption plus autonomous investment decreases from A to B. The total expenditure decreases from $Y_0$ to $Y_1$. We have obtained two points for the AD curve: $P_0$ associated with $Y_0$ and $P_1$ associated with $Y_1$. We can allow more price changes and obtain more data, e.g. $P_2$ associated with $Y_2$ at point C, etc. Collecting all the data, we can draw an AD curve.

This approach is criticized for being inconsistent with the general definition of demand curves and for including the supply-side impact. Colander (1995) used a precocious student as a prop to drive his point home. The student first measured that a price decrease by 2 (e.g. $P_2 - P_1 = 2$) increased the output by 2 (e.g. the size of HG in Fig. 4.37), thus he obtained a slope of AD as $-1$. Since this increase in output is due to an increase in autonomous consumption/investment, the output increase is not affected by income. Then, the student took care of the multiplier effect and measured the increase of total output by 6 (e.g. the

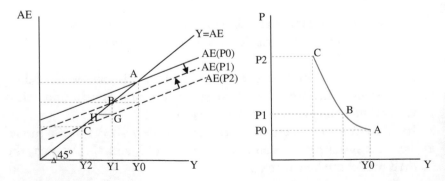

**Fig. 4.37** Deriving the AD curve from the multiplier model

size of $Y_1 - Y_2$ in Fig. 4.37) when the price decreased by 2, so he obtained a slope of AD of $-3$. The student was perplexed: Which slope is correct? Colander concluded that the AD curve of the slope of $-1$ was consistent with the normal definition of a demand curve—the demand change is caused by the price change alone, other things being equal. Following the same reasoning, he thought the AD curve of the slope of $-3$ was not consistent with the normal definition of a demand curve because the AD curve was obtained from the AE/AP model and thus included the income effect or the dynamic multiplier effect. He further argued that the multiplier effect was caused by the interaction between the supply and demand sides, so the AD curve also embodied the supply-side information.

The criticism of inconsistency in the definition of the demand curve largely hinged on the assumption of 'other things being equal'. When we use this phrase, we mean in practice that 'other **relevant** things are equal'. For example, what happened to firms' management structure is irrelevant to the tastes of consumers and thus irrelevant to the demand curve (one may disagree. Later this is to be discussed further), so a change in firms' production behaviours will not affect the demand curve. Because income affects demand, it is indeed problematic to allow income change when deriving a demand curve from an AE/AP model. However, in deriving the AD curve from the AE/AP model, income is viewed as a function of price. In other words, we did not allow income to change exogenously. As such, the income change in the model is a partial effect of price change, so it must be included as part of the price-induced demand. If one excluded this part, like the student did in his first calculation, one has only included the autonomous change in consumption/investment, so the calculation is incorrect because it did not include the full impact of the price change on the demand. Needless to say that, due to the nature of the Keynesian consumption function, the AD curve derived from the AE/AP model is not exactly the same as the way we derived Marshallian (with fixed income) or compensated (keeping consumer's utility constant) market demands, but the reasoning is similar and the derivation is rigorous. Since we can allow different types of demand curves such as the Marshallian and the compensated demand curve, we should also allow a Keynesian demand curve.

The criticism that the AD curve derived from the AE/AP curve contained supply-side information (some people even claim that the derived curve is also an aggregated supply curve) is an unfortunate mis-understanding due to the use of words 'aggregate output' or 'aggregate income'. These two words indeed give the reader the impression that we are dealing with the supply side. However, there is no active supply side in the AE/AP model. Because Keynes assumed excess supply capac-ity in his theory, the supply side just passively satisfied any requirements of the demand side. In the AE/AP model, the 45° line makes aggregate income/output equal to planned aggregate expenditure. This equality means that, even if we are talking about aggregate output or income, effectively we mean planned aggregate expenditure. Although we can use the words 'aggregate income/output' to connect the model to the passive or latent supply side, we must not be fooled by these words: all informa-tion in the AE/AP model comes from the demand side because the sup-ply is purely passive or accommodating up to full-employment output.

## 2. Obtaining the AD curve from the quantity theory of money

The quantity theory of money states that $M = kPY$ or $MV = PY$. If we fix the money supply $M$, we have naturally obtained an inverse relationship between price $P$ and output $Y$ and thus produced a down-ward-sloping curve. The question is: Does this inverse relationship represent a demand curve? Apparently, not all negative relationships between $Y$ and $P$ are a demand function.

The quantity theory of money is essentially an equation indicating that money supply equals money demand. When we fix the money supply $M$, it becomes exogenous, so the theory is about money demand. Except that a small part of money demand is for speculative purposes as well as for money hoarding due to uncertainty, the majority part of the money demand is for transaction. In other words, demand for money is essen-tially for the purpose of purchasing goods, so the demand for money by and large reflects goods demand. As a result, although the quantity the-ory of money is about money demand and money supply, the equation indirectly reflects people's demand for goods, so the negative relationship derived from the theory can indeed be viewed as an AD curve.

The other criticism about this approach is that the AD curve derived from the quantity theory of money is not related to Keynesian economics and thus it contributes little to neoclassical synthesis. The response to this criticism is that any new theory only has a few unique features so it must use a large body of existing knowledge. In order to reflect Keynesian theory, we do not need to derive everything from its key features. Although Keynes did not agree that the velocity $V$ in the quantity theory of money is constant and reject the claim that increase in money supply would cause price hike in the short run, he did accept this theory in general by saying that 'This Theory is fundamental. Its correspondence with fact is not open to question' (Keynes 1923, p. 81). Hence, Keynes would not object to deriving the AD curve from the quantity theory of money. Moreover, the vast majority of economists, including Keynes, accept a downward-sloping demand curve, so the derived AD curve would be consistent with most economic theories, including Keynesian theories.

## 3. Obtaining the AD curve from the IS/LM model

There is no price variable in the IS/LM model, but the price variable is implicitly included in the real money supply $M/P$ embodied in the LM curve. This embedded price provides a way to derive the AD curve.

The IS curve can be expressed as

$$Y = c(Y - T) + G_0 + C_0 + I_0 - b * r \qquad (4.15)$$

Where $Y$ stands for income, $T$ for taxes, $r$ for interest rate, $c$ for propensity to consume, $b$ is a parameter, while $G_0$, $C_0$ and $I_0$ are autonomous level (not affected by income level) of government spending, consumption and investment, respectively.

The LM curve indicates that the real money demand must be met by the real money supply $M/P$:

$$M/P = kY - hr + A_0, \qquad (4.16)$$

Where $M$ is money supply, $P$ is price level, $A_0$ is the base level of real money supply. In an IS/LM model, that the price $P$ in Eq. (4.16) is fixed, so one can obtain the income $Y$ and the corresponding interest rate which balance money supply and demand. If we relax the assumption of a fixed price and use Eqs. (4.15) and (4.16) to eliminate the interest rate $r$, we can obtain an equation about $Y$ and $P$:

$$(1 - c - bk/h)Y = G_0 + C_0 + I_0 - cT - Mbh^{-1}P^{-1} + A_0bh^{-1} \qquad (4.17)$$

Does this equation represent the demand function? The answer is positive. Actually, the LM curve indicated by Eq. (4.16) is an improved version of the quantity theory of money: Eq. (4.16) can be obtained by adding an extra term '$-hr + A_0$' to the equation for quantity theory of money. Applying the same reasoning used in deriving the AD curve from the quantity theory of money, we can be confident that Eq. (4.17) is an AD curve.

Graphically, the AD curve can be obtained by shifting the LM curve along the IS curve (shown in Fig. 4.38). Suppose that the initial equilibrium is at $A$ with price $P_0$, interest rate $r_0$ and income level of $Y_0$. If the price decreases to $P_1$ while the total nominal money supply is fixed, the real money supply increases so the LM curve shifts to the right and produces a new equilibrium at $B$ with a lower price $P_1$, lower interest rate $r_1$, and higher income level $Y_1$. Continuing to shift the LM curve to the right in response to a price drop, we can obtain a new equilibrium point C, and so on. Collecting all data about $P$ and $Y$, we can construct an AD curve that is shown in the right panel.

So far, the aggregate demand curve—Eq. (4.17)—is derived under the condition that the nominal money supply M is unchanged, so the price level for the AD curve is a fixed-money price and thus a real term. If money supply M changes, the AD curve will shift up or down and the associated price change is a pure nominal phenomenon.

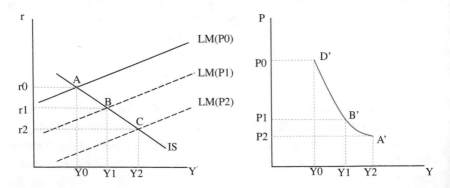

**Fig. 4.38**  Deriving the AD curve from the IS/LM model

One criticism regarding deriving an AD curve from the IS/LM model is that the derived output for the AD curve is not an output demanded but an equilibrium output at the intersection of IS and LM. Thus, the derived AD curve is both an AD and AS curve because it includes not only the demand-side information but also the supply-side information (Moseley 2010). This argument looks reasonable because the points on AD are obtained through the equilibria in the IS/LM model, but the argument results from two misunderstandings which need rectifying. First, any point on any demand curve is an equilibrium point at a given supply. The reason is that we must use different levels of supply to reveal the consumer's willingness to pay in order to construct a demand curve. This necessitates equality of supply and demand at each price the consumer is willing to pay. The way to construct an AD curve by shifting the LM curve is similar to the way of identifying a market demand curve by changing the level of supply. As a result, the obtained AD curve is comparable with the ordinary demand curve. No demand curve contains supply-side information, because a change of supply level or shifting of the LM curve is used only as a tool to reveal demand. Second, there is no active supply side in the IS/LM model. Because of the assumption of excess supply capacity in Keynesian models, the supply side is passively accommodating the change on the demand side. As such, there is no supply-side information in the IS/LM model nor in the derived AD curve.

The other criticism is that the AD curve derived from the IS/LM model is valid only when goods are in excess supply (Barro 1994). The reasoning is that, if the price falls below the market-clearing price, there will be a shortage of goods in the market and thus there will be no equilibrium point in the IS/LM model and thus no data to construct an AD curve. This reasoning is correct but it is an unnecessary concern. One reason is that there is always an excess supply capacity in the IS/LM model due to the Keynesian assumption of excess supply capacity. Thus, if one derives an AD curve from an IS/LM model, the supply capacity is not a concern. The other reason is that while supply capacity is an important tool in deriving a demand curve, a demand curve is not constrained by the capacity of supply. The current supply capacity may be limited but it may increase tomorrow or in the near future, and then we can add the new data to the demand curve. In the end, a demand curve indicates a consumer's tastes and responses to price or quantity change, so it will not be affected by supply-side activities.

The arguments of Barro (1994) and Moseley (2010) have an ideological source from Patinkin (1965), who regarded the AD curve as an equilibrium curve because it is derived from the equilibrium points in the IS/LM model. This idea is actually a misunderstanding of the demand curve. This misunderstanding led to the wrongful claim of interdependency between the demand curve and the supply curve put forward by Rowan (1975), Corden (1978) and Field and Hart (1990). Following Patinkin (1965), Field and Hart (1990) argued that a point on a demand curve indicates an equilibrium between supply and demand so the supply-side is embedded in the demand curve. They further argued that the aggregate demand curve derived under Keynesian assumption implied a supply response of firms that is inconsistent with modern supply theory. Here, we expose the flaws in their arguments.

It is true that every point on a demand curve is an equilibrium between supply and demand, but the points on the demand curve indicate the responses of demand to different supply conditions, rather than the responses of supply (or firms) claimed by Field and Hart (1990). Here, they made a logical mistake. The shift of a supply curve is an instrument to test the response of demand, but Field and Hart (1990) mistook it for a response of the firm's supply to demand. All the supply-side information needed for deriving a demand curve is the availability of supply quantity or price, so the response of supply or the behaviour of suppliers is totally irrelevant to a demand curve. If one considers the response of supply (or firms), the centre of concern is the supply curve (e.g. how to construct a supply curve), not the demand curve, so the charge Field and Hart (1990) laid on the AD curve derived from the IS/LM model is logically flawed. The irrelevance of the supply response in revealing a demand curve can be shown in Fig. 4.39.

Three types of supply curves are shown in Fig. 4.39 to reveal a demand curve. The left panel is a typical Keynesian supply curve—the horizontal supply curve shows that the price is fixed or that there

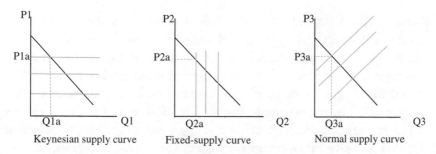

**Fig. 4.39** Using supply curves to reveal demand

is an excessive or unlimited supply at a given price. With this supply curve, the consumer faces a given price and decides how much to purchase, other things being equal. This is exactly what demand means. The middle panel shows the case of a fixed supply. With this supply curve, the consumer faces a given quantity and decides how much they should pay. This reveals the consumer's willingness to pay, which is an alternative way to define demand. The right panel shows a normal upward-sloping supply curve. With this supply, the consumer faces neither fixed price nor fixed quantity, but the intersection between supply and demand reveals how much the consumer wants to buy and at what price. This can fit in either definition of demand. In all three cases, the shift of a supply curve indicates the response of the consumer: buying less when the price is higher (or paying more when the quantity is less). If the interpretation of Field and Hart (1990) is valid that a shift of supply curve indicates a response of firms, not only the Keynesian supply curve but also all other supply curves are at odds with modern supply theory. In the case of the Keynesian supply curve, the firm produces less when the price is higher. In the case of a fixed-supply curve, the firm requires a higher price when the consumer demand is weaker. In the case of an upward-sloping supply curve, the firm either supplies less when the price is higher or asks for higher price when the demand is weaker. Apparently, it is not the slope of the supply curve but the logic of Field and Hart (1990) that is problematic.

Others argued the interdependency of aggregate demand and aggregate supply based on the complexity of an aggregate model.

Rowan (1975) argued that the price-induced change in aggregate supply will affect the aggregate demand through income, i.e. price affects aggregate supply, which in turn affects total output or income. A change in income will then affect aggregate demand. Corden (1978) argued that aggregate demand may be affected by autonomous or parametric shifts in the aggregate supply function. For example, a change in income distribution may affect the aggregate propensity to consume and may also affect investment through changed profit expectations.

It is obvious that an aggregate model for an economy involves many variables and feedback effects. However, in deriving any demand/supply curves, we must hold the condition of other things being equal. With this condition, we can exclude the impact on deriving the aggregate demand curve of the variables exampled by Corden (1978), and in the meantime, we can express the impact of any changes in these variables as a shift of the AD curve.

Although the reasoning of Rowan (1975) is irrefutable, namely a change in aggregate supply will cause a change in income and thus a change in aggregate demand, this statement does not lead us to the conclusion that a change in aggregate supply will affect the aggregate demand curve. Because of the condition of other things being equal when deriving a demand curve, the income level must be kept constant, so the income effect of a change in aggregate supply is excluded when one derives the AD curve. Here, the use of word 'income' may cause some confusion. When we say 'keep income level constant in order to derive an aggregate demand curve', we mean fixed-money income (just like the condition for deriving a Marshallian market demand curve). If one regards it as physical income or real output, one changes the value judgement standard from money value to output itself. This will contradict the meaning of price ($ per output) in the AD/AS model. As we will explain later (when we derive the AD curve from aggregating market demand curves) that, if the money supply and circulation velocity are unchanged, the money income of an economy is unchanged (a change in physical income affects the price level and thus leaves money income unchanged). As such, the aggregate demand curve will not be affected even if a change in aggregate supply affects the physical income and thus affects aggregate demand, because the changed aggregate demand provides another point on the AD curve. For example, an increase in physical income means a higher output level and a lower price level if the money supply and circulation velocity is fixed. This gives another point on the aggregate demand curve.

In short, it is the supply quantity or supply price that is important in revealing the demand of individuals, markets or the economy. The features of a supply response (e.g. the shape or slope of supply curves) are irrelevant, so a demand curve contains no information about a supply response. Although supply capacity is necessary for deriving a demand curve, it is not a component of a demand curve. In other words, sufficient supply capacity is implicitly assumed for a demand curve, so there is no need to restate it explicitly as a condition of a demand curve.

4. Examining alternatives to the AD curve—The DD curve and the HW curve

Based on the Keynes' concept of deficiency of effective demand, the effective demand or aggregate demand may fall short of or exceed total income, so the Eq. (4.15) for the IS curve should be written as

$$Y_D = c(Y - T) + G_0 + C_0 + I_0 - b * r \qquad (4.18)$$

where $Y_D$ indicates effective demand which may differ from real output or physical income $Y$.

From the money demand function—the Eq. (4.16)—we can solve for the interest rate $r$:

$$r = (kY + A_0 - M/P)/h \qquad (4.19)$$

Plugging Eq. (4.19) into Eq. (4.18), we have:

$$Y_D = (c - bk/h)Y - cT + G_0 + C_0 + I_0 - bA_0/h + bM/(Ph) \qquad (4.20)$$

Equation (4.20) is called the HW curve, which is derived by Henry and Woodfield (1985). The HW curve is clearly downward sloping because $Y_D$ and $P$ are in an inverse relationship. Henry and Woodfield have also explained that the HW curve also shows that the income level $Y$ positively affects demand $Y_D$. Everything looks plausible except that the HW curve may not fit into the definition of a demand curve.

To derive a demand curve, one must set the demand equal to the different levels of given supply, i.e. we must find the equilibrium points. As such, to qualify Eq. (4.20) as a demand curve, one must set $Y_D = Y$. This will give an AD curve shown in Eq. (4.17). Henry and Woodfield were also confused about income $Y$. They regarded it as the real income which affects aggregate demand. Based on our definition of different types of income, we know this income $Y$ is actually the physical income or the amount of output which is not a determinant of demand.

Rowan (1975) went further than Henry and Woodfield (1985) to derive the DD curve. With an assumption that the planned supply was fully realized, Rowan derived $Y = Y_S(P)$ based on a production function. This is essentially a positively sloping supply curve, so $Y$ is positively related to price level $P$. Plugging this into Eq. (4.20), he had:

$$Y_D = (c - bk/h)Y_S(P) - cT + G_0 + C_0 + I_0 - bA_0/h + bM/(Ph)$$
$$(4.21)$$

Equation (4.21) is Rowan's alternative aggregate demand curve, called DD curve (Rowan's derivation used a nominal interest rate equation in more general form, so his equation is not exactly the same as Eq. 4.21, but the relationship between $Y_D$ and $P$ is the same). Rowan further argued that the impact of $(c - bk/h)Y_S(P)$ might outweigh the impact of the bM/(hP), so his aggregate demand curve can be positively sloping. However, Rowan's interpretation was a mistake because he changed the concept of real income when he plugged $Y = Y_S(P)$ into the Eq. (4.20).

In Eq. 4.20, $Y$ stands for physical income or real output of an economy, which is equilibrium outcome determined by both aggregate supply and aggregate demand. When he derived the equation $Y = Y_S(P)$ from a production function, he implicitly assumed that $Y$ is determined by supply side only, so what he derived is not an equilibrium outcome and thus is not real output of the economy. Rather, what he derived is simply an aggregate supply function $Y_S = Y_S(P)$. This can easily be seen from their relationship with $P$. Aggregate supply $Y_S$ is positively related to $P$, but the real output of an economy does not have this positive relationship, so it is invalid to plug $Y_S = Y_S(P)$ into Eq. 4.20. By viewing $Y$ as $Y_S(P)$ through equation $Y = Y_S(P)$, Rowan excluded any role of aggregate demand $Y_D$. This is similar to the situation that there is no role of aggregate supply in Keynes's multiplier model. In other words, Rowan made $Y_D$ an accommodating term to take any value required by $Y_S$, and thus, $Y_D$ in Eq. (4.21) actually means $Y_S$. As a result, the DD curve was essentially an aggregate supply function, and the positive slope of the DD curve was of no surprise.

5. Obtaining an AD curve from the direct aggregation of the market demand

To be free from various scepticism and criticism about the AD curve, the best method is to obtain an AD curve by direct aggregation of market demands. Despite the difficulties shown previously, this section will demonstrate that, with some necessary restrictions, aggregation of market demands is not impossible.

We start with the practice of obtaining aggregate price indexes (e.g. CPI). These indexes are generally obtained through weighted average price. For example, the equilibrium outputs or demands for commodity 1 and 2 in a certain period is $Q_1$ and $Q_2$ (for simplicity, we temporarily assume only two commodities in the economy), the prices are $P_1$ and $P_2$, respectively. The value of commodities 1 and 2 and total outputs are $V_1$, $V_2$ and $V$, respectively. We have:

$$V_1 = P_1 * Q_1, \ V_2 = P_2 * Q_2, \ V = V_1 + V_2,$$

Using the value share of each commodity as the weighting, we can calculate the price and quantity of the aggregate output for the economy:

$$P = P_1 * V_1/V + P_2 * V_2/V, \ Q = V/P = V^2/(P_1 V_1 + P_2 V_2)$$

Through this type of aggregation, we can obtain a point on the AD curve. If we can measure the demands for goods 1 and 2 at different sets prices, e.g. at price set (a) we have price $P_{1a}$ and quantity demanded $Q_{1a}$ for good 1, and price $P_{2a}$ and quantity demanded $Q_{2a}$ for good 2; at price set (b) we have $P_{1b}$, $Q_{1b}$, $P_{2b}$ and $Q_{2b}$, using the same method as computing aggregate price indexes like CPI, one can calculate the aggregate price and demand at each set of prices and thus obtain an AD curve.

The key for this aggregation is to find the way to associate a point (e.g. a point determined by $P_{1b}$ and $Q_{1b}$) on the demand curve for good 1 with a point (e.g. a point determined by $P_{2b}$ and $Q_{2b}$) on the demand curve for good 2. In reality, this association is realized through the choices by households and firms in the economy. To aggregate market demands to an AD curve, we need to mimic the reality by applying restrictions on household demand decisions. This can be achieved in a number of ways.

a. Strictly following the conditions for deriving market demand

Since a market demand curve is derived by keeping constant the income level and the prices of other goods, all points on the demand curve for good 1 are associated with the same points on demand curves for other goods, e.g. both the points $(P_{1a}, Q_{1a})$ and $(P_{1b}, Q_{1b})$ on demand curve for good 1 are associated with a point $(P_{2a}, Q_{2a})$ on the demand curve for good 2 and with a point $(P_{3a}, Q_{3a})$ on the demand curve for good 3. To obtain an AD curve, we can choose a market demand curve, associate each point on this curve to the same set of points on other market demand curves, and aggregate prices and quantities as being shown in CPI aggregation. The obtained AD curve is similar to a right shift of the chosen market demand curve and thus will be downward sloping. However, the position of the AD curve can vary, depending on which market demand curve is chosen as the base for aggregation.

b. Applying multiple-stage budgeting

The two-stage budgeting method is popularly used in empirical studies on aggregate demand. In this approach, the researcher allocates some portion of the household budget to two different groups of commodities and thus obtains the quantity and prices for commodities in each group. For the details about composite goods and two-stage budgeting, readers are referred to Nicholson and Snyder (2017) and Blackorby et al. (1978). The method can be shown in a simple two-good case. If we know the market demand curves for two goods and the income spent on them, we have the following three equations:

$$Q_1 = f_1(P_1),$$
$$Q_2 = f_2(P_2).$$
$$P_1 * Q_1 + P_2 * Q_2 = I$$

These three equations include four variables, $P_1$, $Q_1$, $P_2$ and $Q_2$. If the price of one good (e.g. $P_1$) is given, the values for other variables and thus the aggregate price and quantity can be obtained. By assigning different values to the price of the selected good, one can obtain different aggregation points and thus an AD curve.

Figure 4.40 shows the procedure to aggregate market demands to obtain the aggregate demand for an economy. Panel (a): For a given price of good 1, e.g. $P_{1a}$, we can calculate $Q_{1a}$, $P_{1b}$ and $Q_{1b}$. Then, we can find the corresponding points on the market demand curve, e.g. $A_{1a}$ and $A_{2a}$. Panel (b): Although we cannot add $Q_{1a}$ and $Q_{2a}$ directly, we can calculate the value of each good and construct the graphs of the demand value at each price, so we can obtain the new points $A_{1a'}$ and $A_{2a'}$ ($A_{1a'}$ is transformed from point $A_{1a}$, and $A_{2a'}$ is transformed from point $A_{2a}$).

Panel (c): Adding the value of goods 1 and 2, we have the value of aggregate good $V_a$, which is equal to total expenditure or total income ($I$). Moreover, using the calculated values as weighting, we can calculate the aggregate price $P_a$ from $P_{1a}$ and $P_{2a}$. Using $V_a$ and $P_a$, we can produce a point $B_a$. Panel (d): Using $Q_a = V_a/P_a$ to obtain the aggregate quantity $Q_a$, we can have a point $C_a$ on the aggregate demand curve. Choosing another price $P_{1b}$, we can repeat the calculation and obtain another point $B_b$ (with the same value $V_a$ because of the fixed budget allocated) in panel (c) and another point $C_b$ on the aggregate demand curve in panel (d). Choosing a third price $P_{1c}$ and repeating this procedure, we can obtain more points and construct the aggregate demand curve.

At any sets of prices, the total value of spending on goods 1 and 2 is the same (equal to the income allocated). Since total spending is the same in each procedure, for a higher aggregate price $P_a$ (compared with $P_b$), the corresponding aggregate quantity $Q_a$ will be smaller (than $Q_b$). As such, the resulting AD curve will be downward sloping.

For the case of more than two goods, the aggregation can be achieved by repeated two-stage budgeting, i.e. multiple-stage budgeting. First, we aggregate in the way as shown in Fig. 4.40 the demand of any two

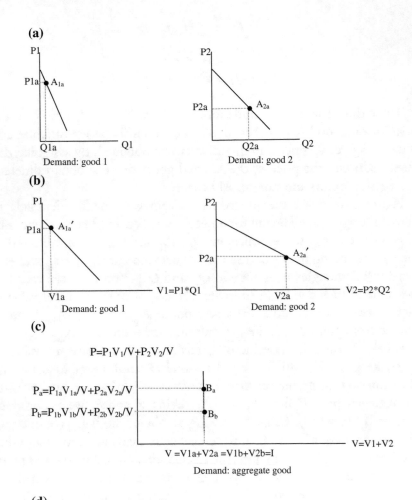

**(a)**

Demand: good 1

Demand: good 2

**(b)**

Demand: good 1     $V1=P1*Q1$

Demand: good 2     $V2=P2*Q2$

**(c)**

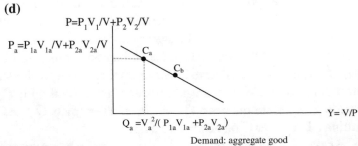

$P=P_1V_1/V+P_2V_2/V$

$P_a=P_{1a}V_{1a}/V+P_{2a}V_{2a}/V$

$P_b=P_{1b}V_{1b}/V+P_{2b}V_{2b}/V$

$V=V1a+V2a=V1b+V2b=I$

$V=V1+V2$

Demand: aggregate good

**(d)**

$P=P_1V_1/V+P_2V_2/V$

$P_a=P_{1a}V_{1a}/V+P_{2a}V_{2a}/V$

$Q_a=V_a^2/(P_{1a}V_{1a}+P_{2a}V_{2a})$

$Y=V/P$

Demand: aggregate good

**Fig. 4.40** Deriving the AD curve using two-stage budgeting

markets to obtain the market demand for the first-stage aggregate good. Second, we aggregate the first-stage aggregate market demand and a third market demand to obtain the market demand for the second-stage aggregate good. Third, we aggregate the second-stage aggregate market demand and a fourth market demand to obtain the market demand for the third-stage aggregate good. Repeating this procedure, we can finally obtain the aggregate demand for the economy. The result of this multiple-stage budgeting approach is similar to that of using a Cobb–Douglas or LES utility function, which allocates spending to each commodity in a proportional or linear fashion.

### c. Using the concept of composite goods

The concept of composite goods is also popularly used in commodity consumption aggregation. This concept is based on the substitution effect between commodities. For close substitutes (e.g. different types of food), their prices tend to move in a similar fashion so one can consider that there is a fixed price ratio among these commodities. In considering commodity prices purely from the demand perspective (e.g. ignore supply condition or assume similar supply functions), if consumer preference does not change but the overall demand increases, the price of all commodities will increase in a similar fashion. As such, all commodities can be viewed as a composite good and a fixed price ratio between different types of commodities can be assumed. Using this assumption, we can demonstrate in Fig. 4.41 the procedure to aggregate market demands to obtain the aggregate demand for an economy.

Panel (a) of Fig. 4.41: for historically determined prices of goods 1 and 2, e.g. $P_{1a}$ and $P_{2a}$, we can obtain from the market demand curves the quantity demanded $Q_{1a}$ and $Q_{2a}$, so we have two points $A_{1a}$ and $A_{2a}$. Panel (b): although we cannot add $Q_{1a}$ and $Q_{2a}$ directly, we can calculate the spending on each good $V_{1a} = P_{1a}Q_{1a}$ and $V_{2a} = P_{2a}Q_{2a}$, so we can obtain the new points $A_{1a'}$ and $A_{2a'}$ ($A_{1a'}$ is transformed from point $A_{1a}$ and $A_{2a'}$ is transformed from point $A_{2a}$). Panel (c): adding the value of goods 1 and 2, we have the value of aggregate good $V_a = V_{1a} + V_{2a}$. Using the calculated values $V_{1a}$ and $V_{2a}$ as weighting, we can calculate the aggregate price $P_a = (P_{1a}V_{1a} + P_{2a}V_{2a})/V_a$. Using $V_a$ and $P_a$, we can

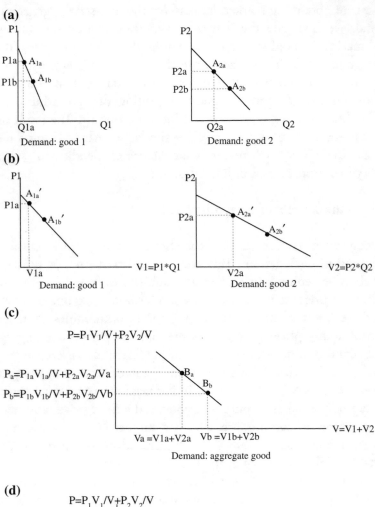

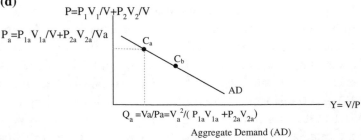

**Fig. 4.41** Deriving the AD curve using composite goods

produce a point $B_a$ in panel (c). Panel (d): the calculated $V_a$ and $P_a$ gives the aggregate quantity demanded $Q_a = V_a/P_a$, and thus the point $C_a$ on the AS/AD curve.

The above procedure is the same as conducting a GDP or CPI aggregation, where the prices and quantities $P_{1a}$, $P_{2a}$, $Q_{1a}$, $Q_{2a}$ are the realized results in history. For aggregation of market demand curves, we need more than one point.

Choosing a price $P_{1b}$ which is less than $P_{1a}$, the constant price ratio for composite goods necessitates that $P_{2b}$ is less than $P_{2a}$, so we have point $A_{1b}$ and $A_{2b}$ in panel (a). Assuming the demand for both goods 1 and 2 is elastic (we will discuss the other cases later), the spending on both goods will be greater than that when prices are $P_{1a}$ and $P_{2a}$, so we have points $A_{1b'}$ and $A_{2b'}$ in panel (b), which lead to the downward-sloping curves in panel (b). Since $P_{1b}$ and $P_{2b}$ are less than $P_{1a}$ and $P_{2a}$, respectively, the aggregated price $P_b$ is less than $P_a$. Meanwhile, the aggregated spending $V_b$ is greater than $V_a$ thanks to the assumption of elastic demand, so we have in panel (c) a point $B_b$ which is below and at the right of $A_a$. This produces a downward-sloping curve in panel (c). Consequently, we have point $C_b$ and a downward-sloping aggregate demand curve in panel (d). If we change the preset value of the price ratio for goods 1 and 2 and repeat the above procedure, we can obtain another AD curve, which can be viewed as a shift of the previous AD curve.

However, at the top part of the linear demand curves for goods 1 and 2, the price elasticity of demand tends to become very small. If the demand for good 1 and/or for good 2 becomes inelastic, the spending may decrease when the prices decrease. This would lead to positively sloping curves in panels (b) and (c), and thus an upward-sloping aggregate demand curve when the aggregate price is small.

Next, we show that, with demand curves of constant price elasticity for goods 1 and 2, the aggregate demand curve is a curve similar to market demand curves, so the aggregate demand curve is downward sloping.

Assume a constant-elasticity demand curve for goods 1 and 2: $Q_1 = a_1 * P_1^{-e1}$, $Q_2 = a_2 * P_2^{-e2}$, (elasticity $e_1 > 0$ and $e_2 > 0$) and a constant price ratio: $P_2 = \lambda P_1$, which $\lambda > 0$. We can calculate $V_1 = P_1 Q_1$, $V_2 = P_2 Q_2$, $V = V_1 + V_2$ and $P = (P_1 V_1 + P_2 V_2)/V$.

$$V = P_1 Q_1 + P_2 Q_2 = P_1 * a_1 * P_1^{-e1} + P_2 * a_2 * P_2^{-e2}$$
$$= P_1 * a_1 * P_1^{-e1} + \lambda P_1 * a_2 * (\lambda P_1)^{-e2}$$
$$= a_1 * P_1^{1-e1} + a_2 * \lambda^{1-e2} P_1^{1-e2}$$
$$= P_1(a_1 * P_1^{-e1} + a_2 * \lambda^{1-e2} P_1^{-e2})$$

$$P = (P_1 V_1 + P_2 V_2)/V = (P_1^2 Q_1 + P_2^2 Q_2)/V$$
$$= P_1^2(a_1 * P_1^{-e1} + a_2 * \lambda^{2-e2} P_1^{-e2})/V$$
$$= P_1(a_1 * P_1^{-e1} + a_2 * \lambda^{2-e2} P_1^{-e2})/(a_1 * P_1^{-e1} + a_2 * \lambda^{1-e2} P_1^{-e2})$$

$$Q = V/P = (a_1 * P_1^{-e1} + a_2 * \lambda^{1-e2} P_1^{-e2})^2/(a_1 * P_1^{-e1} + a_2 * \lambda^{2-e2} P_1^{-e2})$$

This expression is hard to interpret generally, but we can examine two special cases: $e_1 = e_2$ or $\lambda = 1$. If $\lambda = 1$, $P_1 = P_2 = P$, $Q = a_1 * P^{-e1} + a_2 * P^{-e2}$, so the aggregate demand curve is the horizontal aggregation of market demand curve, so it must be downward sloping. If $e_1 = e_2$, $P = P_1(a_1 + a_2 * \lambda^{2-e2})/(a_1 + a_2 * \lambda^{1-e2})$, $Q = P^{1-e1}(a_1 + a_2 * \lambda^{1-e2})^2/(a_1 + a_2 * \lambda^{2-e2})$. The aggregate price level is proportional to market prices and the aggregate demand curve is of the same elasticity as the market demand curves, so it must also be downward sloping. In a general case, it can be shown that the derivatives $dQ/dP_1 < 0$ and $dP_1/dP > 0$, so $Q$ and $P$ will be negatively related, i.e. the demand curve is downward sloping.

d. Utilizing a general equilibrium model

The general solution to obtain an AD curve is to use a general equilibrium model. The model can mimic the behaviour of consumers and firms through mathematic functions and thus obtain the market prices and equilibrium quantities under different supply conditions. The results of a general equilibrium model can be illustrated by the PPF curves on the supply side and indifference curves (IC), as shown in Fig. 4.42.

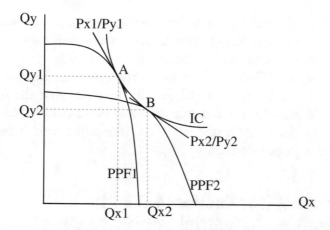

**Fig. 4.42** Obtaining the AD curve from a general equilibrium model

For a closed economy, the equilibrium outcome will be the point when PPF is tangent to an IC, e.g. point A. The tangent point A gives the quantity of goods demanded $Q_{x1}$ and $Q_{y1}$ and the price ratio $P_{x1}/P_{y1}$. Given that the exogenous money income is fixed (e.g. $I = P_{x1}Q_{x1} + P_{y1}Q_{y1}$), we can obtain the monetary price $P_{x1}$ and $P_{y1}$. With given value for $P_{x1}$, $Q_{x1}$, $P_{y1}$ and $Q_{y1}$, we can obtain a point on an AD curve. If the supply-side situation changes to a situation represented by $PPF_2$, we can obtain values for $P_{x2}$, $Q_{x2}$, $P_{y2}$ and $Q_{y2}$, so we can have another point on the AD curve. With changing conditions on the supply side, we can obtain more points and draw the AD curve.

The flexibility of a general equilibrium model means that the resulting AD curve can mimic any situation in reality. However, since many variables in the model can affect the equilibrium outcome, the position and slope of the AD curve depend on these variables and thus can vary considerably.

Different ways of direct aggregation here have their limit. Although the theory of two-stage budgeting and the concept of composite goods are popularly used in empirical research, they are criticized for using unrealistic assumptions. Fixing the demand for other good so as to satisfy the condition of deriving a demand curve is also not realistic. Using a general equilibrium model to derive aggregate demand curve can produce the most realistic result, but this approach is both time-consuming and highly data demanding. Nevertheless, by applying these methods we can demonstrate theoretically that it is possible to derive an AD curve from aggregating different types of market demands. This refutes, once and for all, the claim that the AD curve is totally different from the market demand curve (we refuted this earlier) and the claim that it is invalid to aggregate the market demand to obtain the aggregate demand.

## Appendix 2 (for Section 4.8.3.3): Aggregating Industrial Supply to Obtain a Short-run AS Curve

Based on historical observation on the prices of industrial outputs, we can set up a price ratio for industry 1 and industry 2 and demonstrate in Fig. 4.43 the way to obtain an AS curve by aggregating market supplies. Given the supply curves for these two and the price ratio, once we know one point on market supply of industry 1, e.g., quantity $A_{1a}$ and price $P_{1a}$, we can calculate for industry 2 the corresponding price and quantity, $P_{2a}$ and $Q_{2a}$. Using the prices and quantities $P_{1a}$, $P_{2a}$, $Q_{1a}$ and $Q_{2a}$ in panel (a), we can calculate the revenues $V_{1a} = P_{1a} * Q_{1a}$ and $V_{2a} = P_{2a} * Q_{2a}$, thus we can obtain points $A_{1a'}$ and $A_{2a'}$ in panel (b). Aggregating the total revenue $V_a = V_{1a} + V_{2a}$ and aggregate price $P_a = (P_{1a}V_{1a} + P_{2a}V_{2a})/V_a$, we can obtain the point $B_a$ in panel (c). Calculating aggregate quantities according to $Q_a = V_a/P_a$, we can obtain the point $C_a$ in panel (d).

Selecting another price $P_{1b}$ for industry 1 and using the price ratio set up according to historical observations, we can obtain $P_{1b}$, $P_{2b}$, $Q_{1b}$, $Q_{2b}$ and thus points $A_{1b}$ and $A_{2b}$ in panel (a). Since the industrial supply curve is normally upward sloping, if we set $P_{1b} > P_{1a}$, we have $P_{2b} > P_{2a}$, $Q_{2b} > Q_{2a}$ and $Q_{1b} > Q_{1a}$. Calculating revenues for industries 1 and 2 according to $V_{1b} = P_{1b} * Q_{1b}$ and $V_{2b} = P_{2b} * Q_{2b}$, we will have $V_{1b} > V_{1a}$ and $V_{2b} > V_{2a}$. This produces two points $A_{1b'}$ and $A_{2b'}$ and thus two upward-sloping curves in panel (b). Aggregating point $A_{1b'}$ and $A_{2b'}$ according to $V_b = V_{1b} + V_{2b}$ and $P_b = (P_{1b}V_{1b} + P_{2b}V_{2b})/V_b$,

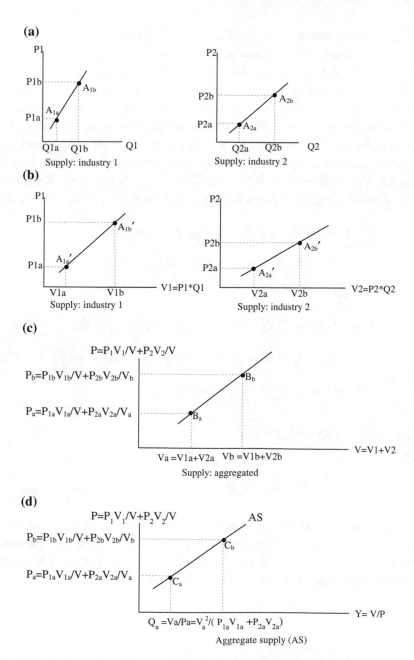

**Fig. 4.43** Deriving the AS curve from short-run industrial supplies

we can obtain the point $B_b$ and an upward-sloping curve in panel (c). Calculating aggregate quantity according to $Q_b = V_b/P_b$, we can obtain the point $C_b$ and an upward-sloping AS curve in panel (d).

It can be proved that, given upward-sloping industrial supply curves, the resulting AS curve is upward sloping. For simplicity, we suppress the intercepts and thus write the industrial supply curves as $Q_1 = a_1 P_1$ and $Q_2 = a_2 P_2$, which $a_1 > 0$ and $a_2 > 0$. The price ratio for industries 1 and 2 can be written as $P_2 = kP_1$, with $k > 0$. Revenue for industries 1 and 2 can be calculated as $V_1 = P_1 Q_1$ and $V_2 = P_1 Q_2$, respectively. Total revenue of two industries $(V)$ and the aggregate price $(P)$ can be written as:

$$V = P_1 Q_1 + P_2 Q_2 = P_1 a_1 P_1 + P_1 a_2 P_2$$
$$= a_1 * P_1^2 + a_2 * (kP_1)^2 = \left(a_1 + a_2 k^2\right) P_1^2.$$

$$P = (P_1 V_1 + P_2 V_2)/V = \left(P_1^2 Q_1 + P_2^2 Q_2\right)/V$$
$$= \left(P_1^2 a_1 P_1 + P_2^2 a_2 P_2\right)/V = \left(P_1^3 a_1 + k^3 P_1^3 a_2\right)/V$$
$$= P_1\left(a_1 + a_2 k^3\right)/\left(a_1 + a_2 k^2\right),$$

Consequently, So we have
Or,

$$Q = V_2/P = P\left(a_1 + a_2 k^2\right)^3/\left(a_1 + a_2 k^3\right), V = P^2\left(a_1 + a_2 k^2\right)^3/\left(a_1 + a_2 k^3\right)^2.$$
$$P_1 = P\left(a_1 + a_2 k^2\right)/\left(a_1 + a_2 k^3\right),$$

Since the parameters $a_1$, $a_2$ and $k$ are all positive, aggregate quantity $Q$ and aggregate price $P$ are positively related, so the resulting AS curve must be upward sloping. If we change the value for parameter $k$, we can obtain another AS curve of a different but still positive slope.

# References

Akerlof, G. A. (1982). Labor Contracts as Partial Gift Exchange. *Quarterly Journal of Economics, 97*, 543–569. https://doi.org/10.2307/1885099.

Akerlof, G. A., & Yellen, J. L. (1985). A Near-Rational Model of the Business Cycle, with Wage and Price Inertia. *Quarterly Journal of Economics, 50*(C), 823–838.

Bailey, M. N. (1974). Wages and Unemployment Under Uncertain Demand. *Review of Economic Studies, 41*(1), 37–50.

Ball, L. (1990). Insiders and Outsiders: A Review Essay. *Journal of Monetary Economics, 26*(3), 459–469.

Ball, L., & Romer, D. (1990). Real Rigidities and the Non-neutrality of Money. *Review of Economic Studies, 57*(2), 183–203.

Ball, L., Mankiw, N. G., & Romer, D. (1988). The New Keynesian Economics and the Output-Inflation Trade-off. *Brookings Papers on Economic Activity, 1988*(1), 1–82.

Barro, R. J. (1994). The Aggregate-Supply/Aggregate-Demand Model. *Eastern Economic Journal, 20*(1), 1–6.

Bernanke, B. S., & Gerlter, M. (1989). Agency Costs, Net Worth and Business Fluctuations. *American Economic Review, 79*(1), 14–31.

Blackorby, C., Primont, D., & Russell, R. (1978). *Duality, Separability, and Functional Structure: Theory and Economic Applications*. Elsevier North-Holland, Inc.

Blinder, A. S. (1991). Why Are Prices Sticky? Preliminary Results from an Interview Study. *American Economic Review, 81*(2), 89–96.

Blinder, A. S. (1994). On Sticky Prices: Academic Theories Meet the Real World. In N. G. Mankiw (Ed.), *Monetary Policy*. Chicago: University of Chicago Press.

Brunner, K. (1968). The Role of Money and Monetary Policy. *Federal Reserve Bank of St. Louis Review, 50*, 8–24.

Chalmers, T. (1808). *Enquiry into the Extent and Stability of National Resources*. Edinburgh: Printed for Oliphant and Brown.

Colander, D. (1995). The Stories We Tell: A Reconsideration of AS/AD Analysis. *Journal of Economics Perspectives, 9*(3), 169–188.

Commons, J. (1934). *Institutional Economics*. New York: Macmillan.

Corden, W. M. (1978). Keynes and the Others: Wage and Price Rigidities in Macro-Economic Models. *Oxford Economic Papers, 30*(2), 159–180.

Diamond, P. A. (1982). Aggregate Demand Management in Search Equilibrium. *Journal of Political Economy, 90*(5), 881–894.

Dunlop, J. G. (1938). The Movement of Real and Money Wage Rates. *Economic Journal, 48*(191), 413–434.

Fischer, S. (1977). Long-Term Contracts, Rational Expectations, and the Optimal Money Supply Rule. *Journal of Political Economy, 85*(1), 191–205.

Friedman, M. (1968). The Role of Monetary Policy. *The American Economic Review, 58*(1), 1–17.

Friedman, M., & Schwartz, A. J. (1963). *A Monetary History of the United States, 1867–1960*. Princeton: Princeton University Press.

Galbraith, J. K. (1952). *American Capitalism: The Concept of Countervailing Power*. New York: Houghton Mifflin.

Galbraith, J. K. (1958). *The Affluent Society*. Boston: Houghton Mifflin.

Galbraith, J. K. (1967). *The New Industrial State*. Boston: Houghton Mifflin.

Garrison, R. W. (2001). *Time and Money: The Macroeconomics of Capital Structure*. London: Routledge.

Glahe, F. R. (1977). *Macroeconomics: Theory and Policy* (2nd ed.). New York: Harcourt Brace Jovanovich.

Gordon, R. J. (1981). Output Fluctuations and Gradual Price Adjustment. *Journal of Economic Literature, 19*(2), 493–530.

Gordon, R. J. (1990). What is New-Keynesian Economics. *Journal of Economic Literature, 28*(3), 1115–1171.

Gordon, R. J. (2003). *Macroeconomics* (9th ed.). New York: Addison-Wesley.

Greenwald, B., & Stiglitz, J. (1993). New and Old Keynesians. *Journal of Economic Perspectives, 7*(1), 23–44.

Hall, R. (1978). Stochastic Implications of the Life Cycle-Permanent Income Hypothesis: Theory and Evidence. *Journal of Political Economy, 86*(6), 971–987.

Hansen, A. H. (1949). *Monetary Theory and Fiscal Policy*. New York: McGrawHill.

Hansen, A. H. (1953). *A Guide to Keynes*. New York: McGraw-Hill Book.

Hayek, F. (1935 [1967]). *Price and Production* (2nd ed.). New York: Augustus M. Kelley.

Heckscher, E. (1935). *Mercantilism*. London: Allen and Unwin.

Henry, K., & Woodfield, A. (1985). Aggregate Demand Curve in Macroeconomic Theory: Some Curiously Antipodean Controversies. *New Zealand Economic Papers, 19*, 21–34.

Hicks, J. R. (1937). Mr. Keynes and the 'Classics': A Suggested Interpretation. *Econometrica, 5*(2), 147–159.

Hobson, J. A., & Mummery, A. F. (1889). *The Physiology of Industry*. London: John Murray.

Horwitz, S. G. (2000). *Microfoundations and Macroeconomics: An Australian Perspective*. London: Routledge.

Kalecki, M. (1943). Political Aspects of Full Employment. *Political Quarterly, 14*(4), 322–331.

Keynes, J. (1923). *A Tract on Monetary Reform*. London: Macmillan.

Keynes, J. M. (1936). *The General Theory of Employment, Interest, and Money*. London: MacMillan.

Keynes, J. M. (1937 [1973]). The General Theory of Employment. *Quarterly Journal of Economics*. Reprinted in Moggridge, D. (Ed.), *The Collected Writings of John Maynard Keynes* (Vol. XIV). London: Macmillan.

Lauderdale, J. (1804). An Inquiry into the Nature and Origin of Public Wealth: And into the Means and Causes of Its Increase. Printed for Arch. Constable & Co.; T.N. Longman & O. Rees.

Littlechild, S. (1990). *Australian Economics*. Aldershot: Edward Elgar.

Lucas, R. (1972). Expectations and the Neutrality of Money. *Journal of Economic Theory, 4*(2), 103–124.

Malthus, T. R. (1820). *Principles of Political Economy*. Cambridge: Cambridge University Press.

Mandeville, B. (1714). *The Fable of the Bees, or, Private Vices, Public Benefits*. London: Printed for Edmund Parker.

Mankiw, N. G. (1985). Small Menu Costs and Large Business Cycles: A Macroeconomic Model of Monopoly. *Quarterly Journal of Economics, 100*(2), 529–553.

Mankiw, G. (2013). *Macroeconomics* (8th ed.). New York: Worth Publishers Inc.

Mankiw, N. G., & Romer, D. (1991). *New Keynesian Economics*. Cambridge, MA: MIT Press.

Marx, K. (1867). *Das Kapital Volume One: The Process of Production of Capital.*

Marx, K., & Engels, F. (1885). *Das Kapital Volume Two: The Process of Circulation of Capital.*

Marx, K., & Engels, F. (1894). *Das Kapital Volume Three: The Process of Capitalist Production as a Whole.*

McCulloch, J. (1856). *A Select Collection of Early English Tracts on Commerce*. London: Political Economy Club.

McCulloch, J. R. (1864 [1965]). *The Principles of Political Economy, with Some Inquiries Respecting Their Application* (5th ed.). New York: Augustus M. Kelley.

McDonald, I. M. (1992). *Macroeconomics*. New York: Wiley.

Menger, C. (1871 [1981]). *Principles of Economics*. New York: New York University Press.

Mill, J. (1808). *Commerce Defended*. London: C. and R. Baldwin.

Modigliani, F. (1944). Liquidity Preference and the Theory of Interest and Money. *Econometrica, 12*(1), 45–88.

Moseley, F. (2010). Criticisms of Aggregate Demand and Aggregate Supply and Mankiw's Presentation. *Review of Radical Political Economics, 42*(3), 308–314.

Muth, J. F. (1961). Rational Expectations and the Theory of Price Movements. *Econometrica, 29,* 315–335.

Myrdal, G. (1930). *The Political Element in the Development of Economic Theory*. Cambridge, MA: Harvard University Press.

Nicholson, W., & Snyder, C. (2017). *Microeconomic Theory: Basic Principles & Extensions* (12th ed.). Cengage Learning.

Owen, P. D. (1987). Aggregate Demand Curves in General-Equilibrium Macroeconomic Models: Comparisons with Partial-Equilibrium Microeconomic Demand Curves. *New Zealand Economic Papers, 21,* 97–104.

Parkin, M. (1986). The Output-Inflation Tradeoff When Prices Are Costly to Change. *Journal of Political Economy, 94*(1), 200–224.

Pasinetti, L. (1981). *Structural Change and Economic Growth*. Cambridge: Cambridge University Press.

Phelps, E. S. (1985). *Political Economy: An Introductory Text*. New York: W. W. Norton.

Phelps, E. S. (1990). *Seven Schools of Macroeconomic Thought*. Oxford: Oxford University Press.

Phillips, A. W. (1958). The Relation Between Unemployment and the Rate of Change of Money Wage Rates in the United Kingdom, 1861–1957. *Economica, 25*(100), 283–299.

Pigou, A. C. (1941). *Employment and Equilibrium: A Theoretical Discussion*. London: Macmillan.

Pigou, A. C. (1943). The Classical Stationary State. *Economic Journal, 53,* 343–351.

Pigou, A. C. (1947). Economic Progress in a Stable Environment. *Economica, NS. 14*(55), 180–188.

Rabin, A., & Birch, D. (1982). A Clarification of the IS Curve and the Aggregate Demand Curve. *Journal of Macroeconomics, 4*(2), 233–238.

Rao, B. B. (1991). What is the Matter with Aggregate Demand and Aggregate Supply? *Australian Economic Papers, 30*(57), 264–277.

Ricardo, D. (1817 [1951–1972]). *On the Principles of Political Economy and Taxation, Vol. 1 of The Works and Correspondence of David Ricardo* (P. Sraffa, Ed.). Cambridge: Cambridge University Press.

Ricardo, D. (1820 [1951–1972]). *Notes on Malthus, Vol. 2 of The Works and Correspondence of David Ricardo* (P. Sraffa, Ed.). Cambridge: Cambridge University Press.

Robbins, L. (1934 [1971]). *The Great Depression.* Freeport, NY: Books for Libraries Press.

Rothbard, M. (1963). *America's Great Depression.* Los Angeles: Nash Publishing.

Rowan, D. C. (1975). *Output, Inflation and Growth: An Introduction to Macro-Economics* (Australian Edition). South Melbourne: Macmillian.

Salop, S. C. (1979). A Model of the Natural Rate of Unemployment. *American Economic Review, 69*(1), 117–125.

Samuelson, P. A. (1948). *Economics.* New York: McGraw-Hill.

Say, J.-B. (1803). *Traite d'Economie Politique* (1st ed.). Paris: Deterville.

Schumpter, J. A. (1934). *The Theory of Economic Development.* Cambridge: Harvard University Press.

Shapiro, C., & Stiglitz, J. (1984). Equilibrium Unemployment as a Worker Discipline Device. *American Economic Review, 74*(3), 433–444.

Sismondi, J. C. L. (1819). *Nouveaus Principes d'Economie Politique* [New Principles of Political Economy]. Paris: Delaunay.

Smith, Adam. (1776 [1904]). *An Inquiry into the Nature and Causes of the Wealth of Nations* (E. Cannan, Ed.). London: Methuen.

Snowdon, B., & Vane, H. (2005). *Modern Macroeconomics: Its Origin, Development and Current State.* Northampton, MA: Edward Elgar.

Spence, W. (1808). *Britain Independent of Commerce.* London: Printed by W. Savage, for T. Cadell and W. Davies.

Sraffa, P. (1960). *Production of Commodities by Means of Commodities.* Bombay: Vora.

Stiglitz, J. E. (1987). The Cause and Consequences of the Dependency of Quality on Prices. *Journal of Economic Literature, 25*(1), 1–48.

Stiglitz, J., Walsh, C., Guest, R., & Tani, M. (2015). *Introductory Macroeconomics* (Australian ed.). Milton, QLD: Wiley.

Tarshis, L. (1939). Changes in Real and Money Wages. *Economic Journal, 49*(193), 150–154.

Taylor, J. B. (1980). Aggregate Dynamics and Staggered Contracts. *Journal of Political Economy, 88*(1), 1–23.

Taylor, J. B. (2000a). Reassessing Discretionary Fiscal Policy. *Journal of Economic Perspectives, 14*(3), 21–36.

Taylor, J. B. (2000b). Teaching Modern Macroeconomics at the Principles Level. *American Economic Review, 90*(2), 90–94.

Veblen, T. (1899). *The Theory of the Leisure Class*. New York: A. M. Kelley.

Veblen, T. (1904). *The Theory of Business Enterprise*. New York: C. Scribner's Sons.

Von Bohm-Bawerk, E. (1889). *Kapital und Kapitalzins. Zweite Abteilung: Positive Theorie des Kapitals*. Innsbruck: Wagner.

Von Mises, L. (1912 [1953]). *The Theory of Money and Credit*. New Haven, CT: Yale University Press.

Weiss, A. (1991). *Efficiency Wages: Models of Unemployment, Layoffs and Wage Dispersion*. Oxford: Clarendon Press.

Yellen, J. L. (1984). Efficiency Wage Models of Unemployment. *American Economic Review, 74*(2), 200–205.

# 5

# A New Theory on Business Cycle and Economic Growth

The phenomenon of the business cycle has a long history and has been one of the most important topics in the economics discipline. As early as 1862, Clement Juglar (1819–1905) identified the business cycle as being about 8–10 years in duration. Jevons (1835–1882) attributed the cause of business cycles to sunspot activities, and later, Moore (1869–1958) developed the Venus theory of trade cycles, in which he attributed business cycles to the intervals that Venus comes between the Earth and the Sun. Despite an enormous amount of empirical and theoretical research done in this area, economists still cannot agree on the nature and causes of business cycles. Classical economists view business cycles simply as large economic fluctuations, so they think that there is no specific cause and no cure for economic recessions. Keynesian economists think the deficiency of effective demand is the cause of economic recession and recommend an increase in government spending and investment as a policy response. Monetarists view business cycles as a monetary or credit problem. Marxists attribute economic crises to income inequality resulting from capitalists' exploitation and advocate class strife to overturn capitalist rule. As economists keep arguing about business cycles, economic recessions continue to haunt mankind.

© The Author(s) 2019
S. Meng, *Patentism Replacing Capitalism*,
https://doi.org/10.1007/978-3-030-12247-8_5

The global financial crisis (GFC) in 2008 is a powerful reminder of the devastating consequence of a recession.

Unable to solve the problem of economic recession, economists focus on long-run economic growth. Solow's exogenous growth model unveiled the important role of technology in economic growth. Later, the endogenous growth model of Romer highlights the importance of human capital. However, economic growth and business cycles are inseparable topics—a large and long economic recession like the GFC must have a considerable impact on economic growth. By forcefully separating economic growth from business cycles, most studies on economic growth solely focus on the production or supply side. The omission of demand-side factors fails to reveal the interaction between the two sides and thus fails to uncover the key mechanism underpinning economic growth.

This chapter attempts to reconnect the demand and supply sides and build a new theory to explain both business cycles and economic growth. In Sect. 5.1, we will discuss the nature of business cycles and examine the existing theories on business cycles and economic growth. Section 5.2 illustrates the new theory in layman's terms while Sect. 5.3 derives the theory by utilizing economic models. Section 5.4 discusses the implications of the new theory. Section 5.5 provides some relevant empirical evidence.

## 5.1   The Nature of Business Cycles and Economic Growth

The nature of business cycles is still contested. An economy may grow very well into a boom, but suddenly the boom will bust and lead to an economic recession. The economy then will gradually recover and grow into a new boom. An economy may repeat this boom-bust cycle, but the time between the consecutive two booms or two busts varies in an unpredictable way. Moreover, the cause of business cycles is largely unknown despite there being a number of hypotheses. In order to uncover the nature of business cycles and economic growth, this section will review the history of economic booms and busts, discuss the features of business cycles and examine the theories concerning the causes of business cycles and economic growth.

## 5.1.1 A Brief History of Economic Booms and Busts

In the long history of social development, there are numerous economic booms and busts, or business cycles. This section reviews some important ones, from which we can gain knowledge of the features and possible causes of economic recessions.

### 5.1.1.1 The Panic of 33 AD

The earliest economic recession can be traced back to the financial crisis of 33 AD in Rome, recorded by the ancient historian Tacitus (56–117 AD). The cause of this recession is generally regarded as the decrease in money supply in both the long run and the short run. According to Frank (1935), Emperor Augustus increased coinage for circulation substantially from 30 to 10 BC, but later he and his successor Tiberius coined very little and were very frugal. Moreover, gold and silver went abroad for paying imports, so the per capita circulation in Italy decreased steadily for forty years.

The long-term decline of money supply is escalated by an acute withdrawal of money by people due to a political act, for which ancient historians like Tacitus and Suetonius provided detailed accounts. The head of the Praetorian Guard, Sejanus, undertook several plots to take the throne, but the emperor Tiberius survived and executed Sejanus in 31 AD. To prosecute the followers and friends of Sejanus, Tiberius revived Julius Caesar's Law of 49 BC on usury and on landowners, which required a certain amount of capital to be invested in Italian land. The law was a wartime measure and had not been enforced for almost 100 years. The re-enactment of this law required that two-thirds of every loan must be invested in Italian land and two-thirds of every loan must be paid off in a short period of time. This led to massive liquidation and distressed sales of real estate, and causing considerable deflation.

The scarcity of liquid assets was exacerbated by the misfortune of a number of firms. One was the firm Seuthes and Son of Alexandria, which lost three richly laden ships in a Red Sea storm and suffered from a sharp fall in the prices of ostrich feathers and ivory. The other was Malchus & Company, which was bankrupted due to a strike among Phoenician workmen and the embezzlement of a freedman manager. These events within firms led to their inability to pay off their loans and thus worsened the balance sheet of banks. It was reported that, when a bank could not pay back a wealthy nobleman his 30 million sesterces deposit, the bank closed. This started a bank run and more banks closed. The crisis even spread to other parts of the Roman Empire. However, the crisis ended quickly after the emperor suspended the revived law, ordered 100 million sesterces from the imperial treasury to be loaned to the neediest debtors with no interest to be collected for three years, and guaranteed a double value of real estate.

### 5.1.1.2 Tulip Mania in 1637

The most famous early economic boom and bust event is the tulip mania in Holland in 1637. Before 1628, The Dutch were great traders but were preoccupied by the fight for liberation from the Spanish. Once they were liberated from Spain, the Dutch focused on business and trade which lead to a thriving economy, creating great wealth from the establishment of global trade, banking, technology and agriculture (Day 2006). Tulips were introduced to Western Europe in the mid-sixteenth century and were viewed as a status symbol (Garber 1989). The new and vivid varieties were particularly in demand and commanded high prices. This attracted investment and trade in tulip bulbs. However, reproducing a new variety of tulip took time. One bulb can produce only two or three offsets a year, and an offset can become a tulip bulb in a year or two, so reproducing bulbs from offsets is slow. Tulips can also be massively produced from seeds, but it takes six or seven years. As such, the trade of new varieties generally involves a future contract. The establishment of a formal futures market in 1636 facilitated substantially the trade in tulip bulbs.

The prices of tulips rose steadily and strongly after their arrival in Holland in the late 1620s. It was recorded that, in 1633, a house in the town of Hoorn was exchanged for three bulbs. The price still kept rising strongly. A type of bulb called Semper Augustus was priced at 5500 guilders in 1633; it was worth 10,000 guilders in January 1637. In the last two months to the price peak in January 1637, the prices increased astonishingly. According to Dash (1999), a tulip called Admirael de Man was bought for 15 guilders and was sold for 175 guilders. A variety called Generalissimo was purchased for 95 guilders and sold for 900 guilders, which is three times the annual wage of an artisan. In early February 1637, the bubble suddenly burst. A bulb worth 5000 guilders in January was sold for only 50 guilders. People who signed future contracts were greatly affected by the price crash, especially people who had borrowed money or mortgaged their houses to engage in the tulip trade. The courts and government of Holland finally ruled in May 1638 that tulip contracts could be annulled upon the payment of 3.5% of the agreed price.

There is little information about the impact of tulip mania. The impact on future contract traders would have been small due to the court ruling. Traders who bought the bulbs at inflated prices would have been affected by hardship because they were unable to sell the bulbs for a similar price. The people being hit hardest were the tulip growers. According to the court ruling, the growers got very little compensation, so they had to bear their own costs while having no prospect that they would be able to sell their bulbs. The impact on the whole economy should have been mild. Otherwise, there would be written records about the impact.

### 5.1.1.3  Mississippi Bubble and South Sea Bubble in 1720

The Mississippi bubble in France and the South Sea bubble in Britain bear similarities. Both were the result of governments' intention to repay the debt of wars, were part of the monopoly right granted by the government, and occurred in the same time period.

In France, the long reign of Louis XIV and the years of wars put the French treasury in a shambles. To fix the budget and repay the debt incurred during the War of the Spanish Succession, the leader of the regents of Louis XV, Duke of Orleans asked for help from a Scottish financier, John Law. In 1716, Law was permitted by the French government to establish a national bank, the Banque Generale, to take in gold and silver and, in return, issue banknotes which were redeemable in official French currency. In 1717, Law acquired a struggling trading company, the Mississippi Company, and was granted a monopoly on the development of France's North American colonies along the Mississippi River. The colonies were considered to have abundant resources, such as gold and silver. In January 1719, the company offered the public shares, which could be purchased by either Banque Generale banknotes or government debt. In December 1719, the share price soared to 10,000 livres from the price of 500 livres per share in January. Since the Banque Generale issued banknotes according to public demand instead of precious metal reserve, the banknotes in circulation increased dramatically in 1719 and caused substantial inflation. In January 1720, some investors decided to take out their profits of shares in the form of gold coins from the bank, but the bank could not fulfil its promise because banks notes were not backed by precious metals. This started a bank run and the share price fell. By November 1720, the shares of the Mississippi Company were worthless and John Law was forced to leave France.

The British government was also embattled with debt due to wars, including the War of the Spanish Succession (1701–1714). Robert Harley, Chancellor of the Exchequer in a government commission, was tasked to finance the debt totalling 9.47 million pounds. A private company, the South Sea Company, was set up to take in debt and issue the shares to the debt holder. The government agreed to pay the company an annuity of 6% interest and granted the company a monopoly to trade with South America. In January 1720, South Sea Company stock was trading at 128 pounds. To stir up public interest, the company directors circulated false information about the success of South American trade and the share price increased to 175 pounds in February, 300 pounds by the end of March, 550 pounds in May, and peaked at 1050 pounds in June. However, investors' confidence started to shake and a sell-off began in early July. The share price decreased to 800 pounds in August and plummeted to 175 pounds in September. The Bubble Act 1720 was introduced to control inflating share schemes. In 1721, a formal investigation was concluded and many of the major players were prosecuted.

The main effect of these bubbles was income redistribution. These bubbles created winners and losers: people who sold shares before the crash of bubbles made a handsome profit while those who did not hand over their hot potatoes quickly had to endure the loss. The governments were the biggest winners—they got rid of a huge amount of debt. The French government had to deal with the high inflation rate after the bubble. However, the impact of these bubbles on the economy was mild. It is hard to find studies on the economic consequence of the Mississippi bubble due to the lack of historical data. In the case of Britain, Hoppit (2002) showed that the South Sea bubble had no direct impact on trade, industry and agriculture.

### 5.1.1.4 The Post-Napoleonic Depression, the Panic of 1819 and the Industrial Revolution

The long Napoleonic wars between France and Britain from 1803 to 1815 caused a wartime boom. Warfare created substantial demand for manufactured goods (e.g. guns and ammunition) and agricultural goods in both countries. Even the USA benefited from the warfare by exporting manufactured and agricultural goods. In Britain, the war acted as a prohibitive tariff to protect its agricultural industry.

As France and Britain settled their differences in 1815, a chronic economic depression ensued. The manufacturing industry plummeted because of the ceased demand and cancelled contracts from the governments. The agricultural industry was also affected negatively. This led to a long depression from 1816 to 1820. In Britain, the return of peacetime meant a large amount of imported agricultural goods. This plummeted the grain prices so landowners and farms went to parliament to request an increase in floor prices set in the Corn Law, which was enacted in 1791. The harvest failure in 1816 also had a negative impact. In the meantime, many soldiers were discharged with no provision. These soldiers contributed significantly to increased unemployment.

The post-war depression was also felt by the USA. During wartime, European nations imported from the USA a substantial amount of industrial and agricultural products. After the war, both France and

Britain reduced military spending and increased their agricultural production, so the demand for US goods reduced sharply. This post-warfare effect was exacerbated by a credit crisis. The USA experienced a westward expansion from 1807 to 1921. The government encouraged people to purchase western land on credit. In 1819, the amount of land purchase reached 3.5 million acres. However, the income of US workers reduced sharply as the Napoleonic war ended, so they could not pay off their loans. This affected the balance sheet of banks, and they were calling in loans and demanding immediate payment. Eventually, this led to the banking crisis and the panic of 1819. The US economy plummeted and did not fully recover until the mid-1820s.

Interestingly, recessions in Britain and the US occurred in the mid of industrial revolution (1760–1840) started in Britain. In the iron industry, the use of coal in iron smelting was started in 1678 by Sir Clement Clerk. In 1709, Abraham Darby used coke to fuel his furnaces. Later, John Wilkinson patented a hydraulic powered blowing engine for blast furnaces in 1757. Henry Cort developed the rolling process in 1783 and the puddling process in 1784, and the puddling process was improved in 1818 by Baldwyn Rogers. In 1828, James Beaumont Neilson developed the hot blast. The development in the iron industry had a substantial impact on other industries.

In the textile industry, the flying shuttle was invented and patented by John Kay in 1733. The spinning jenny was invented by James Hargreaves in 1764 and patented in 1770. Samuel Crompton's spinning mule was introduced in 1779 and Edmund Cartwright developed a vertical power loom in 1785. However, the invention and application of new technology in industry took a long time. These inventions were utilized later by entrepreneurs to increase the efficiency and capacity of the industry.

The technological development in textiles, iron and other industries demonstrated increasing influence from 1820 to 1860. For example, the process of manufacture of iron without fuel invented by Henry Bessemer in 1856 revolutionized the steel industry. Although the industrial revolution bore the name of creative destruction, it eventually transformed an agricultural economy to an industrial economy, lifted Britain and the USA from the recession and into rapid economic growth, improved the living standard of masses of ordinary people and caused the change in social structure and working conditions.

### 5.1.1.5 The Roaring Twenties and the Great Depression in 1929

The roaring twenties started from the recession at the end of World War I (1914–1918). A post-war recession is generally deep because of the sharp decrease in the temporary wartime demand. Unlike the long post-Napoleonian war recession, the post-World War I recession was deep but short (except Germany due to the imposition of the Treaty of Versailles, which forced Germany to pay large amounts of reparations to France and Great Britain). The USA had a brief recession from 1919 to 1920. In European countries, the recession lasted from 1919 to 1923. The quick recovery and the rapid growth in the 1920s were in part related to the invention and mass production of new consumer goods.

Radio technology was proven invaluable in wartime, and it revolutionized communications during the early 1920s. Radio also became the mass broadcasting medium, which demonstrated its value for mass marketing. Automobiles were luxury goods before the war. Henry Ford's application of assembly lines to automobile production produced mass cheap cars for ordinary people. The aviation industry also achieved a milestone. In 1927, Charles Lindbergh succeeded in the first solo non-stop transatlantic flight. Movies and cinemas appeared in the 1910s, but the movies with sound invented in 1923 changed the whole industry. In 1927, 'The Jazz Singer' became a sensation. Warner Brothers released talking movies in 1928, and all-colour, all talking movies in 1929. Inventions on television also had a breakthrough. John Logie Baird transmitted the first long-distance television pictures in 1927 and demonstrated colour transmission in 1928. In the same year, the Scottish biologist Alexander Fleming discovered penicillin.

Stimulated by innovations, many industries flourished, including the automobile industry, the movie industry, the radio industry, the electricity power industry and the chemical industry. Everything seemed cheerful. In 1928, Herbert Hoover, then the Republican Presidential nominee, said: 'We in America today are nearer to the final triumph over poverty than ever before in the history of any land'. However, the stock market suddenly crashed on 29 October 1929. The average value of the stock dropped by nearly 40%. This crash marked the beginning of the longest and toughest depression in history.

The crash of the stock market had considerable implications. Many people borrowed heavily to buy stock on margin with only the stock itself as collateral. As stock price fell, the stock value became lower than the loans, so the stockholders were forced to sell the stock (margin calls), and this caused the share price to fall even further. As a result, the loans could not be repaid, thousands of banks failed and millions of people lost their life savings. During the Great Depression of the 1930s, the unemployment rate in the USA peaked at 25% in 1932–1933, and it took almost a decade for US output to return to its pre-1929 level.

The response of the US government also amplified the negative effect. The Hoover government realized that the economy was slowing down due to overproduction, especially in agricultural goods, so the government's solution was to increase the tariff rate and to expand foreign markets. The rise in the US tariff rate led to retaliation of other countries with their raising tariffs on US goods, resulting in a breakdown of international trade, which increased the severity of the depression in the USA and abroad. In response to the failure of banks, the US government enforced a policy of tighter credit and a suspension of loans from US banks abroad. This policy reduced much-needed liquidity during a recession and led to a sharp drop in commodity and asset prices and, coupled with the rigid gold standard, spread the depression overseas. The government also believed that the economic slowdown was a natural economic fluctuation and adopted a policy of natural recovery.

The depression spread to other countries through its impact on international trade and on investors' confidence. In Australia, export demand and commodity prices fell sharply and unemployment in 1932 reached a record high of 29%. In Canada, industrial production in 1932 fell to 58% of the 1929 level, and unemployment in 1933 reached 27%. In Germany, unemployment in 1932 reached 6 million, accounting for 25% of the workforce. The impact of the Great Depression on Britain and France was relatively milder. Even so, their industrial production decreased by 16.2 and 41.8%, respectively. The impact on other countries can also be shown as a decline in industrial production: Poland 46.6%, Czechoslovakia 40.4%, the Netherlands 37.4%, Italy 33.0%, Belgium 30.6%, Argentina 17.0% and Japan 8.5%.

Recovery started in 1933, but it is generally regarded that the recession lasted until 1939. The sources of recovery are controversial. Commonly cited sources are currency devaluation and monetary expansion, stimulating fiscal policies, and military spending for World War II. An example is the New Deal (a series of programmes, public work projects, financial reforms and regulations) enacted by Franklin Roosevelt between 1933 and 1936. The US output level returned to its long-run trend level in 1942, coinciding with the US declaration of war on Japan in December 1941. After that, the USA entered into a wartime economy.

### 5.1.1.6 Post-World War II Economic Growth and the Stagnation in the 1970s

When World War II (1939–1945) was drawing to a close, prominent economists predicted a deep post-war recession. Paul Samuelson warned that there would be the greatest period of unemployment that any economy had ever faced and suggested the government extend wartime controls. Gunnar Myrdal predicted that there would be a severe post-war economic turmoil which could lead to an epidemic of violence. The reasoning behind these predictions is easily comprehensible: the decrease in government spending for war would press down demand dramatically and thus cause massive unemployment and loss of income. Fortunately, these predictions did not come true. The US economy did experience a short and mild recession after the war. For example, the US GDP in 1947 was about 13% lower than its GDP in 1944, but this was followed by about 25 years of rapid economic growth. The economy around the world experienced a similar golden growth period in the 1950s and 1960s.

What caused the rapid growth during the post-World War II period? Popular answers include repairing wartime damage (reconstruction), rebuilding capital stock and government policies. For example, Milionis and Vonyo (2015) argued that the impact of the prolonged reconstruction process lasted until the mid-1970s. Reichel (2002) concluded that Germany's rapid post-war growth was triggered by the reconstruction effect but was sustained by the German currency reform and the inauguration of the social market economy. Nobel Prize winner Paul Krugman thought a burst of deficit-financed government spending created an economic boom that laid the foundation for long-run prosperity. In his 2009 State of the Union address, US President Obama highlighted the importance of the GI Bill—The Serviceman's Readjustment Act of 1944, which provided educational assistance to service members, veterans and their dependents.

Post-war reconstruction, the US Marshall plan for European recovery, and expansionary fiscal and monetary policies may have had an influence on the post-World War II boom; however, these factors cannot be the fundamental reason for the long golden growth period. The obvious reason is that these factors existed in any post-war period. Why did a long economic boom occur after World War II while a sharp recession followed most wars? The key may lie in technological change. Eichengreen (2007) noted that the late 1920s and the 1930s were a period of instability but also were a period of rapid technological change. Many innovations were developed in the USA in 1920s and 1930s. However, the disruption of World War II delayed the implementation of those innovations. The implementation of these innovations, such as an improved internal combustion engine, television technology, automation technology, mass production methods and personnel-management practices, had a dramatic and long-lasting effect on the post-WWII economy. As the new technologies diffused, the European economies were catching up with the USA. Eichengreen's book demonstrated convincingly the convergence tendency of economic development due to Europe's utilization of an extraordinary backlog of technological and organizational knowledge.

The economic impact of previous inventions in the USA and in Europe started to wear out by the end of the 1960s. There were a number of important inventions during the 1950s and 1960s, such as computers, satellites, digital photographs, laser, global navigation satellite systems and plasma displays. However, these inventions were not mature enough to be implemented by the 1960s, due to the long lag between the creation of inventions and their mass production. The scarcity of applicable innovations led to a weak demand. For example, the growth of the automobile industry and the entertainment industry slowed down significantly because the markets for products like cars and TVs became saturated. This led to a stagnant economy in the 1970s.

Since weak demand is the feature of economic stagnation or economic recessions, the price level generally falls during an economic stagnation. However, the stagnation in the 1970s was associated with a high level of inflation. The causes of high inflation include an expansionary fiscal policy, less stringent monetary policy and energy price shocks. The first two causes may be related to Keynes' theory. In the 1960s, both President Kennedy and President Johnson were influenced by Keynesian economics and intended to make low unemployment the priority, at an acceptable inflation rate. Tax cuts and a number of social programmes were implemented with an aim at creating 'the great society'. This increased government spending dramatically and led to a large government deficit. The US involvement in the Vietnam War also contributed to a heavy government deficit. Government spending and deficit exerted a considerable pressure on inflation. Under the political environment of the Johnson administration, the policy of the Federal Reserve Bank (the Fed) was also softened. Although Chairman of the Board of Governors, William McChesney Martin, was in favour of keeping prices stable, the Fed shifted to a low-inflation policy, which provided a relatively lax monetary environment. Some researchers also argued that the high inflation in the 1970s resulted from the substantial rise in oil prices in 1973–1974 due to the oil embargo imposed by OPEC.

To retain the super inflation in the 1970s, the newly elected Reagan government slashed many social programmes and the Fed raised interest rates. These actions did bring inflation under control, but they sent the economy into recession. The 1980–1982 recession was aggravated by the second oil price shock due to the Iranian Oil Embargo in 1979. However, the recession was over by the end of 1982, and the economy entered another golden growth period in the 1980s and 1990s.

## 5.1.1.7   Japanese Economic Miracle (1945–1990) and the Lost Two Decades (1991–2010)

The post-World War II period witnessed almost 50 years of rapid economic growth in Japan, which is commonly known as the Japanese economic miracle. The miracle can be divided into three periods: the recovery period (1945–1954), the fast growth period (1955–1972) and the steady growth period (1973–1990). The Japanese industrial production in 1946 was about 27.6% of the pre-war level, but it recovered to the pre-war level in 1951 and increased to 350% by 1960. By 1968, the Japanese economy achieved an average growth rate of 10.8% and replaced Germany in becoming the second largest economy in the world. Although Japan was affected by the high energy prices both in 1973 and in the early 1980s due to the world oil price shocks, it maintained an average annual growth rate of more than 3% from 1970 to 1990.

The reasons for the rapid growth of the Japanese economy can be put into three categories. First, the commonly recognized reason is the policies put forward by the Japanese government, including the adoption of the inclined production mode which focused on the production of raw materials like steel, coal and cotton; land reform which forced landlords to sell their land to the government for redistribution to tenant farmers; and the over-loaning practice in the banking system which allowed the Bank of Japan to issue a large amount of loans to industrial conglomerates through city banks to make up the capital shortage in Japan at that time.

Secondly, the post-World War II environment external to Japan also had a significant influence. The Allied Powers required that all production of military materials in Japan be stopped and closed down. This led to the formation of the 1947 Japanese Constitution, which required that Japan give up the right to use any military force forever and relied on the USA for protection from outside force. Being free from the burden of military development, the Japanese government could use all its resources on industrial production. The other important external event was the outbreak of the Korean War. After the USA entered the war, it needed to support logistical operations in Asia and supply firearms and other war materials. Japan stood out and became an important partner of the USA. The large amount of orders and foreign currency from the USA sped up the recovery in Japan.

Thirdly, some researchers realized that the importance of technological diffusion was an essential factor for Japan's economic miracle. For example, Valdes (2003) used a Solow–Swan model to demonstrate that the rapid growth of post-World War II Japanese economy was consistent with the convergence theory, which was based on technological diffusion. Takada (1999) argued that the ability of the Japanese people to imitate and improve on the skills learned was the most important factor for their post-World War II success. These arguments are supported by evidence in Japanese industries such as the steel, automobile and shipbuilding industries.

The above-mentioned factors all played important roles in the Japanese economic miracle, but it is essential to find the fundamental factor or root cause in order to understand the phenomenon in depth and shed light on how to improve the world economic growth in the future. Here, the author argues that technological imitation and improvement were the root cause of the Japanese economic miracle for the following reasons. First, the significant increase in domestic consumption was an important driving force for economic growth. This driving force was underpinned by an increased variety of consumption goods along with activities such as recreation and entertainment, which resulted from technological innovations. Second, given the small population in Japan, exports were essential to maintain a high speed of economic growth for a long period of time. Japanese imitated and improved technologies from other countries so it could compete successfully in the world market. The electronics and car industries are good examples. Japanese TVs and computers have been sold around the world. The excellent fuel efficiency of cars helped the Japanese to capture 21% of the world's automobile market by the mid-1970s. Third, the impacts of other factors were limited and thus were not able to be sustained for a long period of time. For example, the impact of the Korean War phased out when the war ended; the expansionary fiscal and monetary policies could not be extended without limit and had considerable side effects.

The fundamental role of technological improvement and the temporary effect of the other factors in Japanese economic growth can also be shown in the 20 years of economic stagnation following the Japanese economic miracle. The excessive use of expansionary fiscal and monetary policies by the Japanese government had a deadly consequence—the highly inflated asset prices. Compared to the prices of 1985, the commercial land prices increased 302.9% by 1991. When the Bank of Japan started to tighten monetary policy to combat inflation in 1989 and 1990, it caused the burst of the asset bubble. The Nikkei stock index plummeted to half its peak by August 1990. Asset prices began to fall in late 1991. By 1992, the prices of commercial land fell 15.2% from its peak. The crash of stock market and asset prices was followed by a long economic stagnation. The average annual economic growth was only 0.5% for the period from 1991 to 2000, and 0.75% from 2001 to 2010. The inflation rate stayed very low: less than 1% from 1994 to 1999 and −0.5% from 1999 to 2011. In response to the chronic deflation and low growth rate, the government implemented excessive expansionary monetary and fiscal policies in order to stimulate the Japanese economy. The interest rate has been kept below 1% since 1991, and a large government deficit has occurred due to stimulus packages. However, these policies failed to pull the economy out of stagnation.

The two-decade-long stagnation in Japan has triggered a large amount of research trying to uncover the causes and providing a solution. Krugman (1998), Bayoumi (2001), and Hamao et al. (2007) criticized the Japanese monetary policy response allowing deflation or the liquidity trap since the early 1990s. McKinnon and Ohno (2001) blamed the substantial real appreciation of the Japanese currency against the US dollar due to the Plaza Accord of 1985. Saito (2000) and Hayashi and Prescott (2002) argued that Japanese stagnation was mainly due to the low growth rate of aggregate productivity. Hoshi and Kashyap (2004) and Hutchison et al. (2005) attribute the long stagnation to the problems in Japan's financial system. Miyakoshi and Tsukuda (2004) confirmed the one direction causality from Japanese bank lending to the GDP during the lost decade. The outbreak of the GFC triggered a renewed interest on the lost decade in Japan. Schuman (2008) and Onaran (2011) blamed the inefficient 'zombie banks' and 'zombie firms' for the long-lasting stagnation. These zombie banks and firms were created by fund injections from the government and the central bank because they thought these commercial banks and firms were too big to fail. Aloy and Gente (2009) highlighted the declining birth rate that yielded an ageing and contracting labour force. Krugman (2009) regarded the long stagnation as a typical case of liquidity trap and blamed the cozy relationship, and thus the moral hazard problem, between the banks and the firms. Koo (2009) regarded the lost decade as a 'balance sheet recession' triggered by a collapse in land and stock prices. He reckoned that the expansionary monetary policy was not effective because the firms opted to pay down their debts rather than to borrow and invest. Sumner (2011), on the other hand, thought the low interest rate policy of the Bank of Japan was still too tight and thus it caused the prolonged stagnation. By including the performance of China in a model, Tyers (2012) confirmed the explanation of low productivity proposed by Saito (2000) and Hayashi and Prescott (2002).

The above-mentioned explanations have some elements of truth but failed to reveal the key feature and thus the essence of Japanese stagnation. Most studies attributed the cause of the Japanese lost decade to the problems in the financial sector or inappropriate Japanese monetary policies. These issues existed during both economic miracle and stagnation periods, so they are not the fundamental cause of Japanese stagnation. Ageing population happens in other countries and was not associated with economic stagnation, so this factor may be relevant but is not a key factor. Low growth of productivity is typically and not-surprisingly associated with economic stagnation or recessions, but the association does not mean causality. This association did not reveal the reasons for the low growth of productivity and thus for the low level of economic activity during the lost decade.

As with the Great Depression, the key feature of Japanese economic stagnation was stagnant sales, which point to inadequate demand. On the supply side, the Japanese had ample capital accumulated during the economic miracle. Ageing population may have had a negative impact on production, but the high unemployment and the stagnant or even falling wages during the 1990s and 2000s indicated that labour shortage was not a serious issue. On the demand side, the stagnant sales were evidenced by the retailing performance in the 1990s (Dawson and Larke 2004) and indirectly signalled by the deflations during the lost decade. If there were sufficient demand, various sectors would have been able to produce and sell their products, so the economy would have grown despite inefficiency or problems in the financial sectors.

What caused the weak demand? If we attribute it to low income and weak investment during the stagnation period, we go into a causality loop because income is dependent on production, which in turn is dependent on demand, while investment is dependent on future consumption. In other words, since income and investment are endogenous factors, they cannot be the root cause of weak demand. It is also implausible to attribute weak demand to the limits in the human desire for goods and services. Generally speaking, our desire is unlimited, but sometimes, the types of goods and services we want badly are not available due to technology constraint, e.g. a cure for cancer. In this reasoning, the exogenous determinant of demand is the variety of new products. As we have seen, the Japanese imitation and improvement in technology produced products of better quality and more varieties and contributed substantially to the Japanese miracle. As the effect of technological diffusion wore out by the 1980s, the Japanese failed to produce enough applicable innovations in the 1990s. In the meantime, the innovation activity in the USA and Europe grew rapidly. The contrast is manifested by the patent applications statistics during 1987–1997 (Branstetter and Nakamura 2003). As a result, while the USA and other Western countries benefited from the innovations in IT technology, Japan failed to imitate and improve these innovations and thus led to stagnant demand and the lost two decades.

### 5.1.1.8 The Asian Financial Crisis (1997) and the Dot-Com Bubble (2000)

From 1983 to 2007, the US economy, and the world economy in general, experienced a golden growth period, albeit disturbed by recessions such as the 1990 recession in the wake of the oil price shock due to

Iraq's invasion into Kuwait, the 1991 banking crisis in Finland and Sweden, the 1997 Asian financial crisis, the 1998 Russian financial crisis and the 2000 Dot-Com bubble. Among them, the 1997 Asian financial crisis and the 2000 Dot-Com bubble generated global impact, so we will have a closer look at them.

From the 1960s to 1990s, a number of Asian countries experienced rapid economic growth, partly due to the economic policies of these countries and partly due to the diffusion of technology from Western countries. From 1991 to 1996, the average annual GDP growth rates averaged more than 7% for Korea, Indonesia, Malaysia, Singapore, Thailand and Taiwan. Philippines' growth rate was low in the early 1990s, but its rate was more than 5% from 1994 onwards. Corresponding to the high economic growth rates, investment rates as percentage of GDP were as high as 30–40%. The profitability or investment efficiency measured by the ratio of investment rate to output growth rate was also very high. This attracted a large amount of foreign investment. In 1996, private capital flow into South Korea, Indonesia, Thailand, Malaysia and the Philippines reached almost US$100 billion, accounting for one-third of worldwide capital flows. These capital inflows not only helped with economic growth in these Asian countries but also generated potential instability, which was manifested by the very high ratio of debt to foreign reserves for these countries. In 1996, the ratios of debt service plus short-term debt to foreign reserves were 243.31% for Korea, 294.17% for Indonesia, 69.33% for Malaysia, 137.06% for the Philippines and 122.62% for Thailand. Due to the large amount of short-term debt relative to foreign reserves, a negative shock to investors' confidence could cause a capital outflow and lead to a currency crisis.

The potential risk of having a high level of foreign debt was materialized by the pegged or fixed exchange rate regime, which is predominately adopted by most Asian countries. The fixed exchange rate regime provides a safe exit corridor for foreign speculative funds—if foreign speculative investors make a profit in stock or other assets market, they can shift their profit out of the countries safely through currency exchange at the fixed rate. Thus, the fixed exchange rate regime attracts foreign speculative funds and causes large fluctuations in capital inflow and outflow. In the meantime, to maintain a fixed exchange rate, the central bank has to increase the domestic money supply in the face of a net foreign capital inflow. This will inflate the economy. On the other hand, the central bank has to supply foreign currency and reduce domestic currency in the wake of net foreign capital outflow, and this will deflate the economy. Given the high ratio of external short-term debt to foreign currency reserve, it is likely that the foreign reserve is not enough to satisfy the capital outflow. If so, the fixed exchange rate will break down and a currency crisis will occur.

In the years leading to the Asian financial crisis, large amounts of capital inflow into Thailand, Korea, Malaysia and Indonesia led to over-lending to the real estate sector, causing assets bubbles and bad debt in the financial system. The situation was exacerbated by foreign investors' speculative activities in the stock market. When speculative investors were selling currencies of Thailand, Korea, Malaysia and Indonesia, to shift their profits out of the countries, the foreign reserves in these countries were not enough to maintain the fixed exchange rate, so the currency of these countries had to be depreciated heavily. The breakdown of the fixed exchange rate triggered a panic in investment, and investors tried to sell their Asian currencies before their further depreciation. This led to a plummet of the value of the currency of the troubled Asian countries. From June 1997 to July 1998, the currency value decreased by 40.2% in Thailand, 83.2% in Indonesia, 37.4% in the Philippines, 45.0% in Malaysia and 34.1% in South Korea.

The currency crisis also had a large impact on the real economy. The per capita real GDP decreased from 1997 to 1998 was: −15.0% for Indonesia, −11.6% for Thailand, −9.5% for Malaysia, −7.5% for Korea, −6.4% for Hong Kong, −4.6% for Singapore and −2.7% for the Philippines. The time for these countries to recover their GDP to pre-crisis peaks varied between 2 and 7 years, but the recoveries started as early as in 1999 and were fairly rapid, compared to the economic recessions like the Great Depression and the GFC.

The 2000 Dot-Com bubble occurred in the era of the Internet. The World Wide Web was introduced in 1989. With the advances in Internet connectivity, the development of more user-friendly web browsers and better education on the use of the Internet, the usage of the Internet increased significantly from 1990 to 1997. With this new communication platform, doing business online was expected to have great potential: the running cost is low, and the market potential is huge because the Internet could be accessed by anyone in the world. Many young people moved to set up online trading websites with a suffix '.com', while venture capitalists were eager to profit from this investment opportunity. As the Dot-Com companies were put into the hi-tech stock market—Nasdaq through initial public offerings to raise capital, the enthusiasm of investors pushed the share price very high. As a result, the Nasdaq index increased explosively from 751.83 in January 1995 to the peak of 5132.52 in March 2000. In the three years leading to the peak, the Nasdaq index increased by about 300% from 1997 to 2000, causing more than 60% increase in the value in the whole stock market.

However, this extraordinary increase in share prices was not supported by the performance of the Dot-Com companies. Most of these companies gave priority to growth rather than profit and adopted a strategy of 'get big fast' by burning capital on advertisement and by increasing market visibility (i.e. market share) through providing free or discount services or products. The ratio of average share price to earnings for Nasdaq companies increased sharply. It was 25 in 1997, more than 100 in 1998 and more than 200 in 1999 (Perez 2009). Investors eventually realized the doubtful perspective of profitability of Dot-Coms, so the game of earning quick capital gains was over. After its peak in March 2000, the Nasdaq index dropped by 20% in April and May and dropped by 42% from September 2000 to January 2001. The terrorist attack on 11th of September in 2001 caused the stock price to decrease even further. Eventually the stock market started to recover in 2002.

The impact of the Dot-Com bubble on the real economy was modest. The burst of bubble in 2000 caused a slight slowing down of economic growth in the USA—the economic growth rate decreased from 4.7% in 1999 to 4.1% in 2000. Coupled with the 911 terrorist attack in 2001, the US GDP declined by 0.3% from March to November 2001, but it recovered quickly. Although many Dot-Com companies went bankrupt and many venture capitalists lost a fortune, some Dot-Com companies survived the stock market crash. Amazon was a notable example.

It not only survived but became one of the top online shopping companies in the world. Now the large Dot-Com companies such as eBay, Google and Facebook have a remarkable impact on the economy.

### 5.1.1.9 The Global Financial Crisis (2008) and European Debt Crisis (2010)

The world economy experienced a period of low inflation, high growth and modest recessions from the early 1980s to the early 2000s. This period was termed the 'great moderation' by Stock and Watson (2002). The economic growth led by the information technology industry in the 1990s prompted economists to think the economy had entered a new phase: the new economy or knowledge economy. It seemed that large economic recessions and stagnation like the Great Depression in the 1930s and the stagflation in the 1970s were the events of the past. However, the outbreak of the GFC in 2008 and the following European debt crisis in 2010 caused a long and deep recession. By reviewing the events leading to these two crises and their effect, we may shed some light on the causes of the crises and thus find a way to avoid them.

From the 1980s, financial deregulations occurred in the USA. The Garn–St. Germain Depository Institutions Act of 1982 authorized banks to issue short-term deposit accounts with no interest ceilings. The Federal Housing Safety and Soundness Act of 1992 promoted homeownership for low-income and minority groups. The replacement of the Glass–Steagall Act by the Gramm–Bliley–Leach Act in 1999 eliminated the separation between commercial banks and investment banks, so large banks could play multiple roles, e.g. being at the same time a commercial bank, an insurance company, an asset manager, a hedge fund and a private equity fund. The Commodity Futures Modernization Act of 2000 allowed the financial derivatives to be unregulated. In 2004, the US Securities and Exchange Commission relaxed the net capital rule so investment banks could take a higher level of debt. These financial deregulations stimulated the growth of the shadow banking system (Gorton and Metrick 2010) and induced numerous financial innovations, which led to the popularity of various mortgage-backed securities such as Collateralized Mortgage Obligations, Collateralized Debt Obligations and Collateralized Loan Obligation. The risk of these products was not easy to see and thus was hard to evaluate. The deregulation also encouraged risk-taking and a variety of moral hazard problems between bankers, politicians, insurance companies and rating agents (Lin and Treichel 2012).

Another important factor leading to the housing bubble and the GFC was that, from the 1990s onwards, the US Federal Reserve Bank (Fed) adopted a low interest rate policy, especially in the aftermath of a recession. After the 1990–1991 recession, the Fed slashed interest rates from 7% in 1990 to 3% in 1992. In the late 1990s, the interest rate was about 5%. In the wake of the burst of the Dot-Com bubble, coupled with the 911 terrorist attack, the Fed slashed interest rates dramatically from 6% in 2000 to 0.75% in 2002, but increased to 2% in 2003 and early 2004. These extremely low interest rates reduced borrowing cost, stimulated consumption and housing mortgage loans, and thus contributed to dramatically increased housing prices. Considering that the economy had recovered from the 2000–2003 recession, the Fed steadily increased the rate to 6.25% during the period from mid-2004 to late 2006. This rate hike increased the debt service burden, so housing demand subsided and the housing price peaked in April 2006. Sensing that the housing boom was going to end, banks required the subprime mortgage borrowers to pay off debt. This caused intensified mortgage delinquencies, defaults and the drop in housing prices in 2007 and 2008. The burst of the housing price bubble and the growing defaults on mortgages worsened the balance sheet of financial institutions which were exposed to housing loans. In late 2007, a German bank (IKB Deutsche Industriebank) and a British bank (Northern Rock) collapsed. In March 2008, Bear Stearns in the USA filed for bankruptcy and was bought by JPMorgan. In September 2008, financial institutions such as Merrill Lynch and Lehman Brothers and insurance companies such as AIG and HBOS filed for bankruptcy. The collapse of Lehman Brothers—the largest bankruptcy in US history—marked the full eruption of the GFC.

The collapse of Lehman Brothers led to a complete stop of inter-bank lending because of the fear of default. The US government responded quickly by the Troubled Assets Relief Program, using a $700 billion bank bailout to rescue the financial sectors. The Fed reduced its interest rate to almost zero. The crisis in the housing market and the financial sector quickly spread to other sectors through a sharp drop in demand. Retail sales collapsed. In the automobile industry, Chrysler and General Motors filed for bankruptcy, but were rescued by the US Treasury. The stock market also collapsed. The US Dow Jones Industrial Average Index dropped, from its peak of more than 14,000 points in October 2007, to a trough of less than 6600 points in March 2009. At the macro-level, the US quarterly real GDP decreased by 8.9% in the last quarter of 2008. The unemployment rate in the USA increased from 4.6% in 2007 to 10.1% in 2010. The crisis in the USA also spread quickly to other countries. In 2009, GDP fell by 5.2% in Japan, 4.9% in UK, 2.5% in Hong Kong, 4.3% in EU, 6.6% in Hungary, 7.1% in Kuwait, 4.7% in Mexico and 7.8% in Russia. Overall, the world GDP fell by 1.9%.

In the wake of the GFC, the European Sovereign debt crisis erupted. Indirectly affected by the burst of the housing bubble in the USA, many Central and Eastern European banks had asked for a bailout by January 2009. Meanwhile, tax revenues of governments shrank substantially due to the contraction of the economy. To make things worse, some European governments had incurred large deficits which were disguised by accounting procedures in order to circumvent the deficit limit imposed by the EU. These factors caused the blow out of national debt. In the case of Greece, its main industries—shipping and tourism—were hit hard by the GFC. The government budget deficit was reported as 3.7% of 2009 GDP, but later was revised in January 2010 by an EU report to 12.7% of GDP, which was more than four times the maximum allowed by EU rules. The accumulated public debt was as high as 129.7% of Greek GDP in 2009. In April 2010, the Greek government requested an initial loan of 45 billion euro from EU and IMF to cover its financial needs. In May 2010, the Greek government announced a series of austerity measures and secured a loan of 110 billion euros from the EU/IMF. After the implementation of further austerity measures required by the EU, Greece obtained the second bailout fund of 109 billion euro in October 2011. In March 2012, Greece and its private creditors, the EU and IMF reached a debt restructure deal, which effectively wrote off the Greek debt by about 109 billion euros. In November 2012, EU also agreed to prolong debt maturities and lower the interest rate. The Greek economy eventually returned to positive growth in 2014, and the unemployment rate started to decline in 2015.

Other troubled European countries had similar situations. Ireland was the second country which asked for help from the EU and the IMF and was granted an 885 billion euros bailout in 2010. This country enjoyed a government budget surplus in 2007, but similar to the USA, the country had a housing bubble. The burst of the bubble led to a large sum of external debt in order for the government to bail out its banks. Portugal was the third country to ask for a bailout. This country had run a government deficit since the 1974 revolution. The consistent deficit over a long period of time was comparable to the situation in Greece. The low growth rate in Portugal since 2000 due to low productivity enlarged the budget deficit. This situation triggered Moody's Investors Service to lower the rating of Portugal's sovereign bond in the summer of 2010. Portugal's borrowing cost surged, and so the country had to ask for help from the EU and IMF in April 2011. Similar to Ireland, Spain had a government surplus of 2% of its GDP in 2007, but the decrease in tax revenue and the cost of a bank bailout led to a government budget deficit of 11.2% in 2009. In June 2012, Spain asked for and was granted a financial support package of 100 billion euro.

The impact of the European debt crisis coupled with the GFC was dramatic. The data from the World Bank showed that the real GDP decreased by 9.13% in Greece in 2011, 4.03% in Portugal and 2.82% in Italy in 2012, and 5.93% in Cyprus in 2013. According to the European Commission (2013), unemployment rates in 2012 were 25.0% in Spain, 24.3% in Greece, 15.9% in Portugal and 14.9% in Cyprus. The recovery from these crises was very slow. It took four years for the USA, Germany and France to achieve the pre-GFC level GDP. UK and Japan recovered in 2013, Iceland and Ireland in 2014, and Spain in 2017. By 2017, Greece, Portugal and Italy still had not achieved pre-GFC GDP. Interestingly, developing countries in Asia recovered from the crises fairly quickly. China even avoided a recession and maintained fast growth.

This severe negative impact of the GFC and the European debt crisis brought about enormous interest in the search for its causes. For example, Taylor (2008) concluded that frequent monetary excess was the main cause of a housing boom and bust in the USA which led to the GFC. Arner (2009) found that the GFC resulted from unprecedented excessive borrowing, excessive lending and excessive investment incentivized by a range of economic and regulatory factors. Orlowski (2008) argued that the severity of the GFC was influenced strongly by changeable allocations of global savings coupled with excessive credit creation, which led to the over-pricing of various types of assets. Dabrowski (2008) attributed the GFC to the lax monetary policies and financial regulatory failures. Berrone (2008) viewed the GFC as an incentive problem: there was no penalty for managers in the case of the collapse of financial institutions. Crotty (2008) blamed the flawed financial institution and practice—the New Financial Architecture (NFA). Jickling (2009) summarized 26 causes, ranging from bad computer models to financial innovation, most of which concerned flaws in financial regulation. Blundell-Wignall et al. (2008) provided a comprehensive overview. According to them, the macro-policies affecting liquidity, such as the extremely low interest rates in the USA and Japan, the fixed exchange rate in China and the accumulation of reserves in Sovereign Wealth Funds, created a situation like a dam overfilled with flooding water; the faults in the dam—the poor financial regulatory framework—directed the water to mortgage securitization and off-balance-sheet activity and thus led the dam to breaking. Utilizing previous studies on the GFC, Lin and Treichel (2012) tried to find its root cause. Their answer pointed to the financial deregulation aided by lax monetary and fiscal policies in the USA, rather than the currency policy and the export-oriented growth policies in Asian countries.

On the European debt crisis, Lane (2012) thought the crisis originated from the flaws in the design of the EU. He argued that the initial institutional design of the euro increased fiscal risks during the pre-crisis period while the flaws in the design amplified the fiscal impact of the crisis dynamics. He also argued that the restriction imposed by the

EU shaped the recovery period. Lane's solution included a new monetary system focusing on structural budget balance, the creation of a banking union, the introduction of common 'eurobonds' and developing a deeper level of fiscal union. Esposito et al. (2014) attributed the causes of the European debt crisis to Southern Europe's easy access to cheap credit after they joined the EU, the Northern wealthy countries' using debt as an investment vehicle, uncompetitive goods from southern states, and human and social elements. Ruscakova and Semancikova (2016) reviewed a number of empirical studies and claimed that the lax fiscal policy and the application of different monetary policies were the primary sources of the European debt crisis, so they suggested the joint implementation of macro-prudential policies as a solution.

The causes identified above can be put into three categories: first, the lax monetary and fiscal policies, i.e. the low interest rate policy adopted by Greenspan and the high government budget deficit due to expansionary fiscal policy; second, the problems in the financial profession, for example the principal-agent problem due to financial deregulation, the moral hazard problem rose from accounting procedures and the high risks masked by financial innovations; and third, the global saving imbalance, notably the large current account deficit in the USA and Southern European countries, the large current account surplus in Asian countries due to their export-oriented growth policy and/or weak currency policy, and the large trade surplus in Northern European countries due to their high productivity and also due to the convenience of the same currency euro used in the all Eurozone. Although these causes contain elements of truth, they seem not the root cause.

The argument of global external imbalance focuses on the saving made by different countries and thus masks the fundamental fact that lending within and between countries is always made by rich people to poor people through financial intermediates. For example, there was a huge current account surplus in China, but those savings were contributed by rich corporates and individuals. Similarly, the large current account deficit in the USA was due to imports and borrowing from the poor Americans. There were a considerable number of Americans who were rich and lent their money domestically and internationally. Similarly, the impact of lax fiscal and monetary policies highlighted the implications of these policies on lending and borrowing, but they overlooked essential factors embedded in lending/borrowing behaviour. It is valid to argue that the lax monetary and fiscal policies fuelled debt demand and that the financial deregulation and financial innovation caused moral hazard problems. However, these factors only facilitated the lending-borrowing financial flow. The driving forces behind the housing debt and the European government debt are the desire of the poor to borrow and the motive of the rich to lend.

The desire of the borrowers is to maximize utility for any given opportunity. The expansionary fiscal and monetary policies may increase the opportunities for poor households in an unsustainable way; i.e., the financial deregulation and low interest rate policy may have encouraged the poor and Southern European governments to take on unaffordable debt. However, crises will not occur if the lenders choose to be prudent in lending to the needy for their consumption and purchasing housing assets. Why did the lender choose to lend to the poor rather than investing in production? This can be explained by the sales stagnation phenomenon prior to and during the crises— investing in production will make a loss because of weak sales. Why then, do sales become weak and stagnant? It is not because the desire of human beings is limited, but because too many old products are produced while new types of products have not been invented or have not reached the stage of commercial production. This leads us to the root cause of both the GFC and European debt crises—the scarcity of innovations. This reasoning is also supported by the very slow recovery from both crises—advent and implementation of innovations takes time so there is no quick fix for an economic recession accompanied by stagnant sales.

So far, most people still believe that the GFC and European debt crisis are essentially financial issues due to the problems in the financial sector. On the surface, it appears so because the crises started from the debt crisis and ramified through financial links. However, as we will discuss in more details in Sects. 5.1.2.2 and 5.1.2.3, although financial problems may have an impact on the real economy, their impact should be short-lived. The bust of financial bubbles may cause large changes in asset prices, but it changes neither the amount of capital in real term nor the amount of money in the economy (unless monetary policy changes).

Someone is broke during a bust because his/her money/capital is transferred to someone else. If this money or capital can find a profitable investment opportunity (e.g. producing a type of goods highly demanded by consumers), total investment will not reduce. As a result, the economy will not go into a recession or, if it does, it will recover fairly soon.

## 5.1.2  The Features and Essence of a Boom-Bust Cycle

Based on the above historical review, the economic boom-bust cycles are complex. These cycles have different sizes and scopes, unpredictable durations and periods, and are associated with different events such as speculative investment, cheap credit, deflation/inflation, war, natural disaster, trade and domestic policies, etc. In this section, we discuss the features and explore the essence of a business cycle.

### 5.1.2.1  Duration and Stages of Business Cycles

Economists have identified different durations of business cycles from different perspectives. For example, Juglar (1862) identified a business cycle of 7–11 years caused by fixed-investment decisions, Kondratiev (1922) put forward a long wave technological cycle of 45–60 years, Kitchin (1923) claimed a cycle of 3–5 years caused by inventory changes, and Kuznets (1930) proposed an infrastructural investment cycle of 15–25 years. Schumpeter (1939) consolidated different types of cycles and formed a system of cycles, that is short cycles like Kitchin cycles and Juglar cycles within the Kondratiev cycles.

In a study of Juglar cycles, Schumpeter also identified that there are four stages of a business cycle, namely expansion, crisis, recession and recovery. The business cycles are shown as a sine or cosine curve (see Fig. 5.1). Since the size of an economy tends to increase over time, GDP or total output exhibits an obvious growth trend, so the detrended GDP or output is commonly used in studying business cycles. If the economy grows above the trend, e.g. from period 0 to period 1, the economy is in the expansion stage. When the economy

346     S. Meng

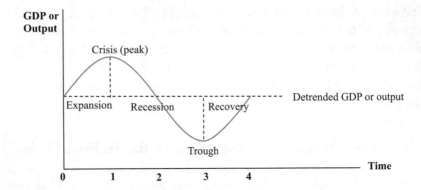

**Fig. 5.1** Stages of a business cycle

has peaked and starts to decline, economic crisis begins and it is followed by a recession, in which GDP or output declines considerably, e.g. from period 1 to period 3. When the decline of GDP or output stops, the recession reaches a trough and then the economy starts to grow to reach the pre-recession level of detrended GDP or output, e.g. from period 3 to 4. This stage is called the recovery stage. Afterwards, the economy will continue to grow to reach a new peak. This marks the beginning of a new expansion stage and a new business cycle. This identification and presentation are generally followed by other economists, although they may give different names and different numbers of stages.

Although a recession is typically defined as from the peak to trough in a business cycle, the term 'economic recession' or 'economic stagnation' is often used loosely to describe a period of slowing down or stagnancy of economic activities. The loose use of the term 'recession' causes significant problems in defining and studying business cycles. This promoted the National Bureau of Economic Research (NBER) to give an official definition: a recession is a significant decline in economic activity spread across the economy, lasting more than a few months, normally visible in real GDP, real income, employment, industrial production and wholesale-retail sales. This definition gives key features of an economic recession, but it also leaves ample room for interpretation, which reflects the disagreement of different schools of economic thought on this phenomenon.

## 5.1.2.2 Business Cycles as a Financial Issue

The cause and essence of a business cycle or economic recession is highly controversial. A number of causes are put forward by economists, such as oversupply of credit, speculative investment, inventory cycle, fixed-investment cycle, institutional or market failure, product life cycle, technological or invention cycle, herd mentality and the collapse of confidence of investors and consumers. Classical economists regard economic recession as a natural fluctuation or even as the result of optimization of consumers. We will discuss the proposed causes of economic recessions in detail when we introduce the major theories on business cycles later. Here, we can shed some light on the essence of economic recessions based on the historical business cycle events described previously.

The displayed economic booms and busts show that these events generally involve crises of the financial sector, e.g. money supply, bank run and bank failure. Are economic recessions essentially a financial problem? To answer this question, we need to discuss three closely related, often confusing, but totally different concepts: money supply, credit supply and capital supply.

The definition of money supply differs from country to country, but the common practice is to have different classifications for money, generally including monetary base, money supply $M_0$ (currency and coins in circulation), $M_1$ (narrow money), broad money $M_2$, $M_3$ and $M_4$. Monetary base is also called money base or base money. It is defined as currency and coins in circulation plus the reserve of a commercial bank held in a Central or Reserve Bank (but in some countries, it is referred to as the former or the latter only). $M_1$ is generally defined as currency and coins in circulation plus a deposit in the checking or transaction account, $M_2$ includes $M_1$ plus savings account deposits and other short-term deposits, $M_3$ is $M_2$ plus long-term deposits, and $M_4$ includes $M_3$ plus other deposits and other liquid assets. The complicated definitions of money supply stem from two complex key functions of money—money as both a measurement of wealth and a medium to facilitate the transactions in the economy. Different definitions tend to reflect the different degrees of the two functions involved. For example, the narrow money $M_1$ is more related to the medium function of money while the broad money $M_4$ is mainly an indicator of liquid assets. However, as will be shown later, these definitions disguise the function of circulation velocity and thus contribute to the confusion about money supply, credit supply and capital supply.

In ancient times, the value of metal money was imbedded in the value of metals. The advent of paper money separated the role of money as a transaction medium and as an asset, but the value of money is backed either by valuable metals or by the authorities. The importance of money as an asset is indicated by the gold standard enforced in the USA until 1933 (or the gold exchange standard in the USA until 1973). Nowadays, the value of paper money is guaranteed by the government—the safest and strongest guarantee in an economy.

Considering that only the amount of currency and coins in circulation ($M_0$) is directly under the control of the central bank, we can call them the core money supply or direct money supply. The other money supply can be called extended money supply or indirect money supply which, as we will see later, is actually a measurement of credit supply. In the case that there is no requirement for commercial banks to reserve a fraction of deposits for issuing credits, the core money means money base, or the total currency and coins the central bank supplied to the economy. On the other hand, if the commercial banks are bound to deposit their reserves into the central bank, the core money should be the total money issued minus the reserves submitted by commercial banks and kept by the central bank.

The difference between money supply and credit supply can be easily seen through the multiplier effect in bank lending. Assume the central bank requires a 10% cash reserve for the lending practice of commercial banks. If the central bank injects $1 million cash to a commercial bank, the commercial bank can lend $0.9 million to custom A. If A spends $0.9 million to purchase goods from B and B deposits $0.9 million back in the commercial bank, the bank can then lend $0.81 million to custom C, and so on, so the total lending the bank can make is:

$$1 * (1 - 0.1) + 1 * (1 - 0.1)^2 + 1 * (1 - 0.1)^3 + 1 * (1 - 0.1)^4 + \cdots$$
$$= 0.9/(1 - 0.9) = \$9 \text{ million}.$$

As such, $1 million increase in money supply leads to a $9 million increase in credit supply. This is called 'money supply multiplier effect'. The lower reserve is required by the central bank, the larger credit supply can be induced by an increase in money supply. If the central bank has no requirement of cash reserve for lending by the commercial bank, the credit supply can increase infinitely. However, the money supply multiplier effect is also constrained by the number of lending transactions the commercial bank are able to make, which in turn depends on the velocity of money circulation or how fast the money circulates, i.e. how many times the money is deposited back in the bank in a certain period. If the circulation velocity is 10 times a year, the money will change hands 10 times and will be deposited back in the commercial bank 5 times, so the credit supply generated by the $1 million cash injection is only $5 million even if there is no reserve requirement from the central bank.

This example highlights the limitation of the complex definitions of the money supply. Since the money has been deposited back in the bank 5 times, the deposits in the bank will be $5 million if no reserve from the commercial bank is required. Depending on what kind of account the deposit is in, the money supply in terms of $M_1$, $M_2$, $M_3$ or $M_4$ would become $6 million, including both the cash injection by the Reserve Bank and the deposits in the bank. As such, the traditionally defined money supply includes the effect of velocity, and thus, the amount of money supply goes hand in hand with the amount of credit supply. This essentially disguises the difference between the money supply and credit supply: only the core money supply $M_0$ is the true money supply; $M_1$ to $M_4$ are actually various types of credit supply. In considering the large difference between core money supply M and credit supplies $M_1$ to $M_4$ (e.g. $M_2/M_0$ in the USA over time is about 9), it is apparent that money supply is only a very small part of credit supply.

Capital supply means the savings made by households and firms. Since savings in real terms mean the produced, but not consumed goods, conceptually speaking, capital and capital supply can be totally unrelated to money and money supply. However, in practice, it is hard to measure and aggregate so many types of saved goods in order to quantify capital, so money becomes a convenient measurement of saved goods in aggregate.

As a result, the money deposited in the savings account is equal to the value of saved goods. Since this value is subject to the variation of prices of saved goods, savings in a bank account mean savings in nominal terms.

Accumulated savings become assets, e.g. unsold goods (inventories), real estate assets and durables. These assets in the real term are independent of any money values, but generally, the value of assets is valued in terms of money and thus becomes a nominal value. The difference between the nominal and real value of assets is assets' price. Generally, asset price indexes are used to measure how the real value and nominal value of assets depart over time. For two special types of assets, financial assets and intangible assets, it is hard to measure their real value because they have no physical quantity and all values are measured by money. Even in this case, we can specify a base-year value as the real value and use asset price indexes to indicate how the real and nominal values of these assets change over time.

Since savings and accumulated savings have both real and nominal values, we should examine capital supply (all savings) and capital demand (investment) in terms of both values. In real value, capital supplied, capital demanded and the real interest rate are determined simultaneously in a supply and demand system, so they are independent of nominal value or any money-related matter. That is why a change in nominal interest rate due to a change in money supply and/or in money demand may not affect capital supply as long as it does not affect the real interest rate. While the money market and credit market may or may not affect the capital market, the capital market has a direct impact on the money market and the credit market. Capital supply cannot affect money supply, but it can affect real and thus nominal interest rates in the money market. Meanwhile, capital supply as various types of saving deposits is the main component of credit supply. On the other hand, capital demand (investment) directly affects money demand and credit demand through the change of circulation velocity. As such, any change in the capital market will reflect in both the credit market and the money market and thus cause a financial disturbance.

From the above discussion, we can easily understand why economic booms and busts generally appear as financial phenomena. Is an economic recession simply a financial crisis or actually a problem in the capital market? It depends on different situations. If it is a pure financial problem (e.g. scarcity of money or credit crunch), the problem would be solved easily by supplying more money or credit, so the economic recession would be short-lived. The panic of 33 AD and the Asian financial crisis in 1997 are good examples of this. Investment bubbles due to unsustainable speculative demand can push the assets/commodity price to an unsustainable level, so they can also be viewed as a financial problem. The examples of this type of boom-busts include the tulip mania, railway mania, South Sea bubble and the Dot-Com bubble. These bubbles were also short-lived. However, many economic recessions lasted a long period of time. This indicates that they may not be a pure financial crisis and thus not a problem confined in the financial sector.

Although the principal and agent problem and the shadow banking system) do impact the whole economy, they are not the root cause of economic recessions. The main role of the financial sector is to provide lubricant to facilitate production and consumption, so the financial sector is not the engine of economic growth. As a result, the severe recession from 2008 to 2013 cannot be attributed solely to the auxiliary financial sector; rather, it indicates some problems in the engine of economic growth.

### 5.1.2.3  Essence of Economic Booms and Busts

Since the disturbance in the money market and credit market is only the outcrop of lasting economic recessions, like the Great Depression and the GFC, what is the essence of an enduring economic recession? If it stems from the problem in the capital market, it is important to reveal the problem and its root cause. In considering the capital market, an economic boom must be related to an increase in capital supply (savings) and/or an increase in capital demand (investment), which lead to a higher level of capital in use and thus a higher level of output. Inversely, an economic bust must be due to a decrease in savings or investment or in both. What causes the change in savings and investment during a boom and a bust? The real interest rate could be a factor, but it is not an exogenous one because savings and investment determine the real interest rate. If both propensity to save and propensity to invest are unchanged, the exogenous determinants of the capital market are the income level for savings and the expected profitability for investment. Since income is an endogenous variable in an economic system, it is determined by other economic variables in the system and thus it cannot be a root cause. As such, we should focus on the expected profitability to obtain the essence of an economic recession.

What causes the expected profitability to change? This stems from the nature of investment—to make a profit from the future selling of products. The expected profitability will be high if one expects good sales in the future, but the expectation will be crushed if further information indicates a pessimistic sales future. It is reasonable that a firm or a number of firms may make mistakes in projecting future sales, but it is not common to see most or all firms in the economy having wrong perceptions. Even if it is possible that all firms make mistakes to overinvest in seemingly profitable projects (e.g. the railway mania and the Dot-Com bubble), later the firms should be able to correct their mistakes by investing in other profitable projects. As such, the economic recession will be over soon because the revised investment on profitable projects will boost the economic growth. However, some recessions last for years or even as long as a decade. The only logical explanation for these persistent recessions is that this massive economy-wide misperception and inability to correct mistakes are due to limited investment opportunities, which result in firms' inability to diversify their investment. As a result, firms are forced to crowd into a few sunny spots, such as tulip production, the stock market, the IT industry and the housing mortgage market. The over-investment in the limited number of sunny spots will lead to a decrease in expected profitability. To make things even worse, some sunny spots have turned out to be not profitable, so the bubble bursts, expected profitability crushes, and investment reduces sharply.

To go one step further, why are there only a few investment opportunities? Why can't the firms invest in products that consumers like so that firms can make a profit in the future? Firms are unable to do so because of technical inability because these products have not been invented yet. This leads us to innovation scarcity. There are many products that have obvious great market potential; however, due to the technological constraint, these products have not been invented and thus cannot be produced. What causes the innovation scarcity? The nature of innovation and inadequacy of patent laws are to blame. These are the focuses of Chapter 6.

To sum up, speculative investment, credit expansion and various fiscal and monetary policies can be associated with an economic boom and bust. These factors alone can cause a relatively short economic fluctuation, but they are not the essence of a lasting economic recession like the Great Depression and the GFC. It is the lack of applicable innovations and thus the lack of new products that leads to the weak demand and thus scarcity of investment opportunity. As such, sales become stagnant and the economy goes into a recession. As applicable innovations are not available in a short period of time, no fiscal or monetary policy can lead to a recovery from the recession quickly. During a recession, the economy is waiting for applicable innovations that will lead to economic recovery and to an economic boom.

## 5.1.3 Patterns of Economic Growth

Economic growth is affected by many factors, so it is of no surprise that the economic growth rates in different countries can differ greatly. The 'little divergence' of economic growth within Europe in 1300–1800 and the 'great divergence' between the West and the East since 1800 have triggered substantial interest and discussions. However, it seems that the economic growth pattern has changed since the mid-twentieth century. The growth of a number of Asian countries has accelerated since the early 1960s, while the growth in the OECD countries has slowed down since 1970. This phenomenon is called the great convergence. Despite the change in the growth trend of the world economy, the economic growth in each country shows a cyclic pattern over time—it is irregularly disturbed by economic recessions and stagnations and followed by recoveries and expansions. This section will discuss the historical patterns of economic growth (cyclic growth, divergence and convergence) as well as the causes and implications of these patterns.

### 5.1.3.1 Cyclic Economic Growth

Economic recessions are a familiar phenomenon for many economies, especially for modern economies. In Sect. 5.1.1, we have seen a number of examples in history. However, any economic recession or stagnation will be over eventually, and the economic growth in the long run has demonstrated a convincing upward trend. Figure 5.2 shows the US real GDP growth from 1948 to 2017. The growth trend over time is obvious, but the growth is not smooth. The shaded periods indicate economic recessions, which occurred repeatedly and with various lengths.

The general explanation of cyclic economic growth is that, although various demand-side or supply-side shocks can cause economic recessions, technology will progress in the long run and this leads to the long-run economic growth trend. This explanation sounds reasonable and also recognizes the important role of technological progress in economic growth. However, the explanation masks the root cause of economic recessions and the mechanism by which technology affects economic growth.

Various shocks can cause economic fluctuations. These shocks include, on the supply side, the oil price shock, natural disasters and wars; and on the demand side, credit shock, income inequality and investment speculations. However, these shocks cannot explain the key feature of

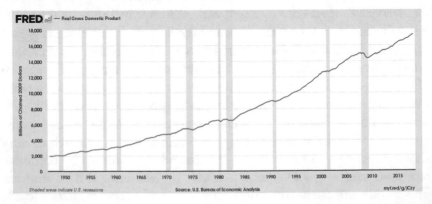

**Fig. 5.2** The US real GDP from 1948 to 2017 (*Source* US Bureau of Economic Analysis, Real Gross Domestic Product [GDPC1], retrieved from FRED, Federal Reserve Bank of St. Louis; https://fred.stlouisfed.org/series/GDPC1, 8 August 2018)

a long economic recession—stagnation in sales. The supply-side shocks have no explanation power on this feature, so they are not the cause of economic recession of plenty. The demand-side shocks can lead to stagnation of sales. For example, a credit expansion followed by credit tightening can lead to a sharp decrease in consumption and investment. Income inequality will lead to the limited purchasing power of the poor and thus stagnation of consumption. Investment speculation will lead to overinvestment in some periods followed by underinvestment when the investment expectation is pessimistic. However, as we discussed previously, these demand-side shocks cannot explain why a recession can last for a long time and why the recession will eventually be over. Mistakes in credit policy can be corrected easily and promptly. Mistakes in investment decisions can be corrected swiftly by reallocating investment to profitable projects. As such, an economic recession caused by credit policy or investment speculation should be short-lived. On the other hand, income inequality can explain the persistence of an economic recession: as long as income inequality exists, the poor have limited purchasing power and thus consumption and economy can stay in a recession forever. However, this explanation is at odds with the obvious positive economic growth trend in the long run.

Both the long-lasting economic recessions and the long-run economic growth trend can be explained by technological progress or innovation. Innovation is the source of technological progress, but the two types of innovation affect economic growth differently. Process innovation can increase the productivity of firms and thus has a positive effect on economic growth; however, the increased production capacity may outpace the growth of consumption and thus lead to sales stagnation and economic recession. On the other hand, product innovation can stimulate consumption and production at the same time, so it is a key driving force in economic recovery and long-run economic growth. The reoccurrence of economic recession suggests that the speed of product innovation has not kept up with the growth of production capacity. If institutional or policy changes (e.g. a thorough revision of patent laws) can speed up product innovation substantially, consumption stagnation and thus economic recessions may be avoided. As a result, the cyclic economic growth pattern may be replaced by a smoother and faster economic growth in the future.

### 5.1.3.2 Divergence of Economic Growth

Regarding the divergence of economic growth, researchers have identified a number of contributing factors. Jones (1981) explained Europe's economic rise came about through a favoured interaction between natural environments and political systems. Pomeranz (2000) argued that Europe's nineteenth-century divergence owed much to the location of coal and the trade with colonized states. Maddison (2007) claimed that the industrial revolution and colonialism brought about the great divergence. Allen (2012) argued that technological progress benefited the rich countries that pioneered it and thus caused the great divergence since 1820. Madsen and Yan (2013) found that culture, contracting institutions and property right institutions were all relevant for growth and development. Using data on international trade, GDP per capita and urbanization, De Pleijt and Van Zanden (2016) confirmed that human capital formation and institutional changes were the drivers of economic growth in the North Sea Area and thus were the cause of little divergence within Europe. Cox (2017) argued that political institutions and economic liberty had a significant impact on the divergence of economic growth in Western Europe before 1800.

These studies have triggered a debate and shed some light on the cause of divergence of economic growth, but no consensus has been achieved on the key factors affecting economic growth. The apparent reason is the complexity of economic growth: a number of social, economic and cultural factors act at the same time to generate an economic growth outcome, so it is difficult to separate their effects. As such, no one can convince others that a particular factor is the key. The way to circumvent the difficulty is to use a natural experiment which occurred in history, for example the separation of Eastern Germany and Western Germany, the separation of South and North Korea, the policy shift in China in 1978 and the dissolution of the Soviet Union in 1991. These natural experiments showed that, for two countries with a similar setting or even for the same country, institutional change or policy change can have a dramatic impact on economic growth. These results tend to indicate that institutions and government policies are a crucial factor in causing the divergence of economic growth. Using this insight from natural experiments, one can easily explain why many resource-constrained countries like Japan, Switzerland and Singapore perform much better than some resource-abundant countries. Using the institution and government policies as a criterion and applying it to various economies in the world, we can notice that, by and large, all advanced economies have a sound and efficient legal and market system. As such, we can draw a conclusion that, although many factors are important for the economic growth of a country, institution and government policies are the key factor.

Identifying institutions or government policies as the key factor is not our ultimate aim because this key factor is the outcome of other factors. What leads to a good or bad institution? The answer is complicated. Culture, technology, history and even the human factor (e.g. strong and powerful characters in leadership) can have substantial influences. For example, the development of the parliamentary system in the UK was driven in part by the monarch's need to raise money through taxes. The American political system was greatly influenced by George Washington. Despite the complexity in the formulation of a political system, it is arguable that the fundamental determinant of an institution is the needs and understanding of the rulers as well as the majority of the public—if the majority are not happy with the institution, they may overthrow the institution through an uprising. The establishment of the patent system as early as in 1474 resulted from the policy makers' realization of the importance of innovation to the economy and of the mechanism of the patent system in stimulating innovation. The establishment of the capitalist system, to a large extent, was due to the acceptance of Adam Smiths' theory of the 'invisible hand'. Without an understanding of the role of competition in achieving efficiency, policy makers might still be under the influence of mercantilism or other interventionism, resulting in great distortion to economic growth. Currently, people cannot predict and are unable to deal with economic recessions because they have not uncovered the fundamental cause of economic recessions and the mechanism to solve this problem.

### 5.1.3.3 Convergence of Economic Growth

Unlike the phenomenon of divergence on economic growth, researchers have reached some consensus on the causes of convergence on economic growth from the 1960s. Although a number of reasons were identified for the accelerated growth of Eastern countries, including globalization, slower population growth in most emerging and many developing economies, the higher proportion of income invested by emerging and developing countries, the fall in communication prices and technological diffusion from the West to the East (Bloom et al. 2002; Dervis 2012; Baldwin 2017), it is generally agreed that the catch-up effect due to technology diffusion is the key factor.

While most economists are happy with the explanation of the rapid economic growth in Asian countries from 1960, the explanation of the slowing down of the Western economy is not satisfactory. The main explanation is based on the balanced growth in a Solow-style exogenous growth model, where the progress of technology is exogenously determined. Due to the less than 100% capital share in the production function, an increase in capital exhibits a diminishing return, so the economy will converge to a balanced growth rate. The assumption of an exogenous technological growth rate was challenged by the endogenous growth model, which allows technological progress to increase the productivity of capital and thus breaks the spell of diminishing return. The endogenous growth model, however, has difficulty in explaining the slowing down of the economic growth rate in developed countries.

Nobody doubts that slow economic growth in developed countries is due to slow technological progress; in other words, the level of technological progress is not enough to sustain a high level of economic growth. Since innovation is the key driver of technological change, the slow technological progress indicates that the speed of innovation in developed countries is too slow. The slow innovation speed can be easily understood considering the two features of innovation. One is the high risk of innovation failure—investment in innovation risks the loss of everything. The other feature is the high possibility of imitation. If an innovation succeeds after numerous failures, it may be imitated by other firms and individuals, so the return on innovation investment is still very low. We cannot change the high-risk nature of innovation investment, but we can establish laws to forbid imitation and thus increase the return to innovation investment. The patent law to encourage innovation activities is a step in this right direction. As we will see in Chapter 6, the design of the current patent system is flawed and thus has failed to encourage enough innovations. Through a thorough revision of the patent system, we may encourage innovation to the greatest extent and sustain economic growth. As such, abundant innovation may lead to fast economic growth in developed countries and thus convergence may cease as a pattern of the world economic growth.

## 5.1.4* Existing Theories on Economic Recession and Growth

The impact of business cycles and the importance of economic growth have prompted a large amount of research in these areas. As a result, a number of theories and models have been formed to explain the causes of business cycles and to uncover the determinants of economic growth. In this section, we examine only the representative ones.

### 5.1.4.1 Underconsumptionists' Explanation

Underconsumptionists view consumption as the purpose, measurement and thus the ultimate determinant of production. In their eyes, economic recessions stem from the moralists' belief of thrift as a virtue, which causes inadequate consumer demand relative to production capacity. The most influential underconsumptionist argument was provided by Mandeville (1723). In his satirical poem 'The Fable of the Bees', Mandeville argued that it was 'the Vices' such as intemperance, luxury and pride of man that led to the high consumption and promoted manufacture and industry. So these vices are the basis for prosperity and public happiness.

The underconsumptionists' view was refuted by moralists. Hutcheson (1750, pp. 61, 66) admitted that a 'small part of our consumption … is owing to our Vices', but he thought an 'equal consumption of manufactures, and encouragement of trade may [exist] without these Vices'.

It is Adam Smith who totally defeated the argument of Mandeville. Smith (1776, p. 321) stated 'That portion which he annually saves, as for the sake of profit it is immediately employed as a capital, is consumed in the same manner, and nearly in the same time too, but by a different set of people'.

However, the reasoning of Smith that savings are immediately invested, and thus consumed, overlooked the fact that investment is dependent on the expected future consumption: without an increase in future consumption, the goods brought about by investment cannot be sold. This point was picked up by Malthus (1836), but Malthus' concern was quickly dismissed by Ricardo (1952) and Mill (1844) because they thought the people's will to purchase was very seldom wanting where the power to purchase existed. Later, the underconsumptionist view was revived and developed by Karl Marx and John Keyes.

While both Keynesianism and Marxism provided some explanation for underconsumption, they failed to gain orthodox status (although some of their elements are absorbed into orthodox economic theory). This may be because its underconsumptionist view is largely rejected by history: it is proven that consumption has kept going up over time and economic recessions have eventually ended. However, one should not discard the whole underconsumptionist's theory without careful reasoning. It is valid to argue that a person's desire for all kinds of goods (i.e. overall consumption) is unlimited and thus overall underconsumption is implausible, but a person's desire for any type of goods may be limited and thus may cause underconsumption or overproduction of any type of goods. This is evident in the phenomenon of market saturation. The paradox between the limited consumption of each type of commodity and the unlimited overall consumption rests on the increasing number of goods: as time passes by, product innovation keeps bringing us new goods and services such as iPhones, driverless cars and space travel. This also explains the paradox between the apparent underconsumption (i.e. stagnation of sales) during an economic recession and forever increasing consumption in the long run. These paradoxes suggest that it is important to study the impact of product innovation on our consumption and thus on economic recessions.

## 5.1.4.2 The Explanation of Classical Economists and the Real Business Cycle Model

Classical economists (Old, New, or Neo) have great faith in the efficiency of market mechanisms and in perfectly competitive markets. They regard economic recessions as large natural economic fluctuations (e.g. Lucas 1975; Kydland and Prescott 1982; Plosser 1989; Prescott 1986). They believe that, if market forces were allowed to operate alone, economic recessions would be temporary or relatively short-lived. Consequently, they argue that government intervention is unnecessary.

The real business cycle model is a representative explanation of new classical economists. Kydland and Prescott (1982) and Long and Plosser (1983) applied rational expectation to develop a real business cycle model, which is further developed by other researchers and acquired the name dynamic stochastic general equilibrium (DSGE) model. This type of model highlights the role of technological shock in business cycles. The curious and controversial results from these models are that business cycles or economic fluctuations are due to economic agents' Pareto-efficient response to technological shocks. In other words, any stage of business (boom, bust, recovery and expansion) cycle is equilibrium; economic recessions are necessary market corrections and thus are efficient and desirable, so government intervention to reduce instability of the economy will reduce welfare. This modelling result is popularly illustrated by a simple Crusoe economy.

By using the story of Robinson Crusoe on a desert island, one can imagine a simple agricultural economy: Crusoe used his labour and necessary tools (a primitive form of capital) to produce crops. If there was unusually good weather compared to the weather Crusoe experienced over the previous years, the productivity of Crusoe would increase dramatically. If Crusoe worked the same hours as before, the output level would be higher than that in previous years. However, it is more efficient for Crusoe to take advantage of the higher productivity resulting from the good weather: Crusoe would work more hours to produce even more crops when the weather was very good. When the weather later becomes bad, he can work less hours and enjoy more leisure time.

This intertemporal labour substitution will lead to larger output fluctuations: the longer working hours and high productivity during good weather lead to a bumpy harvest while the shortened working hours and low productivity during bad weather reduced output to a minimum. As such, the large output fluctuation is actually created by Crusoe's optimal response to change in weather or his productivity.

This model relies on the concept of rational expectation proposed by new classical economists: Crusoe can judge what kind of weather is average, above average and below average. Introducing rational expectation into a neoclassical framework does overcome the reliance on implausible or trivial causes to explain business cycles, such as the money illusion proposed by monetarists based on adaptive expectation. However, even with rational expectation, new classical economists cannot explain satisfactorily the business cycle and economic growth for a number of reasons.

First, new classical economic models inherited Smith's doctrine that saving is invested immediately, which has profound implication in modelling results. Since saving is invested at any given time in the dynamic model, supply always equals demand, so equilibrium is presumed at any time by Smith's doctrine. In other words, any disequilibrium or overproduction is precluded from the model assumption. This contradicts the disequilibrium nature of economic recession manifested by stagnancy of sales, massive unemployment and idle capital even if the interest rate is very low. Precluding any disequilibrium is also incorrect conceptually or theoretically. Equilibrium is a process—market equilibrium price and quantity are achieved from people's reaction to disequilibrium (excess supply or demand), so if one allows the existence of equilibrium, one must allow the existence of disequilibrium. Many factors such as endowment, technological and institutional changes can cause disequilibrium. By ignoring the disequilibrium phenomenon, new classical economists have overlooked an important impact of these changes.

Second, in a real business cycle model, the household utility is determined by both the level of consumption and the amount of leisure time. This would be a plausible specification of utility function if the equilibrium is guaranteed at any point of time: people can choose more consumption or more leisure time dependent on their preference. As discussed above, disequilibrium does exist, especially during a recession.

During disequilibrium, the substitution effect between consumption and leisure time will become an absurd assumption. For example, during a recession, many people were not able to find a job and suffered from hunger and cold. However, according to the utility function in the real business model, people simply substituted consumption with leisure time. Namely, people optimally choose (although they were actually forced to choose) to have more leisure time and less consumption, so they were happy with their unemployment status and enjoyed hunger and cold. Apparently, the efficient and optimal response claimed by new classical economists was presumed in the model and had no relevance to reality.

Third, the illustration of intertemporal labour substitution in the Crusoe economy in Prescott (1986) is not a correct portrait of modelling results. Christopher D. Carroll (2017) presented a concise summary of the model by Prescott (1986) as well as the critiques by Summers (1986). In Prescott's real business cycle model, the household utility was a Cobb–Douglas function of consumption and leisure time. When the price of consumption was normalized to 1, the utility maximization procedure necessitated that the price of leisure at a given time was positively related to consumption and negatively related to the amount of leisure time, namely

$$W_t = \gamma/(1 - \gamma) * C_t/Z_t, \text{ or } W_t * Z_t/C_t = \gamma/(1 - \gamma)$$

where $W_t$ is the price of leisure time, $C_t$ is the amount of consumption, $Z_t$ is the amount of leisure time, and $\gamma$ is the weighting of leisure time in the utility function.

Since $\gamma$ is assumed constant, so the ratio of spending on leisure to spending on consumption is constant over time. This leads to the co-movement between consumption and leisure time. As such, as leisure time increases during bad weather, consumption should also increase. This contradicts common sense: during a recession, most people would cut spending on consumption. Moreover, the argument of the intertemporal labour substitution effect, which is crucial to the real business cycle model, is also not supported by empirical data.

Last, rational expectation may be useful for a model with suitable assumptions, but it is of little use in the long run because our world is full of uncertainty. In the Crusoe economy, new classical economists assume a normal or stable average weather condition, so unusually good weather would trigger an intertemporal substitution effect. In our uncertain world, the weather always changes and nobody knows what the weather will be next year. If Crusoe predicted that the weather next year would be even better, he should work less this year even if the weather is good compared with past experience. Since one does not know the future condition for sure, there is no way one can confidently use rational expectation to conduct an intertemporal optimization. The uncertainty about our lifespan also matters here. Nobody knows exactly how long one can live, so how can one decide how much to save for the leisure time in the future? The safest and most practical way to cope with uncertainty is to work as hard as possible and save as much as possible, no matter if the weather is good or not.

Classical economists also tend to deny or ignore the important features of economic recession highlighted by Keynesian economists. The real business cycle model, for example, explains none of the main features of an economic recession such as stagnant demand, unutilized capital and high unemployment. These features indicate clearly that an economic recession is mainly a problem on the demand side: the economy has plenty of resources (e.g. unemployed capital and labour) and capacity to produce, but the sales stagnate. Classical economists weakly argued that the appearance of oversupply is due to overproduction by high-cost firms or due to a mismatch of production and consumption, and that high unemployment is due to inflexible wages or voluntary unemployment in the short run. Even if these far-fetched arguments were true and thus an economic recession could be viewed as an economic disequilibrium, the question to be answered is why an efficient market allows this disequilibrium to last for years or even for a decade? Classical economists have to admit that either the market is inefficient (so a disequilibrium can last for a decade) or a recession is not simply disequilibrium. Either way, classical economists will contradict their own belief.

By describing economic recessions as natural fluctuations, however, classical economists avoid the task of finding the causes of economic recessions. Instead, they focus on developing economic models and econometric estimations and choose to be indifferent to the economic and psychological damage of a recession on human beings.

It is not surprising that the classical economic solution to economic recession—natural recovery—is equally unpopular with government and the public alike.

### 5.1.4.3  Keynesian Theory of Effective Demand and Multiplier-Accelerator Model

The apparent oversupply or underconsumption of commodities during an economic recession was explained by Keynes (1936) and further by his successors, labelled 'Keynesian economists' (either 'old', 'orthodox', 'new' or 'post'). The main contribution of Keynesian economics to explaining economic recession was shown by the concept of deficiency of effective demand. Keynes attributed this deficiency to decreases in investment. He demonstrated that a decrease in investment would lead to a proportionally greater decrease in output through a multiplier effect. Keynes determined the most important causes of this investment deficiency to be, first, a lack of 'animal spirits' (entrepreneurship), and second, the liquidity preference, or the speculative motive to hold cash in a world characterized by uncertainty (the 'uncertainty argument'). Keynes attributed high unemployment during a recession to the fluctuations of expected profit (or 'marginal efficiency of capital' in Keynes' words), resulting from unstable investment expenditure. The liquidity preference and uncertainty argument were further developed by post-Keynesian economists (e.g. Davidson 1984, 1991), while microeconomic foundations for demand deficiency and unemployment were developed by new Keynesian economists, including the ideas based on real and nominal wage rigidity, price rigidity, efficiency wages, etc. (e.g. Mankiw 1985, 1989; Akerlof and Yellen 1985; Romer 1993).

Harrod, Hicks, Samuelson and others combined the multiplier and accelerator and developed a multiplier-accelerator business cycle model. The principle of accelerator that investment decision is determined by the future increase in demand was put forward by Albert Aftalion (1913) and John Maurice Clark (1917). The multiplier effect that investment increase can lead to a magnified increase in output level was laid out by Kahn (1931) and Keynes (1936). It was Roy F. Harrod (1936) who combined multiplier and accelerator to explain trade cycles. This line of work was continued by Hicks, Samuelson and others. There are many different versions of the model, but the base model can be shown as follows.

The income–expenditure equation at year $t$ can be shown as:

$Y_t = C_t + I_t$, where $Y$ stands for income, $C$ consumption and $I$ investment.

The Keynesian-style consumption function can be written as:

$C_t = C_0 + cY_{t-1}$, where $C_0$ is a constant, and $c$ is the propensity to consume $0 < c < 1$.

As we know previously, the above two equations generate the multiplier effect. The accelerator effect is shown by the following investment equation:

$I_t = v(Y_t - Y_{t-1})$, where $I$ means investment, and $v$ is a positive constant.

Plugging the equations for $C_t$ and $I_t$ into the income–expenditure equation, we have:

$$Y_t = C_{t0} + (c + v)Y_{t-1} - vY_{t-2}$$

This equation is the reduced form of the multiplier-accelerator model, which includes the multiplier parameter ($c$) and accelerator parameter ($v$) and demonstrates the output dynamic over time. Depending on the value of parameter $c$ and $v$, the above equation can generate output cycles of constant amplitude, explosive or damped output cycles and non-cyclic output growth.

Although the multiplier-accelerator model can generate output cycles, it by no means indicates that the model reveals the causes or mechanism of business cycles in reality. The rationale of the accelerator makes sense that investment decision depends on the increase of future consumption, but the model uses the past output growth as an indicator of future consumption change. This effectively assumes that the past patterns will always continue in the future. This is an implausible assumption which makes the model a mechanic and unrealistic tool.

Keynesian economists intuitively identified that the key features of an economic recession are depressed demand and high rates of unemployment. These features were, however, attributed to quite unusual factors, e.g. liquidity preference and the lack of entrepreneurship (by post-Keynesian economists), and wage and price rigidity in an economy (by new Keynesian economists). The question unanswered by Keynesian economists is why these factors exist during a recession but do not exist in an economic boom? In other words, Keynesian economists have not gone far enough to uncover the causes for liquidity preference and the lack of entrepreneurship. Keynesian economists discarded the long-standing assumption of classical economics that perfectly competitive markets exist. By rejecting the existence of Adam Smith's 'invisible hand' (i.e. the efficiency of the market), the solution by Keynesian economists became one of the interventionism.

### 5.1.4.4  Marx's Exploitation Theory

Marx's explanation of economic recession has been given little attention in economics literature, perhaps because of his radical idea of advocating class warfare. Nonetheless, there is an element of truth in the Marxist argument that warrants discussion here. Marxists determine that economic recession is caused by income inequality. Their explanation is based on their observation of the behaviours of capitalists. In order to obtain as much profit as possible, capitalists tend to produce as much as possible and, on the other hand, try to push wages down and raise the rate of surplus value. As a result, the workers are unable to buy up the value they produced and this causes excess supply and inadequate aggregate demand.

Marxists have highlighted the inequality issue. There is little doubt that income distribution inequality plays an important role in economic recessions. While it is clear that income inequality is a contributing factor to the economic recession, this inequality may not be a fundamental factor underpinning recession. Otherwise, the economy would stay in recession or stagnation because the unequal income distribution does not change without a dramatic reform or a change of institution. However, history shows that every recession has moved to recovery and expansion stages. Moreover, in considering the large production capacity in the modern global economy, there is always a possibility for overproduction and thus a deficiency of demand, even if income is equally distributed and everyone has sufficient income to buy what they want. From this point of view, income inequality can aggravate or accelerate a recession, but it is not the fundamental cause.

### 5.1.4.5 Credit Cycle and Speculation Theories

The credit cycle theory is initiated by Fisher (1933) and taken up by post-Keynesian economists and the Austrian school of economic thought. The basic reasoning of this theory is as follows. During an economic expansion, both banks and firms are confident about the future, so banks are willing to provide credits and firms are eager to borrow. The cheap credits lead to the over-indebtedness of firms. As the economy expands and booms, the firms and banks may find that they are overconfident about the future of the economy, so the shake of confidence leads to liquidation. Liquidation in turn leads to distress selling and a fall in the level of prices, profit, net worth of business and output, so the economy goes into a recession, during which banks are pessimistic and are reluctant to lend and firms lose confidence in investment and thus hoard money.

Fisher (1933) acknowledged the complex nature of the business cycle and stated that there may be many forces influencing the cycle; however, he regarded the over-indebtedness during an economic boom and the subsequent debt deflation during a recession as being the primary Factors. In the chain of Fisher's reasoning, the key step is from over-indebtedness to liquidation and distressed selling, which starts the whole sequence of motion. Fisher regarded the alarm of either debtors or creditors or both led to liquidation, but he did not go further to investigate what causes the alarm of debtors and creditors. The alarm or shake of confidence must come from the realization that the expected profitability is not achievable. If this realization comes from a few firms, it will not lead to an economy-wide shake of confidence and thus widely spread liquidation. Even if a number of firms find that they are over-optimistic about their investment but they can find other profitable investment opportunities, the liquidation activity will be accompanied by investment activities, so there will be no money hoarding and no decrease in the price level and output level. By this reasoning, the lack of investment opportunities is the key factor behind a credit cycle.

Speculation theory has a long history and has been used to explain almost all economic bubbles. The basic reasoning of the theory is that the volatile nature of speculation activity can cause the instability of the economy. Minsky (1986, 1992) regarded financial instability was endogenous to capitalist economies. He classified financing to three types: hedge, speculative and Ponzi. He argued that capitalist financial systems had a long-term tendency towards speculative and Ponzi investment, which led to asset price booms and busts. Although Minsky did not think that financial cycles and real cycles were congruent, he stated that both types of cycles might converge at times. The damage caused by speculation activity prior to the Asian financial crisis and to the GFC were recognized by many so they propose reforms of the international monetary system, e.g. a new Bretton Woods agreement, the use of special drawing rights (SDR) at IMF as international reserve currency, or construction of a supranational monetary institution or an international clearing union (Wang 2016).

Speculation activities can cause large economic fluctuations. This is especially true with the establishment of modern financial markets. The invention of the share market has provided a way for firms to raise capital in the primary share market, but it has also created an opportunity for speculation in the secondary share market. A sharp rise in share prices can make numerous millionaires overnight, but a sharp fall also makes many go broke quickly. The creation of the financial derivatives market (e.g. futures and options markets) has provided a channel for firms to hedge various risks, but it also provides more speculative products and increases the leverage of funds. Since speculators in futures and options markets do not need to buy the underlying financial assets, they just need to spend a small amount of money (compared with the value of the underlying financial assets) to buy a contract or a right for purchasing or supplying the assets in the future. The options market separates rights from obligations. This, on the one hand, gives the option buyer the rights with no obligation and, on the other hand, gives the option writer the chance of collecting premiums from buyers while incurring potentially unlimited obligation. If the speculation turns out to be wrong, there are severe consequences to the speculators and ramifications on the market and on the economy.

The impact of speculation in financial markets on the real economy can be shown as the price or wealth effect. When share prices increase, the wealth of shareholders increases and thus they feel richer. As such, they tend to spend more and stimulate the economy. The opposite situation may occur when the stock market crashes. However, this impact is only for the short run. Once the easy money from the stock price windfall is spent in the economy, commodity prices will go up and the purchasing power of the increased wealth shrinks. Once the commodity prices are stabilized in the higher level (or the inflation rate falls), the effect of increased wealth will be cancelled out by higher living costs and people's spending will decrease. This effect is similar to the situation of a substantial increase in the money supply or quantitative easing of monetary policy. During the price adjustment period, people tend to spend hot money as soon as possible to obtain real assets or commodities at lower prices, so there will be a temporary effect on the real economy. Once the price adjustment period completes, the influence on the real economy ceases.

Moreover, the impact of speculation on the economy is generally overstated due to misunderstandings or misinterpretations of the financial market. One such misunderstanding is the failure to differentiate the real value from the nominal value. It is often reported that a stock market crash wiped out many billions or even trillions of dollars of wealth overnight. There were celebrations at New York exchange when Dow Jones or Nasdaq index went above a historical point (e.g. 5000 points for Nasdaq). The salaries of senior managers of many listed financial institutes (e.g. banks) and non-financial companies are tied to the price of the listed stocks of these entities. In fact, the prices of shares indicate the nominal value only. For example, the rise of the share price of the Apple Company increases only the nominal value of the company in its accounting book, and the real value of the company does not change because the people and the real assets of the Apple Company do not change with share prices. This is easily understood by the analogy of housing prices. If the price of a house is doubled due to market factors, its nominal value doubles, but the physical asset is unchanged. As such, the shares index represents only the nominal value of the financial market, and thus, it is not necessarily a true indicator of the underlying economy, and thus, a significant increase in the share price index is by no means worth celebrating. Instead, the stock price index should be used to deflate the nominal value of financial assets, fulfilling a function similar to the CPI or the housing price index.

The misunderstanding of 'market liquidity' and 'capital' also contributes to the overstatement of the impact of speculation activities. During a stock market crash, many people lose money and very few buyers inject money into the market, so the market liquidity becomes very low. However, the amount of capital in the economy does not change. The cash money lost by someone in the market is simply transferred to one or more other participants in the market, so the cash money or liquid capital does not disappear during a stock market crash. It simply changes hands and gets out of the stock market. If this capital is used to invest in other parts of the economy, the crash of the stock market will have minimal impact on the real economy. This reasoning is consistent with the facts during the Asian financial crisis—a large volume of funds exited from Asian nations but flew back into the US economy. There were short-run impacts on Asian countries, but the overall impact on the world economy was mild and relatively brief.

The case of the GFC was different. When the funds exited from housing markets and financial markets, the capital was not reinvested in the economy and this caused a long and deep recession around the world. The funds or capital did not disappear after the outbreak of the GFC, but stayed in liquid form due to the preference of the investors (i.e. liquidity trap in Keynes's term). On the surface, the cause of a liquidity trap is the lack of confidence of investors. Going one step further, one would find that the lack of confidence is due to the scarcity of investment opportunities, which was manifested by the widespread sales stagnation prior to and during the GFC. As such, the underlying cause of the GFC was the sales stagnation in the real economy, rather than the problems or fluctuations in the financial market.

## 5.1.4.6 The Theories of Product Life Cycle and Technological Progress Cycle

The concept of product life cycle is originated by Raymond Vernon. In studying international trade and investment, Vernon (1966) introduced a product cycle concept. He identified three stages of product development: new product, maturing product and standardized product. During the standardized product stage, production is shifted from developed countries to developing countries, so developed countries need to import the product. Vernon's concept was evolved into a product life cycle theory, which generally included four phases of a life cycle: introduction, growth, maturation and decline.

However, to use product life cycle theory to explain business cycles requires that the life cycles of most products in the economy are approximately synchronized—the products are at the same stage at approximately the same time. Otherwise, different stages of various products will average out the life cycle effect and thus will not lead to a business cycle. The innovation cycle or technological progress cycle may help to generate synchronized product life cycle. Based on innovation cycles, Kondratiev (1922) developed a long-wave theory of business cycles which showed the correlation between business cycles and fundamental discoveries implemented in production. He identified three long waves in economic history: the first cycle is about 60 years from 1789 to 1849; the second is about 47 years from 1849 to 1896; and the third is from 1896 to 1920. Schumpeter (1939) further developed the Kondratiev wave theory by introducing a chain of cycles. He claimed the longest Kondratiev cycle consists of 6 Juglar cycles, each of which in turn consists of 3 Kitchin cycles. In supporting Schumpeter's theory, Mensch (1975) investigated the cause of technology stagnation. He distinguished three types of innovations—basic innovations, improved innovations and pseudo innovations—and argued that the lack of basic innovations caused a stalemate in technology.

The product life cycle theory and the long wave theory are innovative approaches in that they demonstrate that the explanations of the business cycle may come from non-economic factors. While it is valuable to emphasize the importance of technology and innovation, these two theories have not linked technology to either the supply or demand side. Although non-economic factors may influence economic performance, they must act through the economic system. Failure to connect non-economic factors to the economic system leads to the failure to uncover the mechanism by which the non-economic factors work. More importantly, the cause of technology or product cycles has not been addressed. Mensch (1975) made a valuable contribution in this direction by identifying that the lack of basic innovations is the cause of technology stagnation, but he did not go further to investigate what caused the shortage of basic innovations. As a result, he was unable to provide a solution to the problem.

### 5.1.4.7* Exogenous Growth Models

The current economic growth theories only consider the supply-side factors; i.e., the output of the economy is generally determined by inputs (e.g. labour and capital) and production technology. In an exogenous growth model, the technology is determined outside of the model. There are mainly three types of exogenous models, the Solow model by Solow (1956), the Ramsey–Cass–Koopmans model by Ramsey (1928), Cass (1965) and Koopmans (1965), and the Diamond model by Diamond (1965).

The Solow model assumes that technology improves labour effectiveness, so its production function takes the following form:

$Y(t) = F[K(t), A(t)L(t)]$, where $Y$ is output or income, $K$ is capital input, $L$ is labour input, and $A$ is the level of technology. They are all changing over time. $A(t)L(t)$ is called effective labour.

With the assumption of constant returns to scale, the above production function can be transformed to:

$$Y(t)/[A(t)L(t)] = F\{K(t)/[A(t)L(t)], 1\}.$$

Letting output per unit of effective labour $y(t) = Y(t)/[A(t)L(t)]$ and capital per unit of effective labour $k(t) = K(t)/[A(t)L(t)]$, we can have a production function in the intensive form:

$$y(t) = f[k(t)]$$

According to the principle of diminishing marginal productivity, the return to capital decreases as the capital per effective labour increases, that is:

$$dy(t)/dk(t) = d\{f[k(t)]\}/dk(t) < 0, \text{ or } y(t)' = f[k(t)]' < 0$$

Assuming labour and technology grow at a rate of $n$ and $g$, respectively, we have:

$$dL(t)/L(t)dt = n \tag{5.1}$$

$$dA(t)/A(t)dt = g \tag{5.2}$$

The capital accumulation is determined by gross investment (or savings) and capital depreciation. Assuming a saving rate of $s$ and a depreciation rate of $\delta$, the capital dynamic can be expressed as:

$$dK(t)/dt = sY(t) - \delta K(t) \tag{5.3}$$

In the form of per effective labour, we have:

$$
\begin{aligned}
dk(t)/dt =& d\{K(t)/[A(t)L(t)]\}/dt \\
=& \{A(t)L(t) * dK(t)/dt - K(t)A(t)dL(t)/dt \\
& - K(t)L(t)dA(t)/dt\}/[A(t)L(t)]^2 \\
=& [A(t)L(t)]^{-1} * dK(t)/dt \\
& - K(t)[A(t)L(t)]^{-1} * dL(t)/L(t)dt \\
& - K(t)[A(t)L(t)]^{-1} * dA(t)/A(t)dt
\end{aligned}
$$

Recall Eqs. (5.1), (5.2) and (5.3), we have

$$
\begin{aligned}
dk(t)/dt =& [A(t)L(t)]^{-1} * [sY(t) - \delta K(t)] \\
& - K(t)[A(t)L(t)]^{-1} * n - K(t)[A(t)L(t)]^{-1} * g \\
=& s[A(t)L(t)]^{-1}Y(t) - \delta[A(t)L(t)]^{-1}dK(t) \\
& - nK(t)[A(t)L(t)]^{-1} - gK(t)[A(t)L(t)]^{-1} \\
=& s * y(t) - \delta k(t) - nk(t) - gk(t)
\end{aligned}
$$

Since $y(t) = f[k(t)]$, we have

$$dk(t)/dt = sf[k(t)] - (\delta + n + g)k(t)$$

The dynamics of capital are determined by the relative size of $sf[k(t)]$ and $(\delta + n + g)k(t)$. Since output level is positively related to capital inputs (i.e. $f[k(t)] > 0$), and marginal productivity of capital is diminishing (i.e. $d\{f[k(t)]\}/dk(t) < 0$), the growth of output per effective labour $f[k(t)]$ is small when $k(t)$ is high. This may lead to $sf[k(t)] < (\delta + n + g)k(t)$ and thus a negative growth of capital (i.e. $dk(t)/dt < 0$). On the other hand, when capital level $k(t)$ is low, $f[k(t)]$ is high enough to satisfy $sf[k(t)] > (\delta + n + g)k(t)$. This leads to a positive growth of capital. As such, the capital level tends to converge to a break-even level at which $sf[k(t)] = (\delta + n + g)k(t)$ and thus leads to a steady-state growth or a balanced growth.

The Solow model was empirically examined by growth accounting and by convergence of growth of different economies in the world. The theoretical foundation for growth accounting can be briefly derived as follows.

Based on the Solow production function $Y(t) = F[K(t), A(t)L(t)]$ and using the traditional assumptions that wage equals marginal product of labour and that capital rent equals marginal product of capital, we have:

$$
\begin{aligned}
dY(t)/dt &= [\partial Y(t)/\partial K(t)] * dK(t)/dt + [\partial Y(t)/\partial L(t)] * dL(t)/dt \\
&\quad + [\partial Y(t)/\partial A(t)] * dA(t)/dt \\
&= v * dK(t)/dt + w * dL(t)/dt + [\partial Y(t)/\partial A(t)] * dA(t)/dt
\end{aligned}
$$

In terms of percentage growth rate, we have:

$$
\begin{aligned}
dY(t)/Y(t)dt &= v * dK(t)/Y(t)dt + w * dL(t)/Y(t)dt + r * dA(t)/Y(t)dt \\
&= v * K(t)/Y(t) * dK(t)/K(t)dt \\
&\quad + w * L(t)/Y(t) * dL(t)/L(t)dt \\
&\quad + r * A(t)/Y(t) * dA(t)/A(t)dt \\
&= \alpha_v * dK(t)/K(t)dt + \alpha_v * dL(t)/L(t)dt + R(t)
\end{aligned}
$$

So, the percentage growth of output can be attributed to the percentage growth of capital, the percentage growth of labour and the Solow residual $R(t)$ due to other factors including technology. The empirical data showed that the residual $R(t)$ is substantial.

The Solow model predicts that an economy will converge to a balanced growth rate. Applying this convergence to different countries in the world, one would conclude that poor economies have lower capital per effective labour and thus tend to grow faster and the growth of rich economies tends to slow down, so the economic growth in the world should also converge. Baumol (1986) examined the economic growth of 16 industrial countries from 1870 to 1979 and found perfect convergence, but this study was criticized by De Long (1988) as spurious due to sample selection bias as well as measurement error. Nevertheless, the history of economic development shows that the developing countries are catching up while the developed countries are slowing down, so the convergence of economic growth is accepted by most economists.

In the Solow model, the saving rate is exogenous and the model lacks a microeconomic foundation. The Ramsey–Cass–Koopmans model advanced the Solow model by including competitive firms and a fixed number of infinitely lived households in the model. The saving rate is endogenously determined by the household consumption decision through maximization of lifetime utility subject to intertemporal budget constraints. The steady state is at the level of capital which stabilizes both consumption and investment. The economy moves towards the steady state through the saddle path or balanced growth path. When government spending is included in the model, it is expected that a permanent increase in government spending will cause a change in investment and consumption while a temporal increase in government spending will lead to a decrease in capital stock and thus an increase in the real interest rate. This prediction is supported by the empirical study by Barro (1987) but was rejected by Weber (2008).

The assumption of infinitely lived households in the Ramsey–Cass–Koopmans model is replaced by an overlapping generation of finite lifespan in the Diamond model. For simplicity, each individual in the model lives for only two periods, so there are only two generations in each period. The results from the Diamond model were similar to those from the Ramsey–Cass–Koopmans model, except that the Diamond model allows for dynamic inefficiency. Since each generation achieves a different level of utility, the total utility must be the weighted sum of different generations. It is impossible to maximize this total utility because the weighting is arbitrary. The results for a decentralized economy from the Diamond model may also be Pareto inefficient: the capital stock on the balanced growth path may exceed the level allowed by the golden rule, so a permanent increase in consumption is possible. Abel et al. (1989) proved that a sufficient condition for dynamic efficiency is that net capital income exceeds investment. Using this criterion to examine the seven major decentralized economies, they concluded that, although dynamic inefficiency is possible in the model, it does not appear in practice.

### 5.1.4.8   Endogenous Growth Models

Although the exogenous growth models can explain the role of technology in economic growth and the convergence of economic development in the world, the models have two major shortcomings. One is that the technology growth rate is viewed exogenously so there is no room for economic policies to improve technology growth. In reality, economic policies such as research and development (R&D) funds will have a positive impact on technological progress. The other shortcoming is that the production function used in the exogenous model assumes constant returns to all inputs (capital and labour). This necessitates a diminishing return to capital. However, experience tells us that, due to the progress of technology, the return to capital in the long run does not diminish. These shortcomings stimulate the rise of endogenous growth theories.

The diminishing return to capital in a Solow-style exogenous growth model is imbedded in the commonly used Cobb–Douglas production function:

$$Y = \mathrm{AK}^{\alpha}L^{\beta}, \quad \text{where } \alpha + \beta = 1.$$

To address the difference between the assumption of diminishing return and the observed non-diminishing return to capital in reality, Frankel (1962) introduced into the production function a parameter $H$—the efficiency of firms, so the production function becomes:

$$Y = \mathrm{AHK}^{\alpha}L^{\beta}, \quad \text{where } \alpha + \beta = 1.$$

He argued that, for an advanced economy, the parameter $H$ is large, so this will offset the diminishing returns to capital. To see this, we can let $H = (K/L)^\beta$, so the production function becomes $Y = AK$, which delivers a constant return to capital.

In a similar approach, Mankiw et al. (1992) introduced human capital ($H$) into the Cobb–Douglas production function and produced an augmented Solow model

$$Y = K^\alpha H^\beta (\mathrm{AL})^{1-\alpha-\beta}, \quad \text{where } \alpha + \beta = 1.$$

The introduction of human capital in addition to physical capital increases the share of capital in the production function, so the return to total capital decreases at a much slower pace. Mankiw et al. showed that the augmented Solow model explained the cross-country data very well.

The other approach to the diminishing return issue is to use a different production function. A simple form of production function exhibits constant returns to capital is the AK production function:

$$Y = AK,$$

where $A$ is a constant.

This production function has an origin from Cassel (1924 [1967]), an economist of the Austrian school of economic thought, but the AK production function was popularized by Roy Harrod (1939) and Domar (1946). They independently developed a model—the Harrod–Domar model, which is a predecessor of the AK model. The basic results of an AK model can be derived as follows.

The accumulation of capital is determined by the investment level and capital depreciation:

$$dK/dt = sY - \delta K.$$

As such, the economic growth rate can be expressed as:

$$g = dY/Ydt = dK/Kdt = (sY - \delta K)/K = sA - \delta.$$

Since $A$ is a constant, $\delta$ is exogenous, so the economic growth rate is exogenous if saving rate s is exogenous.

Romer (1986) and Lucas (1988) used an intertemporal utility maximization to determine household consumption level and thus the saving rate, so the economic growth rate becomes endogenous and the AK model becomes an endogenous growth model. Romer (1986) and Lucas (1988) used a more general AK function and with different foci. Lucas (1988) emphasized human capital accumulation through schooling and specialized human capital through learning by doing. Romer (1986) used a competitive equilibrium model with knowledge as an input that has increasing marginal productivity.

As an alternative to the above approaches, some economists modelled technology changes through innovations and established the 'innovation-based' growth theories. Romer (1990) assumed that aggregate productivity is a function of product varieties, which are created by innovations, so he used the Dixit-Stiglitz-Ethier function for his model.

$$Y = L^{1-\alpha} \int_0^A x(i)^\alpha, \ \ 0 < \alpha < 1$$

where $Y$ is output, $L$ is labour inputs, and $x(i)$ is a flow of intermediate inputs.

Since $0 < \alpha < 1$, the production function exhibits diminishing returns to both labour and intermediate inputs. However, as the variety of intermediate inputs (i.e. $A$ in the production function) increases, the use of each input decreases and thus productivity of inputs will rise.

Keely (2002) used the framework of Romer (1990) to demonstrate an argument put forward by Schmookler (1966) that a key determinant of technological change is the usefulness of new technology. Keely added to the model of Romer (1990) a number of variables, including the number of technological ideas or problems to be pursued $H$, the labour employed in manufacturing sector $L_Y$ and the labour involved in research or innovation activities $L_R$. The growth of the technological level is positively related to the number of technological ideas H and the amount of labour in research:

$$\Delta A = H * \left[1 - \left(1 - v\right)^{L_R}\right],$$

where $v$ is the probability of turning technological ideas to innovations.

Since it is assumed that the number of technological ideas $H$ is mainly determined by capital investment level $I$, Keely formulated three scenarios: $H$ is either a positive function of $I$, or a positive function of both I and the technological level A, or a positive function of I and $L_Y$. Based on this setting, the labour in innovation activities $L_R$ can lead to the growth of the technological level and would have a positive impact on the economic growth rate; however, labour in manufacturing sector $L_Y$ is the source of capital investment which has a dominant influence on technological ideas. Since the $L_R$ and $L_Y$ are constrained by the total labour force, the increase in research labour will increase the probability of transforming ideas into innovation and, in the meantime, it will also decrease capital investment and thus reduce the number of ideas. Consequently, the increase in research labour does not necessarily increase technology growth nor per capita income growth.

Other economists (e.g. Segerstrom et al. 1990; Grossman and Helpman 1991; Aghion and Howitt 1992) built endogenous models by utilizing the concept of 'creative destruction' by Schumpeter (1942) that innovation creates new products which replace old products. The difference between new products and old products is commonly expressed by a quality variable in the models—the new products have higher quality than the old products. A typical production function in a Schumpeter-style endogenous growth is as follows:

$$Y = L^{1-\alpha} \int_0^1 A(i)^{1-\alpha} x(i)^\alpha, \ 0 < \alpha < 1$$

In this production function, each intermediate input has a quality or productivity parameter $A(i)$. With this parameter, the growth rate of output is positively related to innovation activities because innovations can increase the average productivity parameter $A$.

Although endogenous growth models shed light on the factors that affect technological growth, the approach has been challenged on empirical grounds. For example, the endogenous growth models indicate that the long-term economic growth of different countries is likely to differ because of different economic institutions and policies; however, Mankiw et al. (1992), Barro and Sala-i-Martin (1992), and Evans (1996) showed that most countries seem to converge to similar long-run growth rates. The endogenous growth models generally show that R&D funds have an important impact of growth rate, but this has not been confirmed in an economy-wide context (Griliches 1988). The endogenous models relying on monopolistic competition predict that stronger competition laws will negatively affect growth rate. If this prediction is right, we should abolish competition laws and encourage market power. This is against common sense. The Schumpeter-style endogenous models show a scale effect; that is, a larger population leads to a larger market for successful innovation and thus induces a higher rate of innovation. However, Jones (1995) shows that the evidence of the USA and other OECD countries since the 1900s rejected the scale effect.

Facing empirical challenges, proponents of endogenous growth theories can always make a counter-argument or even modify their models and produce results consistent with empirical evidence. For example, the convergence of world economies is explained by technology diffusion and global investment flow from advanced countries to developing countries. Regarding the scale effect, Dinopoulos and Thompson (1998), Peretto (1998), and Howitt (1999) argued that the proliferation of product varieties reduces the effectiveness of R&D and thus offsets the scale effect. They modified their models and produced results consistent with empirical data. Aghion and Howitt (1998) demonstrated a variety of channels through which competition may spur economic growth.

From the above discussion, it can be said that previous studies of economic recession and growth hold some elements of truth, but they fail to provide a full picture of the issues. While some theories focus on the demand side in explaining business cycles, the others look only at the supply side in studying economic growth. Since both supply and demand are important determinants of economic growth and business cycles, ignoring any side will lead to only a partial understanding of economic phenomena. Some models have included both supply/production and demand/consumption sides, for example the real business cycle model, the Ramsey model, the Diamond model and some endogenous growth models. However, the household utility functions used in the models assume unlimited ability to consume, so the demand side exerts no constraint on the system. As such, these models essentially belong to the supply-side approach, but with a demand-side dressing. Since existing theories provide only a partial understanding, they have failed to uncover the fundamental cause of the recession and thus have failed to provide a satisfactory solution. As long as economic recessions continue to occur, the search for answers must also continue. Before we embark on this task, we need to examine the limits in the classical economic framework.

## 5.2   Limits of Classical Economics and Their Implications

Classical economics provides an elegant framework to aid our understanding of economic issues. However, due to a number of limitations, the framework is impotent in explaining economic recessions while explaining very well economic phenomena in normal situations. This section explores the limits and limitations of classical economics.

### 5.2.1  Limits of the 'Invisible Hand'

The classical framework started with the 'invisible hand' theorem initiated by Adam Smith. The theorem states that the competition in the market will drive the long-term price of goods down to production cost so the personal interest of chasing profit will lead to social benefits (i.e. the optimal allocation of resources). This harmony of the capitalist system critically rests on the assumption of competitive markets. If this assumption does not hold, the invisible hand does not work.

The assumption of a competitive market might have been plausible in Smith's time, but it is unrealistic for most economies nowadays. This forms the base for heterodox economists' attack on classical economics. Although heated arguments about whether or not the assumption should be abolished may attract people's attention, a more important task for us is to investigate what causes the different market structures and why market structures change at different stages of economic development. Non-competitive markets may be attributed to many factors, e.g. rules in society, technological barriers and economies of scale. In considering the change of market structures over time, technology and innovation play a crucial role. On the one hand, new technology may create more technological barriers and thus cause more market power (e.g. patent monopoly); on the other hand, the invention of new technology and new products may reduce or even destroy existing monopoly powers (e.g. the impact of the Internet on telephone and telegraph communication). If we can design a set of rules that can maximize the benefit of new technology and inventions, we can improve the competitiveness of markets and thus extend the limits of the invisible hand.

More broadly speaking, the functionality of the invisible hand is fundamentally dependent on the institutional set-up. Without a proper set of rules, a voluntary and mutual beneficial market may not exist, let alone a competitive market. For example, in an anarchical society, violence and criminality may override markets and thus tie up the invisible hand; alternatively, an autocracy government may extend its administrative hand to any markets and prevent the proper functionality of the market force. From this point of view, institutionalists have provided important insights into the limits of market mechanism. Three types of transactions (bargaining, managerial and rationing) proposed by Commons (1934) expressed different ways of allocation of resources in a society. To increase the functionality of the invisible hand, our task is to create and enforce suitable rules in order to increase the degree of market transactions and reduce the amount of other types of transactions. In the case of innovation scarcity, we need to set up appropriate rules to facilitate the formation and functioning of the innovation market.

## 5.2.2 Limits of the Utility Theory

The consumer utility theory is a cornerstone of classical economics. The theory rests on three assumptions which are based on psychological preferences.

One assumption is that the more goods and services an individual consumes, the higher utility he/she can achieve. This assumption is embedded in the utility curves or indifference curves in classical economics textbooks.

Figure 5.3 shows the typical indifference curves. The left panel shows how the utility level increases as the quantity of a type of good consumed increases. As the quantity increases, an additional good generates a smaller amount of utility (i.e. decreasing marginal utility), but the total utility level keeps increasing. This situation will continue even if the quantity of the good approaches infinity (shown by the dashed arrow). The right panel shows the indifference map—the utility levels in the case of consumption of two goods. On each indifference curve, the consumption of the combinations of goods achieves the same level of utility. As the utility curve moves towards top-right, the utility level increases because the consumption of goods increases. This situation will also continue without a limit; i.e., the utility curve can keep on moving towards top-right forever, shown by the dashed arrow in the right panel.

This 'more is better' assumption shown in Fig. 5.3 can be easily understood and works well in ordinary situations, but the assumption should not be extended to extreme cases—unlimited consumption and unlimited utility. Since no one can consume an infinite quantity of goods and no one can have infinite utility, there must be a limit for any consumer even if the limits may differ for different consumers.

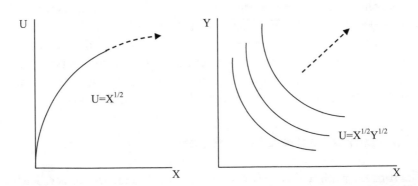

**Fig. 5.3**  Illustration of the utility theory

Considering the finite assumption of each type of good, the indifference map cannot expand infinitely. This limit in utility theory has a direct influence on the limit in a demand function, and thus on the limit in the equilibrium framework. The confined indifference map resulting from consumption limits can also shed light on economic growth and recessions.

Figure 5.4 shows a simple economy with two types of goods. The PPFs (i.e. $PPF_1$ to $PPF_4$) are production possibility frontiers of firms, which indicates the maximum or optimal amount of output combination (i.e. combination of good 1 and good 2) firms can produce for a given amount of inputs (e.g. capital, labour and technology). The most inward PPF (i.e. $PPF_1$) indicates the least amount of inputs available and thus can achieve the least amount of output combination, and vice versa. On the other hand, the indifference curves or indifference map (i.e. $U_1$, $U_2$ and $U_3$) indicates the level of happiness or utility that consumers can achieve. The higher position of indifference curve (e.g. $U_3$) means that more goods can be consumed and thus the higher level of happiness can be achieved.

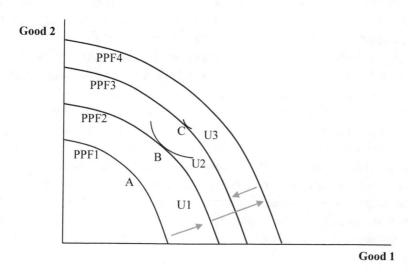

**Fig. 5.4** Impact of consumption ceiling on economic growth

In neoclassical economics, the optimal situation for an economy is that the maximum amount of output produced by firms can achieve the maximum amount of happiness of consumers, so the optimal or equilibrium point for the economy is a point where a PPF is tangent to an indifference curve, for example point A. If the economy has more inputs available, it can produce more output and thus the PPF shift outwards (e.g. a shift from $PPF_1$ to $PPF_2$). The new PPF (e.g. $PPF_2$) will be tangent to an indifference curve at higher position (e.g. $U_2$), so the new equilibrium point with higher amounts of outputs (e.g. $B$) is achieved and the economy grows to a new level. Neoclassical economists assume that there is no limit on consumption so the indifference curves can expand to the right-top corner infinitely. Meanwhile, if the amount of resource or inputs permits, a PPF can always shift outwards to meet an indifference curve at a higher position (i.e. with the higher level of happiness/utility) so the economy can grow unlimitedly. Because the shifts of PPFs are the sole source of economic growth and because the shifts of PPFs is solely constrained by the amount of resources or inputs, in a neoclassical view, economic growth is solely determined by the supply side.

However, based on our previous discussion, indifference curves cannot expand outward infinitely due to the limits on consumption of each good. This is shown by the most outward indifference curve $U_3$ in Fig. 5.4. No more difference curve at the top right can be drawn due to limits on consumption of goods 1 and 2. If a large increase in inputs (e.g. application of a new technology in production, an increase in capital due to increase investment level) pushes the PPF to $PPF_4$, it can meet no indifference curve. This means that firms produce too much to be purchased by consumers. Eventually, the $PPF_4$ has to shift inwards to $PPF_3$ and meet indifference curve $U_3$. This will cause an economic recession. The economy will stay in a recession because it is pointless for firms to product beyond the maximum amount of consumption. The only way to get out of the recession and continue economic growth is to invent more types of goods so as to lift the consumption ceiling.

## 5.2.3 Limits of Intertemporal-Choice Models

The indifference curves of consumption of two goods were extended by Fisher (1930) to explain the consumer choices with two time periods, i.e. the intertemporal choice theory. Figure 5.5 shows the Fisher diagram explaining intertemporal choice.

In the diagram, $Y$, $C$ and $r$ stand for disposable income, consumption and the interest rate, respectively. Given the interest rate $r$, as well as the individual's income $Y_1$ in period 1 and $Y_2$ in period 2, one is able to calculate the intertemporal income as $Y_1 + Y_2/(1 + r)$ in period 1 and $Y_1(1 + r) + Y_2$ in period 2 and, thus, one can derive an intertemporal budget line EF. The intertemporal utility curves $U_1$ and $U_2$ can be drawn as usual. The tangency B is the optimal consumption point. If $Y_1 > C_1$, the individual will choose to save in period 1 and to utilize the saving in period 2. On the other hand, if $Y_1 < C_1$ (as shown in Fig. 5.5), the individual will choose to borrow against the future income and keep the consumption level $C_1$, and to pay back at period 2.

Mathematically, assume that an individual discounts both future income and future consumption by the interest rate $r$ and discounts the future utility by a rate of $\rho$, so the intertemporal budget constraint is:

$$C_1 + C_2/(1 + r) = Y_1 + Y_2/(1 + r),$$

and the utility function is

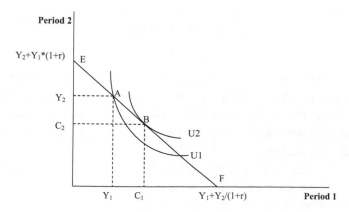

**Fig. 5.5** Fisher's intertemporal choice model

$$U(C_1, C_2) = U(C_1) + U(C_2)2/(1 + \rho).$$

To maximize utility,

$$dU = \frac{\partial U(C_1)}{\partial(C_1)} * d(C_1) + \frac{\partial U(C_2)}{(1 + \rho)\partial(C_1)} * d(C_2) = 0, \text{ or}$$

$$\frac{d(C_1)}{d(C_2)} = -\frac{(1 + \rho)U'(C_1)}{U'(C_2)}$$

Given the interest rate r and intertemporal budget constraint, the change in consumption in period 1 by $d(C_1)$ necessitates an opposite change in consumption in period 2 by $(1 + r)d(C_2)$, namely, $d(C_1 = -(1 + r)d(C_2)$. As such, $\frac{d(C_1)}{d(C_2)} = -(1 + r)$, and we can conclude that

$$U'(C_1) = \frac{(1 + r)}{(1 + \rho)}U'(C_2)$$

For a multi-period case, the intertemporal budget line is

$$\sum_{i=1}^{\infty} \left[(1 + r)^{-i}C_{t+i-1}\right] = \sum_{i=1}^{\infty} \left[(1 + r)^{-i}Y_{t+i-1}\right]$$

Thus, the optimization solution can be expressed by the standard Euler equation:

$$U'(C_t) = \frac{(1 + r)}{(1 + \rho)}U'(C_{t+1})$$

This equation shows that the marginal utility of consumption in two periods is related to the interest rate and the intertemporal discount rate. To maximize the utility of consumption for a lifetime, an individual is inclined to borrow or save in period 1, depending on the value of interest rate and discount rate.

This rationale looks perfect except for the implicit assumption that whether or not one can treat the consumption in different periods in the same way as treating the consumption of different goods in the same period. Fisher treated them in the same way and thus had produced a similar indifference map. In doing so, he implied that the contents of consumption in two periods were different. However, in the absence of innovation, the goods consumed in both periods should be the same, so consumption in period 1 and consumption in period 2 constitute a case of perfect substitution. This means the intertemporal utility curves straight lines (the slope of the line depends on the person's discount rate), shown in Fig. 5.6.

The optimal consumption point in Fig. 5.6 can be either a corner solution or any point on the budget line, depending on the slope of the intertemporal budget and the slope of the linear utility curve. If the slope of the budget line (e.g. budget 1) is greater than the slope of the utility curve, the optimal consumption occurs at $A_2$ in period 2; if the slope of the budget line (e.g. budget 3) is less than the slope of the utility curve, the optimal consumption occurs at $A_1$ in period 1; if the slope of the budget line (e.g. budget 2) is equal to the slope of the utility curve, the budget line will coincide with a utility curve, so the intertemporal consumption can happen on any point on the budget line. Thus, the whole model of intertemporal choice collapses—people tend to consume all at one time period and nothing at the other time period, or people are indifferent to the amount of consumption at each period. Both cases are a meaningless outcome.

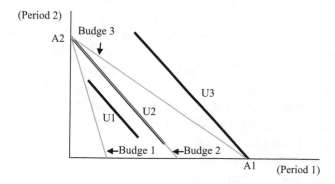

**Fig. 5.6** The collapse of the intertemporal choice model

This result can be interpreted in a more straightforward way. Since the slope of intertemporal utility is determined by the individual's discount rate and the slope of the intertemporal budget line is determined by the interest rate, the saving and borrowing behaviour will be determined by the interest rate and the individual's discount rate: if the two rates are the same, the person will be indifferent in consuming in period 1 or in period 2; if the person's discount rate is higher than the interest rate, he/she will borrow and consume all his/her income in the period and vice versa.

In short, the intertemporal choice model makes sense only when the goods in the two periods are different. Since the change of a variety of goods requires innovation, product innovation is the prerequisite for the intertemporal choice model.

## 5.2.4  Limits in the Production Cost Theory

The behaviour of the production cost depends on different firms and industries. For most firms and industries, as the output increases, the average production cost increases. However, for firms with high upfront fixed costs, the average cost is likely to decrease with the size of output because the fixed cost can be averaged for more outputs. For some new industries, the average production cost may decrease as more firms join in the industries—some costs like advertisement and infrastructure can be shared by more firms. The typical type of production cost of a firm, or an industry presented by classical economists is shown in Fig. 5.7.

Panel (a) shows the short-run production cost of a typical firm. The average cost and marginal cost experience a similar pattern: as the output increases, both costs decrease initially and increase after a turning point. This pattern can be explained by the efficient utilization of fixed capital in the early stage of production expansion and the capital or size constraint in the later stage. Since the marginal cost curve will pass through the lowest point of the average variable cost curve, the firm's supply curve is the part of the marginal cost curve above the average variable cost curve, shown as the broad line in panel (a). The long-run average and marginal cost curves shown in panel (b) are similar to those in the short run with the long-run average cost curve as the envelope of all short-run average cost curves. The long-run production of the firm is

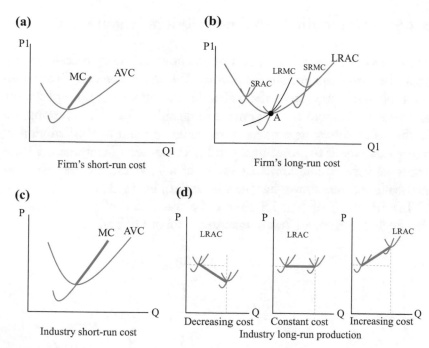

**Fig. 5.7**   Production cost at the firm and industry levels

at point A, which has the lowest long-run average cost. From panels (a) and (b), it is clear that the firm's supply in both the short run and the long run has a minimum supply quantity and a minimum supply price.

The same conclusion can be drawn for a typical industry. The short-run average cost and marginal cost of the industry shown in panel (c) are similar to the firm's short-run costs in panel (a). By the same reasoning, the short-run supply curve of the industry is the part of the marginal cost curve above the average cost curve, which starts from a minimum quantity and price, shown as the broad segment in panel (c). The industry supply in the long run depends on the type of industry: the downward sloping curve for a decreasing cost industry, the horizontal line for a constant cost industry and the upward sloping curve for an increasing-cost industry. However, no matter what type of industry, the long-run supply curves all have a minimum price and quantity. These are the limits of industry supply.

## 5.2.5  Limits in the Partial Equilibrium Framework

The limits of the utility theory and production cost theory necessitate limits in the partial equilibrium framework. The production cost theory indicates that the supply curve of a typical firm or industry is upward sloping and with a minimum starting price and quantity. The finite consumption in the utility theory necessitates a maximum amount of consumption of any goods even if the price of a good becomes zero. Incorporating these limits into the equilibrium framework of a good market, we can draw marshalling crosses shown in panels (a) and (b) in Fig. 5.8.

The panel (a) of Fig. 5.8 shows the usual case of a good market. Although the demand has a maximum limit Qd and the supply has

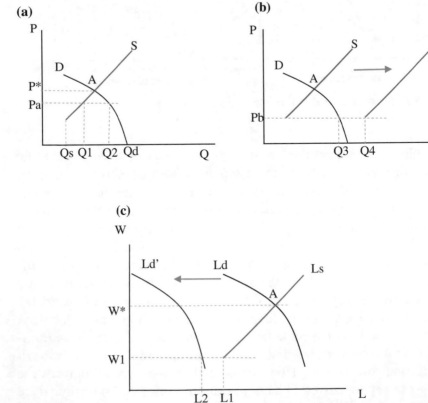

**Fig. 5.8**  (Dis)Equilibrium in goods market and labour market

a minimum requirement Qs, these limits do not affect the equilibrium theory because the supply and demand curves intersect at point A to achieve an equilibrium outcome. If the market price is below the equilibrium price (e.g. $P_a < P^*$), the market is in disequilibrium because market demand will be greater than market supply (i.e. $Q_2 > Q_1$). The excess demand will bid up market price, and thus, the market will move towards the equilibrium point A. In this case, any disequilibrium in the market will be short-lived and the equilibrium will be achieved automatically.

However, it is a different story in panel (b). If the supply curve S shifts to S' due to the increased firm size as a result of capital accumulation, the minimum production quantity increases. At the minimum price level Pb, the minimum supply $Q_4$ is greater than the demand $Q_3$, and there will be an excess supply $Q_4 - Q_3$. Since the minimum supply cannot be reduced and the market demand reaches its limit, this market disequilibrium may be persistent and the market mechanism fails.

A similar situation can happen in the labour market. Like the supply curve in the goods market, the labour supply curve should have a minimum price and quantity. The minimum wage may result from the cost of labour reproduction, e.g. the cost of food and clothes. The minimum quantity may be required by the population in the economy. There should also be a maximum limit on labour demand, which is determined by the limits in goods demand. These features are shown in panel (c) of Fig. 5.8.

In a normal case, the labour supply curve Ls can intersect the labour demand curve Ld at point A, so labour supply and demand are balanced and equilibrium wage $W^*$ is achieved. However, if there is an oversupply in the goods market, labour demand will fall sharply and the labour demand curve Ld may shift dramatically to the left until Ld'. In this case, the minimum labour supply $L_1$ at the minimum wage $W_1$, is greater than labour demand $L_2$. This disequilibrium cannot be eliminated through the market mechanism because the minimum supply of labour and the minimum wage level cannot be reduced.

## 5.2.6* Limits in the General Equilibrium Framework

Although classical economists admit that persistent oversupply in a few markets is possible, they argue that the oversupply in some markets will be cancelled out by the excess demand in other markets and thus oversupply in the total market is impossible. This is the reasoning behind Walrus' general equilibrium theorem, which states that, at any given set of prices, the total excess demands for all markets must be zero. It seems that this theorem eliminates the possibility of general overproduction in an economy. However, the general equilibrium framework based on Walrus' theorem also has its limits for the following reasons.

First, after the primitive stage of economic development, all economies involve money—the prices of all goods are measured by money. Money is essentially a special type of good which can act as a medium of exchange. Although money is not tied to gold or other species anymore, we still need to view money as a special type of commodity because it can purchase any goods the consumer wants and thus generate utility as other goods do. As such, the general equilibrium for a money economy must include the money market. If we apply Walrus' law correctly to a money economy, the excess demand for money in the money market must equal the total excess supply in goods markets; thus, there is always a possibility of general overproduction in goods markets.

Second, even in a barter economy, the general equilibrium theory cannot rule out the possibility of overproduction. The general equilibrium theory deals with supply and demand in markets. A person supplies something to a market in a barter economy only when he/she wants something else from the market, so the supply must equal demand for each person and for the whole economy. However, this requirement of market supply is inapplicable to production. One may produce a large quantity of sheep but only supply a small portion to the market. As a result, a general overproduction does not contradict the general equilibrium theory and is highly possible. A hypothetical example may drive this point home. Consider a simple economy of production of two goods: sweet potatoes and apples. Good weather conditions may bring a bumper year for both products—the output is far beyond the consumption level of the economy. With limited preservation technology in primitive times, the result was an absolute general overproduction: much of both goods has to be left rotting in the fields.

Third, in a capitalist economy, the supply of one type of goods is not due to the demand for other types of goods but for profits, so the assumption of market supply that one sells something to a market only when one wants something else from the market does not satisfy and thus the mechanism of general equilibrium breaks down. To obtain more profits, the capitalists tend to overestimate consumption and supply as much as possible in a competitive market, so oversupply is highly likely. Once the capitalists detect that the demand is weak and profitability drops, they will not use their earned profits to purchase investment goods (otherwise, they will produce more products in future, cause even more oversupply and bring loss to themselves). These uninvested profits are the counterparts of general overproduction in the economy.

Fourth, although people's desire to consume is unlimited, people's ability to consume any type of goods is limited. The consumption limits for each type of goods necessitates an aggregate consumption limit for the economy. With a slow increase in varieties of goods and services due to the slow pace of product innovation, the increase in consumption limits is slow. On the other hand, the production capacity increases rapidly thanks to the accumulation of capital. As a result, general overproduction is a necessary outcome.

Finally, although Walrus' law is generally regarded as the guarantee that economy-wide overproduction is impossible, the conditions for the proofs of Walrus' law show otherwise. Here, we examine three versions of modern proofs of Walrus' law.

The straightforward version of proof is the budget constraint approach: everyone in an economy is bonded by budget, i.e. one's demand or spending is constrained by one's income obtained through supplying goods. We provide and discuss a proof of this type first.

Assume an economy with two goods $X$ and $Y$ and $n$ individuals. For the purpose of generality, we assume each individual supplies and demands for both $X$ and $Y$ (if one demands for $X$ and supplies $Y$, it can be regarded that his/her supply of $X$ and demand for $Y$ are zero). For individual 1, the demand for $X$ and $Y$ is $X_{D1}$ and $Y_{D1}$, supply of $X$ and $Y$ is $X_{S1}$ and $Y_{S1}$, respectively. For individual 2, they are $X_{D2}$ and $Y_{D2}$, and supply of $X$ and $Y$ is $X_{S2}$ and $Y_{S2}$, respectively, and so on. The prices of $X$ and $Y$ are $P_X$ and $P_Y$, respectively.

For individual 1, the net demand for $X$ must be supported by the income from the net supply of $Y$, so we have:

$$P_X(X_{D1}-X_{S1}) = P_Y(Y_{S1}-Y_{D1})$$

or

$$P_X(X_{D1}-X_{S1}) + P_Y(Y_{D1}-Y_{S1}) = 0$$

Define individual 1's excess demand for $X$ and $Y$ as $ED_{X1} = X_{D1}-X_{S1}$, $ED_{Y1} = Y_{D1}-Y_{S1}$, respectively. The above equation can be rewritten as:

$$P_X * ED_{X1} + P_Y * ED_{Y1} = 0$$

The above equation means that the total value of excess demand for individual 1 is zero. Similarly, we can write the total value of excess demand for individual $i$ as:

$$P_X * ED_{Xi} + P_Y * ED_{Yi} = 0 \qquad (5.4)$$

Adding the excess demand value for all $n$ individuals, we have:

$$\sum_{i=1}^{n} P_X * ED_{Xi} \sum_{i=1}^{n} P_Y * ED_{Yi} = 0$$

If we extend the two-good case to the case of $m$ goods, we have:

$$\sum_{i=1}^{n} \sum_{j=1}^{m} P_j * ED_{Xij} = 0$$

The above equation means the total excess demand in the economy is zero for any given set of prices for all $m$ commodities, so we successfully derived Walrus' law.

However, the base for this derivation of Eq. (5.4) is worth more consideration. The budget constraint for individual $i$ indicated in Eq. (5.4) means two things. On the one hand, the demand of individual $i$ is constrained by the income from the supply of goods; in other words, net excess demand must be **non-positive**, namely

$$P_X * ED_{Xi} + P_Y * ED_{Yi} \leq 0 \qquad (5.5)$$

On the other hand, the purpose of individual $i$'s supply is to cater for his/her needs. Putting it differently, he/she supplies according to his/her demand, or supply is constrained by the amount of demand. As such, total demand cannot be less than total supply, or the excess demand must be **non-negative**, i.e.

$$P_X * \mathrm{ED}_{Xi} + P_Y * \mathrm{ED}_{Yi} \geq 0 \qquad (5.6)$$

The double binding required by inequalities (5.5) and (5.6) necessitates the Eq. (5.4). This double binding between supply and demand underpins the proof of Walrus' law. However, in a modern economy, firms supply in order to obtain profits rather than to satisfy demand, so inequality (5.6) does not hold anymore. The inequality (5.5), which is the only valid one in this case, indicates that Walrus' law does not hold and that general gluts are likely.

Since profit is denominated in money, inequality (5.5) and thus Walrus' law are still valid if we include money as one type of good. In this case, however, Walrus' law means the general equilibrium in the combined goods and money market, so general gluts can become a reality if the total demand for money is greater than money supply.

The second version of proof was provided by Walrus himself. Walrus used a hypothetical auctioneer to adjust the prices. The auctioneer may announce a set of prices for $m$ type of goods, and each individual responds to these prices by submitting the quantity he/she wants to purchase or sell. Then, the auctioneer checks the balance between supply and demand. If demand for good $x$ exceeds supply, the auctioneer will increase the price of good $x$ and this will lead to a decrease in excess demand; if the supply of good $y$ exceeds demand, the auctioneer will decrease the price of good $y$ and this will lead to a decrease in excess supply. Repeating this procedure, the auctioneer can obtain a set of prices at which demand equals supply for all goods.

The process of the Walrus auctioneer can generally achieve equilibrium prices, but it depends crucially on two assumptions. One is that an increase in price can reduce excess demand, and the other is that a decrease in price can reduce excess supply. There is no problem with the first assumption, but there are limits for the second assumption. One limit is that the auctioneer cannot reduce the price to zero. If the price is below production cost, the producer will not supply to the market. In this case, the market becomes non-existent because of the withdrawal of supply. The other limit is related to consumption ceiling. Any individual has a limit in consuming any type of goods; e.g., one cannot eat an unlimited amount of ice cream. The implication of this consumption limit is that, once a person reaches his/her consumption ceiling, he/she will not purchase more even if the price decreases. This creates a possibility that demand is less than supply even if the price becomes zero. That is, even if the firms can produce and supply goods at no cost, market equilibrium cannot be achieved due to the cap on demand.

The third version of proof of Walrus' law is based on the Brouwer fixed-point theorem. The theorem states that any continuous function from a closed compact set onto itself will have a fixed point. The theorem can be illustrated by Fig. 5.9.

In a unit simplex where $0 \leq x \leq 1$ and $0 \leq f(x) \leq 1$, as long as $f(x)$ is a continuous function, it must pass through the 45° line $x = f(x)$ when the value of $x$ increases from 0 to 1. To apply the Brouwer fixed-point theorem to prove Walrus' law, we can first normalize the $m$ non-negative prices to a new set of prices through the following equation:

$$p_i = P_i \left/ \sum_{k=1}^{m} P_k \right.$$

The summary of the new set of prices is equal to one, mathematically: $\sum_{k=1}^{m} p_k = 1$. This necessitates that any prices satisfy $0 \leq p_i \leq 1$; namely, they are within the unit simplex. Defining a continuous function:

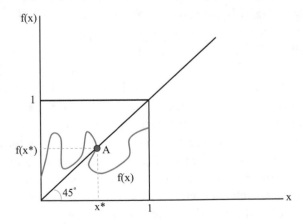

**Fig. 5.9**  Illustration of Brouwer fixed point theorem

$$f(p) = p + k * p * (X_D - X_S) = p + k * p * (ED),$$

where $p$ is a set of prices, $k$ is a set of parameters, $X_D$ is demand, $X_S$ is supply, and ED is excess demand.

This function must pass through the 45° line in the simplex, so we have:

$$p = f(p) = p + k * p * (ED), \text{ or } p * (ED) = 0.$$

This equation means that, at any set of non-negative prices $p$, the sum of the product of $p$ * ED is zero. Thus, Walrus' law is proved.

This proof is elegant, but it proves only the existence of non-negative equilibrium prices. It is possible that, as shown in Fig. 5.10, the function $f(p)$ may pass through the 45° line at the point A (1,1) or at the origin, where some or all goods are free (i.e. $p=0$). These free goods may indicate non-exclusive public goods (e.g. a toll-free bridge), but it may also indicate a case of excess supply. In other words, with a solution of $p=0$, Walrus' law does not exclude the possibility of general gluts.

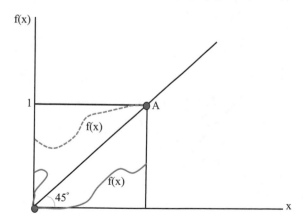

**Fig. 5.10**   Limits of Brouwer fixed point theorem

## 5.2.7  Limits of the AS/AD Model

As shown in Sect. 4.8.3 in the previous chapter, we can aggregate all demand curves and supply curves of the goods markets and obtain an aggregate demand curve and an aggregate supply curve. Due to the limits in demand and supply in the goods market (see Sects. 5.2.2 and 5.2.3), the aggregate demand and supply curves must also have limits. That is, there is a maximum demand limit, a minimum supply quantity and a minimum supply price. With these limits, there is a possibility of general oversupply or underconsumption thanks to capital accumulation and process innovation, shown in Fig. 5.11.

The aggregate demand curve AD in Fig. 5.11 bends towards the horizontal axis and reaches the consumption ceiling at point 'a'. The aggregate supply, AS, is normally an upward slope, but it has a starting point with a positive price level and a positive quantity level, indicating that production of goods has a minimum cost and minimum scale. This setting of AS is an analogy to the standard supply curve (the part of the marginal cost curve above the average cost curve) in microeconomics. Due to the increase in inputs (e.g. labour and capital) and the progress of technology, the aggregate supply curve shifts from AS to AS'. The new intersection point indicates a higher output level with a lower

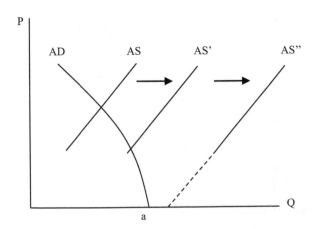

**Fig. 5.11**  The effect of capital accumulation and process innovation

price level. As the aggregate supply curve continues to shift to AS",
however, it is out of the reach of the aggregate demand curve AD.
Consequently, this leads to the overproduction or deficiency of aggregate demand.

Similar to the case of the partial equilibrium, the overproduction
caused by the right shift of aggregate supply curve cannot be eliminated
by market mechanism because of the cap on demand, the minimum
production capacity and the minimum price of goods. However, the
disequilibrium can be solved by introducing production innovations,
shown in Fig. 5.12.

Assuming that capital accumulation and process innovation can shift
the aggregate supply from AS to $AS_1$, causing generally overproduction. However, thanks to product innovation, an increase in the number of goods and services can lift the aggregate consumption limit so
that the aggregate demand curve shifts from AD to $AD_1$. If the speed of
product innovation can match that of capital accumulation and process
innovation, the size of the aggregate demand shift is not smaller than
the size of the aggregate supply shift, so aggregate supply and aggregate
demand can intersect at point B. As a result, general overproduction can
be avoided.

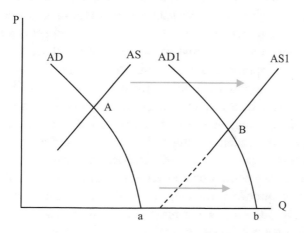

**Fig. 5.12**  The effect of process innovation and product innovation

## 5.3   Rationale of the New Theory

The foregone discussion shows that the existing theories provide views of economic growth and business cycles from different angles, but all these theories have drawbacks. For example, the classical economic framework provides great tools to analyse economic phenomena, but it also has many limitations. In this section, we propose a new theory which overcomes the shortcomings of existing theories and is able to explain both business cycles and economic growth in a consistent way. Before we introduce the new theory, however, it is necessary to recap our refutation of two dominant views in economics so that one can have a chance to appreciate the new theory.

One view is that theories generally include unrealistic assumptions, so they are not as feasible or useful as empirical studies. We have discussed the relationship between empirical and theoretical research in Chapter 2 and have discussed the problems in current empirical research in Chapter 3. Here, we need to emphasize that although empirical research looks more realistic by including all factors in reality and can fit in the research data well, this type of research does not intend or is unable to find the mechanism behind the reality. Mechanism is the most important thing for prediction and for working out solutions for economic problems, so it should be the goal of economic research. The strong point of theoretical research is that it focuses on the main task of uncovering the mechanism that governs economic phenomenon and generates economic data. Although it is impossible for a theory to include all information in reality, it is important that the assumptions of a theory must be reasonable or the proposed condition must exist in reality. The three axioms to be discussed shortly in this section are an effort in this direction. Eventually, it is the synergy of both theoretical and empirical research that improves our understanding and solves the problems in reality.

The other view is that economic growth and business cycles are separate issues, so they should be addressed separately. In the minds of most economists, economic growth is about the long-run growth trend, for which supply-side factors matter, while business cycles are about the fluctuations around the growth trend, which result from demand-side factors in the short run. This view severs the links between long run and short run, between growth trend and fluctuations, and between supply side and demand side. Generally speaking, a short run can evolve into a long run and large economic fluctuations can affect growth trend, so they are not totally separable. More importantly, the factors governing the short run also govern the long run. Capital and technology are essential for long-run growth, but they also play an important role in the short run, e.g. the recessions caused by shortage of capital and other resources due to a number of wars throughout history and by oil shock in the 1970s and 1980s. Similarly, while consumption is important in the short run, it is also important in the long run because it can affect the direction and speed of innovation and technological change (more discussion on this is provided later). Production and consumption (or demand and supply) are two indispensable forces determining the outcome in both the short run and the long run because of the internal link between them: people's desire to consume can be satisfied only through available supply while the purpose of supply is eventually to satisfy demand by consumers. This internal link leads to economic dynamics through the role of innovation. The intention of firms to cut costs in order to make more profit sparks process innovations, which increase production capacity and speed, resulting in an oversupply of commodities of existing types, and leading to economic recessions and stagnation. The saturated appetite of consumers forces firms to innovate products that are wanted by consumers. Successful product innovations will satisfy consumers' desire and stimulate production; hence, the economy gets out of recession/stagnation and grows to a new level. By studying business cycles and economic growth at the same time, this book intends to reflect the interaction between supply and demand and reveal the important role of innovation.

Business cycles and economic growth are a complex phenomenon that involves many factors. The complexity makes it difficult for us to find the essence of the problem. To overcome the complexity, we consider for our new theory a very simple economy of one individual (or household) and one firm. Some economists may disagree with this simplification by saying that this economy has no government and thus precludes any role of monetary and fiscal policies. It is true that the government has no explicit role in this simplest economy, but the role of the government as an institution is implicitly included in this simple economy. To avoid any violation of voluntary exchange (e.g. menace, stealing and robbery), law and order are enforced by the government. This is the foundation of any market economies. Beyond this, the government's role of economic management is not essential for a simple economy. Any income collected by the government will be spent or redistributed to households. Any fiscal or monetary policies are to influence the consumption or saving decisions of the households. As such, the role of the government is implicitly reflected in the behaviour of the household which is included in the model, so it is not essential to add another layer of complication. The rationale for this simplification is that the factors that work in this simple economy must also work in any other more complex economy. Any complex cases can be studied by adding more variables to this simplified case.

A caveat is necessary for our new theory. As shown earlier, existing theories have addressed these factors and thus have merits. The new theory is not an attempt to deny the complexity of an economy or to discard the valid points in existing theories such as the demand/supply equilibrium in classical economics, demand deficiency in under-consumptionist and Keynesian theories, technological changes and the role of entrepreneurs and other institutions in institutionalist theories, income equality in socialist and Marxist theories, and the role of money in the monetarist theory.

In essence, the new theory extends and advances existing theories by uncovering the fundamental factor in a capitalist economy that primarily causes economic recessions and stagnation. For example, Schumpeter highlighted the importance of technology and innovation. He followed Marx's step and suggested that the creative destruction of new technology caused business cycles, but he did not explain what determines the speed of innovations and how exactly innovation cycles cause business cycles. The new theory will explore the characteristics of high risk and low return of innovation investment in determining innovation speed and will illustrate how the interplay of product innovation and process innovation contributes to business cycles. By focusing on the key factor, the new theory can deepen our understanding of the phenomenon of business cycle and economic growth, as well as the existing economic theories.

The new theory rests critically on three simple but revolutionary axioms, from which we can connect production with consumption, uncover the key factor in a modern economy and reveal the dynamics between business cycles and economic growth. In considering readers without an economics background, this section explains the theory in layman's terms but also as accurately as possible. More rigorous derivations based on economic models are provided in Sect. 5.4.

## 5.3.1  Axioms on Consumption, Saving and Investment

The three axioms to be discussed in this section summarize the fundamental but overlooked facts in an economy. While these axioms may look unconventional, they are easily understandable because they are based on real-world experience and logical reasoning. Before we proceed to discuss these axioms, it is important to remind the reader to focus on the simple situation of one individual or one firm as indicated by axioms. The complex situations for an economy will be considered in Sect. 5.6.

**Axiom 1** (consumption satiety): every individual has a satiation point in consumption of any commodity. Consumption over this satiation point offsets utility.

To appreciate this axiom, one must consider a hypothetic case of no resource constraint; i.e., an individual has the ability to obtain whatever he/she wants and at whatever level of quantity. The satiation consumption point thus represents an ultimate or absolute ceiling for an individual's consumption of each commodity. In the case of constrained resources or finance, an individual may not reach satiation, but the satiation point still exists. The constrained resources also imposed a consumption ceiling to an individual. We will discuss this type of consumption ceiling later.

Economists against the idea of consumption satiety or consumption ceiling may argue that the assumption of a consumption ceiling is inappropriate because people's desire for goods and services is unlimited, or simply that consumption increases over time without any ceiling.

This opinion is reflected in the accepted preference and utility theory, which assumes that consumers always prefer greater quantities of any kind of commodity without limit. That is, the greater the quantity of commodities consumed, the higher the utility will be. This view held even for Keynes, who believed in demand deficiency but never thought of consumption deficiency; instead, he attributed the deficiency of effective demand to the deficiency in investment demand.

We should, however, not just think about consumption loosely. Next, we focus on the case of one person's consumption of one commodity, for example one's consumption of ice cream. Most people like ice cream, and generally speaking, the more ice cream consumed, the higher utility will be achieved. Eating too much ice cream will, however, lead to vomiting or stomach pains. Similarly, for harmless commodities like shoes and clothes, one must have a limit on their quantity—given a large number of shoes and clothes, one needs a lot of storage space and a lot of time to find the pair you want to put on. The same can be said for services; e.g., too much massage would damage your skin (due to friction). Listening to music for too long a day would damage your hearing. The World Health Organization estimates more than 1 billion young people are in danger of hearing loss from portable audio devices, including smartphones. Based on a study by Australian Broadcast Company (2018) reported that anyone who uses headphones for more than 90 minutes each day could be jeopardizing their hearing. These examples can be generalized to any goods and services (the author challenges the reader to name a commodity otherwise) and suggest that overconsumption can be a burden for a consumer.

The above examples demonstrate that a satiation point exists for consumption of any commodity (here the concept of consumption is strictly defined as the direct usage of goods or services).[1] If the amount of consumption surpasses the satiation point, the utility from

---

[1]It is arguable that consumption can include commodities purchased for future use or for others' use (e.g. gift, donation or bequeathing). This type of consumption can be treated either as other people's consumption if the commodity is used up, or as other people's savings if the commodity is unused, which will be discussed next.

consumption will decline. To embody the satiation point in the utility function, a parabolic utility function can be proposed:

$$U(x) = 2mx - x^2$$

This utility function is illustrated in Fig. 5.13.

When the amount of consumption of $x$ is less than $m$, the utility achieved increases as the consumption increases. Once $x$ is greater than $m$, however, further increases in consumption will result in lower utility. For a rational consumer, the maximum consumption of $x$ is $m$ and the maximum utility is achieved at point A.

Since each commodity we consume has a satiation point, adding up all amounts of commodities at satiation points, we can have a consumption ceiling for any consumer or household. Adding up all the consumption ceilings of all households in an economy, we can have a consumption ceiling for the economy (we can consider the consumption ceiling per capita to simplify the case of changing population). In short, the one-commodity case may be generalized and applied to the real world of multiple types of commodities (there is no fallacy of composition here because the satiation point can be applied to each commodity and to each household).

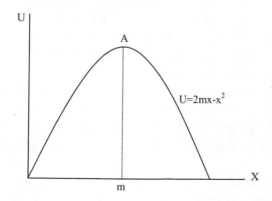

**Fig. 5.13**   A utility function with consumption ceiling

Since an individual needs different types of commodities (and services) to satisfy his or her needs, the following utility function can be used for a general case:

$$U = U(x_1, x_2, \ldots, x_n) = \sum_{i=1}^{n} \alpha_i (2m_i x_i - x_i^2)$$

where $\alpha_i$ is the weight of consuming good $x_i$ in the individual's utility.

The number of commodity type $n$ in the above equation is very important because it is the key to understanding the seeming contradiction between consumption satiety and overall unlimited human desire. A human being cannot consume an unlimited amount of any good, but his/her desire can be unlimited when the varieties of goods are abundant and/or when the varieties increase thanks to the new products being invented over time.

Consumption satiation is in fact not a new concept. The consumption ceiling caused by consumption satiation points has been studied by other researchers such as Fisk (1962), Stent and Webb (1975), Pasinetti (1981), Witt (2001a, b), Andersen (2001), Aoki and Yoshikawa (2002), Saviotti and Pyka (2013), and Chai and Moneta (2014), Murakami (2017), and Saint-Paul (2017). In fact, Pasinetti proposed a satiation hypothesis similar to Axiom 1: 'there is no commodity for which any individual's consumption can be increased indefinitely. An upper saturation level exists for all types of goods and services although at different levels of real income' (Pasinetti 1981, p. 77). He classified goods into three types. One is necessities for physiological needs (e.g. food) which have an Engel curve with expenditure growing slower and reaching a maximum point as income increases; the second is other normal goods which have an Engel curve with expenditure accelerated and then decelerated and reaching a maximum point as income increases; the third one is the inferior goods which have an Engel curve with expenditure increasing to the maximum point and then decreasing as income increases.

This line of work by Pasinetti (1981) was resumed later by other economists. For example, Aoki and Yoshikawa (2002) presented a model showing that economic growth is restrained by saturation of demand for individual consumption good. In their model, they thought innovations were accidental but depended on the prior knowledge. Murakami (2017) represented demand saturation by an S-shaped logistic function and built a model of economic growth through demand creation in the face of demand saturation. He linked the amount of demand creation to R&D investment through the probability of product innovation, or the birth rate of a new product. Saint-Paul (2017) assumed that utility derived from any individual good reached a maximum at a finite consumption level and allowed new products to be introduced into the economy. He showed that a social planner would choose to introduce new products to prevent the economy from converging to a satiated steady state. When a decentralized economy was considered, he found that the market did not deliver the required innovation level to lift the economy from its satiation trap.

The difference presented in Axiom 1 is that all goods are inferior goods at various levels of real income. Moreover, although the implication of satiation phenomenon on economic growth as well as innovation activity is studied by many researchers (e.g. Pasinetti 1981; Saviotti 2001; Witt 2001a, b; Ruprecht 2005), this book contributes more along this line by using Axiom 1 to study business cycles and thus shed more light on the critical role of product innovation.

The nature of Axiom 1 is common sense, which based on historical observations or personal experience and is demonstrated clearly by examples and reasoning provided. That is the reason why the author calls it an axiom. No empirical data are needed for common sense or for an axiom (e.g. the straight line is the shortest between two points), but economists require empirical data for everything. The direct proof for Axiom 1 would be an experiment on anybody using any goods or services, such as those shown in our hypothetical examples like ice cream, shoes, music and massage. If you do not agree with these common-sense examples, you can do experiments on yourself, e.g. listening to music 24 hours on end.

The indirect proof on a large scale can be found in product cycle theory, market saturation phenomenon and the backward Engel curve for inferior goods. However, using this kind of indirect proof needs to be accompanied by careful logical reasoning. For example, an Engel curve is obtained by expenditure data on different commodity groups by different household income groups, so this bears some similarity to the effect of increasing income on an individual, but the collective behaviour of households of different income groups may not be the same as that of one individual when income rises. Moreover, the data are based on commodity groups and income groups, which give the empirical results an indicative or gross nature. For example, a commodity group like 'computer' includes computers of many different brands, speeds, capacity, etc. A person would only need one desktop and one laptop (say this is the satiation point), but the advent of new high-speed computers may induce him to buy one to replace the old one. This example demonstrates that innovation changes satiation point of a commodity group but does not change the satiation point of a specific commodity (e.g. one needs only one computer of the same brand and same model). The income group has a similar issue, and more importantly, the expenditure data may not include sufficiently high income levels (i.e. 'out of sample') at which the satiation point appears. As a result, despite the fact that most goods do not have a backward Engel curve and that Chai and Moneta (2014) showed that the Engel curve can change position and shape over time, these facts do not disprove Axiom 1.

**Axiom 2** (dual role of savings): an individual's savings act both as a precautionary premium and as saved resources.

Saving is traditionally treated as future consumption plus some interest income derived from the saving, so it is normally not included in a utility function. The theory behind this practice is the life cycle theory developed by Modigliani and Brumberg (1954) and Modigliani (1986), and the permanent income theory developed by Friedman (1957). Both theories utilized the intertemporal choice model developed by Fisher (1930). However, these studies assumed away any uncertainty in our life and thus ignored the fundamental function of saving—saving for rainy days, i.e. precautionary saving.

Many researchers have studied precautionary saving (e.g. Leland 1968; Kimball 1990; Weil 1993). These studies recognized the role of uncertainty in saving behaviour, so precautionary saving in the studies is treated as an insurance premium and is determined by intertemporal utility maximization based on the expected utility function. In the current book, however, the author argues that the nature of precautionary saving indicates that savings can generate utility directly and immediately—with savings in hand the person at once feels more secure and thus happier.

In considering the consumption decision under uncertainty, saving plays a dual role. First, saving can be viewed as premium paid to reduce uncertainty, namely the equivalent precautionary premium in Kimball (1990). Second, as usual, savings are saved resources to be used in the future. The utility of savings as a precautionary premium is achieved when the saving action takes place, so this utility is in the current period and thus savings have to be put into the utility function in the current period. The utility of savings as saved resources is achieved when the savings are consumed in the future period, so this part of utility of savings should be expressed as utility of the increased consumption for that period.

Including savings in a utility function is not a novel practice. For example, Howe (1975) treated saving as a good in the current period in a linear expenditure system. In so doing, Howe derived the same extended linear expenditure system as developed by Lluch (1973), who used an intertemporal maximization of the Stone–Geary utility function. However, it is worth noting that the different implications of putting savings into the utility function. In Howe (1975), the saving in the utility function means saved resources, which are equivalent to future consumption, so the uncertainty of the future is completely ignored. In the current book, the savings in the utility function mean precautionary premium to reduce the uncertainty in the future.

The amount of utility generated by the role of savings as a precautionary premium at the present time is positively related to the amount of savings; i.e., the more one saves, the more secure one feels. Since there is no absolute security for any individual, there is no limit on the security feeling from savings action. As a result, there is no ceiling on the intention to save and the concept of 'more is the better' applies here with no constraint. A number of utility functions can be used to describe the nature of saving actions. For simplification, we use a linear function for the utility of saving: $U(\text{Savings}) = \alpha_S * \text{Savings}$. This utility function of saving also allows negative saving (dissaving) to generate negative utility. In considering all factors affecting household utility, the following utility function can be created:

$$U = U(x_1, x_2, \ldots, x_n, \text{Savings}) = \sum_{i=1}^{n} \alpha_i(2m_i x_i - x_i^2) + \alpha_S * \text{Savings}$$

$\alpha_S$ is the weight of savings in the individual's utility.

Some might still object to putting saving into a utility function. One such argument could be that, if savings generate utility and negative savings (i.e. borrowing) generate negative utility, how can you explain why so many households take on debts (e.g. a housing mortgage loan)? People raising this question only see the one type of utility related to savings—the utility generated from saving action. In the case of household debt (dissaving), the current utility related to the action of dissaving is negative, but the dissaving allows the household to consume more in this period, and this generates more utility from the increased current consumption. As long as the utility from the increased consumption outweighs the negative utility from the dissaving action, the total utility increases when the household borrows. As a result, the rising household debt phenomenon does not disprove the validity of putting saving into a utility function.

**Axiom 3** (investment-consumption dependency): to be profitable, the investment of a firm must be used to cater for future consumption.

Investment demand is usually treated as an independent final demand, which has the same status as consumption, but Axiom 3 proposes that investment demand is not an independent one. The logic for investment-consumption dependency is quite simple. The purpose of investment is to make profits. This aim can be achieved only when the products resulting from the investment can be sold to consumers. In other words, to be profitable, investment must be used to cater for future consumption. Consequently, investment today is dependent on the expected consumption growth in the future, or consumption growth potential. This simple logic can be argued in a more rigorous manner.

What determines investment demand? Tobin (1969) noticed that investment demand depends critically on profitability. Further thinking reveals that the profitability is in turn dependent crucially on the final demand.

We think about the case of investment in production first. In this case, the profitability of an investment is determined by both the cost of and the revenue from an increase in production capital. On the cost side, interest rates (or capital rental price), taxation policy, wages and prices of intermediate inputs are important factors. We can use the interest rate as a representative of investment cost. On the revenue side, investment income is achieved through sales of output to both intermediate demand and final demand. Since the sales to intermediate demand are to cater for the final demand, the final demand will be crucial in determining both price and the quantity of sales made.

Similar reasoning can be applied to investment in assets, such as housing, bonds and shares. On the cost side, interest rates (or borrowing rates) are a significant factor. On the revenue side, the sales of assets ultimately depend on the final demand for products. For example, if you invest in company shares or in housing assets, the revenue is determined by the prices of shares or housing prices, which are ultimately determined by the sales of the company's products or by the renting and/or selling of the housing to the final demand.

There are many kinds of final demand such as household consumption, government consumption, exports and investment demand. Household consumption, government consumption and exports simply reflect consumption by different consumer groups, so they can be grouped under a broader definition of household consumption for the purpose of simplicity. As such, there are two types of final demands: the investment demand and the broadly defined household consumption. Since our aim is to discover what determines investment demand, in considering that investment demand cannot be a determinant of itself, the fundamental final demand—household consumption—must be the determinant of investment demand.

The direct link between investment and consumption did not draw much attention from economists. Keynesian economists blamed the lack of investment demand on the lack of animal spirits and on liquidity preference. The traditional investment theory (e.g. the loanable funds theory) proposed by classical economists claims that a flexible real interest rate can *always* make sure that the supply of savings is equal to the demand for investment, thereby producing equilibrium in the capital market. Keynesian economists rejected this theory based on liquidity preference, which rejects the neutrality of money, the core assumption of neoclassical economics.

Actually, even with the assumption of neutrality of money, i.e. without liquidity preference, the loanable funds theory does not hold all the time. This can be shown in Fig. 5.14. Because of precautionary saving, in a general case (i.e. a case does not rely on accumulated savings from the previous years because the household has to maintain a fixed balance sheet in the long run), there are always some positive savings in an economy ($S_0$ shown in Fig. 5.14) even if the interest rate is zero. If the investment decreases dramatically (left shift of investment curve from $I$ to $I''$ in Fig. 5.14), there will be excessive savings although the interest rate is perfectly flexible.

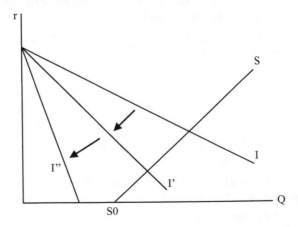

**Fig. 5.14** The possibility of uninvested savings

It should be pointed out that Fig. 5.14 shows only a simplified case but still holds firm for a generality. One may argue that, in the case of having a stock of liquid wealth, the flow of savings can be zero or even negative (dissaving). The first response to this argument is that the positive stock of liquid wealth indicates that the saving flow in the past is positive. Secondly, relying on past savings cannot last forever, so this is a special case in the short run and thus does not invalidate the possibility of uninvested savings in the long run. From this point of view, we must keep the balance sheet unchanged to present a general case.

Some readers may argue that the saving cannot finance the investment in the same period, so the saving-investment balance in Fig. 5.14 is mispresented. Figure 5.14 does not show the time frame in order to simplify the presentation. Considering different periods, the investment in Fig. 5.14 can be viewed as the prospective investment in period 1, which is equal to the investment in period 2, so the investment can balance the savings in period 1. This reasoning is the same as that for the supply/demand schedule in the goods market. Goods must be produced first and then can be consumed, so this involves two periods. We can use the prospective demand to overcome this time difference and produce a supply/demand equilibrium.

Some may also argue that although the interest rate is determined by investment and savings in Fig. 5.14, in reality the interest rate is managed by the Reserve Bank through the control of money supply. This argument mixes real interest rates with the nominal interest rate. The interest rate determined in Fig. 5.14 is the real interest rate while the interest rate stipulated by the Reserve Bank is the nominal interest rate. The Reserve Bank generally adopts a targeting-inflation interest rate policy, i.e. letting nominal interest rate keep up with the inflation rate. This will leave the real interest rate unchanged because the real interest rate = nominal interest rate − inflation rate. Given the unchanged real interest rate, the investment decision will not be affected. In short, it is the real interest rate, rather than the nominal interest rate, that is relevant to the decision on investment and savings.

More generally speaking, investment demand is determined by both the cost side and the revenue side of the investment. On the cost side, the interest rate is a good indicator. On the revenue side, the broadly defined household consumption or more accurately the household consumption growth potential determines investment profitability. The potential of household consumption growth can be represented by the difference between the consumption ceiling and the current consumption level. As such, we can have the firm's aggregate investment demand function:

$$ I = \frac{B}{1+r} \left( \sum_{i=1}^{n} m_i w_i - \sum_{i=1}^{n} c_i w_i \right) $$

where $B$ is a parameter indicating the propensity to invest when no borrowing cost exists, $r$ is the interest rate, which discounts $B$, and $w_i$ is the weighting parameter aggregation, $m_i$ and $c_i$ are the consumption ceiling and current consumption of good $i$, respectively.

The above investment function is very different from the traditional investment function: there is no role for income here. In conventional macroeconomics, investment demand is positively related to income. The reasoning is that higher income may indicate higher future consumption and thus investment becomes worthwhile (again, future consumption is the key for investment). However, this reasoning is problematic during a recession or in considering consumption satiation: higher income enables high purchasing power but does not necessarily lead to higher consumption. To avoid this problem, the investment function here links the future consumption to consumption growth potential, rather than to income. Nevertheless, the impact of income on investment in this book is indirectly reflected on savings, which ration the investment. The impact of income on savings has been taken care of by including savings in the utility function.

Regarding these 3 axioms, one may argue that they are essentially assumptions or hypotheses imposed by the author. What is the empirical evidence for them? The author's response is that these axioms are logical conclusion based on historical observation. Since everyone has a limit in consuming a good in limited time, the Axiom 1—consumption ceiling—is obvious. If one agrees that savings generate both security satisfaction at the present time and consumption satisfaction in the future, the Axiom 2—dual utility of savings—is easily comprehended.

Because the purpose of investment—to make profits—can be achieved only if there is an increase in future consumption, which can exhaust the increased goods and services resulting from investment, Axiom 3— the investment-consumption dependency—is a natural and logical conclusion. We can find many trivial cases to support our common sense, but it is not necessary to pile up the trivial examples to support common sense.

## 5.3.2  Overall Consumption Ceiling for an Economy and Product Innovation

Axiom 1 necessitates a ceiling for an individual's consumption for any commodity. If we add up the consumption ceilings of all commodity types and for all individuals in an economy, we will have an overall consumption ceiling for the economy. This overall consumption ceiling acts as a cap on economic growth. One way to move up this ceiling is to increase the population in the economy—more population means more consumption of all goods in the economy. Population growth may increase the total consumption of the economy; however, the well-being of the average individual may not improve. To improve the welfare of an average person in the economy, we need to move up the per capita consumption ceiling. The only way to lift the per capita consumption ceiling is to increase the types of commodity in the economy. This is the task of product innovation.

Do we have enough product innovation in our economy? Most people may be satisfied by the level of innovation and may cite many new products we are enjoying, e.g. the Internet, smartphones, GPS and drones. But this level of product innovation is still not enough for our society considering the stagnant sales in the economy. If we have adequate product innovations, firms will be busy producing new products that are highly demanded by consumers, and thus, there will not be widespread sale stagnancy. The modern economy today has high pressure to increase sales. This indicates that production capacity is abundant, but innovation is short of supply. Why don't people put more money in innovation than in production? The answer lies in the obstacles to innovation activities. One such obstacle is the high risk of investment in innovation. Innovation by definition involves the creation of something new. Inventors are continuously stepping into uncharted territory, so it is understandable that many successful innovations come only after numerous failed experiments. Although statistics on innovation failure/ success are difficult to obtain, it is widely accepted that a high percentage of investment in R&D is not successful. Moreover, there are two types of innovations: product innovation and process innovation. The goal of the former is to invent new products while the effort of the latter is to improve production efficiency. Compared with production innovations, product innovation has a much higher risk of failure because it normally involves much larger (or more radical) changes and there is much less information available to investors.

The other factor hindering innovation is imitation. Innovation requires hard and intelligent work, takes a long time and requires a great deal of money. Imitating an innovation is, however, fairly easy. For example, software that takes several years and costs millions of dollars to develop may take only a few minutes to copy. Other imitations may be harder and cost more (e.g. re-engineering a medicine), but in considering the numerous failures before a successful invention is created and the correct directions an imitator learns from the successful inventor, the imitation cost will be much lower than the cost of the invention. As a result, the social benefit or positive externality of inventions should be large.

Compared with process innovation, product innovation is much more vulnerable to imitation. Because process innovations are applied to production procedures or machinery, imitating these innovations requires knowledge about the production environment. However, imitating a new product does not require this knowledge. The vulnerability of product innovation to imitation means that the social benefit of product innovation is substantially higher than the benefit to the innovator.

Due to the distinct possibility of innovation failure and the low chances of getting a good return because of imitation, risk-averse investors are reluctant to invest their money in innovations, especially in product innovation. Rather, they prefer to invest in production that has a relatively certain investment return. Therefore, innovation investments, or R&D funds, become scarce, and this leads to scarcity of product innovation activity.

The low speed or scarcity of product innovation means the per capita consumption ceiling increases very slow and thus could become a bottleneck of the economy. This bottleneck shapes the economic growth pattern, which will be discussed in Sect. 5.4.

### 5.3.3  Dynamics Between Economic Recession and Growth

In the absence of product innovation, the gap between the consumption ceiling and the actual consumption level narrows as consumption increases thanks to economic growth. That is, future consumption growth (or consumption growth potential) has to decrease. This decrease will lead to a reduction in investment because investment is positively related to consumption growth potential. The limited increase in consumption and the large decrease in investment will lead to a fall in demand. As demand falls short of production capacity, not all factors of production (i.e. labour and capital) will be utilized and thus the economy enters stagnation or a recession. Facing the stagnation of sales, firms have to put more effort into product innovation. New products eventually appear, and the consumption ceiling is lifted. This increases the consumption potential, so consumption and investment increase, and the economy recovers and expands. Due to the obstacles to product innovation, the speed of innovation cannot catch up with the economic expansion, so the expansion cannot be sustained. As a result, the economy will fall into a recession again and thus form a business cycle. Since the consumption ceilings and innovation time for different new products vary, business cycles have no fixed periods.

Once the economy is back in recession, stagnation of sales will force the firms to engage in innovation activity and a new business cycle will start again. The new business cycle will have a higher output and consumption level. As new business cycles continue to develop at the end of the previous cycle, the economy will keep growing in a cyclical fashion. This cyclical economic growth is underpinned by two mechanisms.

One is that the obstacles to innovation lead to innovation scarcity and thus to economic recessions. The other mechanism is that stagnation of sales will force the firms to innovate and the resulting new products will drive the economy out of recession. However, this cyclic growth pattern may change if we can set up proper rules to stimulate enough innovations.

Although the new theory can explain business cycles well, some readers may still be sceptical. They may question: given that many factors can contribute to business cycles or economic recessions, how can you claim that product innovation is the single most important overarching factor? The key here lies in the definition of business cycle or economic recession. If one regards recessions as a phenomenon that includes a phase of negative growth or a large downward economic fluctuation, it is indeed that many factors are relevant and no one can claim a key factor. Negative economic growth caused by resource constraints (e.g., due to wars, natural disasters, oil price shock, etc.) is easily comprehensible, and the remedy is straightforward. This type of 'recession' is generally short-lived. It constitutes an economic fluctuation but not a business cycle, so we exclude them from 'recession' in order to simplify our discussion.

However, the other type of negative growth is quite different. During the recession phase of a business cycle, economic stagnancy or negative growth is accompanied by abundant commodity and other resources. This type of recession is hard to explain. Marxists and socialists blamed income inequality, Keynes attributed it to the lack of animal spirits and liquidity preferences, post-Keynesian economists and the Austrian School linked it the credit cycle and speculative investment, Schumpeter linked them to technological progress cycle, and institutionalists criticize the capitalist structure. These explanations have some valid elements but have not revealed the key factor.

The key factor is behind the feature of economic recessions—sales stagnancy. On the surface, sales stagnancy indicates limited consumption ability. Since the desire of human beings is unlimited, the seemly limited consumption ability indicates firms have produced too many old products. If firms can create and produce a number of new products liked by consumers, there would not be sales stagnancy. So, the scarcity of product innovation is the key factor we uncovered our new theory.

## 5.4*   An Economic Modelling Approach to the New Theory

The existing theories regarding economic growth and business cycles are based on aggregate macro-models (e.g. Keynes' multiplier model, the AS/AD model, the life-cycle permanent income model, the real business cycle model and the endogenous growth model). An aggregate macro-model has only one good for the consumer (i.e. consumption), so the model is unable to capture the effect of market dynamics and market saturation. On the other hand, there are a number of multi-commodity models, for example product variety models and numerous empirical computable general equilibrium models. However, the household utility functions used in these models are normally the CES function or the LES function. In these functions, there is no limit on the consumption of any goods, so the model cannot reflect consumption situation or market saturation phenomenon. To demonstrate mathematically how the three axioms proposed in the previous section can lead to an economic stagnation and thus to uncover the genesis of economic recessions, this section adopts a general equilibrium approach to build a multi-commodity macro-model, with the three axioms embedded in the consumption function and investment function. For more details, see Appendix 1 at the end of this chapter.

## 5.5   Implication of the New Theory

Starting from the three basic axioms, we derived a new theory of business cycles and economic growth. Based on this theory, in the absence of product innovation, an economy will inevitably go into and stay in a recession or become stagnant. The intertemporal equilibrium model even predicts that the output level and consumption level will keep falling in the long run. The theory also indicates that the only way out of economic recession is to invent new products. These results from the new theory have an important implication on the nature of business cycles and on the ways of improving economic growth. The new theory will also have important implications for the development of the economics discipline: it would help end the long-lasting controversy on general overproduction and unite schools of economic thought.

### 5.5.1 Ending the Arguments on Partial Gluts v.s. General Gluts

Starting from de Laffemas in 1598, the argument on underconsumption or overproduction has lasted for more than 500 years, but orthodox and heterodox economists still cannot achieve an agreement. The new theory reveals the full picture as well as the root cause of this issue and thus could end the long-lasting argument.

Since one of the features of economic recession is the stagnation of sales, nobody denies the problem of overproduction or underconsumption during a recession. However, classical economists consider that this overproduction only occurs in some markets (i.e. partial) while Keynesian and Marxism economists think overproduction occurs in every commodity market (i.e. general gluts). The argument of partial gluts is generally built on Say's law that supply creates purchasing power which underpins demand. It is argued that overproduction in some markets necessitates excess purchasing power, so there must be excess demand or underproduction in other markets. However, this argument is untenable for the following reasons:

First, purchasing power is a necessary but not a full condition for effective demand. According to Axiom 1, when a person reaches his/her consumption ceiling, excess purchasing power will not lead to more consumption. This is unarguably proved by the behaviour of rich people—their demand is much smaller than their purchasing power.

Second, excess purchasing power is by and large driven by people's desire to pursue savings or wealth. It is human nature to accumulate wealth in order to cope with the uncertainty in the future. For this reason, Axiom 2 indicates that there is no limit in a person's desire for savings. As such, excess purchasing power indicates an excess demand for future goods, rather than underproduction in any current markets.

Third, partial gluts by nature are not economy-wide phenomena and thus are not the feature of an economic recession. If most markets are undersupplied, the production for these markets will grow. This leads to a growing economy rather than an economic recession. However, many large partial gluts can lead to general gluts and thus cause a recession. Partial gluts cause some firms to lose profits and thus seek for other profitable opportunities. If some markets are undersupplied, the firms suffering from the partial gluts will flock into these undersupplied markets until all markets are oversupplied. As a result, a recession occurs.

Last, if one insists that the excess purchasing power indicated by partial gluts necessitates an undersupply in some markets, there must be obstacles preventing people from entering these undersupplied markets. Otherwise, the firms in oversupplied markets would find and occupy the undersupplied markets. One such obstacle is the legal barrier, for example a heroin market, an elephant tusks market or a human kidney market. Necessary laws are needed to forbid such markets for the sake of our society and the environment. The other obstacle is a technical barrier. For example, there would be a huge market for space travel or cures of cancer, but so far nobody can provide these products. This obstacle highlights the importance of innovation and technological progress.

Based on the above discussion, the feature of an economic recession is general gluts rather than partial gluts, so the heterodox economists' arguments engender the truth. However, orthodox economists also have a valid point: the level of consumption increases over time and appears to have no limit, so lasting underconsumption or consumption ceiling argument is untenable. The argument between orthodox and heterodox economists highlights the importance of product innovation. Consumption of any type of commodity has a limit, but consumption can increase over time without limit because of the advent of new products thanks to product innovation.

## 5.5.2  Uniting Existing Schools of Economic Thought

It might be controversial and risky to make a claim of uniting different schools of economic thought because people see things from different perspectives and thus have different ideas. However, here the author makes such a claim based on the reasoning that the new theory has included the valid points of existing schools of economic thought and addressed their shortcomings.

The new theory utilizes the classical economic framework, including the assumptions of a competitive market and neutrality of money, utility theory, production theory and equilibrium theory. However, the new theory has overcome the limits in classical theories, notably the limits in utility, production cost, supply and demand. In addressing the limits in classical economics, the new theory has revealed that scarcity of product innovation is the underlying cause behind business cycles.

Underconsumptionists and Keynesian economists would support the new theory because it recognizes and emphasizes the possibility of demand deficiency. Moreover, the new theory overcomes the difficulties faced by these two schools of economic thought. Underconsumptionists could not explain the contradiction between underconsumption and people's unlimited desire. Keynesian economists attribute deficiency of demand to deficiency of investment due to the lack of animal spirits and liquidity preference in an uncertain world, but they cannot explain why uncertainty does not cause a deficiency of demand during economic booms. The new theory advances these two schools of economic thought by uncovering the root cause of demand deficiency— consumption ceiling and innovation scarcity. An insufficient number of innovations cause a consumption ceiling at the aggregate level and thus have a negative impact on future profitability, which in turn leads to a deficiency of investment and an economic recession. When innovation scarcity is eventually replaced by a burst of innovations, the consumption ceiling is lifted and the prospect of future profit becomes bright, so consumption and investment increase and the economy recovers and expands. As such, the conundrum faced by underconsumptionists and Keynesian economists is solved.

Since the new theory adopts the classical assumption of neutrality of money, the theory does not include any elements of the credit bubble theory, which is believed by both monetarists and the Austrian school of economic thought. The reasoning of the credit bubble theory is based on the investment cost: an increase in money supply reduces investment cost, increases the future nominal value of assets and thus creates an investment bubble. This effect can be reflected in the effect of a decrease in the interest rate in the new theory. As money supply increases, the interest rate decreases. This leads to an increase in investment level and an increase in income. Moreover, a change in money supply may create a temporary consumption effect during the price adjustment period: the change in the money supply will lead to the change of the nominal price of all goods, but this process needs time. The new theory recognizes the temporary consumption effect of a change in money supply through the distributional parameter. Since the poor tend to (or have to) spend their income more quickly than the rich, the price adjustment period affects the real income of the poor and rich differently. For example, an increase in money supply will lead to a rise in price, so the real income of the poor increases during the price adjustment period as they spend most of their income before the price rise. As such, the distributional parameter increases, the aggregate consumption ceiling increases, and there is an increase in economic growth potential.

However, the impact of an increase in money supply is temporary. On the one hand, the adjustment period is limited, so the improvement in income distribution and thus the increase in the consumption ceiling will be reversed after the adjustment period. On the other hand, the impact of an increase in money supply on investment is not sustainable because the main driver of investment is not the investment cost but profitable investment opportunities, which are dependent on the availability of product innovations. Evidently, it is not easy to create an investment boom through an expansionary monetary policy during a recession because there is no investment opportunity. As such, a change in money supply may exacerbate the boom-bust cycle but is not the root cause.

People who appreciate the role of technology and innovation in economic growth and business cycles would appreciate the new theory because it highlights the role of product innovation and process innovation. The believers of product cycle theory, technological cycle theory, Schumpeter's creative destruction theory, and the Solow growth model, feel the importance of technology and innovation but have failed to uncover the mechanism by which innovations contribute to business cycles and economic growth. By identifying the mechanism, the new theory is able to suggest a way to increase innovations, avoid economic recessions, and speed up economic growth.

Finally, institutionalists, Marxists, and other socialists would agree with the conclusion from the new theory that the reoccurrence of economic recessions signals that institutional change is inevitable. While most institutionalists have not provided a firm idea about the future society which will replace or reform capitalism, Marxists believe in communist end point and socialists are satisfied with a socialist society. However, they share a common ground in that they have to rely on a planned economy to overcome the shortcomings of capitalism. This ideology effectively rejects the long-established wisdom that the market mechanism is a more efficient way to allocate resources and improve people's living standard. By contrast, the new theory suggests a thorough revision of the patent system to establish a functional and effective patent market, which will lead to a patentist society and an innovative economy.

### 5.5.3 Achieving Faster and Smoother Economic Growth

The analysis based on the static and dynamic models indicates that the only sustainable way to avoid an economic recession is to lift the consumption ceiling and thus the income ceiling over time. Based on this result, we can conclude that the repeated occurrence of economic recessions indicates that the speed of product innovation is not able to lift the cap on income fast enough to avoid a recession.

It is worth mentioning that the links between consumption satiation, innovation and economic growth have been identified by previous researchers (e.g. Pasinetti 1981; Andersen 2001; Witt 2001a, b; Ruprecht 2005), but no one has realized that product innovation is the key to avoiding an economic recession and that it is the engine for economic growth. Also, the importance of innovation to an economy is well recognized by most people. There are numerous studies on product and production innovations. Notable studies include those by Lancaster (1966), Romer (1990), Grossman and Helpman (1991), Aghion and Howitt (1992), Jones (1995), Klette and Kortum (2004), Acemoglu et al. (2013), and

Georges (2015). In these studies, innovations are determined by human capital or R&D funds, so no one has identified or discussed the problem of innovation scarcity and no one realized that innovation scarcity is the root cause of economic recessions. This book has advanced previous studies by uncovering the mechanism and dynamics between innovation and economic growth.

Is there any way to stimulate investment in product innovation so as to avoid economic recessions? Although no one can change the high-risk nature of product innovation, one can balance this high risk with high returns through forbidding imitation. An effort of this kind is seen in the patent laws, but the limitations in current patent laws fail to protect inventors fully. A full discussion of the limitations of patent laws is to be covered by Chapter 6; here, we discuss them only very briefly. On the one hand, current patent laws impose limited durations and a compulsory license rule on patent rights, aiming at moderating the monopoly power of the patentee and at forcing the patentee to implement patented technology. These clauses cause considerable stress for inventors and discourage innovation. On the other hand, the patent laws allow granting exclusive patent licenses, which transfer monopoly power of patentees to licensees so that the patent monopoly power magnifies in the economy. This only aggravates the problem of the abuse of patent monopoly power.

A thorough revision of patent laws can further encourage innovation activity while overcoming the problem of patent monopoly power abuse in a positive way, for example abolishing the time constraint put on patentees, forbidding both the assignment of patent rights and exclusive patent licensing (to avoid the magnification of patent monopoly power), and implementing a non-exclusive patent licensing system. Under this system, anyone could use innovations by applying for a license from the inventor. With the property right of the patent licensing established and clearly defined (infinite duration of patent right) in the new system, a patent market may come into reality and it will automatically channel funds into innovation activities.

In short, the new theory reveals that the lack of product innovation is the root cause that limits consumption, results in economic recessions, and slows down economic growth. Through a thorough revision of the patent law, the properly defined patent right and standardized practice can reward innovators through a patent market, so the speed of product innovation will match the growth rate of production capacity. As a result, the consumption ceiling can be lifted in time, economic recessions can be avoided, and the economic growth can reach its potential.

## 5.6   Empirical Evidence

Empirical evidence is vital for any theories because the credibility of a theory has to be examined by evidence. However, empirical evidence is not always readily available. A good example is Einstein's theory of relativity: Einstein had to rely on thought experiments to build his theory, which was subsequently proven to be correct by empirical evidence. To prove our new theory on business cycles and economic growth, we need detailed data on the consumption of different varieties of specific commodities. There is no such data on the variety of commodities at the macro-level, so there is no direct empirical evidence to prove that product innovation is the key factor underpinning the cyclic economic growth. Nevertheless, in this section, we endeavour to find some indirect evidence supporting the new theory.

A caveat about empirical evidence is also necessary here. As we have discussed in detail in Chapter 3, macroeconomic data do not satisfy the condition of other things being equal, so empirical evidence on business cycles and economic growth can only be indicative—it cannot definitely prove or disprove a theory because of the limitation in data. This is especially true when it comes to the empirical evidence for the 3 axioms on which the new theory is based. As stated in Sect. 5.3, the credibility of these axioms is based on historical observations or personal experience of anyone, so there is no need for any empirical evidence at the macroeconomic level. However, the empirical evidence presented here may reveal a complex picture, which some readers may be interested in.

## 5.6.1   On Consumption Ceiling, Engle Curve, and Market Saturation

Consumption ceiling is one of the important axioms for the new theory. The reasoning and obviousness of this axiom have been demonstrated in Sect. 5.3 by using thought experiments on the consumption by an individual. In this section, we discuss two macroeconomic phenomena—market saturation/production cycle and Engle's law—and examine their implications on the consumption ceiling and the business cycle.

### 5.6.1.1   Engel's Law

Engel's Law was put forward by Ernst Engel (1821–1896) in the 1857 article 'The Consumption-Production Relations in the Kingdom of Saxony'. Engel stated that the poorer a family is, the larger the budget share it spends on nourishment (Engel 1857, pp. 28–29). However, this law was interpreted in different ways later.

The commonly used version of Engel's law is that often used in a microeconomics textbook to explain inferior goods. In this case, Engel's Law is interpreted as that, as a person's income increases, the person's consumption of inferior goods decreases. This is a much stronger version of Engel's Law compared with Engel's original statement. Based on this strong version of definition, one can draw a backward Engel curve for a necessity, shown in Fig. 5.15.

The left panel of Fig. 5.15 shows the consumption of two goods: fast food and eating in restaurants. As income level increases from $Y_1$ to $Y_3$, people having fast food and eating in restaurants both increase. However, as income level increases from $Y_3$ to $Y_5$, people having fast food decreases. Relating the level of income to the amount of fast food consumed, we have an Engel curve shown in the right panel: the positive sloping part from $B_1$ to $B_3$ shows that the consumption of fast food grows with income while the negatively sloping part from $B_3$ to $B_5$ shows a decrease in fast food consumption as income level increases from $Y_3$ to $Y_5$.

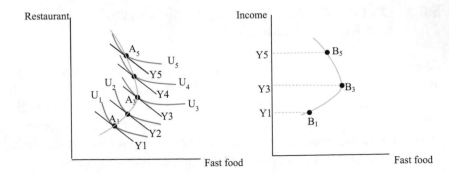

**Fig. 5.15** Deriving the Engel curve

This stronger version of Engel's law and the Engel curve has some empirical support. The 2015 consumer expenditure survey by the US Bureau of Labour Statistics showed that the number of rented dwellings decreases as the household income level increases from $30,000 to $70,000 and above. However, the survey also showed that, for other goods such as food and clothing, health care and entertainment, the amount of consumption keeps increasing as income increases.

A weaker version of Engel's Law is that the consumption of a good or service has an upper limit (saturation level or consumption ceiling) or has a tendency to saturate. This weaker version is easily accepted, but the key issue regarding this version is whether or not any good has an upper limit. Many researchers estimated Engel's law using different methods. For example, log-linear function was used by Working (1943), the log-normal distribution was used by Aitchison and Brown (1954), and the nonlinear functional forms were used by Prais (1953) and Banks et al. (1997). These studies suggested that, for almost all goods studied, the Engel curves exhibit a saturation tendency. Moneta and Chai (2014) used the nonparametric method and the long-run UK household expenditure (1968–2006) to estimate the Engel curves and also find that almost all major expenditure categories exhibit saturation at some point, but the propensity to saturate across different types of expenditure categories varies substantially. They also conclude that the Engel curves shift over time.

Generally speaking, the studies on Engel curves confirmed the generality of consumption ceilings on a wide variety of goods and services. Based on this, Pasinetti (1981) claimed that an upper saturation level exists for all types of goods and services although at different levels of real income. He further postulated that the saturation level will cause industrial structural change: as the income level increases, the old industry will be replaced by new industry to satisfy the shift of consumer preferences. The saturation levels have significant implications for economic growth and business cycles. The industrial structural change predicted by Pasinetti may cause large economic fluctuations or even long-lasting economic recessions. If the consumption of most or all goods and services approach saturation level, economic growth will slow down and eventually cease. The only way to avoid this outcome is to create new products.

However, some people have argued that some high-order goods may not have saturation issues. Menger (1871) observed that economic development led to a new class of higher-order goods, e.g. electricity, which do not directly satisfy consumers' want so do not have a saturation level. Witt (2001a, b) put forward a class of higher-order goods which he labelled 'tools' (e.g. ovens). People do not consume these higher-order goods so these goods do not have a saturation level like food or clothing does. However, the saturation levels of these higher-order goods should also exist because they are determined by the saturation level of the low-order goods which are associated with these high-order goods. Example, a family generally does not require two or more ovens because of the limited family size and outsourced baking activities. Gallouj and Weinstein (1997) and Moneta and Chai (2014) claimed that services are an example of higher-order goods because of the service providers tend to improve service quality and increase variety according to consumers' feedback. This argument involves an interesting point: although improved services are still in the same category as consumption, they become different types of products because the contents of services change thanks to innovations. This argument also highlights the importance of innovations because they are the sources of improved services.

Some researchers have claimed that innovations can be driven by changes in income distribution (Foellmi and Zweimüller 2006) or by firms' reacting to the slowdown in demand (Witt 2001a, b; Moneta and Chai 2014) so the Engel curve can change shape over time and thus lift the saturation level. This claim has a valid point but has overlooked the obstacles to innovation: the high possibility of innovation failure due to the nature of innovation and the high possibility of imitation if the innovation is successful. These two obstacles deter innovations, and thus, the change in income distribution and the slowing down in demand may not enough to spur innovation and change of the shape of an Engel curve.

### 5.6.1.2  Market Saturation/Product Cycle

Market saturation is a common business phenomenon. A prospective new product comes into a market with a high price and low quantity. As time passes by, the quantity of the new product increases and the price drops. Eventually, the market is saturated and the sales stagnate or even decline. The product cycle model describes market saturation in four stages: introduction, growth, maturity, and decline. The empirical support for the product cycle model is demonstrated by a number of researchers, e.g. Vgchartz (2008), Marchand (2016).

Marchand (2016) studied the life cycle in the video games industry. Figure 5.16 adopted from this study shows the yearly installations of various hardware and software of video games from 2005 to 2014. The graph shows a distinguishing product cycle pattern for each product: it starts with very low installations, followed by rapid growth. Then the growth slows down and is followed by a decline in installations.

A similar product cycle pattern can be found in the production of digital cameras and iPods. Figure 5.17 shows the shipments of digital cameras from 1999 to 2014. It demonstrates a textbook version of the 4-stage product cycle. The iPod sales data from Apple Company also show a typical product cycle pattern. It is worth mentioning that the products with life cycle are generally products for final users and most likely durables or semi-durables.

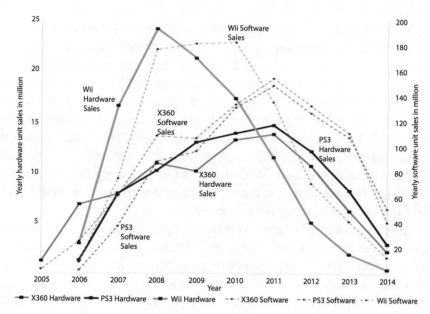

**Fig. 5.16** Life cycle of video games (*Source* Marchand 2016)

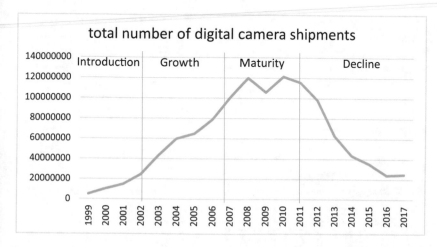

**Fig. 5.17** Life cycle of digital cameras (*Source* Based on data from Camera & Imaging Products Association)

When it comes to raw materials or intermediate products, the product cycle pattern becomes less obvious. Figure 5.18 demonstrates Australian wool production from 1901 to 2011. Although the production demonstrates an obvious cyclic pattern in each sub-period (e.g. 1901–1916, 1916–1946, 1946–1976, 1976–2011), there is a growth trend from 1901 to 1990. Despite the fact that Australian wool production decreased from 1991 to 2011, the decrease may not indicate that the wool production was in its final decline stage. The demand for raw materials depends on the demand for final products. With the advance of new technology and the invention of final products, the new market for invented products may lead to a substantial growth of raw materials.

Although market saturation/product cycle bears the similarity to a consumption ceiling, the linkage between them is complicated by other factors. Examples of product cycle theory are typically durables or semi-durables, which can last for a long time and are not purchased frequently. When most consumers with purchasing power have bought durables or semi-durables, the demand will decrease and the sales will decline. Another contributing factor is the updating or replacement of the old products by new products. For example, the decline stage of iPods may also be due to the fact that iPads have become popular

**Fig. 5.18** Australian wool production cycle (*Source* Based on data from ABS, Historical Selected Agriculture Commodities [cat. no. 7124.0])

and largely replaced iPods. Product cycle and consumption ceilings are generally applicable to specific goods rather than to an aggregate goods category. For example, although each model of video game products exhibits an obvious product cycle pattern, the aggregate product of video games, including all brands and models, may not show a product cycle pattern. This is because the contents of aggregate products may change over time and thus lift the consumption ceiling.

The formation of a 4-stage product cycle is also related to income distribution. The pattern of income distribution may vary from country to country, but it should generally follow the rule that most people have a middle level of income and fewer people have a very high or low income. With this income distribution pattern together with the concept of consumption ceiling, we can demonstrate the formation of a product cycle in Fig. 5.19.

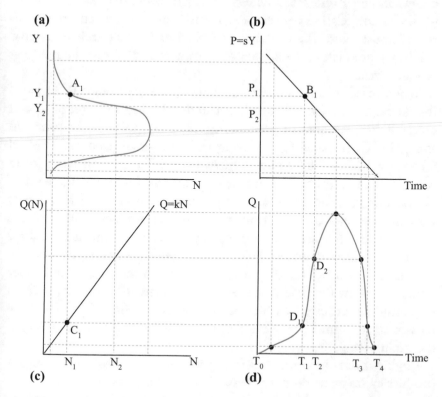

**Fig. 5.19**   Income distribution and product cycle

The income distribution of the economy is shown in panel (a). The long tail at the high-income end indicates the large gap between high-income households and median income households. The length of tails may change depending on different countries. However, the median income is always corresponding to the highest numbers of households. Panel (b) shows the price setting strategy for the new product. At the early stage, the firm sets the high price to cater for the high-income households, so the link between price and income can be set as $P=sY$, where s is the share of income spent on the new product (here we assume one household buys one new product. If buying more, the size of s will change, but the positive relationship between P and Y does not change). For convenience, we let the scale of the price axis in panel (b) be greater than that of the income axis in panel (a), so that the corresponding $Y$ and $P$ can be related by horizontal dashed lines. Panel (c) shows the relation between the number of products and the number of households. The equation $Q=kN$ implies that each household will buy $k$ products at the affordable price. Panel (d) shows the resulting product cycle.

When the firm sets a price $P_1$ at time $T_1$, we have point $B_1$ in panel (b). At price $P_1$, $N_1$ number of households with income level of $Y_1$ will purchase $kN_1$ number of products, so we have point $A_1$ in panel (a) and point $C_1$ in panel (c). Relaying the number of products in panel (c) to panel (d), we have point $D_1$. At period $T_2$, the firm sets a lower price $P_2$. At this price, $N_2$ number of households with income level of $Y_2$ will purchase the products. The previous $N_1$ number of households with income level of $Y_1$ will not purchase the products again because they have already reached their consumption ceilings (this is typical for durables and semi-durables). Following the same procedure, we can find a point $D_2$ in panel (d). Repeating the procedure, we can find other points and draw the product cycle graph in panel (d). From $T_0$ to $T_1$ is the introduction stage, at which sales grow slowly; from $T_1$ to $T_2$ is the growth stage; from $T_2$ to $T_3$ is the mature stage; and from $T_3$ to $T_4$ is the decline stage.

Product cycles have important implications for a business cycle, but product cycles alone do not necessarily cause business cycles. If innovations are plenty and thus the advent of new products is very frequent, it

is very likely that the product cycles of different new products may over-
lap and thus cancel each other out, e.g. the growth stage of one product
is accompanied by the decline stage of another product. As such, product cycles will not lead to a business cycle. However, if innovation speed
is low, product cycles cannot cancel each other out, so product cycles
will lead to business cycle. This is the reasoning behind the technological wave theory.

## 5.6.2 On Investment-Consumption Dependency

One important axiom for the new theory is the linkage between investment and consumption: the amount of investment depends on the
expected future consumption. Depending on the behaviour of investors, the expected future consumption and thus investment level may
be associated with consumption at various periods. If investors have
perfect information about the future and exercise rational expectations,
the investment would be positively correlated with future consumption,
or the lagged investment would be positively correlated with current
consumption. If investors believe future consumption is indicated by
current consumption and immediately change their investment plan,
then current investment would be positively associated with the current
investment. Finally, if investors adopt the adaptive expectation practice, i.e. use the past consumption as the indicator of future consumption, then the lagged consumption would be positively associated with
current investment. Here we use empirical data to examine the investment-consumption dependency axiom. For details, see Appendix 2 at
the end of this chapter.

## 5.6.3 On R&D Investment, Innovations, and Economic Growth

The new theory identifies that innovations are the key driver of economic
growth, so it is of interest to examine the relationship between these two
variables. Also, R&D investment is supposed to have a positive impact
on innovations and thus on economic growth, so we examine in the

section the relationship between R&D investment, innovation, and economic growth. For details, see Appendix 3.

# Appendix 1 (for Section 5.4): An Economic Modelling Approach to the New Theory

To sharpen the focus, the multi-commodity model used in this section is a static one, but a dynamic upgrade is provided for interested readers.

## A Static General Equilibrium Model

The economy in the model consists of one representative household and $n$ representative firms. For simplicity, the government is not included in the model but the function of government is implicitly included in the broad definition of households. Government spending and investment are similar to those for households. The function of government taxation and social welfare influences income distribution, which is reflected in household income distribution. Since income inequality is not the focus of the study, only one representative household is used in the model. This means that income inequality, as well as lending and borrowing, are not explicitly considered in the model.[2] However, they are indirectly included in the income distribution parameter. Lending and borrowing can lead to temporarily more equitable income distribution, so they can be expressed as a change in income distribution parameter. Also for simplicity, a closed economy is assumed, so international trade and finance are not included in the model.

The basic transactions in the model are as follows. The household provides labour and capital to all firms and obtains wages and capital rentals in return. The household also uses its income to purchase goods

---

[2]For simplicity, lending and borrowing are not considered in this paper. Lending and borrowing can delay the problems caused by consumption ceiling but cannot change the nature of the consumption ceiling because debts are required to be paid off eventually. Explicitly including lending and borrowing will not change the results, but will complicate the model.

from firms for consumption purposes and supplies its savings to firms for investment purposes. Under the zero economic profit condition, each firm uses labour, capital and technology to produce a unique type of commodity for the economy, and decides on its requirements for labour, capital, and investment in production. We express the role of each agent in mathematical form.

## Household Consumption and Savings

The ultimate goal of a society is to maximize household utility (other goals such as investment, accumulation and development are parts of household utility in the future). This means that household utility is a crucial part of an economy-wide model. The utility function described in Sect. 5.3.1 requires further modification before it is used in the model. First, since commodity demand includes both consumption demand and investment demand, we use '$c_i$' to replace '$x_i$' in the utility function in the previous section to explicitly indicate consumption demand. Second, we need to consider the fact that there are a large number of households in an economy and that the distributional effect is an important factor in household consumption and utility. It is desirable to develop a multi-household model to include the distributional effect. This would, however, complicate the model and thus interfere with the main purpose of this chapter. Instead, the author adds a distributional effect parameter in the utility function of the representative household. Finally, the varieties of commodities may increase due to product innovation, so $n + \Delta n$ is used to reflect this effect. For simplicity, this study does not model the determinants of product innovation, so $n + \Delta n$ is assumed as exogenous. The new utility function is as follows:

$$U = U(c_1, c_2, \ldots, c_{n+\Delta n}, \text{Savings})$$
$$= \sum_{i=1}^{n+\Delta n} \alpha_i (2\theta m_i c_i - c_i^2) + \alpha_S * \text{Savings} \qquad (5.7)$$

where $\alpha_i$ and $\alpha_S$ are weights for consumption and savings, respectively. $\theta$ is the distributional parameter, $0 < \theta \leq 1$. $\theta = 1$ indicates that every household in the household group has the same level of income. When income distribution is not equal, some households cannot reach their consumption saturation point due to a lack of income support. In other words, their consumption ceilings are practically lowered due to income constraint. This effect is captured by making $\theta < 1$.

The household budget can be expressed as:

$$Y = \sum_{i=1}^{n+\Delta n} P_i * c_i + \sum_{i=1}^{n+\Delta n} P_i * S_i$$

where $c_i$ is consumption of each commodity, $S_i$ is the amount of commodity saved, and $P_i$ is the commodity price.

To obtain aggregate real savings, we need a weighting parameter for aggregation. Letting it be $w_i$, we have aggregate real savings as:

$$\text{Savings} = \sum w_i * S_i.$$

Let $P_s$ be the price of aggregate savings, we have:

$$\sum_{i=1}^{n+\Delta n} P_i * S_i = P_S * \text{Saving}.$$

Defining $\delta_i$ as the share of each commodity saved ($S_i$) in total savings, i.e. $\delta_i = S_i/\text{Savings}$, we can obtain the price for aggregate savings as:

$$P_S = \left( \sum_{i=1}^{n+\Delta n} P_i * S_i \right) \Big/ \text{Saving} = \sum_{i=1}^{n+\Delta n} P_i * \delta_i.$$

As such, the optimal consumption problem for households can be expressed as:

$$\text{Maximize } U = U(c_1, c_2, \ldots, c_{n+\Delta n}, \text{Savings})$$

$$= \sum_{i=1}^{n+\Delta n} \alpha_i (2\theta m_i c_i - c_i^2) + \alpha_S * \text{Savings}$$

Subject to $Y = \sum\limits_{i=1}^{n+\Delta n} P_i * c_i + P_S * \text{Savings}$

Setting up a Lagrangian expression:

$$\ell = U + \lambda \left( Y - \sum_{i=1}^{n+\Delta n} P_i * c_i - P_S * \text{Savings} \right)$$

Using the first order condition we can derive the optimal consumption of good $i$ as follows:

$$c_i = \theta m_i - \frac{\alpha_S}{2\alpha_i} \frac{P_i}{P_S} \tag{5.8}$$

$$\text{Savings} = \left( Y - \sum_{i=1}^{n+\Delta n} \theta P_i m_i + \sum_{i=1}^{n+\Delta n} \frac{P_i^2 \alpha_S}{2\alpha_i P_S} \right) \Big/ P_S \tag{5.9}$$

## Firm's Investment and Unsold Stock

Since the firm's investment decision has an impact on its production, we discuss the firm's investment first. It is assumed that the firm can identify both the consumption ceilings and the impact of the distributional effect on consumption, so the firm can invest a proportion of the perceived consumption growth potential, i.e. a proportion of the gap between the constrained consumption ceiling and the current consumption. Letting the investment demand for commodity $i$ be proportionally related to consumption growth potential and to the propensity to invest after being discounted by interest rates, we have the following investment demand function for each commodity:

$$I_i = \frac{B}{1+r}(\theta m_i - c_i)$$

where $I_i$ is investment demand, $B$ indicates the propensity to invest, $r$ is interest rate, $m_i$ is the maximum amount of consumption on good $i$, $c_i$ is the actual amount of consumption of good $i$.

It is assumed that $0 \leq B \leq 1$. When $B=1$, the firm invests the highest amount in production to produce a maximum amount of goods which will be purchased by the household.

To obtain the aggregate real investment, we need a weighting parameter for aggregation. We use the same weighting $w_i$ as that for aggregating savings because, in the case of the existence of market clearance, the same weighting ensures that the amounts of investment and saving at both aggregate and disaggregate levels are equal, namely, $I_i = S_i$ and $I =$ Saving. As such, we have

$$I = \sum w_i * I_i, \text{ or}$$

$$I = \frac{B}{1+r} \left( \sum_{i=1}^{n+\Delta n} \theta m_i w_i - \sum_{i=1}^{n+\Delta n} c_i w_i \right) \tag{5.10}$$

Based on the investment functions at both disaggregated and aggregate levels, we can calculate the share of each commodity in total investment: $\beta_i = I_i / I$.

It is worth mentioning that, for simplicity, we use the same parameter $B$ for the investment for all commodities so we have the same overall propensity to invest in the total investment demand function. This treatment does not lose generality. If one uses different parameters as the propensity of investment demand for different commodities, the parameter for propensity of overall investment demand will be the weighted average of the parameters for all commodities. The only difference is that calculation of weight average is required to obtain the parameter for the overall propensity to invest.

This investment demand is financed by household savings. The uninvested household saving (the gap between saving and investment demand) equals the unsold stock ($S$) or inventory at firm. Although the firm has not cleared its inventory, it has paid the household the value of the inventory as wages or capital rentals. The household spends money on consumption and saves the rest. Household saving can be divided into two parts: the part invested equals the value of investment goods and the part uninvested equals the value of firm's unsold stock. As such, we have.

$$S = \text{Savings } - I = \sum w_i S_i - \sum w_i I_i$$

In a static model for a closed economy, the unsold stock $S$ in the above equation should be non-negative because investment demand must be financed by savings. However, the unsold stock $S$ can be negative when the economy is open or when the model has multiple periods. The additional finance in this case can come from overseas. In considering the accumulated past savings (e.g. wealth), the savings in a dynamic model can be negative, i.e. dissaving.

## Firm's Input Demand

To depict the firm's production, the following Cobb–Douglas function is used for the purpose of simplicity:

$$x_i = (A_i + \Delta A_i) * L_i^{\gamma_i} * K_i^{1-\gamma_i}$$

where $L$ means labour, $K$ capital, $A$ the level of technology, $\Delta A$ technological changes, and $\gamma$ the share of labour in total inputs.

The optimal production problem can be expressed as:
Minimize Cost $= P_L * L_i + P_K * K_i$
Subject to Output $= x_i = (A_i + \Delta A_i) * L_i^{\gamma_i} * K_i^{1-\gamma_i}$
Setting up a Lagrangian expression:

$$\ell = P_L * L_i + P_K * K_i + \lambda \left[ x_i - (A_i + \Delta A_i) * L_i^{\gamma_i} * K_i^{1-\gamma_i} \right]$$

Using the first order condition we can show the optimal demand for labour and capital as follows:

$$L_i = \left( \frac{x_i}{A_i + \Delta A_i} \right) \left( \frac{\gamma_i P_K}{(1 - \gamma_i) P_L} \right)^{1-\gamma_i} \tag{5.11}$$

and

$$K_i = \left( \frac{x_i}{A_i + \Delta A_i} \right) \left( \frac{(1 - \gamma_i) P_L}{\gamma_i P_K} \right)^{\gamma_i} \tag{5.12}$$

These results link the firm's demand for labour and capital ($L_i$ and $K_i$) to the firm's output $x_i$. More generally, the results show that the factor market is closely related to the commodity market.

## Resource Constraints and Market Clearance Condition

The resources constraint in a closed economy can be expressed as

Savings are not less than investment: $S = $ Savings $- I = \sum w_i S_i - \sum w_i I_i \geq 0$.

Labour supply is not less than labour demand: $L \geq \sum L_i$

Capital supply is not less than capital demand: $K \geq \sum K_i$.

Next, we consider the market clearance condition. The total supply of commodity $x_i$ in the economy is the sum of both the consumed and the unconsumed commodity, namely, $x_{Si} = c_i + S_i$. On the other hand, the total demand for commodity $x_i$ comprises the consumption demand and the investment demand, so that the total demand for $x_i$ can be expressed as $x_{Di} = c_i + I_i$. Thus, the excess demand function for $x_i$ is: $ED_i = x_{Di} - x_{Si} = I_i - S_i$.

The conditions for market clearance require that, for each commodity $i$, $ED_i = x_{Di} - x_{Si} = I_i - S_i = 0$ or $I_i = S_i$. Aggregating all commodities, we have market clearance condition: $\sum w_i I_i = \sum w_i S_i$, or $I = $ Savings.

In a traditional general equilibrium model, investment is always equal to savings because neoclassical economics simplistically and idealistically assumes that all savings are invested, so $I = $ Savings is guaranteed by presumption. With this guarantee, a general equilibrium is achievable at any time: if $I_i \neq S_i$ for some or all commodity types, the price mechanism will kick in and adjust any difference between investment and savings in all commodity types.

However, in our static model, investment is determined by the consumption growth potential (the difference between current consumption and the maximum consumption) while saving is determined by the utility maximization procedure, so there is no guarantee that total investment equals total savings—only the resource constraint (Savings $\geq I$) is applied to savings and investment. As a result, the general equilibrium is not guaranteed: if Savings $> I$, the price mechanism cannot work out the solution for $S_i = I_i$ because in this case there is a positive net saving and thus an overall oversupply in the economy.

## Static Result Interpretation: A Demand-Side Perspective

At this point, we assume the prices in the static model are fixed so that we can derive some intuitive but essential results from the model. For an economy to grow without a recession, the total supply of a commodity must be cleared by the market, i.e. excess demand for any commodities must be nonnegative: $\mathrm{ED}_i = x_{Di} - x_{Si} = I_i - S_i \geq 0$, or $\mathrm{ED} = \sum_{i=1}^{n+\Delta n} I_i w_i - \sum_{i=1}^{n+\Delta n} S_i w_i = I - \text{Saving} \geq 0$.

Recalling the investment equation (Eq. 5.10), we can express the condition to avoid a recession as:

$$\mathrm{ED} = \frac{B}{1+r} \sum_{i=1}^{n+\Delta n} \theta m_i w_i - \frac{B}{1+r} \sum_{i=1}^{n+\Delta n} c_i w_i - \text{Saving} \geq 0$$

Plugging the consumption equation (Eq. 5.8) and the saving equation (Eq. 5.9) into the above inequality, we have:

$$\frac{B}{1+r} \sum_{i=1}^{n+\Delta n} \theta m_i w_i - \frac{B}{1+r} \sum_{i=1}^{n+\Delta n} \left( \theta m_i w_i - \frac{\alpha_S P_i w_i}{2\alpha_i P_S} \right)$$
$$- \left( Y - \sum_{i=1}^{n+\Delta n} \theta P_i m_i + \sum_{i=1}^{n+\Delta n} \frac{P_i^2 \alpha_S}{2\alpha_i P_S} \right) \Bigg/ P_S \geq 0$$

$$Y \leq \sum_{i=1}^{n+\Delta n} \theta P_i m_i + \sum_{i=1}^{n+\Delta n} \frac{B P_i \alpha_S w_i}{2(1+r)\alpha_i} - \sum_{i=1}^{n+\Delta n} \frac{P_i^2 \alpha_S}{2\alpha_i P_S} \qquad (5.13)$$

This inequality shows that, to avoid a recession, the household income must be below a certain level! To allow income to increase without a ceiling, one may increase $\theta$ (i.e. improving the equality in income distribution) or increase $B$ (the propensity to invest) or decrease $r$ (the interest rate), but the effect of these efforts is limited because the maximum value of both $\theta$ and $B$ is 1 and the minimum value of $r$ is 0. The only way to allow the income level to increase unrestrictedly is to increase $\Delta n$, i.e. inventing new products. Since $\theta * m_i$ is much larger compared with $P_i\alpha_s/2\alpha_i P_s$ (for an economy, the consumption ceiling $m_i$ is generally very high compared with other items here), when $\Delta n$ increases, the increase in the first term on the right-hand side will outweigh the increase in the third term and thus the cap on $Y$ will be lifted.

Household supply of labour and capital is determined by household willingness to obtain income, which in turn is determined by consumption and savings. So, the household will supply the amount of labour and capital to produce the amount of output of good $i$ that is equal to the sum of the consumed and the saved by the household. In this reasoning, we substitute $x_i = c_i + S_i$ into Eqs. (5.11) and (5.12) and obtain the amount of labour and capital supplied for the production of good $x_i$:

$$L_{Si} = \left( \frac{c_i + S_i}{A_i + \Delta A_i} \right) \left( \frac{\gamma_i P_K}{(1 - \gamma_i) P_L} \right)^{1 - \gamma_i} \tag{5.14}$$

$$K_{Si} = \left( \frac{c_i + S_i}{A_i + \Delta A_i} \right) \left( \frac{(1 - \gamma_i) P_L}{\gamma_i P_K} \right)^{\gamma_i} \tag{5.15}$$

On the other hand, the demand for labour and capital is determined by final demand $c_i + I_i$. Therefore, substituting $x_i = c_i + I_i$ into Eqs. (5.11) and (5.12) we have the labour and capital demand functions:

$$L_{Di} = \left( \frac{c_i + I_i}{A_i + \Delta A_i} \right) \left( \frac{\gamma_i P_K}{(1 - \gamma_i) P_L} \right)^{1 - \gamma_i} \tag{5.16}$$

$$K_{Di} = \left( \frac{c_i + I_i}{A_i + \Delta A_i} \right) \left( \frac{(1 - \gamma_i) P_L}{\gamma_i P_K} \right)^{\gamma_i} \tag{5.17}$$

Excess demand in the factor market is the sum of the excess demand for labour and capital in producing each commodity, namely:

$$\text{ED}_L = \sum_{i=1}^{n+\Delta n} \text{ED}_{Li} = \sum_{i=1}^{n+\Delta n} L_{Di} - \sum_{i=1}^{n+\Delta n} L_{Si} \qquad (5.18)$$

$$\text{ED}_K = \sum_{i=1}^{n+\Delta n} \text{ED}_{Ki} = \sum_{i=1}^{n+\Delta n} K_{Di} - \sum_{i=1}^{n+\Delta n} K_{Si} \qquad (5.19)$$

Substituting Eqs. (5.14) to (5.17) into the Eqs. (5.18) and (5.19) and utilizing the saving share $\delta_i$ and investment share $\beta_i$, we have:

$$
\begin{aligned}
\text{ED}_L &= \sum_{i=1}^{n+\Delta n} \left(\frac{I_i - S_i}{A_i}\right)\left(\frac{\gamma_i P_K}{(1-\gamma_i)P_L}\right)^{1-\gamma_i} \\
&= \sum_{i=1}^{n+\Delta n} \mu_i I_i - \sum_{i=1}^{n+\Delta n} \mu_i S_i = \sum_{i=1}^{n+\Delta n} \mu_i \beta_i I - \sum_{i=1}^{n+\Delta n} \mu_i \delta_i S
\end{aligned}
\qquad (5.20)
$$

$$
\begin{aligned}
\text{ED}_K &= \sum_{i=1}^{n+\Delta n} \left(\frac{I_i - S_i}{A_i}\right)\left(\frac{(1-\gamma_i)P_L}{\gamma_i P_K}\right)^{\gamma_i} \\
&= \sum_{i=1}^{n+\Delta n} v_i I_i - \sum_{i=1}^{n+\Delta n} v_i S_i = \sum_{i=1}^{n+\Delta n} v_i \beta_i I - \sum_{i=1}^{n+\Delta n} v_i \delta_i S
\end{aligned}
\qquad (5.21)
$$

where

$$\mu_i = \left(\frac{\gamma_i P_K}{(1-\gamma_i)P_L}\right)^{1-\gamma_i} \bigg/ (A_i + \Delta A_i), \quad v_i = \left(\frac{(1-\gamma_i)P_L}{\gamma_i P_K}\right)^{\gamma_i} \bigg/ (A_i + \Delta A_i)$$

The above two equations show that the excess demand for both labour and capital is the difference between weighted total investment and weighted total savings. This is very similar to the excess demand function in the commodity market. Thus, if the investment demand is greater than savings, there is an excess demand in commodity markets, there will be an excess demand for labour and capital. The size of excess demand in different markets will, however, differ due to the weights. Since the demand for primary factors closely links to the demand for commodities, the reasons for excess supply in the commodity market can also explain the excess supply in the factor market.

## Static Result Interpretation: A Supply-Side Perspective

The above demand-side approach gives us an intuitive picture of commodity and factor markets. However, this picture is not a high-resolution one because the prices for the commodities and for factors are set as exogenous. In fact, the prices in the model are related to each other and they can change when the economy goes from disequilibrium to equilibrium, so we allow the prices be endogenous in this section. In determining the prices of commodities and factors, we can link the production side (or supply side) to the consumption side (or demand side).

First of all, by using Eqs. (5.11) and (5.12), we can obtain the price linkage between labour and capital.

$$P_L = \frac{\gamma_i}{(1 - \gamma_i)} \frac{K_i}{L_i} P_K$$

Based on the zero economic profit assumption, the cost of producing $X_i$ will determine the price of $X_i$, so,

$$P_i X_i = P_L L_i + P_K K_i = \frac{\gamma_i}{(1 - \gamma_i)} \frac{P_K K_i}{L_i} L_i + P_K K_i = \frac{P_K K_i}{(1 - \gamma_i)}$$

or,

$$P_i = \frac{P_K K_i}{(1 - \gamma_i) X_i}$$

Normalizing the price level of the economy to 1, we have

$$1 = \frac{\sum_{i=1}^{n+\Delta n} P_i X_i}{\sum_{i=1}^{n+\Delta n} X_i} = \frac{P_K \sum_{i=1}^{n+\Delta n} \frac{K_i}{1-\gamma_i}}{\sum_{i=1}^{n+\Delta n} X_i} \quad \text{or}$$

$$P_K = \frac{\sum_{i=1}^{n+\Delta n} X_i}{\sum_{i=1}^{n+\Delta n} \frac{K_i}{1-\gamma_i}} = \frac{\sum_{i=1}^{n+\Delta n} (A_i + \Delta A_i) L_i^{\gamma_i} K_i^{1-\gamma_i}}{\sum_{i=1}^{n+\Delta n} \frac{K_i}{1-\gamma_i}},$$

And thus,

$$P_i = \frac{K_i}{(1 - \gamma_i)X_i} \frac{\sum_{i=1}^{n+\Delta n} X_i}{\sum_{i=1}^{n+\Delta n} \frac{K_i}{1-\gamma_i}}$$

$$= (1 - \gamma_i)^{-1}(A_i + \Delta A_i)^{-1}L_i^{-\gamma_i}K_i^{\gamma_i} \frac{\sum_{i=1}^{n+\Delta n}(A_i + \Delta A_i)L_i^{\gamma_i}K_i^{1-\gamma_i}}{\sum_{i=1}^{n+\Delta n} \frac{K_i}{1-\gamma_i}} \quad (5.22)$$

Based on this $P_i$, we can easily calculate Ps as $P_s = \sum_{i=1}^{n+\Delta n} P_i\delta_i$. Plugging $P_i$ and $P_s$ into Eq. (5.8), we can assess (albeit a bit complicated) the impact of a change in inputs ($K$ and $L$) on consumption.

The excess supply of commodities is the difference between commodities saved and commodities invested, i.e.

$$ES_i = S_i - I_i = S\delta_i - I\beta_i$$

$$= \delta_i\left(Y - \sum_{i=1}^{n+\Delta n} \theta P_i m_i + \sum_{i=1}^{n+\Delta n} \frac{P_i^2\alpha s}{2\alpha_i P_S}\right)\bigg/ P_S$$

$$- \frac{B}{1+r}\beta_i\left(\sum_{i=1}^{n+\Delta n} \theta m_i w_i - \sum_{i=1}^{n+\Delta n}\left(\theta m_i w_i - \frac{P_i\alpha s w_i}{2\alpha_i P_S}\right)\right)$$

$$= \delta_i\left(Y - \sum_{i=1}^{n+\Delta n} \theta P_i m_i + \sum_{i=1}^{n+\Delta n} \frac{P_i^2\alpha s}{2\alpha_i P_S}\right)\bigg/ P_S - \frac{B}{1+r}\beta_i\sum_{i=1}^{n+\Delta n} \frac{P_i\alpha s w_i}{2\alpha_i P_S}$$

The excess commodity supply will reduce the commodity supply in the next period, so the factors contributing to the excess commodity supply will not be employed in the next period. This causes unemployed labour and unutilized capital in the factor markets.

To avoid economic stagnation, it is necessary that there is no overall excess supply, i.e. $\Sigma ES_i \leq 0$. Summarizing all $ES_i$ and using the fact that $\Sigma\delta_i = 1$ and $\Sigma\beta_i = 1$, we have:

$$\sum_{i=1}^{n+\Delta n} ES_i = \left(Y - \sum_{i=1}^{n+\Delta n} \theta P_i m_i + \sum_{i=1}^{n+\Delta n} \frac{P_i^2\alpha s}{2\alpha_i P_S}\right)\bigg/ P_S - \frac{B}{1+r}\sum_{i=1}^{n+\Delta n} \frac{P_i\alpha s w_i}{2\alpha_i P_S} \leq 0$$

$$Y \leq \sum_{i=1}^{n+\Delta n} \theta P_i m_i + \frac{B}{1+r}\sum_{i=1}^{n+\Delta n} \frac{P_i\alpha s w_i}{2\alpha_i} - \sum_{i=1}^{n+\Delta n} \frac{P_i^2\alpha s}{2\alpha_i P_S}$$

This is the same results as Eq. (5.13) derived for the income ceiling from the demand-side perspective. However, the prices for commodities and for savings in the above equation are determined by the amount of capital and labour inputs through Eq. (5.22). Considering this, the ceiling on income is not fixed.

To be more accurate, the income derived above is measured by money so it is a nominal income. In real terms, we must leave out price $P_i$ in above equation, so the real income in the equilibrium should be the goods consumed and invested, i.e.,

$$Y_{\text{real}} = \sum_{i=1}^{n+\Delta n} \theta m_i + \frac{B}{1+r} \sum_{i=1}^{n+\Delta n} \frac{\alpha_S w_i}{2\alpha_i} - \sum_{i=1}^{n+\Delta n} \frac{P_i \alpha_S}{2\alpha_i P_S} \quad (5.23)$$

In this equation, the prices of commodities (and thus the prices of savings, labour and capital) only affect the negative item, so the fixed ceiling still exists for real income.

Equation (5.23) can also be used to demonstrate a case of government tax policy. Although there is no government in the model, the effect of a lump sum tax can be demonstrated indirectly. If the government imposes this tax and distributes it to poor households, the distributional parameter $\theta$ in Eq. (5.23) increases with other things being equal, and this leads to an increase in real income. This is consistent with intuition: income redistribution to improve equality will encourage consumption and thus stimulate economic growth. If the government use the tax to boost budget balance (i.e. tax revenue is unspent), this indicates an increased preference to save, so $\alpha_S$ increase while $\alpha_i$ decreases in Eq. (5.23). The consequence of this is an increase in the absolute size of both the positive investment effect (the second term at the right of the equation) and the negative saving effect (the last term at the right of the equation). The enlarged negative saving effect results from the decreased current consumption level due to government tax; and the enlarged positive investment effect is due to the increased gap between the consumption ceiling and the decreased current consumption. The overall effect of this lump sum tax depends on the relative size of changes in both the saving effect and the investment effect. Generally speaking, the investment effect is smaller because the value of

$B/(1 + r)$ is less than one, so the overall effect would be a decrease in real income.

## A Recursive Dynamic Model

The model presented so far is a static one, but it can be easily upgraded by adding a time frame and by considering dynamics in technology, capital, and wealth.

Since wealth is accumulated savings, so the wealth dynamic can be described by equalling $W_{t+1}$ (wealth in time $t+1$) to $W_t$ (wealth in time $t$) plus savings in time $t$.

$$W_{t+1} = W_t + \text{Savings}_t$$

The technological and capital dynamics for industry $i$ can be expressed as
$A_{i,t+1} = A_{i,t} + \Delta A_{i,t}$,
where $\Delta A_{i,t}$ is the change of technology in industry $i$ at time $t$.
$K_{i,t+1} = (1 - \delta)K_{i,t} + I_{i,t}$,
where $I_{i,t}$ is the change of capital in industry $i$ at time $t$, $\delta$ is the depreciation rate.

The household utility function at time $t$ can be written as:

$$U_t = U(c_{1,t}, c_{2,t}, \ldots, c_{n+\Delta n_t}, \text{Savings}_t)$$
$$= \sum_{i=1}^{n+\Delta n} \alpha_{i,t}(2\theta_t m_{i,t} c_{i,t} - c_{i,t}^2) + \alpha_{S,t} * \text{Savings}_t$$

This utility function is subject to a budget:
$W_t + Y_t \geq \sum P_{i,t} c_{i,t} + P_{S,t} \sum S_{i,t}$.
The optimal production problem can be expressed as:
Maximize
$x_{i,t} = A_{i,t} * L_{i,t}^{\gamma_i} * K_{i,t}^{1-\gamma_i}$,
subject to production cost $= P_{L,t} * L_{i,t} + P_{K,t} * K_{i,t}$.
The investment demand function is as follows:

$$I_t = \frac{B_t}{1 + r_t} \left( \sum_{i=1}^{n+\Delta n} \theta_t m_{i,t} w_{i,t} - \sum_{i=1}^{n+\Delta n} c_{i,t} w_{i,t} \right) \qquad (5.24)$$

In a similar fashion to the static model, the household uninvested savings equal the unsold stock (i.e. inventory) at firm, which are calculated as the difference between savings and investment demand,

$$S_t = \text{Savings}_t - I_t$$

Investment demand is financed by savings; any uninvested savings (or stocks) will be accumulated as wealth; and any excess investment demand over savings will be drawn from wealth.

The above equations transform the static model to a recursive dynamic model. This recursive model can generate the equilibrium for each period. The results from the previous period will have an influence on the results in the next period. For example, $I_t$ (the investment in time $t$) will affect $K_{t+1}$ (the capital in time $t+1$), which in turn will affect a number of variables in time $t+1$ such as output level ($x_{t+1}$), savings ($S_{t+1}$), consumption ($c_{t+1}$), and investment ($I_{t+1}$). However, the mechanism determining the equilibrium or disequilibrium in each period is the same as in the static model.

## An Intertemporal General Equilibrium Model

The above recursive model is unable to determine either the intertemporal equilibrium or the optimal time path for the economy, so an intertemporal model is needed. Since an intertemporal equilibrium model with multi-commodity is very complex, we have to be content with a Ramsey/Solow-style one-commodity model, but the essence of including multi-commodity in the static model—consumption ceiling—will be reflected in the intertemporal model. Moreover, to reduce the number of variables, we eliminate the variable labour by measuring capital, output, utility, and consumption in per capita term. In a traditional Ramsey/Solow model,[3] all savings are assumed to be invested. This assumption is implausible thus it has to be relaxed. The three axioms featured in the static model will also be used in the intertemporal equilibrium model.

Since we are considering an intertemporal equilibrium model, we have to use continuous time, which is different from the discrete time

---

[3]For detail of Ramsey/Solow model, see Ramsey (1928), or more recent books on macroeconomics or mathematical economics such as Turkington (2007), Romer (2013).

in the recursive model. Also, because all variables in the intertemporal model are measured in per capita terms, we use the lower case for most variables so as to differentiate them from the aggregate variables, e.g. using $k$ for capital, $c$ for consumption, and $s$ for savings.

In per capita term, the function can be written as:

$y = a * f(k)$,

where $a$ is technology, $k$ is capital per worker, and $f(k)' > 0$.

The household utility function in per capita terms can be written as:

$u(c) = \alpha_C * (2mc - c^2) + \alpha_S * (af(k) - c)$,

where, $\alpha_C$ is the weighting for utility from consumption, $\alpha_S$ is the weighting for utility from saving, $m$ is consumption ceiling, $c$ is actual consumption, $c \leq m$.

The investment per capita is proportional to the gap between the consumption ceiling and the actual consumption level, so it can be written as

$i = b(m - c)$,

where $i$ is investment and $b$ is an interest-rate-discounted investment propensity parameter, $b = B/(1 + r)$, $0 < B < 1$; as before, $B$ is the propensity to invest, $r$ is the interest rate.

Based on this investment demand function, the per capita capital dynamics can be written as:

$k' = \Delta k = b(m - c) - \delta k - nk$,

where $\delta$ is the capital depreciation rate and $n$ is the growth rate of population (or labour force).

The dynamics of per capita assets are determined by uninvested household saving or unsold stock (inventory) at the firm:

$s' = \Delta s = \text{Savings} = y - c - i = af(k) - c - b(m - c)$,

where $s$ stands for stock.

Considering a time preference rate (i.e. future discount rate) of $\theta$, the optimal control problem can be expressed as:

Maximize

$V = \int_0^\infty u(c)e^{-\theta t} dt$

Subject to

$s' = af(k) - c - b(m - c)$, $k' = b(m - c) - \delta k - nk$,

$s(0) = 0$, $s(\infty) \geq 0$, $k(0) = 0$, $k(\infty) \geq 0$, and $0 \leq c(t) \leq m$.

The standard Hamiltonian function is

$$H = u(c)e^{-\theta t} + \lambda_1 \left[ af(k) - c - b(m - c) \right] + \lambda_2 [b(m - c) - \delta k - nk].$$

From this Hamiltonian function, it is easy to see that if we impose a condition that $\lambda_1 = \lambda_2$, then the model collapses to the traditional Ramsey/Solow model.

The current-value form of Hamiltonian function is

$$H_c = u(c) + \eta_1\left[af(k) - c - b(m - c)\right] + \eta_2[b(m - c) - \delta k - nk],$$

where $\eta_1 = \lambda_1 e^{\theta t}$, and $\eta_2 = \lambda_2 e^{\theta t}$.

The necessary condition for an optimal solution is that, at each $t$,

(i) $\partial H_c/\partial c = 0$
(ii) $\eta_1' = -\partial H_c/\partial s + \theta\eta_1$.
(iii) $\eta_2' = -\partial H_c/\partial k + \theta\eta_2$.
(iv) $s' = \partial H_c/\partial \eta_1$
(v) $k' = \partial H_c/\partial \eta_2$
(vi) $s(0) = 0, s(\infty) \geq 0, k(0) = 0, k(\infty) \geq 0, 0 \leq c(t) \leq m$.

Condition (ii) gives: $\eta_1' = -\partial H_c/\partial s + \theta\eta_1 = \theta\eta_1$. The obvious solution for $\eta_1$ is $\eta_1 = e^{\theta t}$.

Condition (i) gives $u' + \eta_1(-1 + b) + \eta_2(-b) = 0$, or $\eta_2 = \eta_1 (b - 1)/b + u'/b$.

Considering $\eta_1 = e^{\theta t}$ and $u = \alpha_C * (2mc - c^2) + \alpha_S * (af(k) - c)$, we have $\eta_2 = e^{\theta t}(b - 1)/b + (2\alpha_C m - 2\alpha_C c - \alpha_S)/b$, and thus $\eta_2' = \theta e^{\theta t}(b - 1)/b + 2\alpha_C c'/b$.

Condition (iii) gives:

$$\begin{aligned}
\eta_2' &= -\partial H_c/\partial k + \theta\eta_2 = -[\eta_1 af(k)' - \eta_2(\delta + n)] + \theta\eta_2 \\
&= -\eta_1 af(k)' + \eta_2(\delta + n + \theta) \\
&= -e^{\theta t}af(k)' + \big[e^{\theta t}(b - 1)/b \\
&\quad + (2\alpha_C * m - 2\alpha_C * c - \alpha_S)/b\big](\delta + n + \theta).
\end{aligned}$$

Based on the above two equations, we have:

$$\begin{aligned}
\eta_2' &= \theta e^{\theta t}(b - 1)/b + 2\alpha_C * c'/b \\
&= -e^{\theta t}af(k)' \\
&\quad + \big[e^{\theta t}(b - 1)/b + (2\alpha_C * m - 2\alpha_C * c - \alpha_S)/b\big](\delta + n + \theta), \text{ or,}
\end{aligned}$$

$$\begin{aligned}
c' &= -0.5\theta e^{\theta t}(b - 1) - 0.5e^{\theta t}abf(k)' \\
&\quad + 0.5\big[e^{\theta t}(b - 1) + (2\alpha_C * m - 2\alpha_C * c - \alpha_S)\big](\delta + n + \theta), \text{ or,}
\end{aligned}$$

$$c' = -\ 0.5e^{\theta t}\left[abf(k)' + (1-b)(\delta + n)\right]$$
$$+ (\alpha_C * m - \alpha_C * c - 0.5\alpha_S)(\delta + n + \theta) \qquad (5.25)$$

Conditions (iv) and (v) gives the growth rate of stock and capital respectively:

$$s' = af(k) - c - b(m - c) \qquad (5.26)$$

$$k' = b(m - c) - \delta k - nk \qquad (5.27)$$

These conditions will be used in the next section.

## Result Interpretation: A Dynamic Perspective

The mechanism in the recursive model is essentially the same as that in the static model, so the equilibrium and disequilibrium in each period in the recursive model will be very similar to those in the static model. However, the investment function (5.24) plays a key role in patterns of economic growth. As the investment at time $t$ is proportional to the potential of consumption growth (i.e. the gap between current consumption and consumption ceiling), the change of potential of consumption growth leads to cyclic the investment behaviour. When the gap between the current consumption and the consumption ceiling is large, investors perceive the high potential of increase in sales in the future and thus invest more in production. This leads to a large increase in aggregate demand and pushes the economy into the boom phase. As the gap between the current consumption and the consumption ceiling gets smaller (assuming no new product is invented), stagnancy or very slow growth of consumption is coupled with an investment decrease. This leads to a decrease in aggregate demand and thus an economic recession.

During a recession, the firm cannot find a chance to increase sales because the consumption level is close to the consumption ceiling. Under this circumstance, firms have no choice but to invest in research and innovation, hoping to invent a new product. Once the innovation succeeds, the new product will lift the consumption ceiling and the gap between the current consumption and the consumption ceiling increases. As a result, both consumption and investment increase and the economy enters a recovery phase which is followed by an expansion phase. This cyclical growth will continue as long as there is no mechanism to stimulate product innovation.

From a different perspective, the intertemporal equilibrium model in the previous section can demonstrate the same point: a steady or balanced growth is not achievable in the long run if product innovation cannot keep pace with production growth.

We start with a task to find out the steady-state conditions as well as the optimal economic growth path. At the steady state, the growth of stock and capital becomes zero. Using Eqs. (5.26) and (5.27), we have:

$$s' = af(k) - c - b(m - c) = 0$$

$$k' = b(m - c) - \delta k - nk = 0$$

Combining the above two equations, we have:

$$k' = af(k) - c - \delta k - nk \tag{5.28}$$

Setting $k' = 0$, we have

$$c = af(k) - (\delta + n)k \tag{5.29}$$

Equation (5.29) defines a $k' = 0$ curve.

The condition for optimal consumption is $\partial c/\partial k = 0$. This gives the same golden rule as in the Ramsey/Solow model:

$$af(k)' = (\delta + n) \tag{5.30}$$

At a steady state, the growth of consumption must also be zero. Setting Eq. (5.25) to zero, we have:

$$\begin{aligned} c' = &- 0.5e^{\theta t}\left[abf(k)' + (1 - b)(\delta + n)\right] \\ &+ (\alpha_C * m - \alpha_C * c - 0.5\alpha_S)(\delta + n + \theta) = 0 \end{aligned} \tag{5.31}$$

Equation (5.31) defines a $c' = 0$ curve.

Combining Eqs. (5.29) and (5.31), we can solve for the steady state $E(c^*, k^*)$. However, this steady state will not be steady if the consumption ceiling ($m$) is fixed: due to the term $e^{\theta t}$, $c'$ will reduce over time. In other words, to maintain a steady state, $m$ has to increase over time.

If the steady state is optimal, it must coincide with the optimal consumption point.[4] We apply the golden rule by substitute Eqs. (5.29) and (5.30) into Eq. (5.31),

$$c' = -0.5e^{\theta t}(\delta + n) + (\alpha_C * m - \alpha_C * c - 0.5\,\alpha_S)(\delta + n + \theta) = 0 \tag{5.32}$$

This gives: $c = m - 0.5e^{\theta t}(\delta + n)/[\alpha_C(\delta + n + \theta)] - 0.5\alpha_S/\alpha_C$ (5.33)

The condition to achieve a steady consumption is:

$$dc/dt = dm/dt - 0.5\theta e^{\theta t}(\delta + n)/(\delta + n + \theta) = 0, \ \text{or,}$$

$$m = 0.5e^{\theta t}(\delta + n)/[\alpha_C(\delta + n + \theta)] + \text{constant} \tag{5.34}$$

These results can be shown in Fig. 5.15. Panel (a) shows the space view of the phase diagram. The $k'=0$ curve is given by setting $k'=af(k) - c - (\delta + n)\,k=0$, i.e. equation (5.29). Based on Eq. (5.28), $\partial k'/\partial c=-1<0$, $k'$ and $c$ move in opposite directions, i.e. as $c$ increase, $k'$ decrease. This necessitates $k'>0$ within the $k'=0$ curve and $k'<0$ outside the $k'=0$ curve. Putting it differently, $k$ will increase within the $k'=0$ curve and $k$ will decrease outside the $k'=0$ curve. This movement is indicated by the solid arrows accompanied by letter '$k$' (Fig. 5.20).

The steady state is at $E$ ($c^*$, $k^*$), which is the intersection of the $k'=0$ curve and the $c'=0$ curve. For simplicity, we assume that the steady state is at optimal. Since $c'=0$, we have $c=c^*$ (this assumption can be relaxed and the analysis is similar, but the graphs will be more complicated).

---

[4]A steady state can happen at any capital level, i.e. multiple steady states. In this case, we can obtain at each capital level a consumption level at which $c'=0$. The analysis for each steady state is similar. The assumption of an optimal steady state simplifies the analysis.

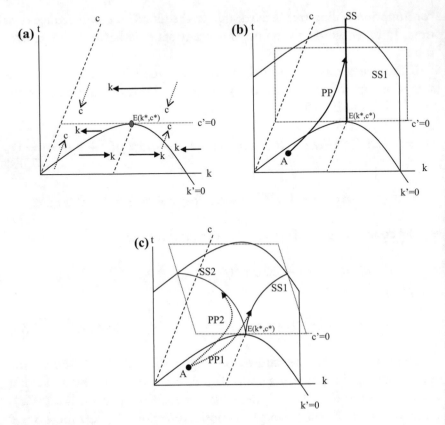

**Fig. 5.20** Economic growth in an intertemporal optimization model

According to Eq. (5.32), $c'$ and $c$ move in opposite directions. As a result, $c' > 0$ when $c < c^*$, and $c' < 0$ when $c > c^*$. In other words, $c$ will increase when $c < c^*$, and $c$ will decrease when $c > c^*$. The movement of $c$ is indicated by the dotted arrows accompanied by letter '$c$'.

The phase diagram for $c$ and $k$ indicates that the economy can converge to a steady state E either (a) when $c < c^*$, $k < k^*$, and within the $k' = 0$ curve, or (b) when $k > k^*$ and outside of the $k' = 0$ curve.

Panel (b) of Fig. 5.15 shows the evolution of the economy over time. $k' = 0$ is now a curve space and $c' = 0$ is a vertical plane at $c = c^*$. The tangent line of these two spaces SS shows the steady state of the economy. An economy at point A can reach the steady state SS through a path PP.

However, the $e^{\theta t}$ term in Eq. (5.34) shows that the existence of the steady state is conditional on the lifting of the consumption ceiling over time. If the consumption ceiling ($m$) is fixed, the term $e^{\theta t}$ in Eq. (5.32) necessitates that $c$ will decrease over time. As a result, the $c' = 0$ plane will be inclined to the $k$ axis and intersect with $k' = 0$ space on two curves $SS_1$ and $SS_2$. The economic equilibrium will evolve along either curve. The time path of economic growth is shown as the dotted arrow $PP_1$ or $PP_2$. Since the $c$ and $k$ on either $SS_1$ or $SS_2$ will change over time, there is no steady state—the consumption will decrease continuously.

In short, the intertemporal equilibrium model of single-commodity with consumption ceiling shows that, thanks to the fixed consumption ceiling, the economy will not reach a steady state—the consumption will keep falling in the long run. To reach a steady state, the consumption ceiling must keep increasing. Since the consumption of any commodity has a fixed ceiling according to Axiom 1, the only way to increase the consumption ceiling for the economy is to increase the variety of commodities. That is, product innovation is the key to reaching a steady state, or balanced growth.

# Appendix 2 (for Section 5.6.2): On Investment-Consumption Dependency

Our new theory is rested on an assumption that the amount of investment depends on the expected future consumption. Here we use empirical data to examine the investment-consumption dependency axiom.

The standard econometric approach is to identify all factors related to private investment and use them to do a multi-regression. However, there are two shortcomings with this approach. One is that it is impossible to include all relevant variables and exclude irrelevant variables. The other is that nobody knows the correct function form for regression. Most econometricians simply assume a linear or log function for convenience. Thus, this approach is subject to data distortion if irrelevant variables are included or if the function form is incorrect. Since we are interested only in the correlation between investment and consumption, we use two variable regressions to avoid data distortion.

This approach implies that the impact of other variables is negligible, so the results from this approach, just like results from other econometric approaches, are only indicative.

First, we examine the linkage between investment and consumption by using the 1929–2017 yearly US private consumption and private investment data, which are freely available from the website of the Bureau of Economic Analysis (BEA). The regression shows that private consumption is highly correlated with investment in all periods, i.e., current, lagged and leading investments. For example, the result of regressing of the current private investment 'investpriv' on current private consumption 'consump' using Stata software is shown in Fig. 5.21.

The low standard error and high $t$-value show the extremely high significance of the consumption variable. The $R$-squared value 0.9795 is close to the maximum value of 1, which indicates the extremely high power of the consumption variable in explaining the behaviour of private investment. The results of regressing lagged investment on current consumption and the results of regression of leading investment on current consumption are not displayed here but they are very similar to those in Fig. 5.21. These results let us wonder about the types of investors' behaviours: rational expectation, adaptive expectation, or adjusting investment immediately according to current consumption?

Actually, this regression suffers from a serious defect. Both private investment and consumption demonstrate a growing trend (see Fig. 5.22). This growing trend may be caused by other factors, e.g. the

```
. regress investpriv consump
```

| Source | SS | df | MS | | | |
|---|---|---|---|---|---|---|
| | | | | Number of obs | = | 89 |
| | | | | F(1, 87) | = | 4162.78 |
| Model | 8.0159e+13 | 1 | 8.0159e+13 | Prob > F | = | 0.0000 |
| Residual | 1.6753e+12 | 87 | 1.9256e+10 | R-squared | = | 0.9795 |
| | | | | Adj R-squared | = | 0.9793 |
| Total | 8.1835e+13 | 88 | 9.2994e+11 | Root MSE | = | 1.4e+05 |

| investpriv | Coef. | Std. Err. | t | P>|t| | [95% Conf. Interval] | |
|---|---|---|---|---|---|---|
| consump | .2448048 | .0037943 | 64.52 | 0.000 | .2372633 | .2523463 |
| _cons | 40593.77 | 18767.27 | 2.16 | 0.033 | 3291.787 | 77895.74 |

**Fig. 5.21** Results of regressing US investment on consumption

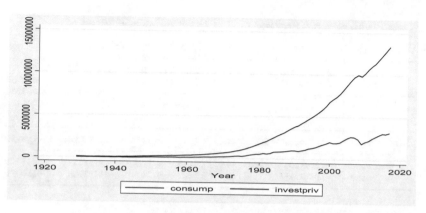

**Fig. 5.22** The US consumption and investment from 1927 to 2017

increased size of the US economy over time. Any two time series with the same trend will have a strong positive correlation but will indicate no causality between them.

To avoid the problem of the common trend, we use the first difference (i.e. the yearly change) of private investment and consumption to conduct regression. The regression results are showing in Fig. 5.23.

In Fig. 5.23, we used a prefix of $d$ for all variables (e.g. dinvestpriv, dconsump) to indicate that they are differenced values. The results show that, although the correlation between differenced investment and differenced consumption are very significant in all three regressions (showing by the high $t$-value, low standard error and very low $p$-value), the $R$-squared reduced significantly, compared with those in Fig. 5.21. For example, the $R$-squared for differenced investment and consumption (0.3752) decreased by more than half, compared with the value of $R$-squared of 0.9795 in Fig. 5.21. This is because we have excluded the misleading correlation due to a growth trend. Nevertheless, the reduced values of $R$-squared in Fig. 5.23 are still reasonably high considering that we omitted other variables which may affect investment. The $R$-squared value for the regression involved in the differenced lagged investment (dinvestlag1) and differenced lagged consumption (dconsulag1) are 0.2245 and 0.0774, which are much smaller than 0.3752. This indicates that the correlation between the differenced current

```
. regress dinvestpriv dconsump
```

| Source | SS | df | MS |
|--------|-----|-----|-----|
| Model | 3.4306e+11 | 1 | 3.4306e+11 |
| Residual | 5.7118e+11 | 86 | 6.6416e+09 |
| Total | 9.1423e+11 | 87 | 1.0508e+10 |

| | |
|---|---|
| Number of obs | = 88 |
| F(1, 86) | = 51.65 |
| Prob > F | = 0.0000 |
| R-squared | = 0.3752 |
| Adj R-squared | = 0.3680 |
| Root MSE | = 81496 |

| dinvestpriv | Coef. | Std. Err. | t | P>\|t\| | [95% Conf. Interval] | |
|-------------|-------|-----------|---|---------|---------------------|---|
| dconsump | .3647104 | .050746 | 7.19 | 0.000 | .2638307 | .46559 |
| _cons | -18881.78 | 11595.48 | -1.63 | 0.107 | -41932.83 | 4169.26 |

```
. regress dinvestlag1 dconsump
```

| Source | SS | df | MS |
|--------|-----|-----|-----|
| Model | 2.0204e+11 | 1 | 2.0204e+11 |
| Residual | 6.9780e+11 | 85 | 8.2094e+09 |
| Total | 8.9984e+11 | 86 | 1.0463e+10 |

| | |
|---|---|
| Number of obs | = 87 |
| F(1, 85) | = 24.61 |
| Prob > F | = 0.0000 |
| R-squared | = 0.2245 |
| Adj R-squared | = 0.2154 |
| Root MSE | = 90606 |

| dinvestlag1 | Coef. | Std. Err. | t | P>\|t\| | [95% Conf. Interval] | |
|-------------|-------|-----------|---|---------|---------------------|---|
| dconsump | .281281 | .0566987 | 4.96 | 0.000 | .1685488 | .3940131 |
| _cons | -8139.25 | 13029.85 | -0.62 | 0.534 | -34046.07 | 17767.57 |

```
. regress dinvestpriv dconsulag1
```

| Source | SS | df | MS |
|--------|-----|-----|-----|
| Model | 7.0656e+10 | 1 | 7.0656e+10 |
| Residual | 8.4179e+11 | 85 | 9.9034e+09 |
| Total | 9.1244e+11 | 86 | 1.0610e+10 |

| | |
|---|---|
| Number of obs | = 87 |
| F(1, 85) | = 7.13 |
| Prob > F | = 0.0091 |
| R-squared | = 0.0774 |
| Adj R-squared | = 0.0666 |
| Root MSE | = 99516 |

| dinvestpriv | Coef. | Std. Err. | t | P>\|t\| | [95% Conf. Interval] | |
|-------------|-------|-----------|---|---------|---------------------|---|
| dconsulag1 | .1716621 | .0642677 | 2.67 | 0.009 | .0438806 | .2994435 |
| _cons | 11653.53 | 14228.42 | 0.82 | 0.415 | -16636.38 | 39943.44 |

**Fig. 5.23** Results of regressing changes in yearly investment and consumption

investment and the differenced current consumption is much higher, so we may conclude that investors tend to adjust their investment immediately based on the current consumption situation.

However, this explanation is still not rigorous. One reason is that both the change in current consumption and in current investment may result from other common factors (e.g. a change in GDP level will affect consumption and investment in a similar way), so the correlation between differenced current consumption and differenced current investment may not indicate any causality between them. The other reason is related to the yearly data. One year is a long time frame for investors. If investors adjust their investment decision in the later part of the year according to the consumption data in the earlier part of the year, the yearly data would mask this investment behaviour.

To avoid the shortcoming in the above regression, we employ the quarterly US consumption and investment data from 1947Q1 to 2017Q4, provided by the Federal Reserve Bank of St Louis. Again, we use the first-differenced data to avoid high correlation caused by the trend of time series. The results of regressing current private investment, investment lagged by 1 period and investment with 1-period lead on private consumption are shown in Fig. 5.24.

The regression results show the significance of consumption in all cases. This is not surprising because we only have one explanatory variable here. The $R$-squared values suggest that the correlation between the leading investment and current consumption are significantly higher than those other two cases (regression between current investment and current consumption, and regression between lagged investment and current consumption). By trying a different number of lags and leads, we find the highest $R$-squared value between current consumption and investment with a lead of one quarter shown in Fig. 5.24 is highest. These empirical results tend to suggest that investors engage in adaptive expectation so the consumption lagged by one quarter has a strong influence on investment decisions.

(10 vars, 288 obs)

. regress dinvestpriv dconsump

| Source | SS | df | MS |
|---|---|---|---|
| Model | 91123.9735 | 1 | 91123.9735 |
| Residual | 405717.634 | 282 | 1438.71502 |
| Total | 496841.608 | 283 | 1755.62406 |

| | |
|---|---|
| Number of obs | = 284 |
| F(1, 282) | = 63.34 |
| Prob > F | = 0.0000 |
| R-squared | = 0.1834 |
| Adj R-squared | = 0.1805 |
| Root MSE | = 37.93 |

| dinvestpriv | Coef. | Std. Err. | t | P>|t| | [95% Conf. Interval] | |
|---|---|---|---|---|---|---|
| dconsump | .3518771 | .0442142 | 7.96 | 0.000 | .2648453 | .4389089 |
| _cons | -5.117453 | 3.092895 | -1.65 | 0.099 | -11.20554 | .9706378 |

. regress dinvestlag1 dconsump

| Source | SS | df | MS |
|---|---|---|---|
| Model | 76710.1792 | 1 | 76710.1792 |
| Residual | 415193.183 | 281 | 1477.55581 |
| Total | 491903.362 | 282 | 1744.33816 |

| | |
|---|---|
| Number of obs | = 283 |
| F(1, 281) | = 51.92 |
| Prob > F | = 0.0000 |
| R-squared | = 0.1559 |
| Adj R-squared | = 0.1529 |
| Root MSE | = 38.439 |

| dinvestlag1 | Coef. | Std. Err. | t | P>|t| | [95% Conf. Interval] | |
|---|---|---|---|---|---|---|
| dconsump | .3232794 | .0448666 | 7.21 | 0.000 | .234962 | .4115968 |
| _cons | -4.043633 | 3.144055 | -1.29 | 0.199 | -10.23252 | 2.145258 |

. regress dinvestlead1 dconsump

| Source | SS | df | MS |
|---|---|---|---|
| Model | 140101.481 | 1 | 140101.481 |
| Residual | 356567.459 | 281 | 1268.92334 |
| Total | 496668.94 | 282 | 1761.23737 |

| | |
|---|---|
| Number of obs | = 283 |
| F(1, 281) | = 110.41 |
| Prob > F | = 0.0000 |
| R-squared | = 0.2821 |
| Adj R-squared | = 0.2795 |
| Root MSE | = 35.622 |

| dinvestlead1 | Coef. | Std. Err. | t | P>|t| | [95% Conf. Interval] | |
|---|---|---|---|---|---|---|
| dconsump | .4382252 | .0417055 | 10.51 | 0.000 | .3561303 | .5203202 |
| _cons | -9.090165 | 2.905294 | -3.13 | 0.002 | -14.80907 | -3.371263 |

**Fig. 5.24** Results of regressing changes in quarterly investment and consumption

# Appendix 3 (for Section 5.6.3): On R&D Investment, Innovations, and Economic Growth

R&D investment is supposed to have a positive impact on innovations and thus on economic growth, so we examine the relationship between R&D investment, innovation, and economic growth. Considering the complications caused by many variables explained in Appendix 2, we use two variables models and focus on the correlation between the pairs.

We use the real GDP as an indicator of the performance of the economy. The 1959–2015 US real GDP data is obtained from the BEA. The US R&D data from 1959 to 2007 are also available from the innovation account on the BEA website. The number of innovations is indicated by the number of patent applications and patent approvals. The data on US patent application and patent approvals from 1963 to 2015 are obtained from the US Patent and Trademark Office. Since all data display positive trends, we use the first-differenced data (changes over previous year) to avoid misleading high correlations.

Figure 5.25 shows the results of regressing different periods of real GDP on the number of patent applications with US origins. Since we are dealing with first-differenced data for all variables, we omit the prefix '$d$', which were used in Figs. 5.23 and 5.24. The $R$-squared value of 0.3444 for the regression between current GDP (gdp) and current US patent applications (patappus) indicates the high relevance between these two variables. However, this correlation cannot be interpreted as the positive contribution of innovation (indicated by patent application) to economic growth (indicated by GDP) because there is a significant time gap between patent application and implementation of the innovation. On the contrary, it is more likely that the condition of the economy has an influence on patent application: a better economic performance indicated by a higher GDP means that the patent applicants have more resources to file patent applications.

Due to the time lag between the patent application and the implementation of patent technology, we can examine the impact of patent technology on economic performance by regressing current GDP on lagged

```
. reg gdp patappus

    Source |       SS          df       MS          Number of obs  =       44
           |                                        F(1, 42)       =     22.06
     Model | 6.2214e+11          1  6.2214e+11      Prob > F        =    0.0000
  Residual | 1.1843e+12         42  2.8198e+10      R-squared       =    0.3444
           |                                        Adj R-squared   =    0.3288
     Total | 1.8064e+12         43  4.2010e+10      Root MSE        =    1.7e+05


       gdp |      Coef.   Std. Err.       t     P>|t|     [95% Conf. Interval]

  patappus |    16.1615    3.44068     4.70    0.000     9.217928     23.10508
     _cons |   250380.4   28763.41     8.70    0.000     192333.5     308427.3
```

```
. reg gdplead2 patappus

    Source |       SS          df       MS          Number of obs  =       42
           |                                        F(1, 40)       =      7.48
     Model | 2.6207e+11          1  2.6207e+11      Prob > F        =    0.0092
  Residual | 1.4005e+12         40  3.5014e+10      R-squared       =    0.1576
           |                                        Adj R-squared   =    0.1366
     Total | 1.6626e+12         41  4.0551e+10      Root MSE        =    1.9e+05


  gdplead2 |      Coef.   Std. Err.       t     P>|t|     [95% Conf. Interval]

  patappus |   11.37827   4.158996     2.74    0.009     2.972621     19.78391
     _cons |   288758.1   32078.39     9.00    0.000     223925.3       353591
```

```
. reg gdplead5 patappus

    Source |       SS          df       MS          Number of obs  =       39
           |                                        F(1, 37)       =     23.17
     Model | 5.5548e+11          1  5.5548e+11      Prob > F        =    0.0000
  Residual | 8.8685e+11         37  2.3969e+10      R-squared       =    0.3851
           |                                        Adj R-squared   =    0.3685
     Total | 1.4423e+12         38  3.7956e+10      Root MSE        =    1.5e+05


  gdplead5 |      Coef.   Std. Err.       t     P>|t|     [95% Conf. Interval]

  patappus |   17.63578   3.663417     4.81    0.000     10.21299     25.05857
     _cons |   293908.2   27138.03    10.83    0.000     238921.4     348895.1
```

**Fig. 5.25** Results of regressing changes in US GDP and patent applications

patent application numbers or, alternatively, by regressing leading GDP on current patent application numbers. As leads of GDP increase, we found the $R$-squared value decreases; e.g., the $R$-squared value for GDP of 2 leads (gdplead2) is 0.1576, which is less than half of that for regression of current GDP on current patent application numbers. However, as the number of leads of GDP increases to 5 (i.e. gdplead5), the $R$-squared value increases to 0.3851, which is even higher than that from regressing of current GDP on current patent application numbers. When we increase the number of leads for GDP, the $R$-squared value starts to decrease again. The high $R$-squared value for GDP of 5 leads cannot be explained as the impact of GDP on patent applications, so it tends to indicate a causality from patent application to GDP. Namely, the patent applications have a significant impact on real GDP, but with about a 5-year lag.

Next, we examine the impact of R&D investment on innovation by regressing the various periods of US-origin patent application numbers on R&D investment in the USA. The regression results are shown in Fig. 5.26.

The $R$-squared value of 0.3247 between current total R&D (rndtotal) and patent application numbers (patappus) indicates that the two variables are highly correlated. However, this correlation may not indicate any causality between these two variables because the causality running from R&D investment to patent application requires a significant time gap; meanwhile, it is implausible that the number of patent applications will instantly affect R&D investment. The correlation in the current period is most likely caused by a common factor: good economic conditions mean the firms have more money to invest in research and also have more money to file patent applications.

As the leads for patent applications increase, the $R$-squared value decreases sharply. For example, the $R$-squared value for patent application of 4 leads (patappusf4) becomes as little as 0.0593. However, as the leads increase further, the $R$-squared value peaks at 0.3016 when the lead number is 6 (patappusf6), and then it starts to decline again. This result may be interpreted as the impact of current R&D on the number of innovations with a 6-year lag.

However, Fig. 5.27 shows the results of regressing different periods of real GDP on total R&D investment in USA. The very high $R$-squared value of 0.7156 indicates a very strong correlation between current GDP (gdp) and current R&D investment (rndtotal). Again, this correlation is more likely due to the impact of GDP on R&D investment. R&D investment cannot affect GDP instantly, so the only plausible

```
. reg patappus rndtotal
```

| Source | SS | df | MS | | | |
|---|---|---|---|---|---|---|
| | | | | Number of obs | = | 44 |
| | | | | F(1, 42) | = | 20.19 |
| Model | 773389858 | 1 | 773389858 | Prob > F | = | 0.0001 |
| Residual | 1.6085e+09 | 42 | 38297813.9 | R-squared | = | 0.3247 |
| | | | | Adj R-squared | = | 0.3086 |
| Total | 2.3819e+09 | 43 | 55392977.8 | Root MSE | = | 6188.5 |

| patappus | Coef. | Std. Err. | t | P>\|t\| | [95% Conf. Interval] | |
|---|---|---|---|---|---|---|
| rndtotal | .5448821 | .1212524 | 4.49 | 0.000 | .3001849 | .7895793 |
| _cons | -817.0505 | 1415.863 | -0.58 | 0.567 | -3674.377 | 2040.276 |

```
. reg patappusf4 rndtotal
```

| Source | SS | df | MS | | | |
|---|---|---|---|---|---|---|
| | | | | Number of obs | = | 48 |
| | | | | F(1, 46) | = | 2.90 |
| Model | 169297937 | 1 | 169297937 | Prob > F | = | 0.0954 |
| Residual | 2.6873e+09 | 46 | 58418752.9 | R-squared | = | 0.0593 |
| | | | | Adj R-squared | = | 0.0388 |
| Total | 2.8566e+09 | 47 | 60777884.5 | Root MSE | = | 7643.2 |

| patappusf4 | Coef. | Std. Err. | t | P>\|t\| | [95% Conf. Interval] | |
|---|---|---|---|---|---|---|
| rndtotal | .2454729 | .1441962 | 1.70 | 0.095 | -.044779 | .5357248 |
| _cons | 1768.013 | 1613.14 | 1.10 | 0.279 | -1479.069 | 5015.095 |

```
. reg patappusf6 rndtotal
```

| Source | SS | df | MS | | | |
|---|---|---|---|---|---|---|
| | | | | Number of obs | = | 48 |
| | | | | F(1, 46) | = | 19.87 |
| Model | 1.0101e+09 | 1 | 1.0101e+09 | Prob > F | = | 0.0001 |
| Residual | 2.3386e+09 | 46 | 50840020.8 | R-squared | = | 0.3016 |
| | | | | Adj R-squared | = | 0.2864 |
| Total | 3.3487e+09 | 47 | 71248993 | Root MSE | = | 7130.2 |

| patappusf6 | Coef. | Std. Err. | t | P>\|t\| | [95% Conf. Interval] | |
|---|---|---|---|---|---|---|
| rndtotal | .5995862 | .134518 | 4.46 | 0.000 | .3288154 | .870357 |
| _cons | -403.9479 | 1504.869 | -0.27 | 0.790 | -3433.092 | 2625.197 |

**Fig. 5.26** Results of regressing changes in patent applications and investment

. reg gdp rndtotal

| Source | SS | df | MS | | | |
|---|---|---|---|---|---|---|
| Model | 1.5067e+12 | 1 | 1.5067e+12 | | | |
| Residual | 5.9892e+11 | 46 | 1.3020e+10 | | | |
| Total | 2.1056e+12 | 47 | 4.4800e+10 | | | |

| | | |
|---|---|---|
| Number of obs | = | 48 |
| F(1, 46) | = | 115.72 |
| Prob > F | = | 0.0000 |
| R-squared | = | 0.7156 |
| Adj R-squared | = | 0.7094 |
| Root MSE | = | 1.1e+05 |

| gdp | Coef. | Std. Err. | t | P>|t| | [95% Conf. Interval] | |
|---|---|---|---|---|---|---|
| rndtotal | 23.15745 | 2.152703 | 10.76 | 0.000 | 18.82428 | 27.49062 |
| _cons | 101721.9 | 24082.54 | 4.22 | 0.000 | 53246.26 | 150197.6 |

. reg gdplead3 rndtotal

| Source | SS | df | MS | | | |
|---|---|---|---|---|---|---|
| Model | 6.2316e+11 | 1 | 6.2316e+11 | | | |
| Residual | 1.2605e+12 | 43 | 2.9314e+10 | | | |
| Total | 1.8837e+12 | 44 | 4.2811e+10 | | | |

| | | |
|---|---|---|
| Number of obs | = | 45 |
| F(1, 43) | = | 21.26 |
| Prob > F | = | 0.0000 |
| R-squared | = | 0.3308 |
| Adj R-squared | = | 0.3153 |
| Root MSE | = | 1.7e+05 |

| gdplead3 | Coef. | Std. Err. | t | P>|t| | [95% Conf. Interval] | |
|---|---|---|---|---|---|---|
| rndtotal | 20.48169 | 4.442292 | 4.61 | 0.000 | 11.52296 | 29.44043 |
| _cons | 168289.9 | 39664.74 | 4.24 | 0.000 | 88298.37 | 248281.5 |

. reg gdplead9 rndtotal

| Source | SS | df | MS | | | |
|---|---|---|---|---|---|---|
| Model | 8.0250e+11 | 1 | 8.0250e+11 | | | |
| Residual | 6.3983e+11 | 37 | 1.7293e+10 | | | |
| Total | 1.4423e+12 | 38 | 3.7956e+10 | | | |

| | | |
|---|---|---|
| Number of obs | = | 39 |
| F(1, 37) | = | 46.41 |
| Prob > F | = | 0.0000 |
| R-squared | = | 0.5564 |
| Adj R-squared | = | 0.5444 |
| Root MSE | = | 1.3e+05 |

| gdplead9 | Coef. | Std. Err. | t | P>|t| | [95% Conf. Interval] | |
|---|---|---|---|---|---|---|
| rndtotal | 31.17885 | 4.576881 | 6.81 | 0.000 | 21.90521 | 40.45249 |
| _cons | 167042.2 | 33788.76 | 4.94 | 0.000 | 98579.63 | 235504.7 |

**Fig. 5.27**   Results of regressing changes in US GDP and R&D investment

explanation is that firms may have more money to invest during good economic times. As the number of leads for GDP increases, the $R$-squared value decreases.

Figure 5.27 also shows that the $R$-squared value for GDP of 3 leads (gdplead3) decreased to 0.3308, less than half of the $R$-squared value for current GDP on current R&D. Increasing the GDP leads further, we find that the $R$-squared values started to increase and peaked at 0.5564 when the lead number is 9 (gdplead9). This high $R$-squared tends to indicate that the impact of R&D on the GDP has a lag of about 9 years.

Although the above interpretation is plausible, there may be alternative explanations about the results. It may be argued that the correlation within the time series (i.e. autocorrelation) may be responsible for the high correlation between patent application numbers and GDP of 5 leads, between R&D investment and patent applications of 6 leads, and between R&D investment of GDP of 9 leads.

The correlogram tests indeed indicate that there are autocorrelations for each time series. Judged by the 95% confidence level, the autocorrelations are significant between the current GDP and the GDP of 1–3 lags, between the current R&D and the R&D of 1 or 2 lags, and between the current patent applications and the patent applications of 5 or 6 lags. The autocorrelations between the current patent applications and those of around 5 lags may have contributed to the high $R$-squared value for regression between the current patent applications and the GDP of 5 leads, and it may also have contributed to the correlation between the current R&D and the patent applications of 6 leads. However, the autocorrelation of 1–3 lags in time series GDP and the autocorrelation of 1–2 lags in time series R&D can lead to a correlation between current R&D and the GDP of maximum 5 leads, so they cannot explain the high $R$-squared value between current R&D and GDP of 9 leads. A more plausible explanation is as follows. R&D investment has a positive impact on innovations with about 5-year lags. Meanwhile, innovation indicated by patent applications have a positive impact on real GDP after about 5 years. Consequently, R&D influences GDP through innovation after about 10 years.

# References

Abel, A., Mankiw, N., Summers, L., & Zeckhauser, R. (1989). Assessing Dynamic Efficiency: Theory and Evidence. *Review of Economic Studies, 56,* 1–20.

Acemoglu, D., Akcigit, U., Bloom, N., & Kerr, W. (2013). *Innovation, Reallocation and Growth* (NBER WP 18933).

Aftalion, A. (1913). *Les crises périodiques de surproduction.* Paris: Rivière.

Aghion, P., & Howitt, P. (1992). A Model of Growth Through Creative Destruction. *Econometrica, 60*(2), 323–351.

Aghion, P., & Howitt, P. (1998). *Endogenous Growth Theory.* Cambridge, MA: MIT Press.

Aitchison, J., & Brown, J. A. C. (1954). A Synthesis of Engel Curve Theory. *Review of Economic Studies, 22*(1), 35–46.

Akerlof, G., & Yellen, J. (1985). Can Small Deviations from Rationality Make Significant Differences to Economic Equilibria? *American Economic Review, 75*(4), 708–720.

Allen, R. (2012). Technology and the Great Divergence: Global Economic Development Since 1820. *Explorations in Economic History, 49,* 1–16.

Aloy, M., & Gente, K. (2009). The Role of Demography in the Long Run Yen/USD Real Exchange Rate Appreciation. *Journal of Macroeconomics, 31,* 654–667.

Andersen, E. (2001). Satiation in an Evolutionary Model of Structural Economic Dynamics. *Journal of Evolutionary Economics, 11*(1), 143–164.

Aoki, M., & Yoshikawa, H. (2002). Demand Saturation-Creation and Economic Growth. *Journal of Economic Behaviour and Organization, 48*(2), 127–154.

Arner, D. (2009). *The Global Credit Crisis of 2008: Causes and Consequences* (AIIFL Working Paper No. 3).

Australian Broadcast Company. (2018). *Are Headphones Damaging Young People's Hearing?* https://www.abc.net.au/news/2018-06-06/headphones-could-be-causing-permanent-hearing-damage/9826294.

Baldwin, R. (2017). *The Great Convergence: Information Technology and the New Globalization.* Cambridge: Belknap Press.

Banks, J., Blundell, R., & Lewbel, A. (1997). Quadratic Engel Curves and Consumer Demand. *The Review of Economics and Statistics, 79*(4), 527–539.

Barro, R. (1987). Government Spending, Interest Rates, Prices, and Budget Deficit in the United Kingdom, 1701–1918. *Journal of Monetary Economics, 20,* 221–247.

Barro, R. J., & Sala-i-Martin, X. (1992). Convergence. *Journal of Political Economy, 100*, 223–251.

Baumol, W. (1986). Productivity Growth, Convergence, and Welfare: What the Long-Run Data Show. *The American Economic Review, 76*(5), 1072–1085.

Bayoumi, T. (2001). The Morning After: Explaining the Slowdown in Japanese Growth in the 1990s. *Journal of International Economics, 53*, 241–259.

Berrone, P. (2008). *Current Global Financial Crisis: An Incentive Problem* (IESE Occasional Paper, OP-158).

Bloom, D. E., Cannin, D., & Sevilla, J. (2002). *Technological Diffusion, Conditional Convergence, and Economic Growth* (NBER Working Paper No. 8713).

Blundell-Wignall, A., Atkinson, P., & Lee, S. (2008). *The Current Financial Crisis: Causes and Policy Issues*. Financial Market Trends, Vol. 2008/2. Paris: OECD.

Branstetter, L., & Nakamura, Y. (2003). *Is Japan's Innovative Capacity in Decline?* http://www.nber.org/papers/w9438.

Carroll, C. D. (2017). *The Prescott Real Business Cycle Model*. http://www. econ2.jhu.edu/people/ccarroll/public/LectureNotes/DSGEModels/ RBC-Prescott/.

Cass, D. (1965). Optimum Growth in an Aggregative Model of Capital Accumulation. *Review of Economic Studies, 32*, 233–240.

Cassel, G. (1924 [1967]). *The Theory of Social Economy*. New York: Augustus M. Kelley.

Chai, A., & Moneta, A. (2014). Escaping Satiation Dynamics: Some Evidence from British Household Data. *Journal of Economics and Statistics, 234*(2/3), 299–327.

Clark, J. M. (1917). Business Acceleration and the Law of Demand: A Technical Factor in Economic Cycles. *Journal of Political Economy, 25*, 217–235.

Commons, J. (1934). *Institutional Economics*. New York: Macmillan.

Cox, G. W. (2017). Political Institutions, Economic Liberty, and the Great Divergence. *The Journal of Economic History, 77*(3), 724–755.

Crotty, J. (2008). *Structural Causes of the Global Financial Crisis: A Critical Assessment of the 'New Financial Architecture'* (University of Massachusetts Working Paper 2008–14).

Dabrowski, M. (2008). *The Global Financial Crisis: Causes, Channels of Contagion and Potential Lessons*. CASE Net work E-Briefs 7, https://www. cesifo-group.de/DocDL/forum4-08-focus5.pdf.

Dash, M. (1999). *Tulipomania: The Story of the World's Most Coveted Flower and the Extraordinary Passions It Aroused*. New York: Three Rivers Press.

Davidson, P. (1984). Reviving Keynes's Revolution. *Journal of Post Keynesian Economics, 6*(4), 561–575.

Davidson, P. (1991). Is Probability Theory Relevant for Uncertainty? A Post Keynesian Perspective. *Journal of Economic Perspectives, 5*(1), 129–143.

Dawson, J., & Larke, R. (2004). Japanese Retailing Through the 1990s: Retailer Performance in a Decade of Slow Growth. *British Journal of Management, 15*(1), 73–94.

Day, C. (2006). Paper Conspiracies and the End of All Good Order: Perceptions and Speculations in Early Capital Markets. *Entrepreneurial Business Law Journal, 1*(2): 286.

De Long, J. B. (1988). Productivity Growth, Convergence and Welfare: Comment. *American Economic Review, 78*(5), 1138–1154.

De Pleijt, A., & Van Zanden, J. (2016). Accounting for the 'Little Divergence': What Drove Economic Growth in Pre-industrial Europe, 1300–1800? *European Review of Economic History, 20*, 387–409.

Dervis, K. (2012). World Economy Convergence, Interdependence, and Divergence. *Finance and Development, 49*(3), 10–14.

Diamond, P. A. (1965). National Debt in a Neoclassical Growth Model. *The American Economic Review, 55*(5), 1126–1150.

Dinopoulos, E., & Thompson, P. (1998). Schumpeterian Growth Without Scale Effects. *Journal of Economic Growth, 3*, 313–335.

Domar, Evsey. (1946). Capital Expansion, Rate of Growth, and Employment. *Econometrica, 14*(2), 137–147.

Eichengreen, B. (2007). *The European Economy Since 1945: Coordinated Capitalism and Beyond*. Princeton, NJ: Princeton University Press.

Engel, E. (1857). Die Produktions- und Consumtionsverhältnisse des Königreichs Sachsen, Zeitschrift des Statistischen Breaus des Kniglich Schischen Ministeriums des Innern 8 and 9.

Esposito, M., Chatzimarkakis, J., Tse, T., Dimitriou, G., Akiyoshi, R., Balusu, E., et al. (2014). *The European Financial Crisis: Analysis and a Novel Intervention*. https://scholar.harvard.edu/files/markesposito/files/eurocrisis.pdf.

European Commission. (2013). *Unemployment Statistics*. http://epp.eurostat. ec.europa.eu/statistics_explained/index.php?title=File:Unemployment_rate,_2001-2012_%28%25%29.png&filetimestamp=20130627102805.

Evans, P. (1996). Using Cross-Country Variances to Evaluate Growth Theories. *Journal of Economic Dynamics and Control, 20*, 1027–1049.

Fisher, I. (1930). *The Theory of Interest*. New York: Macmillan.

Fisher, I. (1933). The Debt-Deflation Theory of Great Depressions. *Econometrica, 1*(4), 337–357.

Fisk, E. (1962). Planning in a Primitive Economy: Special Problems of Papua New Guinea. *Economic Record, 38,* 462–478.

Foellmi, R., & Zweimüller, J. (2006). Income Distribution and Demand-Induced Innovation. *Review of Economic Studies, 63*(2), 187–212.

Frankel, M. (1962). The Production Function in Allocation and Growth: A Synthesis. *American Economic Review, 52,* 995–1022.

Friedman, M. (1957). *A Theory of the Consumption Function*. Princeton, NJ: Princeton University Press.

Gallouj, F., & Weinstein, O. (1997). Innovation in Services. *Research Policy, 26,* 537–556.

Garber, P. (1989). Tulipmania. *Journal of Political Economy, 97*(3), 535–560.

Georges, C. (2015). *Product Innovation and Macroeconomic Dynamics, Manuscript*. Hamilton College. http://academics.hamilton.edu/economics/cgeorges/product-innovation-and-macro-dyn.pdf.

Gorton, G., & Metrick, A. (2010). Regulating the Shadow Banking System. *Brookings Papers on Economic Activity, 2,* 261–312.

Griliches, Zvi. (1988). Productivity Puzzles and R&D: Another Nonexplanation. *Journal of Economic Perspectives, 2*(4), 9–21.

Grossman, G. M., & Helpman, E. (1991). *Innovation and Growth in the Global Economy*. Cambridge, MA: MIT Press.

Hamao, Y., Mei, J., & Xu, Y. (2007). Unique Symptoms of Japanese Stagnation: An Equity MarketPerspective. *Journal of Money Credit and Banking, 39,* 901–923.

Harrod, R. F. (1936). *The Trade Cycle*. Oxford: Oxford University Press.

Harrod, Roy F. (1939). An Essay in Dynamic Theory. *The Economic Journal, 49*(193), 14–33.

Hayashi, F., & Prescott, E. (2002). The 1990s in Japan: A Lost Decade. *Review of Economic Dynamics, 5,* 206–235.

Hoppit, J. (2002). The Myths of the South Sea Bubble. *Transactions of the RHS, 12,* 141–165.

Hoshi, T., & Kashyap, A. (2004). Japan's Financial Crisis and Economic Stagnation. *Journal of Economic Perspectives, 18*(1), 3–26.

Howe, H. (1975). Development of the Extended Linear Expenditure System from Simple Saving Assumptions. *European Economic Review, 6,* 305–310.

Howitt, P. (1999). Steady Endogenou Growth with Population and R&D Inputs Growing. *Journal of Political Economy, 107,* 715730.

Hutcheson, F. (1750). *Reflections upon Laughter, and Remarks upon the Fable of the Bees.* Glasgow: Printed by R. Urie for D. Baxter.

Hutchison, M., Ito, T., & Westermann, F. (2005). *The Great Japanese Stagnation: Lessions for Industrial Countries* (EPRU Working Paper Series 2005-13).

Jickling, M. (2009). *Causes of Financial Crisis* (Congressional Research Service (CRS) Report for Congress No. 7-5700).

Jones, E. L. (1981). *The European Miracle. Environment, Economies and Geopolitics in the History of Europe and Asia.* Cambridge: Cambridge University Press.

Jones, C. (1995). R&D-Based Models of Economic Growth. *Journal of Political Economy, 103*(4), 759–784.

Juglar, C. (1862). *Des crises commerciales et de leur retour périodique en France, en Angleterre et aux Etats-Unis.* Paris: Guillaumin et Cie, second edition 1889.

Kahn, R. F. (1931). The Relation of Home Investment to Unemployment. *Economic Journal, 41*(162), 173–198.

Keely, L. (2002). Pursuing Problems in Growth. *Journal of Economic Growth, 7,* 283–308.

Keynes, J. M. (1936). *The General Theory of Employment, Interest, and Money.* London: MacMillan.

Kimball, M. (1990). Precautionary Saving in the Small and in the Large. *Econometrica, 58*(1), 53–73.

Kitchin, J. (1923). Cycles and Trends in Economic Factors. *Review of Economics and Statistics, 5*(1), 10–16.

Klette, J., & Kortum, S. (2004). Innovating Firms and Aggregate Innovation. *Journal of Political Economy, 112*(5), 986–1018.

Kondratiev, N. D. (1922). *The World Economy and Its Conjunctures During and After the War* (in Russian). Moscow: International Kondratieff Foundation.

Koo, R. (2009). *The Holy Grail of Macroeconomics—Lessions from Japan's Great Recession.* New York: Wiley.

Koopmans, T. (1965). *On the Concept of Optimal Economic Growth, in the Econometric Approach to Development Planning.* Amsterdam: North Holland.

Krugman, P. (1998). It's Baaack: Japan's Slumpand the Return of the Liquidity Trap. *Brookings Papers on Economic Activity, 2,* 137–205.

Krugman, P. (2009). *The Return of Depression Economics and the Crisis of 2008.* New York: W. W. Norton.

Kuznets, S. (1930). *Secular Movements in Production and Prices: Their Nature and Their Bearing upon Cyclical Fluctuations.* Boston: Houghton Mifflin.

Kydland, F. E., & Prescott, E. C. (1982). Time to Build and Aggregate Fluctuations. *Econometrica, 50*(6), 1345–1370.

Lancaster, K. (1966). Change and Innovation in the Technology of Consumption. *American Economic Review, 56,* 14–23.

Lane, P. (2012). The European Sovereign Debt Crisis. *Journal of Economic Perspectives, 26*(3), 49–68.

Leland, H. (1968). Saving and Uncertainty: The Precautionary Demand for Saving. *Quarterly Journal of Economics, 82*(3), 465–473.

Lin, J., & Treichel, V. (2012). *The Unexpected Global Financial Crisis— Researching Its Root Cause* (Policy Research Working Paper WPS5937). The World Bank.

Lluch, C. (1973). The Extended Linear Expenditure System. *European Economic Review, 4,* 21–32.

Long, J. B., & Plosser, C. (1983). Real Business Cycles. *Journal of Political Economy, 91*(1), 39–69.

Lucas, R. E. (1975, December). An Equilibrium Model of the Business Cycle. *Journal of Political Economy, 83*(6), 1113–1144.

Lucas, R. E. (1988). On the Mechanics of Economic Development. *Journal of Monetary Economics, 22,* 3–42.

Maddison, A. (2007). *Contours of the World Economy, 1-2030 AD.* Oxford, UK: Oxford University Press.

Madsen, J., & Yan, E. (2013). *The First Great Divergence and Evolution of Cross-Country Income Inequality During the Last Millennium: The Role of Institutions and Culture* (Discussion Paper 14/13). Department of Economics, Monash University.

Malthus, T. (1836 [1964]). *Principles of Political Economy, Considered with a View to Their Practical Application.* New York: A. M. Kelley.

Mandeville, B. (1723). *The Fable of the Bees: Or, Private Vices, Public Benefits* (2nd ed.). London: Printed for Edmund Parker.

Mankiw, N. G. (1985, May). Small Menu Costs and Large Business Cycles: A Macroeconomic Model of Monopoly. *The Quarterly Journal of Economics, 100*(2), 529–537.

Mankiw, N. G. (1989). Real Business Cycles: A New Keynesian Perspective. *Journal of Economic Perspectives, 3*(3), 79–90.

Mankiw, N. G., Romer, D., & Weil, D. N. (1992). A Contribution to the Empirics of Economic Growth. *Quarterly Journal of Economics, 107,* 407–437.

Marchand, A. (2016). The Power of an Installed Base to Combat Lifecycle Decline: The Case of Video Games. *International Journal of Research in Marketing, 33*(1), 140–154.

McKinnon, R., & Ohno, K. (2001). The Foreign Exchange Origins of Japan's Economic Slump and Low Interest Liquidity Trap. *The World Economy, 24,* 279–315.

Menger, C. (1871). *Grundsätze der Volkswirthschaftslehre.* Wien: Wilhelm Braumüller.

Mensch, G. O. (1975). *Stalemate in Technology. Innovations Overcome the Depression.* Cambridge: Ballinger.

Milionis, P., & Vonyo, T. (2015). *Reconstruction Dynamics: The Impact of World War II on Post-War Economic Growth.* https://www.aeaweb.org/conference/2016/retrieve.php?pdfid=235.

Mill, James. (1844). *Elements of Political Economy* (3rd ed.). London: Henry G. Bohn.

Minsky, P. (1986). *Stabilizing an Unstable Economy.* New Haven: Yale University Press.

Minsky, P. (1992). *The Financial Instability Hypothesis* (Working Paper No. 74). Levy Economics Institute.

Miyakoshi, T., & Tsukuda, Y. (2004). The Causes of the Long Stagnation in Japan. *Applied Financial Economics, 14,* 113–120.

Modigliani, F. (1986). Life Cycle, Individual Thrift, and the Wealth of Nations. *American Economic Review, 76,* 297–313.

Modigliani, F., & Brumberg, R. (1954). Utility Analysis and the Consumption Function: An Interpretation of Cross-Section Data. In K. Kurihara (Ed.), *Post-Keynesian Economics.* New Brunswick: Rutgers University Press.

Moneta, A., & Chai, A. (2014). The Evolution of Engel Curves and Its Implications for Structural Change Theory. *Cambridge Journal of Economics, 38*(4), 895–923.

Murakami, H. (2017). Economic Growth with Demand Saturation and Endogenous Demand Reaction. *Metroeconomics, 68,* 966–985.

Onaran, Y. (2011). *Zombie Banks: How Broken Banks and Debtor Nations Are Crippling the Global Economy.* Hoboken: Wiley.

Orlowski, L. (2008). *Stages of the 2007/2008 Global Financial Crisis: Is There a Wandering Asset-Price Bubble?* (Economics-eJournal Discussion Paper No. 2008-43).

Pasinetti, L. (1981). *Structural Change and Economic Growth: A Theoretical Essay on the Dynamics of the Wealth of Nations.* Cambridge: Cambridge University Press.

Peretto, P. F. (1998). Technological Change and Population Growth. *Journal of Economic Growth, 3*(4), 283–311.

Perez, C. (2009). The Double Bubble at the Turn of the Century: Technological Roots and Structural Implications. *Cambridge Journal of Economics, 33,* 779–805.

Plosser, C. I. (1989). Understanding Real Business Cycles. *Journal of Economic Perspectives, 3*(3), 51–77.

Pomeranz, K. (2000). *The Great Divergence: China, Europe and the. Making of the Modern World Economy.* Princeton, NJ: Princeton University Press.

Prais, S. J. (1953). Non-linear Estimates of the Engel Curves. *The Review of Economic Studies, 20*(2), 87–104.

Prescott, E. C. (1986). Theory Ahead of Business Cycle Measurement. *Carnegie-Rochester Conference Series on Public Policy, 25,* 11–66.

Ramsey, F. (1928). A Mathematical Theory of Saving. *Economic Journal, 38,* 543–559.

Reichel, R. (2002). Germany's Postwar Growth: Economic Miracle or Reconstruction Boom? *Cato Journal, 21*(3), 427–442.

Ricardo, D. (1952). *The Works and Correspondence of David Ricardo* (Vol. 6). In P. Sraffa & M. Dobb (Eds.). Cambridge: Cambridge University Press.

Romer, P. (1986). Increasing Returns and Long-Run Growth. *Journal of Political Economy, 94,* 1002–1037.

Romer, P. (1990). Endogenous Technological Change. *Journal of Political Economy, 98*(5), S71–S102.

Romer, D. (1993). The New Keynesian Synthesis. *Journal of Economic Perspectives, 7*(1), 5–22.

Romer, D. (2013). *Advance Macroeconomics* (4th ed.). New York: The McGraw-Hill.

Ruprecht, W. (2005). The Historical Development of the Consumption of Sweeteners: A Learning Approach. *Journal of Evolutionary Economics, 15*(3), 247–272.

Ruscakova, A., & Semancikova, J. (2016). The European Debt Crisis: A Brief Discussion of Its Causes and Possible Solutions. *Procedia—Social and Behavioral Sciences, 220,* 339–406.

Saint-Paul, G. (2017). Secular Satiation (CEPR Discussion Paper 2017–18). https://halshs.archives-ouvertes.fr/halshs-01557415/document.

Saito, M. (2000). *The Japanese Economy.* Singapore: World Scientific.

Saviotti, P. (2001). Variety, Growth and Demand. *Journal of Evolutionary Economics, 11,* 119–142.

Saviotti, P., & Pyka, A. (2013). The Co-evolution of Innovation, Demand and Growth. *Economics of Innovation and New Technology, 22*(5), 461–482.

Schmookler, J. (1966). *Invention and Economic Growth*. Cambridge, MA: Harvard University Press.

Schuman, M. (2008, December 19). Why Detroit Is Not Too Big to Fail. *Time*. http://content.time.com/time/business/article/0,8599,1867847,00.html.

Schumpeter, J. (1939). *Business Cycles: A Theoretical, Historical, and Statistical Analysis of the Capitalist Process*. New York and London: McGraw-Hill.

Schumpeter, J. A. (1942). *Capitalism, Socialism and Democracy*. New York: Harper.

Segerstrom, P. S., Anant, T. C. A., & Dinopoulos, E. (1990). A Schumperterian Model of the Product Life Cycle. *American Economic Review, 80,* 1077–1091.

Smith, A. (1776 [1904]). *An Inquiry into the Nature and Causes of the Wealth of Nations*. In E. Cannan (Ed.). London: Methuen.

Solow, R. (1956). A Contribution to the Theory of Economic Growth. *The Quarterly Journal of Economics, 70*(1), 65–94.

Stent, W., & Webb, L. (1975). Subsistence Affluence and Market Economy in Papua New Guinea. *Economic Record, 51,* 522–538.

Stock, J. H., & Watson, M. W. (2002). *Has the Business Cycle Changed and Why?* (Vol. 17). NBER Macroeconomics Annual 2002.

Summers, L. H. (1986). Some Skeptical Observations on Real Business Cycle Theory. *Federal Reserve Bank of Minneapolis Quarterly Review, 10,* 23–27.

Sumner, S. (2011). Why Japan's QE Didn't Work. *The Money Illusion*. www.themoneyillusion.com/why-japans-qe-didnt-work/.

Takada, M. (1999). *Japan's Economic Miracle: Underlying Factors and Strategies for the Growth*. www.lehigh.edu/~rfw1/courses/1999/spring/ir163/Papers/pdf/mat5.pdf.

Taylor, J. B. (2008). *The Financial Crisis and the Policy Responses: An Empirical Analysis of What Went Wrong* (NBER Working Paper No. 14631). http://www.nber.org/papers/w14631.

Tobin, J. (1969). A General Equilibrium Approach to Monetary Theory. *Journal of Money Credit and Banking, 1*(1), 15–29.

Turkington, D. (2007). *Mathematical Tools for Economics*. Australia: Blackwell.

Tyers, R. (2012). Japanese Economic Stagnation: Causes and Global Implications. *Economic Record, 88*(283), 517–536.

Valdes, B. (2003). An Application of Convergence Theory to Japan's Post-WWII Economic Miracle. *Journal of Economic Education, 34*(1), 61–81.

Vernon, R. (1966). International Investment and International Trade in the Product Cycle. *The Quarterly Journal of Economics, 80*(2), 190–207.

Vgchartz. (2008). *Video Games, Charts, News, Forums, Reviews, Wii, PS3, Xbox360, DS, PSP*. http://www.vgchartz.com.

Wang, J. (2016). *The Past and Future of International Monetary System*. Singapore: Springer. http://dx.doi.org/10.1007/978-981-10-0164-2.

Weber, E. J. (2008, March). *The Role of the Real Interest Rate in U.S. Macroeconomic History*. Available at SSRN: https://ssrn.com/abstract=958188 or http://dx.doi.org/10.2139/ssrn.958188.

Weil, P. (1993). Precautionary Savings and the Permanent Income Hypothesis. *Review of Economic Studies, 60*(2), 367–383.

Witt, U. (2001a). *Escaping Satiation: The Demand Side of Economic Growth*. Berlin: Springer.

Witt, U. (2001b). Learning to Consume—A Theory of Wants and the Growth of Demand. *Journal of Evolutionary Economics, 11*(1), 23–36.

Working, H. (1943). Statistical Laws of Family Expenditure. *Journal of the American Statistical Association, 38*, 43–56.

# 6

# A New Patent System to Usher in an Innovative Economy

The previous chapter showed that the lacking of new types of products resulting from innovation scarcity is the genesis of economic recessions. In other words, the repeated occurrence of economic recessions indicates that the current laws or systems which are responsible for stimulating innovations, fail to encourage innovation effectively. Although all intellectual property laws (e.g. copyright act, designs act, and laws on trademarks, performers' right, and plant breeder's right) are relevant to innovation, arguably patent laws play a major role. Through examining the theories and practice of the patent system, this chapter demonstrates that a new design can transform the current patent system to a new one, a system which can encourage innovation activity effectively and efficiently and thus help avoid economic recessions.

## 6.1 Features of Innovation Activities and Incentives to Innovate

Innovation or invention (invention generally refers to large innovation, but they are of the same nature so we use the words interchangeably here) is to create something new, so special ability or talent is required.

© The Author(s) 2019
S. Meng, *Patentism Replacing Capitalism*,
https://doi.org/10.1007/978-3-030-12247-8_6

There is an old argument: the impetus to innovate is the tendency or interest of innovators to create new things, so any other incentives will not induce more innovation activities. It is true that even if given a million dollar award, a person who has no interest or talent in innovation will not initiate any innovation. However, innovation is not cost free. At least, it requires the time of the innovators. Moreover, innovation activity often involves experimentation and testing, also incurring financial cost. Thomas Edison, an invention genius, is a case in point. In his early career, he put all his income into experiments and even incurred debt to develop a chemical vote-recorder, but he was unable to sell the vote-recorder to anyone. He learnt a lesson from this and determined never to invent anything unsalable. Without the proceeds from his later inventions, Edison would not have been able to build his research laboratory and conduct numerous experiments. Financial gains are often the motive for inventions, such as the discovery of anaesthetic gases and the invention of many modern medicines.

It may be argued that financial incentives can be given by a market: the inventor can sell his/her invention or produce the invented products for profit. However, an invention market without special laws or regulations generally fails to generate enough incentives due to the features of innovation investment. One feature of innovation is the high possibility of failure or high risk to investment in innovation. There are two types of innovations: product innovation and production innovation. The goal of the former is to invent new products while the effort of the latter is to improve production efficiency. Compared with production innovations, product innovation has a much higher risk of failure because it normally involves much larger (or more radical) changes and there is much less information available to product investors.

The other feature of innovation is its large positive externality resulting from innovation imitation. Although allowing for imitation can speed up the application of invention and this seems a good outcome for the economy, it greatly reduces return to the innovator. Facing high risk of innovation failure (zero return rate) and low return when innovation is successful, innovators are discouraged to conduct innovation activities. As a result, fewer inventions will appear in the future and everyone is worse off.

There are generally three ways to avoid undersupply of inventions: public produce, government-subsidized private produce, and rewarding inventors through a patent act. The public innovation institution solution is criticized as breaking the economic link between the innovation and its use—it is evident that public research is shown to be insensitive to technological requirements in the marketplace (Nelson 1990; Block 1991). Subsidized private innovation suffers from the difficulty of monitoring and the high cost of administration. These two solutions are even more inefficient if we consider the high risk, and large social benefit, of innovation. We demonstrate it by a graph.

Figure 6.1 is drawn in such a way as to reflect the features of innovation activity. Due to the high possibility of innovation failure, innovation activities are very costly. This means that the marginal cost of innovation is very high or, to be more precise, the initial marginal cost (from innovation failure to success) is very high. MC in Fig. 6.1 shows the marginal cost of innovation while MC' shows a firm's marginal cost after being offset by government subsidies. On the other hand, social return on innovation is enormous due to the large positive externality of

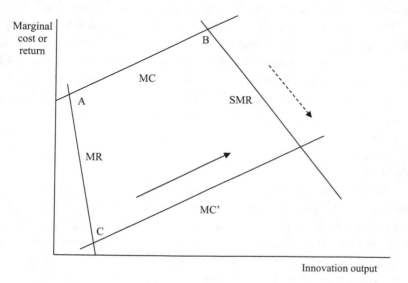

**Fig. 6.1**   Different ways to stimulate innovation

innovation, while the private return to the innovator is very low due to the low cost and easiness of innovation imitation. MR in Fig. 6.1 is the marginal return to the inventor, and SMR is the social marginal return. They differ considerably due to the high externality of innovation.

Without a policy to stimulate innovation, the high level of MC and the low level of MR for the innovation firm necessitate that only a very small number of innovations occur at point A. The public innovation institute solution means that the government will produce innovation according to the marginal cost MC and the social marginal cost SMC, so a higher innovation output can be achieved at point B. However, this higher innovation output requires the government to pay the very high innovation cost of MC and this will lead to a huge social burden. In the case of the subsidized private produce solution, the government will pay a great deal of subsidy $(MC - MC')$ so the firm's marginal cost shifts from MC to MC'. Despite the much lower marginal cost curve MC', the innovation output increases only very mildly at point C due to the large divergence between the private and social return to innovation. Considering that there are numerous innovations the government has to support and that not all innovations will have market success, the government's spending on stimulating innovation may be massive.

The mechanism and the performance of the patent system are quite different. Theoretically, if imitation is eliminated by 'perfect' patent protection (i.e. patent laws are perfectly monitored and enforced without a time limit), the social marginal revenue curve SMR becomes the firm's marginal revenue curve, i.e. MR shift to SMR. With MC as the firm's marginal cost curve, the best outcome at point B is achieved without any increase in government spending. In comparing the three ways of stimulating innovation, the patent system is the most effective and efficient.

The effectiveness of the patent system can also be viewed from the perspective of property rights. Like the intellectual property rights protected by copyright laws, innovation is the intellectual property of inventors. The efficiency of a clearly defined property right is well explained by Coase (1960), and the necessity for privatizing the property right of an invention was demonstrated by Kitch's prospect theory

(Kitch 1977). According to Coase's theorem, clearly defined property rights and obligations, combined with bargaining and negotiations between the parties concerned, can often ensure a more efficient allocation of resources than the allocation through government intervention or by the provision of public goods. If an innovation is legally defined as the intellectual property of the inventor, then others have to pay for using it. As a result, the social return and private return equalize and the problem of underinvestment of innovation is solved.

From the above discussion, it appears that the patent system is superior to the other methods and is effective and efficient in stimulating innovation. However, the patent system has the drawback of limiting the application of patent technology. The social welfare loss due to this drawback will be discussed later, but here we can provide a brief statement that, considering the high chance of innovation failure and all the costs of successful and numerous failed innovations, the social welfare loss due to patent monopoly should be well below the cost of government produce or government subsidy. Due to the importance of the patent system, in the next section we briefly introduce the development of the patent system.

## 6.2   Development of the Patent System

The word 'patent' originates from the Latin word 'litterae patentes', meaning an open letter. The patent system can be traced back to medieval guild practices in Europe. To raise revenue, medieval European monarchs frequently sold some of their right and privileges, for example, monopoly over trade in specified commodities. The rights and privileges are conferred through an open letter. With a royal seal, the open letter served as the proof of rights conferred.

Patents granted to inventions can be dated back to thirteenth and fourteenth centuries in various European countries. It is generally agreed that the first informal patent system regarding invention was developed in Renaissance Italy. The first general statute on patent was enacted by the Venetian State in 1474, which stated:

we have among us men of great genius, apt to invent and discover ingenious devices; and in view of the grandeur and virtue of our city, more such men come to us every day from diverse parts. Now, if provision were made for the works and devices discovered by such persons, so that others who may see them could not build them and take the inventor's honour away, more men would then apply their genius, would discover, and would build devices of great utility to our commonwealth. Therefore:

BE IT ENACTED that, by the authority of this Council, every person who shall build any new and ingenious device in this City, not previously made in our Commonwealth, shall give notice of it to the office of our General Welfare Board when it has been reduced to perfection so that it can be used and operated. It being forbidden to every other person in any of our territories and towns to make any further device conforming with and similar to said one, without the consent and license of the author, for the term of 10 years. (cited in Reid 1993)

However, the principle that monopoly should be granted only for innovators was first laid down by Francis Bacon in 1602. In 1623, this principle was adopted by the British Parliament in the Statute of Monopolies. Section 6 of the Statute authorized 'letters patent and grants of privilege for the term of fourteen years or under, hereinafter to be made, of the sole working or making of any manner of new manufactures within this Realm, to the true and first inventor of such manufactures, which others at the time of making such letters patent shall not use'. About a half-century later, the patent system was enacted in the USA by Article 1, Section 8 of the 1787 constitutional provision, which stated that 'The Congress shall have Power … To Promote the Progress of Science and useful Arts, by securing for limited Times to Authors and Inventors the exclusive Right to their respective Writings and Discoveries'. These statutes form the foundation of the modern patent system.

The patent system was first internationalized in 1883, when 11 countries ratified the 'Paris convention of the international union for the protection of industrial property' (The 'Paris Convention'). The Paris Convention has subsequently been amended on numerous occasions and now has approximately 175 member countries. In the meantime, other international conventions or treaties have been established.

Notable examples include the Patent Cooperation Treaty of 1970, the European Patent Convention of 1973 and the Patent Cooperation Treaty of 1978. The World Intellectual Property Organization (WIPO) was formed in 1974 as a specialized agency of the United Nations. In 1977, the European Patent Office was established. The Agreement on Trade-Related Aspects of Intellectual Property Rights (TRIPS) was ratified in 1994. These international developments helped to harmonize the patent system of each country.

## 6.3   Patent System—A Double-Edged Sword

From Sect. 6.1, we see that the patent system is more effective and efficient in stimulating innovations than other policies such as public produce and subsidized private produce. This section focuses on the downside of the patent system. The patent system provides the patentee or the innovator with an exclusive right or monopoly power of producing the product he/she invented. This right or monopoly power is a double-edged sword. The right forbids other people from producing the patented products and thus gives the innovator the chance to benefit from his/her innovation. As a result, the patent system can encourage innovation and thus benefit society. The other side of the coin is that this exclusive right of producing invented goods limits the application of invention and thus reduces the benefit to society. It is often claimed that the James Watts' patent on the steam engine delayed the railway system for many years, that the patent of the Wright brothers on flight control systems hindered the development of the aviation industry for decades, and that the telecommunication industry took off only when the telephone patent by Alexander Graham Bell expired about 18 years after his invention. In a modern patent system, the patentee can issue licences to other producers and increase application of patented technology, but the high licence fees impose extra cost and thus are obviously an obstacle to the diffusion of patented technology.

Monopoly power granted to the patent is also subject to abuse, i.e. the patentee can use the monopoly power to his/her advantage at the expense of society. One type of abusing patent monopoly power is called 'tie-in'. That is, apart from the royalties and other payment for using the patented technology, the patent holder may insert in the patent licensing agreement other conditions to the advantage of the patent holder. For example, the licensee may be required to purchase materials from the patentee at a high price or the patentee may give the licensee only a very small production quota. The General Electric Company used this practice to maintain its dominant market power in the incandescent lamp industry in the early 1900s.

General Electric was formed in 1892 through the consolidation of the two largest electrical-goods manufacturers in the USA: Edison General Electric Company and the Thomson-Houston Electric Company. General Electric inherited a large number of patents in incandescent lamp production and used patent licensing as a tool to make a pricing agreement with Westinghouse Electric & Manufacturing Company and a federation of small companies named the National Electric Lamp Company. This triggered an antitrust suit by the Department of Justice in 1911. After that, General Electric changed tactics in licensing. Except for the royalty payment, it only gave a small quota for specified types of incandescent lamps. As a result, it maintained over 60% of the market share until the other antitrust action in 1924.

Another type of abuse of patent monopoly power is called patent suppression, or the sleeping patent, where the patentee or the exclusive patent licence holder deliberately neither uses nor licenses a patent for various reasons. In this case, no patented product is produced and thus consumers are totally denied the benefit of invention. McGee (1966), Blair (1972), and Saunders (2002) included and discussed a number of examples of patent suppression. Here, the author illustrates the downsides of patents with a few examples.

The most easily understood case of patent suppression is the nonuse of patent due to the fixed cost or existing investment in production. In 1908, there was a case in USA: Continental Paper Bag Co. v. Eastern Paper Bag Co. Continental Paper Bag Company made substantial investment in machines. When there was a patent that could revolutionize the paper bag manufacturing machine, the company obtained this patent and shelved it for two reasons. First, the existing machines could not be improved or replaced without great cost, so the company could not use the patent. Second, the company has to purchase the patent to prevent others from using it. If the company did not obtain the patent, the competitor may obtain the patent and had an advantage in production. When a competitor—Eastern Paper Bag Co.—used the patent technology, Continental Paper Bag Co. filed a lawsuit against Eastern Paper Bag Co. for patent infringement. The defendant argued that the plaintiff was holding the patent in nonuse and refused to license the patent, but the US Supreme Court judged that 'it is the privilege of any owner of property to use or not use it, without question of motive', so Continental Paper Bag Co. won the lawsuit.

The sleeping patent phenomenon could happen even if there is no significant increase in production cost. This can be demonstrated by the strategy of the Standard Oil Company around the 1930s and by the long delay in introducing a safer cigarette.

The main business of Standard Oil was petroleum, so the company had no interest in chemical production. However, in order to eliminate the competition from synthetic fuels and coal lubricants posed to the oil industry, Standard Oil started collaboration with the German conglomerate I.G. Farben in 1929. Standard Oil gained the ownership of I.G. Farben and obtained a monopoly right of using the hydrogenation process patents outside Germany, but it did not use these patents. In 1931, I.G. Farben discovered a synthetic oil product Paraflow, which can lower the temperature at which oil will stop flowing. Standard Oil obtained the exclusive right and started to produce Paraflow in 1932. Later a new product called Santopour was discovered, which had similar function to Paraflow, but was more effective and economical. Standard Oil obtained the exclusive patent right to Santopour, but retired the production of Santopour quickly (because its efficiency will reduce its sales and thus the profit of the company) and quietly to market Paraflow only.

The other example of sleeping patents due to protecting the existing products is a safer cigarette known as 'XA'. As early as the 1960s the researchers at Liggett & Myers Company discovered the way to remove most of the carcinogenic agents in cigarette smoke and thus a safer cigarette could be produced. The company obtained a patent on the XA cigarette but suppressed the patent. One reason is that marketing a safer cigarette might imply that the existing cigarettes were not safe and thus affect company business. The other reason is that the company was bound by industry agreement not to disclose negative information on smoking and health. Only when the tobacco industry admitted the carcinogenic effect of smoking, did Liggett and Myers finally introduce the XA cigarette to the market in 2001.

In the case of sequential invention, patents can also be suppressed by previous patents. An example in medicine demonstrates the case. Amgen Company was producing a bio-engineered version of erythropoietin (EPO) based on a patent. EPO can effectively encourage the development of red blood cells and thus is vital for anaemic patients. However, the treatment is expensive because EPO cannot stay in the body for long before it is excreted in urine. In 1997, a scientist at the Lawrence Berkeley Laboratory developed and patented a binder material that can increase the level of EPO by 10 to 50 times. This would reduce the treatment cost and benefit more patients. However, when the invention was offered to drug companies, no one took up the offer. Amgen was not interested in the offer because it was concerned that the invention would decrease the sales of EPO. Other drug companies were reluctant because they had to obtain both the EPO and the binder patent licences. The expectation is that the patent licence on EPO would be hard to obtain because it is controlled by the Amgen Company.

In a modern economy, most industries have numerous patents. It is said, in a device like the smartphone, there may be more than 250,000 active patents involved. Moreover, to reduce the high risk of patent infringement, many companies acquire more patents not for applying the patent technology but for use as weaponry for litigation and counter-litigation purposes. This results in a large number of sleeping patents.

Due to the double-edged sword nature and complex implication of the patent system, in 'An Economic Review of the Patent System' prepared for the US Senate subcommittee on patents, trademarks and copyrights, Machlup (1958, p. 80) gave a ambiguous answer: 'If we did not have a patent system, it would be irresponsible, on the basis of our present knowledge of its economic consequences, to recommend instituting one. But since we have had a patent system for a long time, it would be irresponsible, on the basis of our present knowledge, to recommend abolishing it'.

## 6.4*  Current Patent System Design—A Compromise

When the patent system was first established, patents were regarded as rewards to innovators and the purpose of the patent system was to encourage more invention to be used by the public, so the power of monopoly production was given to inventors only for a limited time. This tradition not only continues but also emphasizes due to the realization of the double-edged sword nature of patent monopoly power. Currently, all countries have adopted a balanced approach in designing their patent systems. That is to put a time limit (patent duration) and a scope limit (patent breadth) on patent monopoly power. To press the patentee to apply his/her patent, a provision called 'compulsory license rule' has also been adopted by many countries. The balanced approach is apparently a compromise: more restrictions on patent monopoly will limit the cases of abusing patent power but, in the meantime, it will reduce the returns to the patentee and thus lead to less incentive to innovate.

However, some economists deny the compromising nature of the balanced approach based on the belief that there must be an optimal trade-off point in patent duration and breadth. That is, there is an optimal patent length and breadth which can maximize the net benefit. Economic models are built to demonstrate the optimal point. The claim of optimal patent length and breadth in turn justifies the balanced approach to the patent system design. This section has been devoted to proving that there is only a very slim chance that an optimal patent length and breadth might exist. For details, see Appendix at the end of this chapter.

# 6.5    Review of Arguments Against the Patent System

Although the positive impact of the patent mechanism is generally accepted by most people, the downside of the patent system, especially the cases where patent monopoly power is abused, causes many people to rail against the patent system. Some even suggest that the patent system is beyond repair and thus should be abolished. We put the arguments against the patent system into three groups and discuss them in this section.

## 6.5.1    Argument 1 – Natural Incentives Are Sufficient to Stimulate Innovation

For the proponents of this argument, the natural incentives to innovate often refer to market structure, competition, lead-time advantages, trade secrets and imitation costs. For example, innovators have lead-time advantage which is enough to motivate innovators and allows them to recoup the research cost; the trade secret is more efficient to protect the innovator and is more often used than patents; the high imitation cost acts as a natural protection for inventors; moderate economic rents and market competition can stimulate innovation greatly. Gort and Klepper (1982) studied the life cycles of a number of industries and formulated a five-phase life-cycle theory: monopoly, entry, equilibrium, shake-out and maturity. The industry starts with one or few firms (monopoly phase), followed by a phase of rapid entry of new firms (entry phase). The next is a period of a relatively stable number of firms (equilibrium phase), followed by the considerable reduction in the number of firms (shake-out phase) and a small number of dominant firms are left in the market (maturity phase). Gort and Klepper found that innovation rates are highest in the entry phase and equilibrium phase when there are considerable numbers of firms staying in the market. Less innovation occurs in the other phases because firms face less threat of competition. As a result, Gort and Klepper concluded that competition can stimulate innovation. Later, the empirical study by Aghion et al. (2005) claimed that innovation is greatest when firms earn moderate rents. Among these claims, market structure and competition are the most complex case, so we use them as an example to show if moderate market competition can generate enough incentive to innovate.

Before we move on to discuss how effectively market pressure can stimulate innovation, it is beneficial to clear up two issues. First, the existence of effective and efficient natural incentives does not necessarily mean that there are enough incentives to innovation. There is no harm in having a patent system as an additional incentive provided that the system has a net positive impact. Why do we need an additional incentive? This is related to the question: What level of innovation is enough? Based on the discussion in Chapter 5, sufficient innovations are needed to avoid an economic recession. The repeated occurrence of economic recessions shows that the current level of innovation is not enough so we need more incentives to innovate.

Second, correlation and causality are different concepts. The empirical studies on competition and innovation can claim a correlation between competition and innovation, but not causality that competition stimulates innovation. For example, from Gort and Klepper (1982) we can infer only that high innovation rates are associated with the entry and equilibrium phases when a large number of firms are in the market. This may indicate either competition stimulates innovation, or innovation promotes competition (with innovations as an entry weapon, more firms can enter and stay in the market). Moreover, innovations tend to appear in a cyclic fashion and thus may be the cause of the industry cycle in Gort and Klepper (1982). The cyclic advent of innovations comes from the fact that innovation requires sufficient foundational knowledge and technology. Otherwise, innovation is only imaginary. For example, Leonardo da Vinci intended to invent a flying machine but it stayed only on the paper. When the foundational knowledge is mature, a number of innovations may emerge in a certain period of time. The next innovation boom needs to wait for the higher foundational knowledge in the future. History shows that innovations tend to cause industrial structural changes, so innovation cycles may lead to industrial cycles.

The market structure argument includes two threads. One thread (e.g. Thomas 2001) claims that the patent system leads to market concentration, which hampers competition and causes deadweight loss. However, some empirical studies (e.g. Miller 2009) show that there is no direct relationship between market concentration and patent activities. If the patent system does contribute to market concentration, we should assess this impact through different perspectives. First, we must consider the deadweight loss of patent monopoly with the benefit of patent monopoly—stimulating innovation. Second, we must consider the negative impact in terms of the dynamics of the patent system. As we will see in Sect. 6.6, the current patent system has many defects, and these can be avoided by extensive reform. The contribution of patents to market concentration might come from the defects of the current patent system and thus they can be avoided or, at least, decreased in the future.

The other thread of the argument is that the oligopolistic market can generate moderate competition which can stimulate innovations to the highest level so that the patent system is not necessary (e.g. Scherer 1970; Needham 1975; Mansfield et al. 1981; Levin et al. 1987, 2004; Baumol 2002; Pretnar 2003; Aghion et al. 2005; Bessen and Meurer 2007). The reasoning behind the argument is as follows. The perfectly competitive market cannot stimulate innovation because fierce competition diminishes profit margin so the firms cannot accumulate enough R&D funding for invention. Monopoly markets eliminate competition so the firm is satisfied with monopoly power and thus has no motivation to invent. Oligopolistic firms face moderate competition and earn moderate economic rent, which can be used to fund innovation activity. With moderate economic rent, oligopolistic firms have to innovate just to maintain their market position, so the entrants to the oligopoly market can spur incumbents not to rest on their laurels (Aghion et al. 2005). They may also bring diverse knowledge that can increase the rate of innovation success (Bessen and Meurer 2007). As such, oligopolistic competition is the best condition for innovation.

It is also argued that many factors in an imperfect market economy have provided sufficient inducement to innovation, such as the lead-time advantage, the learning curve, reputation, product differentiation and transaction cost (Dasgupta and Stiglitz 1988; Moir 2008; Schacht and Thomas 2005).

Market pressure may be an incentive to innovate, but the question to be answered is: Does market pressure induce enough innovation? If innovation is so attractive to firms and the market economy can provide enough innovation incentives or innovation pressure, a patent system is indeed unnecessary. However, this argument overlooks an important feature of innovation: innovation investment is of high risk, consequently, the majority of innovations end in failure. The impact of high innovation investment risk can be shown in Fig. 6.2.

In Fig. 6.2, DD is the demand curve faced by the firm and MR is its marginal revenue curve. When the firm makes a decision not to be

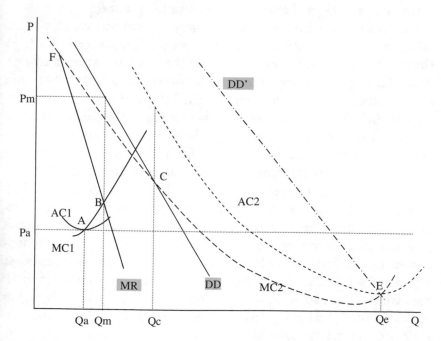

**Fig. 6.2** Innovation decision of a firm

involved in innovation activity, $AC_1$ and $MC_1$ are the average cost curve and the marginal cost curve, respectively. When the firm decides to make a process innovation in order to reduce the production cost, $AC_2$ and $MC_2$ are the new average and marginal cost curves. As the innovation involves very high initial cost, at the beginning both average cost $AC_2$ and marginal cost $MC_2$ are much higher than the $AC_1$ and $MC_1$ in the no innovation scenario, respectively. However, both $AC_2$ and $MC_2$ decrease as demand and thus production size increase. They eventually become very low, shown as point E at quantity $Q_e$.

We consider the no-innovation case first. In a competitive market, the firm will produce a quantity of $Q_a$ at the lowest average cost at point A where $AC_1 = MC_1$, and the market price is $P_a$. Numerous firms in the market can satisfy the market demand DD. If the market is a monopolized by one firm, the firm will produce a quantity of $Q_m$ at point B where $MC_1 = MR$ and charge a price of $P_m$. In the case of oligopoly (i.e. a few firms in the market), the firm will produce a quantity between $Q_a$ and $Q_m$ and charge a price between $P_a$ and $P_m$.

Now assume the firm decides to engage in process innovation (i.e. being an innovative firm) in order to reduce production costs. The high chance of innovation failure necessitates that the innovation might be successful only after many failures. This necessitates a high initial sunk cost and thus a decreasing marginal cost $MC_2$ and a decreasing average cost $AC_2$. If the market demand is very high (e.g. a demand curve of DD′), the firm can indeed enjoy a low production cost at point E (assuming competitive behavior of the firm). This is the best outcome for the firm and for the consumer. However, market demand is generally unable to keep up with the extremely high initial fixed cost. This situation is the same as that faced by a natural monopoly. The low demand indicated by DD necessitates that, in a competitive market case, the firm has to produce a quantity of $Q_c$ at point C at a production cost much higher than $P_a$, but the firm will make a loss because, at any quantity less than $Q_e$, its $AC_2 > MC_2$. If the firm is a monopoly in the market, it will set $MC_2 = MR$, which leads to a significantly higher price at point F than at $P_m$. In either the competitive market or monopoly cases, this substantially higher production cost and price make the innovation firm less competitive and thus to have a higher possibility of making a loss.

In the case of product innovation, the situation becomes worse because the much higher chance of failure for product innovation means a much higher marginal cost. Meanwhile the unknown market demand for a new type of product indicates a higher uncertainty. As indicated by the low application rate of patents, most product innovations have little or no market potential. Even if an innovation has a good market prospect, the initial market size tends to be very small. In considering this, the high sunk cost necessitates an even higher chance for the firm to make a loss if it decides to engage in product innovation.

Generally speaking, being innovative or not is a dilemma for firms. On the one hand, any non-innovating firm will be faced with considerable erosion of its market share (Baumol 2002), so it will die eventually (Freeman and Soete 1997). On the other hand, the high possibility of innovation failure may incur a loss for the innovating firm, and this may let it die sooner. So the decision to innovate or not, to some degree, is the choice between to die quickly and to die slowly. Given the risk aversion nature of human beings, it is more likely the firm will choose the latter unless it is facing the immediate pressure of dying sooner—a sharp decrease in sales. Even if the firm is forced to choose an innovation strategy, it is very likely that, to avoid dying sooner, it will choose the kind of innovation with a low risk—normally a small and unimportant innovation, and leave the risky and hard task to someone else. By this reasoning, even if the innovation has the potential to be very beneficial and all firms are facing innovation pressure, the high possibility of innovation failure may cause the level of innovation activity to fall behind and drag down economic growth.

The market innovation pressure argument also has the following flaws. First, this argument relies on market imperfection. Although it is thought of as the feature of a knowledge economy (Pretnar 2003) or a modern economy (Baumol 2002), there is no theoretical reason and empirical evidence to show that the oligopolistic market structure does and will dominate the economy. For example, the US data showed that although the percentage share of assets for the top 100 manufacturing corporations increased in some period of time, it decreased at other times (Scherer and Ross 1990). The stochastic determinants approach (e.g. Hart and Prais 1956) even claimed that the market structures observed at any moment in time are the result of pure historical chance. The main determinants of market structure are thought of as the economies of scale and government regulations, but there is no evidence that these determinants will change towards encouraging market concentration. Innovation could be concentration-increasing (Nelson and Winter 1982), but Geroski and Pomroy (1990) found a strong tendency for market concentration to fall during the 1970s when innovative activity was vigorous.

Secondly, being innovative is a possible outcome of oligopolistic competition, but not a definite one. The game-theory models about the competitive behaviour of oligopolies (e.g. Dasgupta and Stiglitz 1988; Reinganum 1982; Fudenberg et al. 1983) show that, while some competitors tend to attempt to win the race, the knowledge spillovers also encourage firms to engage in waiting games. If most firms choose to be the followers so as to avoid the risk of innovation failure, the level of innovative activity may be too low.

Thirdly, the incentive firms may be just innovative enough to keep their market position and fail to generate enough technological progress for economic growth. For example, Rosenberg (2004, p. 2) noticed, 'in the United States today, there are more than 16000 firms that currently operate their own industrial research labs, and there are at least 20 firms that have annual R&D budgets in excess of USD 1 billion'. But he concluded that R&D funds are not enough due to uncertainties and their attendant financial risks deriving from a variety of sources (he has listed seven).

In short, mild market concentration may help induce innovation, but this incentive may not be enough due to the high risk of innovation investment. Even with the help of the current patent laws, the inability of the market to produce enough innovation is repeatedly demonstrated by the frequently happening financial crises and economic recessions which, as explained in Chapter 5, result from scarcity of product innovation.

## 6.5.2  Argument 2 – The Political Economy of the Patent System Prohibits a Successful Patent Reform

It is a general consensus among both proponents and opponents of the patent system that the current patent system has many problems and thus can be improved substantially. However, Landes and Posner (2004) and Scherer (2009) raise the issue of the political economy of the patent system and cast doubt on patent reform. Boldrin and Levine (2012) even claim that the patent system should be abolished because political economy pressure makes it impossible to accomplish the required patent reform. Their proposition of political economy of the patent system has some elements of truth but does not depict the whole picture.

There are two central arguments concerning the political economy of the patent system. One is that the patent system is designed and operated by interest groups while consumers (i.e. the public) are excluded. The other is that, since patenting is a technical subject, the interests of voters are not well represented and this causes regulatory capture—the regulators act in the interests of the regulated, instead of the broader public. These two arguments are discussed in turn.

For the first argument, it is true that interest groups may have heavy influences on patent system designs and administration, but this is not unique to the patent system. Like other laws, patent law is enacted by parliament, operated by the government and arbitrated by courts. Other laws and regulations are also subject to lobbies by interest groups. In essence, the political economy of the patent system is the political economy of any other laws and regulations. We cannot abolish all laws and regulations because of lobbying by interest groups. This also applies to the patent system.

The claim in the first argument that the political economy of the patent system excludes consumers, or the public, is misunderstanding and thus misleading. Three agents are involved in the patent system: the patent holder, the producer and the consumer. Here the consumer means the public who are the final users of the patented products, while the producer is a firm who needs patent permission to produce the patented product. If the patent is licensed or assigned to the producer, the patent holder and the producer form one group to extract monopoly rents from the consumer. In this case, the consumer is explicitly included in the patent system and, if there is any lawsuit (e.g. antitrust action), the consumer will be directly involved or will be represented by the government. On the other hand, if the producer has allegedly produced the patent products without a licence, there will be a lawsuit between the patent holder and the producer. In this case, the producer is sided with the consumer because the consumer does not need to pay the patent monopoly rent if the producer wins the lawsuit. Consequently, although the consumer is not involved in the lawsuit directly, the interest of the consumer (the public) as the final user is directly related to the producer through the market price of patented products and thus indirectly represented in the courts by the producer or the defendant. The operation of this system is the same as that of any other law or court system, for example, it is unnecessary for the broader public to be represented in a criminal court or a civil court.

The second argument that the patent system is highly technical is indeed an issue in the forefront of the patent system. The highly technical nature of the patent system may impose some challenges in operating the system, especially when the system is not mature or standardized. However, this should not be a fatal problem for the system. As explained previously, the public is indirectly represented in the court so the public is not required to have a full understanding of technical issues. The representatives of the patent owner, the representatives of the patent user and the judge (or the jury) are directly involved in the patent system so they do need to have a thorough understanding of the patent technology involved. It is reasonable to believe that there are qualified personnel, who can understand the technicalities in the patent, to fill these roles. Eventually the public can understand the technicalities by diffusion of knowledge, experience, education or by learning from mistakes. Broadly speaking, although more specialty is embedded in the patent system, the technical nature is a common factor in specialized courts such as those dealing with property rights, intellectual property rights, the military and marriage, so it is not unique to the patent system.

In short, it is unrealistic to assume that a patent system is designed by impartial legislators and administrated by a benevolent government, so the political economy of the patent system has a valid point.

Albeit having some specialty, the political economy of the patent system is essentially the same as the political economy of any other laws and regulations. The dynamics of different players (i.e. the parliament, the patent office, and the representatives of the patent owner and of the patent user) in the patent system can overcome the political economy problem, so it is an overreaction to advocate abolishing the patent system. Nevertheless, the approach of the political economy of the patent system highlights the issues in the patent system. These issues can be solved through a thorough reform of the patent system.

## 6.5.3 Argument 3 – The Patent System Has Net Negative Effects

The opponents of the patent system generally belittle or even deny the positive effect of the patent system but emphasize the negative effect. As a result, the patent system is viewed as having a large net negative effect and thus should be abolished. The arguments of net negative effects of patents are sometimes made based on statistic empirical studies (e.g. Bessen and Meurer 2008). The statistical approach will be discussed in Sect. 6.5.4. The discussion in this section is confined to logical reasoning and historical case studies.

There are a number of arguments against the positive effect of the patent system, including: (1) the oldest but still often proffered argument is that innovation stems from the interest or genius of inventors. The reward of patent can neither change people's interests nor make people become more talented and, thus, has little impact in stimulating innovation; (2) the motivation of innovation is a market demand. When people need something, inventors will respond by solving the problem, with or without the potential to obtain a patent. Very few people invent just to get a patent; (3) the first move advantage is also a strong incentive for innovation; (4) imitation of innovation is not that easy and cost saving, so innovators can obtain enough profit from their innovation and thus have enough incentive to innovate; (5) people tend to buy or even put a premium on authentic products supplied by inventors because people believe that original creators have more knowledge in the area and thus have better solutions and better quality products. If the imitator has done a better job than the inventor and has more market share, it is also a good outcome because it promotes competition and benefits consumers; (6) the disclosure benefit from the patent system is negligible. Because the patentee intends to cover as much information as possible, patent description is generally written very broadly and thus produces very little benefit to downstream innovation; and (7) patentees receiving the benefit of patents may no longer actively engage in innovation activities. Examples of old companies having a large number of patents but producing very few innovations are often used to illustrate the ineffectiveness of the patent system in encouraging innovation.

These arguments have some elements of truth, but generally fail to provide a picture of the overall impact of the patent system. There may be a few examples to support each claim, but these examples are not a general case in reality. If one has read histories of industries such as the oil industry, the electrical lamp industry, the electronic industry, the pharmaceutical industry and the information communication industry, one can perceive not only the negative impacts of patent monopoly but also the stimulus provided by patents to innovation, the patent races among innovators, and the urgency of firms to obtain patents. Since many of the arguments listed above come from the same themes, the following response is not in a one-to-one fashion.

First of all, the reasoning for the positive impact of the patent system on innovation is clear and sound. Patents give the innovator a power to obtain profit. This power is a reward and an incentive to inventors and thus must have positive impacts on innovation. Who can deny the positive impact of the Nobel Prize in stimulating scientific research? The patent monopoly power has an impact similar to but much more powerful than the Nobel Prize because the potential reward from the patent is much greater. Overall, the mechanism of the patent system also requires no source of financing, incurs no administration cost in allocating rewards and, more importantly, ensures that the rewards are directly linked to the contribution of the invention. As such, the patent system is an effective and efficient rewarding system to stimulate innovation.

Some people may label the above reasoning as 'faith-based' reasoning, or 'over-simplified' reasoning. However, since the logical reasoning is correct and the reward mechanism works, the positive impact must exist in a general case, or at least in our 'oversimplified' case. Because the real world is so complicated by a large number of factors, we need the ability to grab the basic factors to simplify the cases. This is our way to uncover the mechanisms, form theories and find solutions. To understand the more complex cases in the real world, we can add more factors to our basic case. For example, there are other incentives to innovate such as interest and talent in innovation, market demand and the first-mover advantage. These incentives are important, but the patent system can also add an important additional incentive to innovation. This point is convincingly illustrated by Abraham Lincoln when he said that patents 'added the fuel of interest to the fire of genius'.

Second, the incentives due to the interests of inventors and from market demand and market structure are not enough to generate sufficient inventions, so innovation needs extra stimulus. This is evident by the repeated occurrence of economic recessions. As demonstrated in Chapter 5, the scarcity of innovation is the fundamental cause of economic recessions, so more incentives are necessary to encourage innovation so as to avoid recessions. Innovation needs extra stimulus also because of the nature of innovation: the high cost of conducting innovation and the high possibility of innovation failure. Conducting innovation is costly and a successful invention often comes from numerous failures. Edison's first successful electric carbon lamp came from hundreds of experiments by himself. In considering other people's failures, from which Edison learned some lessons, the number of innovation failures is even higher. The total cost of all the successful and unsuccessful experiments is very high and thus justifies more incentives for innovation.

Third, copying or imitation of an innovation is relatively easy, of small cost, and extremely harmful to inventors. Depending on the types of innovation, some (e.g. a software program or a formula for a medicine) are easier to copy or imitate while some (e.g. an improvement in a machine and an innovation of a process) are relatively harder. However, compared with innovation, imitation is much easier and incurs much less cost. Otherwise, the imitator would prefer to invent rather than to imitate and thus imitation activity would not exist. When comparing the cost of imitation with the cost of invention, one should consider not only the cost of successful innovation, but also the unsuccessful trials by the lucky successful innovator and by other unlucky ones.

Since the cost of imitation is much lower than that for invention, imitation activity occurs even if we have a patent law. The harm of imitation to inventors is obvious because imitation severely affects the return to the inventor. This harm is ultimately the harm to society— more people are discouraged from undertaking invention activities and fewer inventions can be available for consumers. The harm of imitation is also well explained by the theory of public goods and is demonstrated in the long history of patent system development. If imitation had no harm to inventors, the patent system would not be proposed by our ancestors and would not exist today. The consumer's preference to buy authentic products at a premium from original creators may help the inventors to a certain degree but, one has to admit that price is the most powerful tool in a market economy. In considering the high cost and uncertainty of invention, the price competition essentially gives an advantage to imitators and thus harms and discourages inventors.

Fourth, the examples of less-innovative old companies reflect the pattern of patent activities in the course of development of a firm or an industry. However, using these examples to demonstrate the ineffectiveness of the patent system is putting the cart before the horse. A new firm or new industry tends to be more innovative. Market pressure and market demand play an important role, but the stimulus from the patent system is also crucial. This is evident by the firms' racing to develop and obtain patents in the development of many industries. Without patent protection, fewer firms or individuals would be willing to undertake innovation activity and/or start a new business requiring inventions. As the firm or industry approaches the mature stage, they become less innovative due to the lack of new ideas, complacency in its market position or preoccupation in routine production. The large amount of revenue from patents may also play a role in complacency of the firm. However, a less-innovative old firm receiving a large sum of patent royalty does not mean that the patent system is ineffective in stimulating innovation. The benefit of patents to the old firm is the reward for its previous innovations and these rewards can encourage other firms to innovate. In other words, the patent system is effective if it can encourage any firms to innovate, so stimulating further innovations from the patentee is not a necessary condition for an effective patent system.

Finally, disclosure information provided by the patent system has at least some positive effect and has no negative effect. The main purpose of the patent system is to stimulate innovation. The disclosure requirement in patent application contributes to this end through aiding downstream innovation. It is true that patent applicants tend to provide as little information as possible in order to have some additional protection on their inventions, but the applicants must provide adequate information required by the patent office. Even a broad description of innovation ideas may be useful (e.g. showing the right direction) for downstream inventors. The bottom line is that the patent office forbids applicants from providing misleading information, so the disclosure information generally has no negative impact.

There are many arguments and examples regarding the negative effect of the patent system. The common ones are: (1) patents often lead to high prices of products and they decrease the welfare of consumers. The outrageously high price of medicines and treatment (e.g. the treatment of HIV) often triggers public outcry against the patent system; (2) patents are a main obstacle of competition and thus are against the antitrust law and against the spirit of a free market; (3) at the expense of consumers, patents generate much more economic rent than required to recoup the research cost; (4) patents negatively affect downstream innovation, so the patent system hinders rather than stimulating innovation; (5) sleeping patents are a waste of valuable resource. The large number of defensive sleeping patents in the modern economy highlights the wastes; and (6) patent trolls who use patent litigation to threaten many businesses are putting in danger many businesses and becoming the enemy of society.

The above-mentioned concerns about patents are valid arguments, at least to certain degrees. A patent system is essentially a monopoly system. The negative impact of monopoly on a market economy is well known and indisputable. However, when considering the negative impact of a patent system, one needs to compare it with the positive impact—the patent system stimulates innovations. That is the wisdom behind Bacon's argument that a monopoly should be granted only to innovations. Moreover, we also need to differentiate between the negative effect built in patent mechanisms from that of the current patent system—some negative effects may stem from the defects in the patent design which can be overcome through patent reforms. With these considerations in mind, we will discuss the arguments related to the high prices of patent products, sleeping patents, downstream innovations and patent trolls.

The excessive price argument (e.g. Singh 2001; Oxfam 2001; Nielsen and Samardzija 2007) states that the monopolistic position of patent holders lets them to require an excessive price for the granting of a patent licence. This argument is very appealing, but it is not well backed by theories. It is claimed that companies in the high-technology sector have been asserting that they are under constant pressure of possible lawsuits that threaten to shut them down (Nielsen and Samardzija 2007). The argument is even more appealing when the medication patent is involved. The reports are empathic and emotive that patients in developing countries are not able to pay the price of patented medicine and were left to die. It is unfortunate that patients have to pay the price of market imperfection, but the other side of the story is that the patients would not have any hope at all if the medicine had not been invented. If we abolish the patent system or impose a compulsory licence on this patent, the current patients can definitely benefit from cheaper drugs but, as a result, the patients in the future will suffer because the drugs or cure for a disease will be found after a very long delay due to less incentives. This is a trade-off between now and the future. To protect the benefit of both future and current patients, the government can step into negotiate an agreement with the patentee to obtain a licence for production of the drug for the patients.

The sleeping patent phenomenon can also be viewed as an extreme case of excess patent price—an infinite patent price: no matter how much the patentee will be paid, he/she will not accept the offer. This phenomenon to a large extent may result from malfunction of a patent market due to the defects in the patent system design. In a market economy, even a monopoly will produce enough output to maximize profit, instead of setting an infinite price and making no profit. However, the sleeping patent phenomenon indicates that the current patent market does not function well for society. In a well-functioning patent market, firms would not be able to acquire an important patent very cheaply so they would not just put aside an expensive patent. They would prefer to sell it or license it at an acceptable market price so as to maximize profit.

The defects in the patent design may be responsible for an ill-functioning patent market. For example, the limited patent durations impose a life sentence on patents and cause much distortion in the later years; the exclusive patent licence and patent assignment magnify monopoly power and reduce patent demand. Similarly, the exclusive patent licence and patent assignment also play a role in causing excessive prices of patent products because these arrangements allow patent monopoly power to be used by producers other than the patent holder. Using this monopoly position, producers can demand a high-profit margin and thus cause high prices. Since the producers directly face the final demand, the profit margin could be much higher than the royalty paid to the patent holder (especially when the final demand is inelastic in the case of medical patents).

The defects in patent system design may also be responsible for the negative impact of downstream innovations as well as the excessive price of patented products. The definition of patent in current patent laws gives the patent holder sweeping exclusive rights in making, using and selling the patented products. This definition prohibits everyone except the patent holder from using the patent without permission, so downstream innovators have to obtain permission (pay royalty) before conducting innovations on an existing patented product. If the patent system is designed in a way which only forbids distribution or commercial use of patented products, the downstream innovators can freely use patents in their innovation activity, so the negative impact of the patent system on downstream innovations can be avoided.

The most appealing scenario is the phenomenon of patent speculators or patent 'trolls'.[1] As the number of patents increases dramatically in the modern economy, some individuals or companies initiate and provide services to patentees: to find out and litigate patent infringements for patentees, or to negotiate a solution for patentees and the infringing companies. Some service providers send out letters indiscriminately to infringing and non-infringing companies demanding a large royalty from firms or threatening a patent lawsuit or injunction, so the service providers have acquired the name 'patent trolls'. In this case, the patent laws seem to be in the wrong hands, deter economic growth and do nothing to stimulate innovation. However, apart from the extreme practices, the services provided by so-called patent trolls are actually legal and useful for enforcement of the patent law. As shown by Barker (2005), not all patent trolls are bad for the economy.

Objectively, if the practice of so-called patent trolls is legal, we should not blame their activity. If the threatened firms have not infringed patent rights, they do not need to worry about the threats. The anger towards patent trolls may result from the low-quality patents issued by patent offices. For example, due to lack of manpower and other resources in the US patent office, and also possibly due to the lack of recognition of the importance of patents, the patent office has approved many patents which are non-innovative (i.e. not qualifying for the 'non-obvious' condition in patent law) and have caused significant impact on normal business. An example is the 'one click' patent. Threat to litigate this kind of patent infringement triggers widespread anger. If the patent approval is improved and if the patent trolls follow lawful practice, the service provided to patentees is useful to patentees and beneficial to the patent system and to society at large.

In summary, a rational person or a balanced argument generally acknowledges both the positive and the negative impact of the patent system, but the question remains as to whether the positive effect or the negative is greater. The empirical evidence on this is mixed. Some studies show considerable positive effect of the patent system. For

[1]For more information about patent trolls, please see F.D. Ferrill (2005), Woellert (2005), and Varchaver (2001).

example, Taylor and Silberston (1973), Mansfield et al. (1981) and Mansfield (1986) showed that a significant percentage of innovations are induced by patents in most industries, especially in the pharmaceutical and chemical industries. Firestone (1971), Griliches (1990), and Thomas (1999) found that a large proportion of patents were worked and renewed. Khan and Sokoloff (1993) and Arora et al. (2003) concluded that patented innovation makes a significant contribution to firms. Bloom and Reenen (2002) and Hall et al. (2007) found that patents stock forms an important part of market value of large firms in UK and other European countries.

On the other hand, some studies claim patent policy does not confer clear net economic benefits. For example, Granstrand (1999) claimed that the patent system has been neither necessary nor sufficient for technical and/or economic progress at country and company levels historically; Encaoua et al. (2006) suggested that patents should not be used as a default policy choice to stimulate investment in innovation. Boldrin and Levine (2004) suggest that patents should be used only when the innovation investment is large and indivisible. Using an evolutionary model, Winter (1993) shows that both R&D investment and total surplus would be higher without patents. Dutton (1984) and Mandeville (1996) both concluded that the most balanced patent policy was a weak patent system. Jaffe (2000) reports that widespread unease that the costs of stronger patent protection may exceed the benefits. Bonatti and Comino (2011) found that, if government has the opportunity to subsidize R&D and if imitation takes time, the social welfare is higher in the absence of patents. Moir (2013) concluded that patent policy is a very blunt instrument and is effective only in a narrow range of circumstances.

A word of caution is necessary for these empirical studies. The statistical approach based on probability theory is used for most studies on patents. As explained in Chapter 3, time series data cannot satisfy the condition of probability law unless one can claim that the model includes all relevant factors and has a correct function form. Since no one can confidently make either claim, the results from a statistical model can only be indicative. That is, one cannot claim a net positive or a net negative effect of a patent system based on only a few empirical studies. The detailed discussion about empirical studies on patents is provided in the next section.

## 6.5.4 Argument 4 – Statistical Studies Indicate Poor Performance of the Patent System

One might hope empirical research could provide hard evidence showing if the patent system has a net positive or net negative effect. However, due to the availability and limitation of data in this area, statistical studies are unable to answer this question categorically. These studies often show mixed or conflicting results. Nevertheless, by reviewing these studies, overall impressions about the performance of the patent system can be obtained.

Empirical studies on the patent system are very diverse. They may use different methods of research, have different scopes of research, and emphases on different aspects of impact. We classify and review the empirical studies mainly based on the aspects of impact of the patent system. In addition, the research on the impact of changes in patent policy will be reviewed at the end of the section.

1. Empirical research on the social cost of the patent system

The studies on the social cost of patent monopoly are scarce, but it is not hard to find studies on consumer welfare loss due to the imperfect competition in the market. Harberger (1954) is the first to attempt to estimate the cost of monopoly powers. Based on the data used in Epstein (1934)—a sample of 2046 corporations in 73 manufacturing industries in USA for the period of 1924–1928, Harberger estimated the resources misallocation due to market power in the manufacturing industry was about $1.2 billion, or 4% of the total resources in the manufacturing industry. The loss of social welfare was about $59 million, or about 0.1% of the national income.

Using different data and varied methods, other researchers obtained much larger estimates. For example, Scherer (1970) argued that the more reasonable assumption would raise the estimation of the US deadweight loss to 0.5–2.0% of the gross national income. His own estimation of the redistribution loss of consumers due to monopoly profits was about 3.0% of US GNP. Comanor and Leibenstein (1969) studied a sample of the consumer-good manufacturing sector for the

1954–1957 period and estimated monopoly overcharge to consumers to be 5.5–6.0%. Kamerschen (1966) estimated the welfare losses from monopoly in the US economy as 8% of GNP. Parker and Connor (1979) concluded the welfare loss was 3–6% of the US GNP; Jenny and Weber (1983) had a result of 0.85–7.39%.

Recently, Katari and Baker (2015) studied the social cost of mis-marketing of five patented drugs in USA. The accumulative cost of both increased morbidity and mortality is estimated at $382.4 billion over the 14-year period from 1994 to 2008. This gives a social cost of over $27 billion per year caused by mis-marketing of five drugs. Compared with the total research spending of the whole pharmaceutical industry of less than $25 billion per year, Katari and Baker concluded that the damage due to marketing abuse of patent monopolies is much larger than the value of research induced by patents.

The above research confirms the negative economic impact of monopoly but, generally speaking, the social cost of monopoly is small relative to GNP. The study by Katari and Baker seems an exception. This study suggested that the social cost of monopoly looks very high in the pharmaceutical industry; however, the social cost of mis-marketing of brand drugs is only indirectly related to patent so the cost cannot be fully attributed to patent monopoly. More importantly, the estimates from the above studies are gross in nature due to the availability of data. For example, many strong assumptions are used in calculating the deadweight loss, e.g. the elasticity of demand, the type of inefficiency and the method of scaling up the sample estimation to the industry and to the economy. A key assumption on the calculation of cost of drug mis-marketing by Katari and Baker (2015) is subjective. Since the assumptions used directly affect estimation results, a change of these assumptions will lead to a substantial change in estimation results. The criticisms and caveats about the limitations due to data availability are well documented in these studies.

2. Research on the social benefit of the patent system

The social benefit of the patent system is in the areas of stimulating innovation and disseminating innovation information. There are many indicators which may directly or indirectly relate to the social benefit of the patent system. For example, the percentage of innovations and R&D investment induced by patents, patenting rate, patent usage data and patent renewal data, and imitation cost and lead time. We group the studies by the indicators they employed.

On innovations stimulated by patents, Taylor and Siberston (1973) conducted a survey on British firms which showed that the patent-induced innovation is 8% for all industries and is about 65% for the pharmaceutical and fine chemicals industries. Using the data for 48 product innovations in four industries in the north-west of the USA, Mansfield et al. (1981) found that patents induced about 90% of innovations for the pharmaceutical industry and about 20% for the chemical, electronics and machinery industries. Based on the random sample of 100 firms from 12 industries in the USA, Mansfield (1986) concluded that patent-induced innovation is 60–65% for the pharmaceutical innovations, 30–38% for the chemicals industry, 15–17% for the machinery industry, 12% for fabricated metal products and 4–11% for electrical equipment.

Moser (2002) studied the inventions displayed by different nations at the World Fair of 1851 and of 1876. She concluded that nations with patent systems were no more innovative than others. Nations with longer patent terms were no more innovative than nations with shorter patent terms. These findings suggested zero social benefit from the patent system. However, she also found that innovation in countries without a patent system was centred on industries having strong trade-secrecy protection. This shows the patent system benefits at least some industries. Jaffe (2000), Jaffe and Lerner (2004), Lerner (2002), and Boldrin and Levine (2008) examined a number of empirical studies and found that the link between patent protection and the number of innovations is weak.

On the impact of the patent system on economic growth, Gould and Gruben (1996) used an index to measure a country's strength of patent protection. In their model, the patent index had a positive coefficient but it was not statistically significant. Park and Ginarte (1997) used a similar model but with an additional measure of general property right named 'market freedom'. They found that the market freedom index had positive and statistically significant effects on economic growth, but the intellectual property rights index had a negative but statistically insignificant coefficient.

On R&D investment induced by patents, Taylor and Silberston (1973) found that patent protection was responsible for 60% of R&D in the pharmaceutical industry, 15% for the chemical industry and 5% for the mechanical engineering with a negligible amount for the electronics industry. Kanwar and Evenson (2003) found that higher patent protection leads to higher research and development spending as a fraction of GDP. However, Ginarte and Park (1997) found that lagged R&D is positively correlated with subsequent intellectual property rights strength. This suggested that R&D investment has an impact on the patent system rather than the other way round.

On patenting statistics, Dutton (1984) showed that hundreds of inventors did patent and many obtained multiple patents. Based on a 1983 Yale survey, Levin et al. (1987) suggested a low patenting rate because the R&D managers regarded patents as less effective than other means in obtaining returns on their R&D investments. For new products, lead time (first in market), learning curve and sales/marking are more important than patents in obtaining a return. For new processes, lead time, learning curves, sale/service and secrecy are more important than patents. Arundel and Kabla (1998) found that the majority of inventions are not patented. On average, large European firms applied for patents on only 36% of product innovation and 25% of process innovations, but pharmaceutical firms applied for patents on 79% of products. Cohen et al. (2000) used both the Yale survey and the Carnegie Mellon survey. While the study claimed an increase in the importance of secrecy after 1982, it suggested that patents may be more central for larger firms and that a sizeable minority of industries counted patents as a major mechanism of appropriation. Moser (2005) studied the information on the invention exhibition at the 1951 Crystal Palace World Fair and found that only 11% of British inventions were patented.

On patent usage and renewal, most studies indicated that the percentage of patent renewal is small and the usage of patent information is not encouraging (e.g. Scherer et al. 1959; Mansfield 1986; Levin et al. 1987; Cohen et al. 2000; Lemley 2001; Blonder 2005).

However, some empirical studies show otherwise. Firestone (1971) found that 50% of Canadian patents were used to support production, 30% of UK patents were worked, 49% of patents in large firms and 71% patents in small firms were worked in USA. Griliches (1990) reported that 41–55% of 1957 US patents were used commercially, up to 71% of patent were used among small firms, about half of granted patents were renewed beyond year 10, and about 10% to the maximum limit. Thomas (1999) estimated the percentage of US patents renewed to their full term was nearly 40%. This study was consistent with Pakes and Schankerman (1984), which found that nearly 60% of US patents were not renewed to full term.

On imitation cost and lead time, Mansfield et al. (1981) found that on average imitating an industrial artefact costs 65% of the original innovation cost and took 70% of the original time. Levin et al. (1987) showed that imitation saved around 50% of the cost of the original R&D, and took at least 6–12 months. These studies indicate a significant amount of cost and time are needed to imitate, so a patent law is not crucial to protect the benefit of inventors; in other words, the benefit of patent protection to inventors is not that high.

To sum up, the research results on the social benefit of the patent system are mixed. Some reports show large impacts of the patent system but other results are insignificant. When considering the results from the above research, we have to be mindful of the limitation of the data and of the method used. For example, the sample size in Taylor and Siberston (1973) was small, so the generalized conclusion for the industry might not be very reliable. Moser (2002) relied on the data of World Fair events. The number of innovations each country displayed on the World Fair may not be of the same proportion of the total innovation for each country, so counting the number of innovations in the World Fair cannot be seen as a conclusion as to which country is more innovative. The patent index and intellectual property right index used in Gould and Gruben (1996), Park and Ginarte (1997), and Lerner (2002) are high aggregate numbers. The aggregation includes many factors and the way of aggregation is subject to the researcher's judgement (e.g. the weighting used). This will introduce subjective elements into the estimation. All econometric estimations should be interpreted as a correlation rather than causality. Even the finding by Ginarte and Park (1997) that R&D spending might determine the strength of intellectual property right was Granger causality, not true causality.

## 3. Research on private returns of patents

Due to the difficulty in measuring social cost and social benefit, the studies in these areas are often criticized as being inaccurate. Under this circumstance, many researchers focus their study on the return of patents to firms. The studies on private return of patents are obviously important to firms, but they are also useful for society as a whole. If firms can earn a significant amount of return from patents, they will have enough incentive to innovate, so the patent system will be very effective in stimulating innovation and society will benefit from this. On the other hand, if the private returns on patents are not significant, the patent system will be ineffective and not beneficial to society.

On the revenue to firms or inventors brought by patents, Khan and Sokoloff (1993) conducted a historical study on the important inventions in the USA during 1790–1846 and found that many of the famous inventors did make profits from patented inventions. Using the data from the 1994 Carnegie Mellon survey, Arora et al. (2003) found that when firms do patent, the patents yield a positive and often substantial return. Toivanen and Vaananen (2012) studied the returns to inventors and found that inventors get a temporary reward of 3% of annual earnings for a patent grant and, for highly cited patents, a longer lasting premium of 30% in earnings three years later.

A large number of researchers studied the private value of patents. Bloom and Reenen (2002) examined the impact of patents on the market value of large UK firms and found that patents affect the firms' market value positively. Hall et al. (2007) studied the private value of patents and R&D in European firms during the period 1991–2004 and found that Tobin's q (the ratio of the market value to historical or book value of a company's assets) is positively and significantly associated with R&D and patent stocks and that the software patents are more valuable than ordinary patents. Shapiro and Hassett (2005) estimated intellectual property constitutes 33% of corporate assets, amounting to about $5 trillion. They obtained this estimate by assuming that all corporate expenditures on R&D, software and databases constituted intellectual property.

Based on the surveys on patentees' willingness to sell, Harhoff et al. (1997) conducted an inventors' survey on the full-term German patents of the 1977 applications and confirmed the high-value tail distribution estimated by Lanjouw et al. (1996). Harhoff et al. (1997) tried different distributions to fit the data and found that the log-normal distribution generated the best fit for the patented invention value data. Gambardella et al. (2005) estimated the determinants of patent value in France, Germany, Italy, the Netherlands, Spain and the UK. The study was based on a large PatVal-EU inventors' survey, which collected more than 9000 responses from 27,000 questionnaire submissions. Using the same methodology by Lanjouw and Schankerman (2004), the study constructed a patent value index and the estimation indicated that the most important determinants of patent value were characteristics of the individual inventors, rather than the characteristics of the firm or the characteristics of the location. This supported the argument that the invention business is more about human capital and talented individuals. Gambardella et al. (2006) used the PatVal sample of firm patents to study the determinants of patentees' willingness to license in Europe. They concluded that firm size was the most important factor while patent breadth, patent value and patent protection also had an impact.

Some researchers obtained the value of patent by studying the data on patent renewal behaviour. Using the patent renewal value and a distribution from a previous survey of inventors, they calculated the median and mean value of patents. Barney (2002), Bessen (2006a), Putnam (1996), and Serrano (2005) studied US patents and estimated a mean value of patents of $47,456–$188,355. Baudry and Dumont (2006), Gustafsson (2005), Lanjouw (1998), Schankerman (1998), Pakes (1986), Pakes and Schankerman (1984) studied patents in Europe and estimated a mean value of $1656–31,704.

Some also used the market value of a patent by employing a model to estimate the contribution of patents to the total market value of the firm, controlling the other sources of value such as physical assets and goodwill. Notable studies include Cockburn and Griliches (1988),

Megna and Klock (1993), Hall et al. (2005), Bessen (2006b). They estimated a mean patent value of $119,000–$370,000.

The above studies estimate only the revenue or gross profit from patents. Bessen and Meurer (2008) used the event study methodology to estimate the average annual cost of patent litigation as $14.9 billion during 1996–1999. Compared with the estimated patent value of $4.4 billion in 1991, they concluded that the cost of patent is larger (or at least not smaller) than the patent value.

Bessen and Meurer (2008) tried to improve the accuracy of empirical studies in a number of ways. First, they confined their research question to the net benefit of patents to a firm. Second, they tried to establish an upper boundary for patent rent and a lower boundary for costs related to patent, which would provide a buffer for their estimations. Third, they tried different ways (e.g. patent renewal information and market value calculations) to estimate consistent patent revenues. Last, they tried to identify and overcome the bias in their estimation, for example, the use of distribution or patent renewals rather than the patent renewal fee in estimating patent value.

However, there still are many factors which can contribute to inaccuracies, e.g. the name-matching software for preparation of the data set may be made as effective as possible but it still caused about 10% error in their testing. Although Bessen and Meurer (2008) criticized that patent value studies based on surveys on patent owners' willingness to sell mixed the value of a patent with the underlying value of the technology, they also adopted the log-normal distribution derived from these surveys. Since what these surveys revealed was the patent owners' willingness to sell, the distribution derived from these surveys cannot be interpreted as the distribution of patent values (the patent owner may underestimate or overestimate the value of patents). In the end, Bessen and Meurer justified their use of log-normal distribution by saying that log-normal distribution was used by most patent renewal studies. Although it is common that econometric studies often assume a type of distribution without sufficient justification, the inability to justify the use of distribution shows the unreliable nature of the econometric method.

Bessen (2006b) explained the way to obtain the market value of patents. He admitted that the theoretical model made several strong assumptions. For example, it is assumed that the firms return rate on capital could not be greater than 10–20%, that the link between commercial success and patenting was weak, and that the total patent rents of a firm were a concave function. The theoretical model also relied on the Taylor approximation. The data used for estimation were obtained through a name-matching software, and were also subject to various treatments. For example, the sample of 1% tails of Tobin's q was trimmed. R&D was calculated assuming 15% annual depreciation rate and an 8% pre-sample growth rate. The patent stock was calculated using a 15% depreciation rate. The rival firms' patent stocks were calculated using a technology distance measure. These assumptions or treatments may be appropriate for the study, but what can be seen here is that there is a high margin of error.

The estimation of patent cost in Bessen and Meurer (2008) was based on case studies and a survey of patent lawyers. The estimation details were given in Bessen and Meurer (2007). Although the legal costs (e.g. attorneys' fees) are estimated based on data from Westlaw and from the survey by the American Intellectual Property Law Association, the main cost of patent litigation came from the estimation of business costs—the impact of filing litigation on the return rate of the firm's stock. Even if it is assumed that there is no problem in data and estimation, the estimated cost may be not only inaccurate but also flawed for two reasons.

One is that the use of the stock return rate as an indicator of patent litigation cost is inappropriate. It is true that a decrease in stock return indicates the damage of patent litigation and damage can be interpreted as a cost. However, the return rate of a firm's stock is not the rate of return to the firm—the former can fluctuate markedly while the latter does not. Apparently, the stock return rate is subject to speculation by investors. Moreover, the loss due to a decrease in the stock return rate belongs to shareholders rather than to the company that issued the shares. The company may still hold a large amount of shares but it generally will not sell its shares in response to a patent litigation. As a result, there is little loss to the company even if the stock return rate drops. In short, the decrease in the stock return rate is a damage to the stock but not a damage or cost to the firm, so it cannot be interpreted as the cost of the patent to the firm.

The other reason is that the use of the return rate as a foundation to calculate litigation cost requires extremely high accuracy of estimation because of the large base of a firm's stock. For example, Bessen and Meurer (2007) estimated a mean of a litigation-induced change in stock return or cumulative abnormal return (CAR) for all alleged infringers is −0.50% with a standard error of 0.16%. This result is of 1% significance. However, if allowing for a one-standard deviation to obtain an interval estimation (this is a common practice in econometrics), we have a change in the return rate of −0.50%±0.16%, or within the interval of [−0.66%, −0.34%]. This means that the point estimation produced by Bessen and Meurer (2007) can increase or decrease by ±0.16/0.50 = ±32%. When the estimated return rate is applied to the large market value of the stock, this level of variation will have a large impact on the calculated patent cost.

From the above discussion it is seen that, even though Bessen and Meurer (2008) intended to do accurate estimations, their estimations not only have a large margin of error but are potentially flawed. As a result, their estimation results are not reliable and their conclusion of a net negative private return of patents cannot be substantiated.

## 4. Research on the impact of changes in patent policies

One difficulty of empirical research is that too many factors are involved in a study so it is hard to assess the impact of a particular factor. This is also true in the study of the impact of the patent system. To overcome this difficulty, some researchers focus on the impact of natural experiments, i.e. the effect of the patent system before and after a change in patent policy is made.

Scherer and Weisburst (1995) and Challu (1995) studied the effect of 1978 Italian legislation which allowed the patenting of medicine and concluded that there is no acceleration of R&D in the Italian medicine manufacturing industry after legislation. Sakakibara and Branstetter (2001) examined the effect of the 1988 Japanese patent law reform that increased the patent scope in Japan. Their study showed that patent reform did not lead to an increase in either R&D spending or innovation output. Bessen and Hunt (2004) investigated the effect of a change of attitude towards software patenting in the 1980s and the 1990s and found that the number of software patents increased dramatically. Interestingly, they found that the firms that acquired relatively more software patents tended to reduce their level of R&D spending relative to sales. Based on the 1990–2002 data of 15,207 software companies, Lerner and Zhu (2007) studied the impact of software patent shifts and found that the increased reliance on patents was correlated with growth in measures such as sales and R&D investment.

Some researchers conducted multi-country studies. Lerner (2002) studied the impact on patenting activity of the changes in the patent systems in 60 countries over 150 years and found that foreign inventors increased their patenting but domestic inventors actually patented at an unchanged or even lower rate after the change. Using the same data set, Lerner (2009) investigated the impact of patent policy changes on innovation. He concluded that strengthening of patent protection had no positive impact on innovation and that enhancing patent protection was less effective when patent protection was already strong. Qian (2007) studied the changes in pharmaceutical patent coverage in 26 countries from 1978 to 2002. She found that changes of strengthening patent coverage for pharmaceuticals alone did not increase domestic innovation. However, she found these changes had positive effects on innovation in more developed countries with greater educational attainment and more market freedom. She also found that, at high levels of patent strength, additional strengthening measures actually decreased innovation. This seems to support the claim that a weak patent system stimulates innovation effectively.

The approach of studying patent policy change has its merit, but this approach cannot eliminate all confounding factors which might affect results. For example, Bessen and Hunt (2004) found that firms that acquired more software patents tended to reduce the ratio of R&D over sales. This result cannot be presented as software patents negatively affecting R&D spending. Since a firm is constrained by its budget, if it spends more to purchase patents, it has to cut funding on research, so budget constraint plays a key role here. Bessen and Meurer (2008) claimed that a large sample size with a number of countries over many years should be limited. This claim was based on the belief that positive confounding events tended to be offset by negative confounding events, but there was no evidence that this belief was reliable. Moreover, the inclusion of a large number of countries and the spanning of a long period of time will involve more variables both cross country and over time. This will lead to heteroscedasticity and autocorrelation problems and thus undermine the estimation results.

The estimation methods themselves may also make empirical results unreliable. Here we use the study of Lerner (2009) to illustrate this point. Lerner identified 177 events of patent policy change in 51 out of 60 nations and examined the impact of these events on patent filings. Britain was used as a reference country because its patent office had tabulated the national identity of the patent applicants and because its patent policy is relative constant. The changes in patent filing of foreign entities in the country undertaking policy change are used as an index of propensity to seek patent protection; the changes in patent filing in Britain by country undertaking policy change are viewed as the impact of policy change on stimulating innovation; and the changes in domestic patent filing in the country undertaking policy change are viewed as the combined results of propensity to file patent applications and increased innovations. These assumptions are reasonable.

However, to examine the impact of the events of patent policy change, Lerner used the changes in patent filing two years before and after the events. This short time frame may be good to examine the impact of patent policy change on the propensity to seeking patent protection, but is definitely insufficient to gauge the impact on innovations. If the increase in patent filing results from an increase in innovations caused by patent policy change, we need a longer time frame because innovation takes time. Furthermore, Lerner included weighted or unweighted average growth rates of patent filings in 10 nations to produce an 'adjusted' difference. This practice implicitly assumed that the country of patent policy change shared the same average trend of other countries. This is an implausible assumption and leads to significant estimation errors because the growth trend of each country can be significantly different from others.

Lerner also included in estimation dummy variables, 'positive patent policy change', 'strong protection prior to change' and 'strong protection x positive change'. He found that both 'positive patent policy change' and 'strong protection prior to change' have a positive impact on innovation, but the interaction dummy, 'strong protection x positive change', is significantly negative. Thus, he concluded that enhancing patent protection in a country with strong patent protection already was less effective. To reach this conclusion, we need an assumption that the independent variables are not correlated with each other (i.e. there is no multicollinearity in the model). This assumption apparently does not hold because the interaction dummy 'strong protection x positive change' is based on its components 'strong protection prior to change' and 'strong protection x positive change'. As a result, the conclusion is untenable.

Even if we ignore the confounding factors and assume that the estimation results are valid and precise, these valid results only show the impact of patent policy changes rather than on the patent system itself. For example, Sakakibara and Branstetter (2001) found Japanese patent reform did not affect innovation output, but this finding did not mean that the patent system itself is ineffective. One must be very careful in interpreting the implication of empirical results.

Based on the above review on empirical studies, we can conclude that these studies are useful in that they can give an impression or the magnitude of impact of the patent system, but the results from these studies are only indicative due to the issues in the data and in econometric methods. Since it is hard to find data on the number of innovations in an economy, some studies use the number of patents as proxy while some studies use the amount of R&D. These proxies do not necessarily have high correlation with the number of innovations. The time series data (e.g. the amount of R&D each year) may be affected by numerous factors (including unknown factors), so it is impossible for the researchers to claim that the results accurately reflect the effect of the patent system. The poor results from the survey data may also result from the shortcomings of the 'current' patent system, not from the principle of the patent system. For example, the reasons for low usage of patent information, as recognized by many researchers, are largely the short-comings in the patent registration process such as the incomprehensible legalistic language (Mandeville et al. 1982), the obscurity of patent titles (Murphy 2002), the ineffective patent classification (Desrochers 1998), and the low novelty and inventiveness criteria for issuing patents (Lunney 2001, Jaffe and Lerner 2004).

Because of the complication in data, one should not put too much faith in empirical results. On the contrary, if the logic is well grounded and straightforward, one should believe the logic because it is supported by the evidence in our lifetime. For example, our experience and economic theory tell us that printing substantially more money will lead to inflation. However, if we regress money supply with CPI, we may find money and CPI are not correlated well or even are negatively correlated due to many other contributing factors. In this case, it is more reliable to believe our common sense. When it comes to the effectiveness of the patent system, one can be sure that patent protection must have a positive impact on encouraging invention because the system gives the inventor high reward. This logic is correct and thus believable. The non-supportive results, on the other hand, might result from complication of data and thus are not reliable.

We also need to apply careful logical reasoning even for well-documented facts. Based on the survey of R&D managers by Cohen et al. (2000), Boldrin and Levine (2013) claimed market pressure (or first-mover advantage) is the main driver of innovation because more than 50% of managers think lead time is important in earning a return on innovation and less than 35% of managers (excluding the pharmaceutical and medical instruments industries) indicate that having patents is important. This claim is invalid because the survey did not indicate that lead time is more important than the perspective of filing a patent. Even if the survey does say so, the conclusion that first-mover advantage is the main driver of innovation is still invalid, because what the survey shows is the opinion of R&D managers who, as the representatives of the producers, largely rely on production rather selling patent licences to earn a return. If anyone had surveyed professional inventors, one can imagine that the answer must be that the patent is the main driver of invention.

## 6.6    Impact of the Patent System—Evidence from the Electric-Lamp Industry

The studies on the patent system are generally based on either models of specific conditions or empirical data of many limitations, so the conclusions from these studies cannot establish that the patent system has net positive or net negative effect. In other words, the question of the net effect of the patent system cannot be answered definitely by theoretical studies (there is no theory which can categorially show the net effect of the patent system is positive or negative), or anecdotal case studies (a few or even a number of cases cannot be generalized to the overall effect or to the majority of patents), or even statistical studies (the drawback of statistical studies on patents has been discussed in the previous section). The author suggests a systematic study of histories of different industries, including electronics, electricals, IT, pharmaceutical, transportation and even the traditional agricultural industries. From the history of an industry, the overall positive impact of the patent system can be shown by the influence of patented technologies and by the firms' behaviours regarding patents. Based on the excellent book by Bright (1949), the author briefly reviews the development of the electric-lamp industry, from which the reader can easily grasp the impact of the patent system as well as the power of patent monopoly.

Due to technological constraints, the traditional man-made light source in the very early days was gas or fuel lighting until continuous electric arc lighting was discovered in 1802 by Humphry Davy. Battery technology was improved greatly by a number of inventions around 1840, notably the inventions of the 1836 Daniell battery, the 1839 Grove battery and the 1842 Bunsen battery. Economic and reliable batteries provided a foundation for electrical lighting and generated enormous interest in electric arc lighting. However, the arc lamps at this early stage were cumbersome, complicated and expensive.

The discovery of electromagnetic induction in 1831 independently by Michael Faraday in UK and Joseph Henry in USA had an important impact on the electric-lamp industry, but the impact of this discovery was realized only in the early 1870s due to three important advances in the electricity generator. The first one was the 1845 dynamo by Charles Wheatstone, who replaced the permanent magnet with an electromagnet and used an electromagnetic field. Around 1867, Samuel Alfred Varley, Charles Wheatstone and the Siemens brothers (Werner von Siemens and Carl Wilhelm Siemens) independently used dynamos to energize the electromagnetic field. The third important advance was the use of ring winding by Zenobe-Theophile Gramme and Friedrich von Hefner-Alteneck in the early 1870s. With these advances, the dynamos generated more efficient and uniform electricity currents, and these restored the interest in arc lighting.

At this time, the importance of patents had not been shown because the inventions were too primitive—they were still far away from being used in daily life. In other words, these inventions were short of demand. However, the situation changed shortly thereafter.

The first permanent installation of arc lighting for general lighting purposes took place in the 1873 Gramme workshop in Paris. By 1877, arc lighting was popularly used in Europe as street and other outdoor lighting and as lighting in factories, but the arc-lighting device was complicated. Charles F. Brush made the dynamo system much simpler and more reliable. In 1877, Brush signed a patent licensing contract with the Telegraph Supply Company of Cleveland and he made the first Brush arc lamp in the same year. Later he made a series-wired arc-lighting system and other inventions. There are other inventors who improved the arc lamp and patented their inventions. These inventions interested capitalists and they formed companies to manufacture the apparatus. To increase sales, these manufacturing companies supported the formation of local companies to install electric light by giving them the patent licences and supplying equipment.

Despite the successful commercialization of arc lamps, the drawbacks of these lamps limited their future. Although arc lamps are very cost effective, they are generally too bright for a private room. Arc lamps also need a high voltage system, large currents and frequent adjustments. These disadvantages necessitated the development of incandescent electric lighting.

The first demonstration of incandescence induced by electricity is generally regarded as also by Humphry Davy in 1802.

The development of batteries around 1840 also aroused interest in lighting by incandescence. In 1841, the first patent on an incandescent lamp was granted to an Englishman, Frederick De Moleyns. In 1845, the British government also granted the young American J.W. Starr a patent on the incandescent lamp. In 1852, M.J. Roberts made a vacuum lamp with graphite illuminant in a pear-shaped glass bulb. In 1854, H. Golel made a lamp with a carbon thread as illuminant in a sealed vacuum bulb, but he took out no patents and published no papers and made no public claim on his invention. In 1859, Moses G. Farmer made a platinum-strip lamp to be operated in the open air. Various other incandescent lamps were made by others. These early lamps generally used platinum, iridium or carbon conductors, which become incandescent as the current passed through them. These incandescent lamps burned only for a short time because of the reaction of oxygen in the air. There were attempts to obtain a vacuum in the bulb or to fill the bulb with nitrogen, but the results were unsatisfactory due to the limited techniques in making a high-quality vacuum. Moreover, lamp design and assembly techniques were also not good enough. By 1860, these shortcomings led to the abandonment of incandescent lighting. However, the invention of a superior mercury vacuum pump changed the situation. The pump invented by Hermann Sprengel in 1865 solved the air-exhaust problem. The British scientist Sir William Crookes used and perfected the pump to exhaust the air in a glass bulb in 1875. The invention and improvement in the pump led to the return of interest in incandescent lighting.

With the perceivable demand and the protection from patent laws, many inventors made various lamps after 1875. For example, in 1878, Hiram S. Maxim made a lamp with a graphite illuminant in hydrocarbon vapour, while St. George Lane Fox-Pitt patented a lamp which was made of loops of platinum–iridium wire in a glass tube filled with nitrogen or air; in 1879, Moses G. Farmer made and patented a lamp with a graphite rod in a glass bulb (exhausted or nitrogen-filled), while William E. Sawyer and Albon Man made a nitrogen-filled horseshoe-type lamp with a carbon rod. Joseph W. Swan made a vacuum lamp with a very slender carbon rod and made a successful demonstration in 1878, but he was slow to take out patents and later was disadvantaged by his mistake. It was Thomas A. Edison who made the greatest contribution to incandescent lighting. From 1877, he experimented with a number of materials as illuminants. In 1879, Edison and his men tried hundreds of different forms of carbon and finally, on October 19, they succeeded in carbonizing a piece of cotton sewing thread by heating it to very high temperature without the presence of oxygen. The carbon-filament lamp burned for almost two days. On November 4, Edison applied for patents in America, Britain, Canada, France and other countries to cover his cotton-thread filament lamp. The American patent No. 223,898 was granted in January 1880. This is a basic patent which was later proven to be very powerful. On 11 December 1879, Edison patented a similar lamp which used carbonized paper. He formed the Edison Electrical Lamp Company late in 1880 and publicly demonstrated the lamp on New Year's Eve. Edison continued to search for the best carbon filament until he found the bamboo filament in 1894.

After the success of inventing the incandescent lamp, Edison worked hard in perfecting the incandescent lighting system, including supporting the central-station generation and distribution of electricity, improving the efficiency of the dynamo from 50% to almost 90%, inventing and patenting the system of multiple distribution of electricity current, selecting the voltage for his system and the metre to measure the energy consumption of each consumer. He formed a new company, the Edison Lamp Company, to manufacture lamps patented by the Edison Electrical Light Company.

Edison's success prompted a number of companies to manufacture electric lamps. Their production in the early years was relatively small, but their output was increasing. More importantly, many of the competitors' lamps were similar to those patented by the Edison's company. The company decided to take legal action. Between 1885 and 1901, the Edison's company and its successors spent the large sum of about $2 million on over 200 infringement suits under its lamp and lighting patents. Defendants of the lawsuits might have spent a similar amount. The lawsuit results were mixed, but the 1891 and 1892 victory of the Edison's company on the basic patent No. 223,898 against the US Electric Lighting company established the monopoly position of the Edison's company and its successor General Electric.

The General Electric obtained injunctions against the producers and users of competing and infringing lamps. The Beacon Vacuum Pump & Electrical Company of Boston tried to avoid an injunction by claiming the priority of the invention by Heinrich Gobel, but the judge ruled that the evidence supplied was insufficient to invalidate Edison's patent. Within a short period of time of injunction, lamp plants of many companies had to be closed. Other companies closed their plants for fear of General Electric's legal action against them. The expiration of the Edison basic patent in 1894 saw a dramatic increase in competition. As a result, the price of lamps dropped sharply. For example, the price of 16 candlepower lamps was $1.00 from 1880 to 1886, $0.50 in 1893, $0.18–$0.25 in 1895 and $0.12–$0.18 in 1896.

As the companies in the electric-lamp industry were fighting in lawsuits and in production, the effort for improving incandescent, gas and arc lighting continued. Largely due to the attempt to get around the basic Edison patent, the search for non-carbon filaments was intensive. In 1883, F.G. Ansell tried aluminium, magnesium and oxidation of calcium on carbon filaments; in 1886, Max Neuthel was issued a German patent for a filament of magnesium and porcelain clay saturated with 'platinum-iridium' salt. In 1889, Turner D. Bottome took out a patent on a process for making a composite carbon and tungsten filament and another patent on carbon and molybdenum. In 1890, Lawrence Poland obtained an American patent on an iridium-filament lamp. After losing the lawsuit against the Edison basic patent, the Westinghouse company employed Alexander Lodyguine to develop a non-infringing lamp. Lodyguine attempted to coat carbon, platinum and other metallic cores with various metals, but was not successful. In 1892, F.M.F. Cazin obtained the first of his series of patents on the incandescent lamp. In 1889, Thomson-Houston Electric Company brought in Rudolf Langhans to develop a substitute for the carbon filament and he obtained a patent in 1894 for his process of chemically combing carbon, silicon and boron in varying combinations. These attempts of finding a better substitute for carbon filament were largely unsuccessful due to the constraints of technology at the time.

Meanwhile, the invention of the Welsbach gas mantle in 1883 reduced the cost of gas lighting by two-thirds so gas lighting became cheaper than electric lighting. However, the technical problem such as the cost, efficiency and life of the mantle prevented its wide use until the early 1890s. In arc lighting, William Jandus applied in 1886 and was granted in 1891 a patent on an arc lamp with an enclosed chamber, which not only increased the life and energy efficiency of the lamp, but also made the arc lamp suitable for indoor lighting. Compared with gas lighting and arc lighting, the carbon-filament incandescent lighting had a very low energy efficiency.

Due to the competition from the gas and arc lighting, the search for a non-carbon filament continued. The advances in inorganic chemistry, especially of the rare earths, and the 1892 invention of the electric furnace by Henri Moissan improved the understanding of heavy metals. The electric furnace had the advantage over the gas furnace because the former could obtain a much higher temperature, with controlled atmospheres. With the new knowledge and a new tool, many non-carbon lamps were made. In Germany, Walther Nernst used a small rod of refractory metallic oxides as an illuminant to make a lamp in 1897; in Austria, Carl Auer von Welsbach used a squirting process to make osmium filaments in 1899 and commercially sold the lamps in 1902. This lamp was patented in Europe and in the USA. In Germany, a new type of incandescent lamp was developed by Werner von Bolton and O. Feuerlein who worked in the Siemens and Halske Company. In 1901, they began to study rare metals such as vanadium, niobium and tantalum. In 1902, they applied for and later obtained German and other patents to cover the process of making pure and ductile tantalum. In 1905, the new Siemens & Halske lamp was on the market in Europe. In 1906, the General Electric Company and the National Electric Lamp Company acquired exclusive rights to make and sell tantalum lamps in the USA. For this exclusive right, the two companies paid $250,000, in addition to royalties on all tantalum lamps sold. These lamps were expensive because of the rareness of osmium and tantalum.

The high melting point and cheap price of tungsten attracted the attention of inventors in Europe, but tungsten is too brittle to be made into a fine wire for electric lamps. A number of inventors tackled this problem. Alexander Just and Franz Hanaman developed two chemical processes for making tungsten into fine wire. Hans Kuzel developed a slightly different process to make tungsten wire. Fritz Blau and Hermann Remane of the Austrian Welsbach Company developed a process of sintering the pressed metallic powder. Von Bolton at Siemens & Halske developed purified ductile wires of different metals, including tungsten. All inventors applied for patents in a number of countries.

In the USA, the research for new filaments prompted the General Electric Company to set up a research laboratory in 1900 to improve the carbon filament. Although a patent on improved carbon filament was applied for in 1904 and was granted in 1909, it was proven to have little commercial success. General Electric commenced research on tungsten filaments when the potential of the tungsten filament lamp was apparent, but it was too late to make a contribution to the non-ductile tungsten filament. Given the great potential of tungsten filament lamps, General Electric was eager to secure an exclusive right to produce and sell them. However, all patent applications were still pending. Not knowing which application would be successful, General Electric started to buy all American patent applications on tungsten filament lamps from all European inventors. This costs the company about $760,000, but it was proven later as an extremely profitable investment. A few years later, the laboratory at General Electric made an important contribution to ductile tungsten filaments. In 1912, William D. Coolidge applied for an American patent for the process of producing ductile tungsten filaments through repeated heating and hot swaging. The American patent was granted in 1913, and the patents in other countries were granted in subsequent years.

The patents helped General Electric to maintain its market power and its position as industry leader. General Electric also enhanced its market power through other methods. One was to enter a general cross-licensing arrangement with the Westinghouse Electric & Manufacturing Company—the second largest company in the industry. The other method was, by providing new working capital, to obtain control over the National Electric Lamp Company—the federation of the small companies in the lamp industry. In this way, the General Electric reduced market competition considerably but, eventually in 1924, led to antitrust action.

The big disadvantage of incandescent lighting was its low energy efficiency. Even the best performance of ductile tungsten filaments was 32.7 lumens per watt in large aviation lamps. However, this disadvantage of incandescent lighting was outweighed by its safety, simplicity and reliability. With technological advances, the safety, simplicity and reliability of other electric lamps become comparable, so the low energy efficiency of incandescent lamps became a predominant issue and most incandescent lamps were replaced by electric-discharge lamps.

The discovery of the electric-discharge light has a long history. First, we consider the development of electric-discharge non-fluorescent lamps. In 1683, Otto von Guericke obtained light from the discharge of a primitive static-electricity machine. In 1700, Newton and Hawksbee found that the interior of the 'exhausted' glass globes would glow as the electrical charges were built up. In 1856, a German artist and glass-blower, Geissler originated the electric-discharge tube which gave off light for a while until the vacuum in the tube deteriorated. The first UK patent was issued in 1862 for using Geissler tubes filled with various gases and vapours for signalling and for lighting buoys. In 1866, a British patent on battery-operated nitrogen-filled Geissler tubes for buoy lighting was given to Adolphe Miroude. Up to 1896, the electric-discharge tubes were energy efficient but had a very short life. D. McFarlan Moore made an important contribution to the electric-discharge tubes by devising an automatic valve. The valve allowed gas to flow into the tube if the pressure became too low. This improved the life of the tube dramatically. Peter Cooper Hewitt invented the mercury-vapour lamp in 1901. Georges Claude demonstrated his first neon sign at the Grand Palais in Paris in 1910 and later he patented his invention and made a commercial success in the USA. In the early 1930s, the sodium-vapour lamps and the high-pressure mercury-vapour lamp were invented and entered the market.

The other technological advance in the electrical lighting industry was the development of fluorescent lamps. The discovery of fluorescence occurred as early as 1602 when an Italian cobbler discovered that certain rocks glowed after receiving light. In 1852, Sir George G. Stokes discovered that ultraviolet light induced fluorescence in various substances. The first electric fluorescent lamp was made by Alexandre

Edmond Becquerel in 1859. Thomas A. Edison also applied in 1896 and was granted in 1907 a patent to cover a short vacuum tube with an inside calcium-tungstate coat. In 1923, Jacques Risler introduced fluorescent powders into neon-type tubes and patented his invention in 1925. Three engineers, Friedrich Meyer, Hans Spanner and Edmund, at the Rectron company in Berlin made the first low-voltage gaseous-discharge device in 1926 and applied for a German patent in the same year and an American patent in the next year. Their lamps entered the market in the early 1930s. In 1927, Albert W. Hull at the Schenectady Research Laboratory of General Electric invented and patented a discharge device which permitted the use of mercury vapour with pressure ranging from one to one thousand microns. However, General Electric did not realize the potential of fluorescent lighting until 1935.

Once General Electric realized that fluorescent lighting was the future of the lighting industry, it put a lot of effort into developing a practical fluorescent lamp. This lamp was placed on the company's special listing in 1937 and was first commercially produced for the New York and San Francisco World's Fair in 1938. Despite the substantially higher efficiency of the fluorescent lamp, the company was reluctant to produce fluorescent lamps on a large scale for a number of reasons. First, the complacency of its market position made it hard to adopt a new direction. Second, the company had vested interest in preventing fluorescent lamps from replacing its existing production of incandescent lamps. Third, the company has a cosy relationship with the electricity utility companies. The high efficiency of fluorescent lamps meant lower consumption of electricity and this worried the electricity utility companies.

The competition from a small lamp company named Hygrade Sylvania Corporation accelerated the entrance of the fluorescent lamp into the US market. As the largest of General Electric B licensees, Sylvania was unhappy with the very small quota of incandescent lamps allowed by General Electric and saw fluorescent lighting as a chance to gain a greater market share. The engineers of Sylvania had experimented since 1931 and made a fluorescent lamp in 1934, but the lamp had a short life so Sylvania abandoned the experiments. In 1935, the news that General Electric had successfully developed a fluorescent lamp spurred Sylvania into resuming its experiments. In 1938, Sylvania started a production line of fluorescent lamps, backed by its own Cox patent on the method of applying fluorescent powders and the exclusive licences under the two patents of the Raytheon Manufacturing Company.

In the spring of 1940, General Electric offered Sylvania a new B licence covering fluorescent lamps but Sylvania declined the offer. In May, General Electric instituted a patent infringement suit against Sylvania, which denied the allegation and instituted a countersuit of patent infringement. The lawsuit was long and complicated. The court hearings started in 1942. By then the antitrust action against General Electric initiated in 1939 had not finished so General Electric did not come to the court with a 'clean hand'. On this basis, the antitrust division of the US Federal Government intervened and filed a new complaint against General Electric. In 1944, the judge declared that both patents put forward by Sylvania were invalid in their entirety and that, with the exception of the broadened claims, two patents of General Electric were valid and had been infringed by Sylvania. However, the judge delayed the final decision until the antitrust action was finished. In 1948, the court found that General Electric and its licensees had violated the Sherman Act. In 1953, the remedial order required that all lamp patents be dedicated to the public without compensation. By 1956, nine compulsory licences had been issued by General Electric.

From this review of the electric-lamp industry, we can tell the vital role of innovations and patents as well as the powerful impact of patents, both positive and negative. Although many patents have been granted during the course of industry development, only a small number of them have been proven to have high economic value. The very high economic value of some patents brought considerable market power and super profit for the patent holder and the exclusive licensees. The high economic returns of a small number of patents inspired inventors and thus stimulated new innovations. This can be shown by the awareness and participation of the patenting of both the independent inventors and inventors employed by companies in the electric-lamp industry. The reader can apply the systematic historical approach to other industries of interest to appreciate the importance of innovations and patents.

## 6.7     Towards a New Patent System

Based on the previous discussion, the double-edged sword nature of the patent system leads to the heated debate about its effectiveness and efficiency. The currently dominant balanced approach to patent design is a compromise. More monopoly power stimulates more innovation but will cause a higher deadweight loss while less monopoly power does the opposite, hence a balanced approach to designing patent system will reduce both positive and negative effects. Based on this approach the current patent law imposes limitations on patent duration and breadth in a hope to solve the problems caused by patent monopoly power, but the consequence is that the problems remain while the incentives to innovate decrease. An optimal patent system needs a totally different approach: to increase the monopoly power to the inventor but also to decrease the social cost. This seems an impossible task. However, after we have clarified the purpose of the new patent system and illustrated the rationale for the new patent system design, we will find a feasible way to transform the current patent system to achieve this task.

### 6.7.1     Purpose of the New Patent System

The purpose of the current patent system is twofold: to stimulate innovation of new technology and to promote the diffusion of new technology. These two objectives are conflicting tasks. In order to stimulate innovation, the system must confer on the inventor strong monopoly power which, unfortunately, will hinder the diffusion of patented technology. On the other hand, patent monopoly power should be reduced so as to encourage diffusion of new technology but, in so doing, the stimulus for the inventor is greatly reduced. Bundling these two tasks as the dual objectives of the patent system compromises the effectiveness and efficiency of the system.

Besides the dual-objective issue, it is also beneficial to clarify some common misunderstandings regarding the purpose of the patent system. One such misunderstanding is that an optimal patent system should allow the inventors just to recoup their research cost for the invention. The rationale behind this argument is as follows. Allowing the inventor to recoup the innovation cost would provide enough incentive and is fair to the inventor; meanwhile, so allowing the inventor to obtain a profit higher than the innovation cost will cause unnecessary loss of consumer welfare. The flaw in this argument is that it overlooks an important feature of innovation activity—the high possibility of innovation failure. A successful innovation results from numerous failures by the ultimately successful inventor and by many other unlucky ones so, from the point of view of society, the cost of an innovation should include the costs of all successful and unsuccessful innovators. Moreover, the chance of success is unknown when one starts innovation activity, so the uncertainty faced by the inventor also needs to be factored into the cost calculation.

More importantly, fairness is not the main concern of the patent system. It is mainly about stimulating more innovations for society. From this point of view, the more rewards to the inventor, more people being attracted to innovation activities, so the better outcome for society. The purpose is necessary to give high rewards to the innovator because innovation is the bottleneck of our economic growth. As we discussed in Chapter 5, the repeated occurrence of economic recession indicates that our society severely lacks innovations, so we need to provide the highest possible rewards to innovators in order to save our economy from economic recessions.

The other misunderstanding regarding the purpose of the patent system is that the system should provide a balance between patent monopoly and competition. It is well known that competition is essential for a market economy and that a monopoly will cause social welfare loss. Patent right is essentially a monopoly power, so encouraging competition seems a task in conflict with patent rights. As a result, this reasoning leads to the compromising nature of the current patent system. However, if we change our perspective in considering patent monopoly and competition, we may have a desirable new solution. That is, if we can confine patent monopoly power to encouraging innovation activity only, the production activity can still remain highly competitive. In this way, patent monopoly and competition working in different spaces cause no conflict.

In considering the crucial needs of our economy for innovation and the high uncertainty of innovation activity, the focus of the new patent system should be to stimulate innovation activity. Since this purpose conflicts with the task of diffusion of new technology, the latter should not be the task of the new patent system. This seemingly single-minded approach is necessary because the main objective of the patent system should not be compromised by other tasks (e.g. improving competition and technology diffusion). On the condition that the objective of stimulating innovation can be achieved, other considerations can be addressed by suitable design of the patent system. In order to reduce the negative effect of the patent system, patent monopoly power should not be applied to production activity. This can be achieved by forbidding exclusive patent licences. If we can fulfil the purpose of the new patent system, it would be a great outcome for society.

## 6.7.2  Rationale for Designing the New Patent System

The mechanism of the current patent system is to allow inventors to profit from their inventions by prohibiting others from producing patented products. Traditionally, inventors will profit by producing and selling the exclusive products they have invented, so inventors are rewarded through their engagement in production activity. However, this type of reward system is suitable only for the technologically underdeveloped period in history, during which both invention and production were relatively simpler, took less time and required less specialty.

In modern times, both inventing and producing patent goods may take a long time and require substantial capital, knowledge and effort. Moreover, managing the production process may not be the strength of the patentee, whose strength is more likely to be on the invention of new products. As a result, the patent system provides various types of patent transactions such as transfer of patent rights, assignment of patent rights, and exclusive and non-exclusive patent licensing. These transactions are consistent with the definition of patent rights in the current patent system. However, not all these transactions are beneficial to society.

Transfer of patent rights involves a change of patent ownership, so the monopoly power of the patent will be transferred but will not be amplified. Non-exclusive patent license does not involve any change of monopoly power of the patent. These two transactions will not lead to an enlarged social cost because the monopoly power does not increase. However, it is a different story when it comes to the other two types of patent transactions.

Assignment of patent rights enables the monopolistic patent rights to be transferred to many people other than the patent holder, so the monopoly power is magnified but, in the meantime, the benefit to inventors has become very limited. Although the assignment of a patent right brings some revenue to the patent holder currently, the patent right signed out reduces the chance for the patent holder to profit through the granting of patent licences in the future. Exclusive license does not transfer the monopoly power of the patent right, but this practice enables the licensees to have monopoly power in producing patent products in different regions, so the patent monopoly power is also magnified. Exclusive licences also limit the benefit to the patent holders as they are unable to grant any licences in the regions where exclusive licences have been granted.

In short, magnified patent monopoly in both cases (i.e. assignment of patent right and exclusive patent licences) necessitates a magnified deadweight loss (or social cost) but has no significant positive impact on stimulating invention, so the magnified patent monopoly in both cases is undesirable. As a result, both exclusive patent licences and assignment of patents should be banned in order to avoid the unnecessary social cost.

Since the practice of both exclusive patent licences and assignment of patents originates from the definition of patent right, the patent right needs to be redefined in the new patent system. The key to the new definition is to distinguish innovation activity from production activity, i.e. to separate the role of the innovator and the role of the producer. Although there are many links between innovation activity and production activity, in the modern economy these two types of activities are separable specialized activities.

If the new definition of patent right can restrict the application of the patent monopoly right to production activity, it will not cause social welfare loss in the production of patented goods, and thus the downside of the patent system is overcome (or at least decreased substantially). Although patented products will still have much higher price due to patent license fees, the higher price paid by consumers is fully awarded to patent holders, so it is fully used for stimulating innovation. As a result, society will benefit much more from the innovations in the future, so the higher prices for patented products are justified. Under this condition, it is desirable to increase the patent monopoly power to stimulate innovation. Thus, the restrictions on patent length and breadth should be abolished. Patent right transactions should also be standardized to reduce transaction cost and to foster a functional patent market.

### 6.7.3 Necessary Reforms to Establish the New Patent System

The rationale for designing the new patent system forms the foundation of patent reform. The undesirable amplification of monopoly power stems directly from the broad definition of patent right in the current patent system, so a new patent system must start with redefining 'patent right'. Several other changes to current patent laws are also necessary. These are now discussed in turn.

1. Redefining patent right

Current patent laws grant patentees the exclusive right to implement their innovations. For example, in the USA, patent rights are defined as the rights granted to inventors by the Federal Government, pursuant to its power under Article I, Section 8, Clause 8, of the US Constitution, that permit them to exclude others from making, using or selling an invention for a definite, or restricted, period of time. Similarly, the TRIPS Agreement—Article 28.1 gives patent holder the exclusive right 'to prevent third parties not having the owner's consent from the acts of: making, using, offering for sale, selling, or importing'. This broad definition is subject to interpretation and can lead to the abuse of patent monopoly. Under this patent rights definition, patentees are able to grant all kinds of licences and sign all sorts of agreements, or even refuse to grant a licence without giving a reason. Moreover, this definition rewards invention through granting exclusive production so it is the role of producer that is rewarded, rather than the role of the inventor. In other words, this definition of patent rights encourages monopoly in production. For example, some inventors do end up being rewarded very highly, such as the creators of Microsoft, Apple and Google, but their rewards come from their roles in producing new products rather than their roles as inventors. The implication of this rewarding mechanism is that, in order to maximize the benefit of an invention, the inventor has to become a producer or a manager.

To prevent the abuse of patent monopoly power and to reward invention directly, patent right should be redefined as 'the right to exclude others from distributing or commercially producing an invention unless they obtain non-exclusive permits from the patent holder'. Putting it differently, patent right is the exclusive right of patentees to grant non-exclusive licences for distribution or commercial production of patented products. This new definition clarifies and confines patent rights, so there would not have many obstacles from society (most people worry about patent monopoly power). However, the definition indicates a radical change in the patent system and requires a revision of both the national patent law and the TRIPS agreement. For example, the TRIPS Agreement—Article 28.2 gives the patent holder the right to assign the patent. This would not be allowed in the new patent system.

This redefinition of patent rights has a number of implications.

First, under the new definition, monopoly power stays with the patent holder and this makes the patent system more effective in stimulating innovation. The current patent system permits exclusive patent licence and patent assignment so the patent monopoly power can be transferred to different agents so they all can obtain economic rents from monopoly power. Under the new definition, only non-exclusive patent licences are allowed, so only the patentee has the monopoly right. As a result, any economic rents from the monopoly power are awarded to the patent holder. If the patent holder is the inventor of the patented product, this reward to the inventor will encourage him/her and other inventors to conduct more experiments and result in more innovations. Even if the patent holder is not the inventor (e.g. the patent is sold to a company by the inventor), the strengthened monopoly power of the patent holder, thanks to the new definition, is likely to increase the market value of the patent before the patent transaction is made. Thus, the new definition will increase the patent value and thus benefit the inventor. Since the inventors are rewarded greatly under the new definition, the new patent system will be effective in encouraging innovation.

Second, the new definition helps to overcome the problems of abusive use of the patent monopoly power such as excessively high patent prices and sleeping patents. The abuse of patent rights under the current patent system has outraged the public and worried experts. The phenomenon of abuse of patent rights to a large extent is related to exclusive patent licence and patent assignment. This can be demonstrated by the outrageously high patent prices in some cases. In the current patent system, exclusive licence and patent assignment are used, so the patent holder has only a limited number of chances to make a deal (e.g. one cannot sell two exclusive licences in the same region). This one-off nature of exclusive patent licence prompts the patent holder to ask for extremely high prices. It is also arguable that the price of an exclusive licence includes the price of some patent monopoly power to be granted to licencees, so its price is much higher than a non-exclusive licence. Patent assignment also involves partial transfer of monopoly power from the patent holder, so the price of patent assignment is also very high.

If exclusive licences are banned in the new patent system, patent holder will be less cautious about patent licence price because he/she has more chances to sell the licence again and to set at another price later according to market responses. To maximize profit, the patent monopoly will still set a high price, but not an extremely high price because a too high price will lead to a very limited number of buyers and thus reduced profits. As a monopoly, the patent holder needs to price a suitable price to maximize profit. This reasoning is consistent with the situation in the current and past monopoly markets, e.g. the iPhone market, the oil market and the electricity market. The prices in these markets are high, but not excessively high, so we should expect that similar situation would happen in a functioning patent market.

In short, the redefinition of patent rights gives the patent holder unlimited chances to sell their licences and creates a thicker patent market. With a functioning market, an excessive price is unlikely.

Third, the new definition does not allow for monopoly in the production of patented products and thus greatly reduces the social cost of the patent system. Social cost is a key factor in the currently heated debate on patents. The social cost of a patent directly stems from the monopoly in production of patented production. Under the new definition of patent right, only non-exclusive patent licences can be issued, so nobody has a monopoly power in producing patented goods. Since the production of patented goods will be under the condition of a competitive market, producers are unable to add extra margin for themselves (otherwise they will lose customs to other producers who supply at a lower price). The cost of non-exclusive licences will be added to the price of patented goods, but this cost is counted as social cost of the patent licensing market, which will be overcompensated by increase innovations in the future. Since producers charge no extra cost, there is little or no social cost in the production process.

It is arguable that the patent holder could have monopoly power in production by offering an outrageously high price for the patent licence, at which nobody will take up a patent licence. This kind of situation is possible when the new patent system first becomes effective, but it is unlikely when patent holders are used to the new patent system and their behaviour becomes more rational. The reasoning is as follows. The goal of the patent holder is to make a profit either by engaging in patent licensing activity or by monopoly production. If a rational patent holder intends to have monopoly power of production rather than to license the patented product, this must indicate that the profit from the monopolistic production of the patented product is higher than that from granting licences. If the licence fees or licence royalties are higher than the profit (and fixed cost in some cases) from producing the patented products, why would an inventor bother to take significant amount of time, money and effort to produce patented products, which is not the strength of an inventor, involves market and imitation risks? A rational patent holder would prefer licensing to producing. Irrational behaviour may exist when a policy is just established, but rational behaviour will dominate afterwards. Eventually, the market will determine the price of patent licences.

Fourth, the new definition separates the role of innovator from the role of producer and encourages specialization. Under the new definition of patent rights proposed here, innovation is rewarded directly through the right to grant patent licences, rather than indirectly through producing patented products. Separating the two activities of invention and production allows the innovation/research department to be an independent, self-financed entity, and thus leads to the more efficient use of specialized human capital.

Last, the new definition may also help to solve the problem of sequential innovation. Based on the current definition of patent right, even with a research exemption, a downstream invention cannot proceed without patent permissions being obtained from the upstream inventor because the patent invention is clearly for commercial purposes (i.e. to invent and sell patent licences) and thus is not qualified for research exemption in the current patent system. Scotchmer (1991) suggested a prior agreement between upstream and downstream inventors to divide the receipts from selling the patent licences to producers, but this suggestion is neither feasible nor efficient because of the high cost of negotiation between inventors when market information about the patent licences is very limited.

However, if the new definition of patent right is in place, the downstream inventor does not need to obtain a licence from any upstream patentee for conducting invention because the downstream inventor is not commercially producing the products patented by the upstream inventor. As such, the downstream innovation is not affected by upstream patents. If the downstream invention is successful and the inventor wants to sell the patent, the downstream inventor does not need to obtain a licence from the upstream inventor either. Instead, it is the production firms that need to obtain patent licences from all inventors involved in the new products. In this way, the rewards to upstream and downstream inventors are determined by the negotiation between inventors and producers, or by the market value of their inventions. The prior agreement between upstream and downstream inventors, as suggested by Scotchmer (1991), becomes unnecessary.

It is worth mentioning that there are still issues in licensing incremental inventions. Because the producer needs to obtain patent licences of both basic and incremental inventions separately from different inventors, this will increase the transaction cost. The author envisions that the market will work out a way to improve efficiency in this area. For example, there could be patent licensing agencies who have better information about both the supply of and demand for patents (just like a real estate agent has more information on housing prices and rental prices). These patent licensing agencies may obtain the rights to license patents on behalf of all basic and incremental inventors.

Another possibility for improving the efficiency in incremental patent licensing is the use of patent pools. The current patent pools have both benefits and drawbacks, which are realized by many researchers (e.g. Priest 1977; Carlson 1999; Merges 1999; Shapiro 2000; Gilbert

2002; Lerner et al. 2003). We will discuss the drawbacks of patent pools later, but here we focus on the benefit of patent pools in overcoming the problem of 'patent thicket'—the high transaction cost due to a large number of overlapping patents or fragments of patents. The high transaction cost may prompt some companies to form a patent pool by obtaining all basic and incremental patents necessary to produce a patented product. The companies can sell patents separately to the producer according to the market prices, or sell a bundle of patents necessary for production. This type of patent pool can save producers the large accumulation of costs from separate negotiations.

## 2. Forbidding both patent assignment and exclusive patent licences

Both patent assignment and exclusive patent licences must be proscribed because they magnify the monopoly power in production and thus cause large social costs. It may be argued that patent laws should not ban the granting of exclusive licences, because this imposes constraints on the way the patentees cash in on their inventions. However, this is a common misperception. Normally it is the producer who desires an exclusive licence because the producer likes the monopoly power of setting the price for the patented product, i.e. the exclusive patent licence actually is in favour of the licensee. Moreover, the exclusive licence actually imposes greater constraint on the patentee. After granting an exclusive licence, the value of the patent will decrease substantially because the patentee is no longer able to grant another licence in the same region. If exclusive licensing is banned, licensees are no longer able to seek an exclusive licence. This means the patentee has unlimited opportunities to grant licences later and thus the patentee has the potential to make more profit.

Banning exclusive licences does place restriction on patent holders who prefer income today, e.g. one could not become rich overnight by selling one expensive exclusive licence. However, the purpose of the patent system is to stimulate innovations, rather than satisfying the desire of innovators. The total returns to innovators is the key for encouraging future innovations. This restriction on the patent holder is more than compensated by both private and social benefits, e.g. unlimited patent licensing opportunities, reigning in extremely high licence price, preventing extension and spread of patent monopoly power, reducing transaction costs.

Banning exclusive licences can encourage patent licensing activities. The patent holder can grant an unlimited number of non-exclusive licences and he/she may be less cautious/hesitant at deciding the licence price, which can be changed later according to market demand. Banning exclusive licences also has the potential to reduce the patent licence price (a non-exclusive licence will be much cheaper than an exclusive licence) and thus to reduce the prices of patented products due to more competition in producing patented products. The reduced patent licence price will not affect the income of the patentee negatively because of the increased number of non-exclusive patent licences to be granted. On the contrary, non-exclusive licence may increase the patentee's income. The patentee can set the patent licence price according to market demand so as to maximize his/her profit. The reduced prices of new products, thanks to the reduced licence fee and increased competition in production, will also considerably benefit consumers.

Banning patent assignment and exclusive patent licences may also help to solve the problem of patent thicket (an overlapping set of patent rights) and reduce the social cost of patent pools. Patent pools in history have been used to monopolize an industry, e.g. the patent pool in the sewing machine industry in 1856 and the one in the electrical lamp industry in 1892. Government also used patent pools to develop an industry. For example, in 1917 the US government forced the two major owners of key airplane patent, the Wright Company and the Curtiss Company, to form a patent pool—the Manufacturers' Aircraft Association. Patent pools have the benefit of reducing negotiating cost and transaction cost, but too many patent pools will be destructive and potentially hold up an industry, so it is desirable to reduce the number of patent pools in any industry.

A large part of monopoly power of a patent pool stems from patent assignment and exclusive licences. For example, by a cross-licensing agreement, two or more companies can share their patents and form a group to discriminate against companies outside the group. Because patent assignments and exclusive patent licenses provide the receiver with patent monopoly power in different regions, the tentacles of monopoly can reach all corners of the world. This increases the problem of patent thicket and thus the numbers of patent pools. If patent assignment and exclusive patent licences are banned, the only way to form a patent pool is to buy up all relevant patents. Because the monopoly right in the new patent system belongs only to the patent owner and thus one patent can contribute to only one patent pool, banning patent assignments and exclusive patent licences will reduce the number of patent pools to a great extent.

3. Standardizing patent transactions

Under current patent laws, a patent transaction is implemented by an unregulated contract. Given the open-ended nature of these contracts, many inappropriate contents may be included and thus may lead to various abuses of patent monopoly, such as the tie-in problem—the practice where the patent holder inserts some clauses (e.g. requiring the licensee to purchase the goods nominated by the patent) in a licensing contract so as to extend their monopoly power. In order to avoid various abuses of patent monopoly, the new patent law needs to simplify and standardize patent licensing.

Patent licensing under the new patent system would be very simple. Even a statement 'I grant xxx the right to use the method/technology of patent No. xxx from xxx date to xxx date' would be sufficient. The style of a patent licensing agreement may vary from country to country, but the key is that standard licence agreements should allow contents only related to the patent rights, such as permission for the use of patent technology in return for a royalty or annual fee.

With a standardized patent licence agreement only involving granting the right of using patent technology, an open price for patent licence becomes possible, so buying a patent licence in the future can be as easy as ordering a good or service in a shop or online. This will greatly reduce the patent transaction cost.

In reality, in order to implement the patented technology, the licensee may require technical assistance from the patentee. However, this kind of cooperation or supporting agreement does not involve the right to use the patented technology, so it can and should be included in a separate agreement. The advantage of using a separate technical support agreement is that any attempt to extend patent monopolistic power will be delinked from patent licensing and thus will eventually fail.

The simplicity of patent licensing would also greatly reduce the transaction costs and thus improve market efficiency. High transaction costs and information asymmetry (i.e. one side has more information than another side) are features of current patent transactions. The complexity of current patent licensing and the extended monopoly power of the patentees are important contributors to high transaction costs. Standardized patent licences can reduce the negotiation cost and prevent the patentee from using information asymmetry to their advantage. The separation of the patent licensing agreement and the technical support agreement can reduce the information asymmetry problem in the technical support agreement because the licensees are given the chance to obtain more information from other technical support service providers than from the patentee.

4. Prolonging the duration of patent rights infinitely

Due to concerns regarding the negative effects of patent monopoly, current patent laws grant only temporary patent rights. The duration of patent right varies from country to country and depend on different kinds of inventions, but these durations have been harmonized through the introduction of the Unified Patent by the TRIPS agreement. Currently, the maximum patent duration is generally 20 years, with medical patents being 5 years longer. The duration for an innovation patent or utility model varies from country to country, but is much shorter than 20 years. For example, the Australian innovation patent is valid for up to 8 years.

As demonstrated in Sect. 6.4, an optimal finite duration of patent protection exists only in some rare cases where the social cost associated with the patent monopoly grows faster than its social benefit and, generally speaking, an infinite length of patent protection will maximize the social benefit of the patent system. With the exclusive licence being banned under the new patent system, the social cost of patent protection would be reduced substantially, as the deadweight loss in manufacturing patented products would be eliminated because the production of patented products becomes competitive. Consequently, the concern about the negative effects of the patent system is unwarranted and the limit on patent duration should be removed. It is also worth noting that, even though the patent law may provide an infinite patent duration, the effective life of a patent is limited. As the patented technology is outdated or the patented product has been phased out of the market, the patent loses its value and disappears naturally.

Limit in patent right duration should be removed also because it is a major obstacle to a functioning patent market. Limited duration means that the property right of a patent is temporary only. When the patent is close to its maturity, the patent value approaches zero, so the temporary property right imposes a great distortion on the patent market. It is an open secret that many companies are exploiting expired patents and are waiting to exploit the nearly expired patents. The removal of limit on patent duration can help to form a thicker, less-distorted, and thus more effective, patent market.

One may sceptical about infinite patent duration because of worrying about that, after a certain number of years, the marginal cost to society may outweigh the marginal benefit of stimulating innovation. Without knowing the specific functions, it is hard to predict how marginal incentive effect (social benefit) and marginal social cost of patents behave and thus it is impossible to quantify them. However, as shown in the Appendix at the end of this chapter, the incentive effect and social cost are directly related to the returns to inventors (higher reward to inventors produces higher incentive effect and also higher social cost), so we can expect that marginal incentive effect and marginal social cost behave in a similar fashion, or at a similar degree. In other words, there is no reason to believe one effect to increase or decrease at a significantly larger degree than the other. As such, if the net social benefit is positive today, it is most likely that it will also be positive at any future time. As long as the net future income is positive, present value of future income (i.e. the discounted income based on discount rate) does not lead to zero marginal effect. The only probability to have zero marginal effect is that the marginal social cost grow faster than marginal social benefit. As explained earlier, this is unlikely.

There may be difficulties to legislate an infinite duration of patent right in some countries with a constitutional law system. For example, the US Constitution requires that copyrights are time limited. But this constitutional requirement may be circumvented by the n-1 term of copyright suggested by copyright scholars. In practice, the patent duration can be extended substantially (e.g. 50, 100 or 200 years) as the first step. If the new patent system is proven to be very successful and essential, the constitution will need to be changed.

## 5. Abolishing the compulsory licence rule

The compulsory licence rule has been adopted by many countries to address abuses of patent rights. For example, while Article 48(3) of Patents Act 1977 in UK lists different kinds of abuse of patent monopoly, Article 48(1) stipulates:

> At any time after the expiry of 3 years, or of such other period as may be prescribed, from the date of the grant of the patent, any person may apply to the comptroller on one or more of the grounds specified in subsection (3) below: (a) for a licence under the patent, (b) for an entry to be made in the register to the effect that licences under the patent are to be available as of right or (c) where the applicant is a government department, for the grant to any person specified in the application of a licence under the patent.

The compulsory licence rule is an extreme case of limited patent duration. Although this rule is rarely used in reality, it exerts potential pressure on patentees in their pricing of patent licences. As Reid (1993, p. 132) realized, 'There have been relatively few compulsory license applications in recent decades… nevertheless they have probably wielded an influence wider than might have been expected from the paucity of the case law. The background threat of a compulsory license application is a potent lever in the hands of a person applying to a patentee for a voluntary license on reasonable terms'. In this regard, it is still worthwhile to discuss the compulsory licence rule here.

The purpose of compulsory licence rule is benign, but it is based on the belief that all monopolies are bad and thus need control.

Motivation of having a compulsory licence comes from the worry about negative use of patent rights, especially the high patent licence price. As being argued previously, the patent licence price will be high but will not be extremely high. A high price is necessary to stimulate innovations so it is desirable. Since exclusive licences and assignment of patent rights are banned in the new patent system, the profit maximization behaviour of a monopoly will not lead to extremely high licence price, so patent monopoly in a revised system is not a concern.

The compulsory licence rule is unnecessary under the new patent system because abuse of patent rights can be avoided in other ways. For example, the tie-in problem can be fixed by standardizing patent licensing, while the excess prices of patent rights and patent burying can be solved by a functioning patent market. Even so, some may argue that the compulsory licence rule is still required to safeguard the implementation of some very important patents such as generic invention patents and patents on pharmaceutical and pollution control equipment. This caution is understandable, but the problem should be solved using methods that will not hinder the patent market, such as government involvement in production of certain goods after the government has purchased the relevant patent licences. If less protection is provided for these extremely important innovations by employing the compulsory licence rule, the public may gain some benefit in the short run, by having cheaper patent licences and thus cheaper prices for new products. However, the public will be worse off in the long run, as fewer innovations of this type will be made in the future.

The compulsory licence rule is not only unnecessary but also very harmful. It wrongly imposes the task of implementing and diffusing patent technology on inventors. This distracts the energy of inventors from continuing invention activity into the implementing of the invention, a job more appropriately suited to the entrepreneur. More importantly, the extremely short period of time stipulated by the compulsory licence rule greatly disadvantages inventors in the negotiation of patent licence prices. If a patent right does not command a high price due to the pressure of limited time, further innovation activity becomes unattractive. As a result, inventors or potential inventors will try to find a more attractive career.

Tandon (1982) argued that, if the government is able to provide an optimal price for patent licences, the compulsory licence rule is of no harm. The problem with this argument is that nobody can detect the optimal price except the market. If the 'optimal price' designated by the government is only a fraction of the market price for the patent, this compulsory licence rule will greatly discourage invention. In considering its damaging effects on the patent system, the compulsory licence rule should therefore be abolished.

## 6. Improving the quality of patent

For various reasons, many patents have been approved even though they were not essentially innovative, for example, Amazon's 'one click' patent. These kinds of patents are of low quality, or 'low height' as termed by Foster and Breitwieser (2012). A non-innovative patent has a net negative impact because it imposes a social cost (patent monopoly power) without any social benefit (i.e. no effect on stimulating invention). A non-innovative business patent may also affect the normal business of a wide range of firms.

An implicit case of low-quality patents is an exaggerated patent breadth. The optimal breadth of patent protection should be the one that matches the true breadth of the patent innovation. It is the responsibility of the patentee to make sure in their application that the claimed breadth of their patent is the same as the true breadth of their innovation. However, there is a tendency for patent applicants to exaggerate the patent breadth of an innovation so that some of the claimed breadth is not a true innovation. Since the exaggerated part of patent breadth imposes a social cost (patent monopoly power) without any social benefit, it entails net social cost and is undesirable.

The consequence of low-quality patents is partly reflected in the anger of business people towards patent trolls. Technical speaking, patent trolls are business people who profit by helping the patent holder enforce their patent rights. There may be a few outliers, but the majority of their business is legitimate because it is done within the legal framework. However, with a relaxed standard of patent approval, non-innovative business models may be patented. The enforcement of these low-quality patents by patent trolls affects many businesses and thus generates much resentment.

Low-quality patents may result from a number of factors. First, the lack of resources in patent offices. This appears to be the case of the US patent office in the 1990s before Congress increased the budget of the US patent office. The budget shortfall of the patent office can be eventually financed by increasing the patent application fee and the patent maintenance cost. Second, the low quality of patent officers. Patent assessment involves a wide range of specialties, so the patent officers have to be experts or need to employ experts in relevant areas. This problem can be solved by proper training, new recruitment and an increase in budget. Third, the lack of understanding of the significance of approving a patent. A high-quality patent has a great positive impact on the economy and on invention while a low-quality patent has a great negative impact on the economy but has little impact on stimulating invention. However, these impacts are less obvious due to the ineffectiveness of the current patent system. As a result, the impacts are often underestimated. With a new definition of patent right, the impact of approving a patent may become more noticeable. Finally, the lack of a rigorous system preventing the granting of low-quality patents. Currently there is no mechanism to hold the patent office accountable for low-quality patents. Although the decision of the patent office can be challengeable in the court system, the invalidation of a patent in court does not affect anyone in the patent office. A proper mechanism needs to be established to hold the patent office accountable for patent quality.

The low patent quality exposed the issues of institutional setting and funding for a country to manage its patent system. Currently the patent office is a part of government organization and funded by taxpayers. With a rapidly increasing number of patents applications and the large number of experts required to assess patent applications in different areas, this organization setting and funding model becomes increasingly inadequate. Due to the technical nature of patent assessment, the responsibility of the patent assessment is likely to be taken up by an independent professional body, and the patent office as a government organization approves or disapproves patents based on patent assessment. Since patent applicants receive services from the patent office and patent holders benefit from patents, the funding for the patent assessment should eventually be covered by patent application fee and maintenance fee.

7. Widening the patent protection scope

For historical and practical reasons, current patent laws only protect practical inventions. Technically speaking, fundamental discoveries are patentable, it is impractical to grant a patent right because the function of a fundamental discovery is mainly to contribute to humanity's knowledge or ideas base. Legal scholars unanimously claim that knowledge and ideas are not patentable (Pretnar 2003). Currently neither patent laws nor copyright laws protect the knowledge gained from scientific discoveries and theoretical breakthroughs, so theoretical and fundamental research is severely underfunded. It heavily depends on government funding because few private companies are interested in it. This funding model in turn causes issue in determining the right of a discovery. Since the discoveries are funded by the public, i.e. the government, the public should have the right to use the knowledge obtained and thus the discovery should be free to the public.

However, scientific discoveries and theoretical breakthroughs are the foundation of practical inventions and have widespread indirect influence on the economy. It is possible and desirable to impose a fundamental-research tax on patent applications or on patent trading transactions. This could be one of many ways to protect the source of the invention, and thus to guarantee an injection of adequate innovations into the economy.

8. Enhancing international coordination in patent protection

Due to globalization, a patent innovation can easily cross national boundaries, but the patent law of one country can protect only the benefit of inventors within that country and become powerless beyond its boundaries. It is arguable that the country thoroughly revising its patent law first will enjoy an enriched pool of innovations and thus stimulate its economic growth the most. However, producers in this country will initially be unfairly disadvantaged, because countries without a strict patent law will be able to utilize these innovations at a reduced cost, or even obtain them free. To protect the benefit of inventors on a global scale, and to give all companies an equal footing in producing patented products, it is highly desirable and fundamental to enhance international coordination in revision and implementation of patent laws.

That being said, the importance of international coordination does not mean that a single country could not lead the way in patent reform to establish a new patent system. The development of the patent system provides an example of good international coordination. The patent system was first established in industries countries in Europe. Then, the system is internationalized in 1883 through the Paris Convention and eventually the WIPO was established in 1974. The agreement on TRIPS was incorporated into the charter of the World Trade Organization (WTO) in 1994 and this resulted in uniform patent standards throughout the membership of the WTO. This experience can be utilized in the globalization of the new patent law so as to stimulate global innovation and thus boost the world economy.

This chapter reviews the theories and arguments on the patent system and suggests a thorough reform to establish a new patent system. The role and impact of the patent system have been heatedly debated. In theory, the patent system is the arguably most efficient way to stimulate innovation, compared with other methods such as public produce and government-subsidized private produce. The arguments against the patent system are largely centred on the net social cost of patent monopoly, i.e. whether or not the social benefit of stimulating innovation is greater than the social cost of patent monopoly charging high prices for patented goods. The chapter shows that empirical results on the impact of the patent system are mixed. The chapter also highlights the indicative nature of empirical results, due to the problems in data and in methodologies used.

Although the current patent system cannot be said as the main policy for stimulating innovation, the chapter places the patent system at the centre of innovation/invention activities because the system is crucial for a functioning innovation market. The importance of the patent system can be shown by the following reasoning. The drawbacks of other policy options (e.g. public produce and government-subsidized produce) mean that these options can play only a minor role. Although the natural incentives (e.g. market pressure faced by firms, lead time for innovators and steep learning curve for the imitators) and natural protection method (e.g. trade secrets) play an important role, they work in a natural way and thus cannot be used as an effective policy. On the other hand, the patent system has a great potential because it contains a market mechanism: it stimulates innovation by directly linking the benefit for the innovator with the performance of the innovation. If the problem of social cost of patent monopoly and other issues can be addressed adequately, the patent system can become more popularly used.

For example, some industries may prefer trade secrets and other means of protection. This may be due to the nature of industries, but it may be also due to unsatisfactory or insufficient protection provided by the current patent system (e.g. trade secrets can protect more than 20 years and, arguably, with higher certainty and at lower costs). In the former case, patent reform does no harm; in the latter case, patent reform will provide a better alternative and thus stimulate innovations. If the patent system can absorb the advantages of other means of protection (e.g. long duration, reduced transaction costs, more certainty in patent protection), industry or people preferring other means of protection may turn to patents, so the revised patent system would become the main means of stimulating innovations.

On the theoretical ground, the chapter reviews and rejects the balanced approach. The design of the current patent system is based on a balanced approach—encouraging innovations by giving innovators a limited period of monopoly rights. Nordhaus (1967) supported this approach and claimed that there is an optimal patent duration. Later, others claim that there is a trade-off between patent duration and patent breadth. This chapter shows that these claims are based on flawed models and further demonstrates that it is very unlikely that trade-offs exist. As a result, the balanced approach is actually a compromised approach because limited patent duration reduces the monopoly rights of the patent holder but also reduces the incentives for future innovation. The chapter suggests a new approach for designing the patent system: banning both exclusive patent licences and assignment of patents to restrict patent monopoly power while prolonging patent duration infinitely to avoid disturbances in patent markets. Following this approach, the chapter proposes a number of patent reforms to establish a new patent system.

# Appendix for (Section 6.4): Current Patent System Design—A Compromise

When the patent system was first established, patents were regarded as rewards to innovators and the purpose of the patent system was to induce more invention to be used by states, so the power of monopoly production was given to inventors only for a limited time. This tradition not only continues but also was emphasized due to the realization of the double-edged sword nature of patent monopoly power. Currently, all countries have adopted a balanced approach in designing their patent systems, that is to put a time limit (patent duration) and a scope limit (patent breadth) on patent monopoly power. To press the patentee to apply his/her patent, a provision called 'compulsory license rule' has also been adopted by many countries. The balanced approach is apparently a compromise: more restrictions on patent monopoly will limit the cases of abusing patent power but, in the meantime, it will reduce the returns to the patentee and thus lead to less incentive to innovate.

However, some economists deny the compromising nature of the balanced approach based on the belief that there must be an optimal trade-off point in patent duration and breadth. That is, there is an optimal patent length and breadth which can maximize the net benefit. Economic models are built to demonstrate the optimal point. The claim of optimal patent length and breadth in turn justifies the balance approach to the patent system design. This section has been devoted to proving that there is only a very slim chance that an optimal patent length and breadth might exist. As a result, the balanced approach is not an appropriate one for designing the patent system.

## Fallacies in Studies on Optimal Patent Length and Breadth

We start with discussing the work by Nordhaus (1967), which is regarded as the pioneering study on optimal patent systems and has largely been accepted and followed by subsequent researchers.

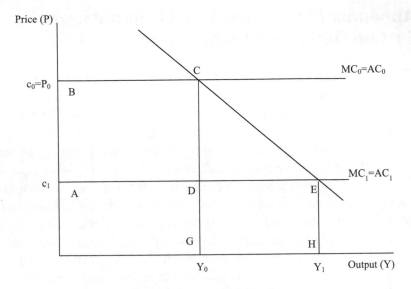

**Fig. 6.3** The run-of-the-mill case illustrated by Nordhaus

Nordhaus extended the model developed by Arrow (1962) to investigate optimal patent duration in the context of small (run-of-the-mill) and of drastic or radical innovations. For the convenience of the reader, Nordhaus's model on small innovation (i.e. the run-of-the-mill case) is reproduced here.

Nordhaus employed a model consisting of a linear demand and constant marginal and average cost under a perfectly competitive market, shown in Fig. 6.3. Without an invention, the firm produces an output $Y_0$ at point C where the marginal cost curve $MC_0 = AC_0$ intersects the demand curve CE, so the market price is $P_0 = c_0$. An invention reduces the production cost to $MC_1$, so the market price is $P_1 = c_1$ the firm produces an output $Y_1$ at point E where the marginal cost curve $MC_1 = AC_1$ intersects the demand curve. The demand curve CE is described by the demand function $Y = a - dP$.

The royalty returns from inventions is given by:

$r = P_0 - P_1 = c_0 - c_1 = B(R)$, where $R$ is the amount of research.

Here Nordhaus introduced an 'invention possibility function $B(R)$' and assumed that the first and second derivatives of $B(R)$ satisfy that $B'(R) > 0$ and $B''(R) < 0$. This implies that, as $R$ increases, the royalty $r$ also increases but at a diminishing pace. The profit from the invention is

$\pi(R) = \int_0^T Y_0 r * e^{-\rho v} dv - sR$

where $s$ is the unit cost of research.

The profit maximization requires $\pi'(R) = 0$, or

$$\pi'(R) = \int_0^T Y_0 B'(R) * e^{-\rho v} dv - s = \frac{Y_0 B'(R)(1 - e^{-\rho v})}{\rho} - s = 0.$$

Let $\varphi(T) = 1 - e^{-\rho v}$, we have $B'(R)\varphi Y_0 - s\rho = 0$.

The net social benefit (NSB) from the invention is

$$W = \int_0^\infty Y_0 r * e^{-\rho v} dv + \int_T^\infty 0.5(Y_1 - Y_0) r * e^{-\rho v} dv - sR,$$

since $Y_1 - Y_0 = -d(P_1 - P_0) = rd$, we have $W = \frac{rY_0}{\rho} - \frac{r^2 d(1-\varphi(T))}{2\rho} - sR$.

Maximizing $W$

subject to $B'(R)\varphi Y_0 - s\rho = 0$,

we can have

$\varphi(T) = \frac{2B'^2(dB+Y_0)}{2B'^2(dB+Y_0) - B''B^2 d}$.

To normalize the model by setting $Y_0 = P_0 = 1$, $d$ is converted to an elasticity.

We have:

$$\varphi(T) = \frac{dB + 1}{dB\left(1 + \frac{\sigma}{2}\right) + 1}, \text{ where } \sigma = -B''B/B'^2 \qquad (6.1)$$

Or $T = -\frac{1}{\rho}\log(1 - \varphi)$.

Since it is assumed that $B > 0$, $B' > 0$, $d > 0$ and $B'' < 0$, so we have $\sigma > 0$.

Because $d$ and $B$ both are also assumed to be positive, Nordhaus (1967) concluded that $0 \leq \varphi \leq 1$, $0 \leq T \leq \infty$. This means that there is a positive finite optimal patent length.

Nordhaus' research appears to provide sound mathematical proof that there exists a finite optimal patent length, at least for the run-of-the-mill case. The conclusion of finite optimal patent length seems also plausible because it is supported by the decreasing return assumption which is consistent with the general assumption used in economics. However, the model of Nordhaus (1967) has two major defects which invalidate the conclusion of having a finite optimal patent length.

First, the model assumes no uncertainty. This is highly inappropriate for modelling innovation. Innovation means to attempt something totally new, so uncertainty is the key feature of innovation investment or innovation activities. Nordhaus (1972) did acknowledge this limitation and introduced a risk premium on invention to modify the conclusion. However, this risk premium did nothing to make plausible his key assumption for maximizing the profit from invention: $r = B(R)$. This assumption links return or royalty to the amount of research. However, this link does not exist. Due to the uncertainty of research outcome and of the demand for new products, some research may not be successful and some new products may have no demand. As a result, setting $r = B(R)$ is problematic in the first place.

Nordhaus (1967) not only set $r = B(R)$ but also further assumed 'decreasing rate of return to innovation research', i.e. $B > 0$, $B' > 0$ and $B'' < 0$. This further assumption was crucial to Nordhaus' conclusion. As we saw previously, this assumption resulted in $\sigma > 0$ and thus $0 \leq \varphi \leq 1$ and $0 \leq T \leq \infty$, necessitating a finite optimal patent length $T = -(1/\rho)\log(1 - \varphi)$. Although the assumption of 'decreasing rate of return' is widely used for firm's production and for research in macroeconomics, this assumption is implausible for studying firms' innovation, which is the focus of Nordhaus' model. A large number of innovation attempts fail in the end, so the return to research is zero (i.e. $B' = 0$). Moreover, it is not uncommon that some research costs a large sum but has a very low return while some low-budget research turns out to be highly profitable. In this case, the royalty is negatively related to the amount of research (i.e. $B' < 0$). For some inventions, the return rate may be positively related to invention cost, but there is no basis for the assumption of decreasing return.

Second, even if all the assumptions in Nordhaus' model were correct, his conclusion that 'Since $\sigma$ is positive, we know that the optimal life (of a patent) is a finite, positive period' (Nordhaus 1967, p. 9) is not correct since he ignored 'the horrible second-order condition' (Nordhaus 1967, p. 7) for the constrained maximization problem. Using Nordhaus' social welfare function, we can derive the following second-order condition.

Social welfare function in his study (Nordhaus 1967, p. 7) is as follows:

$$W = rY_0/\rho + r^2 d[1-\varphi]/(2\rho) - sR.$$

Since he defined $r = B = B(R)$, the welfare function becomes:

$$W = BY_0/\rho + B^2[1-\varphi]d/(2\rho) - sR.$$

The relevant derivatives to obtain the second-order condition are:

$$f_1 = \partial W/\partial \varphi = -B^2 d/(2\rho),$$
$$f_2 = \partial W/\partial R = B'Y_0/\rho + BB'(1-\varphi)d/\rho - s,$$
$$f_{11} = \partial f_1/\partial \varphi = 0,$$
$$f_{12} = \partial f_1/\partial R = -BB'd/\rho,$$
$$f_{22} = \partial f_2/\partial R = B''Y_0/\rho + \left(B'^2 + BB''\right)(1-\varphi)d/\rho.$$

The second-order condition for a true optimization requires:

$$f_{11}f_2^2 - 2f_{12}f_1f_2 + f_{22}f_1^2 < 0$$

Substituting the first and second derivatives into the above condition we have,

$$f_{11}f_2^2 - 2f_{12}f_1f_2 + f_{22}f_1^2 = 0 - 2\left[-BB'\,d\,/\rho\right]$$

$$\left[-B^2d/(2\rho)\right]\left[B'Y_0/\rho + BB'(1-\varphi)\,d/\rho - s\right]$$

$$+ \left[B''Y_0/\rho + d\left(B'^2 + BB''\right)(1-\varphi)/\rho\right]\left[-B^2d/(2\rho)\right]^2$$

$$= -\left(B^3B'd^2/\rho^3\right)\left[B'Y_0 + BB'(1-\varphi)d - s\rho\right]$$

$$- \left(0.25B^4d^2/\rho^3\right)\left[B''Y_0 + \left(B'^2 + BB''\right)(1-\varphi)d\right]$$

Plugging into above equation Nordhaus' results (1967, p. 6) from the firm's profit maximization condition, $s\rho = B'\varphi Y_0$, we have:

$$f_{11}f_2^2 - 2f_{12}f_1f_2 + f_{22}f_1^2 = -\left(0.25B^3d^2/\rho^3\right)\left[4B'^2Y_0(1-\varphi) + 4BB'^2(1-\varphi)d\right]$$

$$- \left(0.25B^3d^2/\rho^3\right)\left[BB''Y_0 + \left(BB'^2 + B^2B''\right)(1-\varphi)d\right]$$

$$= -\left(0.25B^3d^2/\rho^3\right)\left[(1-\varphi)\left(4B'^2Y_0 + 5BB'^2d + B^2B''d\right) + BB''Y_0\right] < 0.$$

By definition $B$, $d$, $\rho$ are all positive, so the above inequality requires the content in the square bracket to be positive. Following Nordhaus' convention to let $Y_0 = 1$ so that d represents demand elasticity in absolute value. We have the following second-order condition:

$$1 - \varphi > -BB'\left/\left(4B'^2 + 5BB'^2d + B^2B''d\right)\right., \text{ or}$$

$$\varphi < 1 + BB''\left/\left(4B'^2 + 5BB'^2d + B^2B''d\right)\right. \qquad (6.2)$$

Since $\varphi = 1 - e^{-\rho T}$, the above equation can be rewritten as:

$$T < T\text{max} = -(1/\rho)\,\ln\left[-BB''\left/\left(4B'^2 + 5BB'^2d + B^2B''d\right)\right.\right] \quad (6.3)$$

This second-order condition means that the limited duration of patent protection ($T < T$max) is a prerequisite for the true optimal length of patent protection provided by Nordhaus. Since it is assumed $B > 0$ and $B'' < 0$, the numerator in the 'ln' function is positive, i.e. $BB'' > 0$.

Considering the denominator in the 'ln' function, $4B'^2 + 5BB'^2d$ is positive, but $B^2B''d$ is negative. If $B^2B''d$ is sufficiently large to reduce remarkably most of the positive value of $4B'^2 + 5BB'^2d$, there is a possibility that in the 'ln' function the numerator is greater than the denominator. This will lead to a positive value for the 'ln' function and thus a negative value '$T$max'. In this case, the second-order condition requires that the optimal patent length must be negative, so the conclusion of a finite positive patent length may not be true.

One can grasp this issue more readily by assigning some numbers to variables in the second-order condition and in Nordhaus' solution. Following Nordhaus (1967), the author assumes an innovation response function $B=R^\alpha$ and uses the numbers $\alpha = 0.1$, or $\sigma = 9$, suggested by Nordhaus. For convenience of calculation, the author further assigns $d=0.5$, $\rho = 0.1$ and $R=1$, so we have $B' = \alpha R^{\alpha-1} = \alpha = 0.1$, $B'' = \alpha(\alpha - 1)R^{\alpha-2} = 0.1 * (0.10 - 1) = -0.09$. Substituting these numbers into Eqs. (6.2) and (6.3), we have the second-order condition: $1 - \varphi > -(-0.09)/0.02 = 4.50$, or $T < -10*\ln 4.5 = -15.04 < 0$. This suggests a requirement of negative length of patent protection for an optimal solution.

However, based on Eq. (6.1) we have Nordhaus solution of an optimal length of patent protection: $\varphi = (dB + 1)/[dB(1 + \sigma/2) + 1] = -1.5/3.75 = -0.4$, or $T = -10*\ln(1 - \varphi) = -10*\ln 0.6 = 5.12$. Apparently, this solution is not a true optimal patent duration because it does not satisfy the second-order condition. Believing that his results were valid, Nordhaus had great difficulty in explaining the counter-intuitive implication of his results—the optimal length of a patent for more important innovations should be shorter.[2] There were also other implausible implications arising from the Nordhaus model. Some of these implications were identified by Domar and Stiglitz (1969). Since the optimal

---

[2]Nordhaus (1967) admitted that it was almost universally agreed that more important inventions should have longer patent lives because their development poses a higher risk to the inventor and those funding their work, and requires longer periods of development. To explain his implausible results, he overlooks the greater social welfare of more important inventions and exaggerates its deadweight loss by saying 'in general more important inventions involve larger second order effects and thus should have shorter lives'.

length of patent derived by Nordhaus is false, it follows that the belief that there is a trade-off point in the length of patent protection is not tenable.

Patent protection has two dimensions, length and breadth, so some researchers think that there may be an optimal trade-off between these dimensions. From the early 1990s, a number of researchers have set out to identify this trade-off, using various definitions of patent breadth.

Gilbert and Shapiro (1990) defined patent breadth as the flow rate of profit available to the patentee, or the ability of the patentee to raise price. Using this concept, they maximized social welfare subject to the optimal reward available to the patentee. They concluded that the optimal patent was of an infinite length, with patent breadth adjusted to provide the required reward for innovation. In other words, a long and narrow patent is preferred.

Klemperer (1990) defined patent breadth as the distance between the patented product and the products in a product space that competing firms can sell without infringing the patent. By minimizing the social cost per dollar of profit to the patentee (or the ratio of social cost to the patentee's profit), Klemperer concluded that when all consumers have identical transport costs, infinite length narrow patents are optimal. However, when all consumers have identical reservation prices for the most-preferred product variety, a short-lived patent with infinite width is optimal.

Gallini (1992) links patent length and breadth to imitation costs. She defined patent breadth as the flow of profits earned by the innovator during patent life, which was similar to the definition by Klemperer (1990). Imitation cost is assumed to be positively related to patent breadth, while incentive to imitate is determined by the return to the imitator during the length of patent protection. By minimizing the discounted deadweight loss plus profits lost to imitation, while letting the return to the innovator equal imitation cost, Gallini claimed that the optimal patent policy consists of broad patents with patent lengths adjusted to achieving the desired reward to the patentee. Horowitz and Lai (1996) defined an innovation as quality of goods along a quality ladder and assumed a limit-pricing strategy by the market leader. They also claimed an optimal patent length both for stimulating innovation and for increasing social welfare.

Denicolo (1996) studied optimal patent length and breadth in the context of many firms racing to establish a patent. He defined patent breadth as the fraction of the technological knowledge of a patent that is not freely available. Assuming that the losers of the patent race can profit from the post-innovation equilibrium, Denicolo maximized the total social welfare subject to the condition that the returns to all firms in the patent race are equal to the equilibrium research and development (R&D) investment level. He derived a number of optimal patent policies under different circumstances. However, he concluded that generally, the less efficient the competition in the product market, the more likely it is that broad and short patents are socially optimal.

The aforementioned studies used a method very similar to Nordhaus (1967). That is, a micro- or firm-level approach to maximizing social welfare (or minimizing social cost) subject to the optimal return to the patentee (or to all firms in a patent race). This approach confines the research to a specific case involving invention and patents. Even if the conclusions drawn from this approach are correct, they are applicable only to the cases specified. In other words, their research loses generality because it does not answer the question: Is there a finite positive optimal patent length and breadth for the majority of firms or for a typical firm?

Moreover, this approach needs very detailed restrictive modelling settings or assumptions regarding the firm's behaviours. Many conclusions directly stem from or heavily rely on various assumptions. Gilbert and Shapiro (1990) assume social welfare ($W$) becomes increasingly costly as patent breadth increases ($W' < 0$ and $W'' < 0$), so it is not surprising that their conclusion is in favour of long but narrow patents. For Klemperer (1990), when the ratio of the social cost to the patentee's profit $r(w)$ is minimized at zero breadth of a patent ($w = 0$), it is assumed that the social cost will increase faster than the profit to the patentee as the breadth of the patent increases. Thus it is natural that a narrow patent is preferred in this case and a wider patent preferred otherwise. Due to the assumptions of Gallini (1992) regarding imitation cost and incentive to imitate, the long patent duration encourages imitation and thus is not desirable, while a wide patent breadth will increase imitation costs and thus deter imitation. So, it is understandable that her study is in favour of broad and short-lived patent protection.

Horowitz and Lai (1996) assumed that a limit-pricing strategy is used by the leading firm, which led the firm to innovate or file a patent when the previous patent is expiring. This led to their conclusion on the frequency of innovation, the rate of innovation and the welfare-maximizing patent lengths. Denicolo (1996) made different assumptions about the behaviour of social cost of patents under different scenarios and, accordingly, reached different conclusions.

However, these assumptions may not be plausible or may not be the general case for the majority of firms. We have already seen that Nordhaus' assumption of 'diminishing return to research' is not plausible for studying innovations. Gilbert and Shapiro (1990) assumed that net social welfare is increasingly costly as the profit to the firm increases. This assumption is based on the reasoning that broader patents confer the firm greater market power that generates higher deadweight loss. This reasoning ignores the positive-side impact of patent breadth. A broad patent indicates a high impact innovation that will contribute more to the economy. Moreover, the higher profit to the firm, due to the broader patent breadth, can provide more stimulus to basic innovation. Similarly, the assumptions on firms' behaviour in Klemperer (1990), Gallini (1992), Horowitz and Lai (1996), and Denicolo (1996) cannot be said as a general case for a typical firm. Once the assumptions are proven implausible or not general, the value or the applicability of the research based on the assumptions is discounted greatly.

To sum up, although these previous studies on optimal patent protection make a worthwhile contribution to this issue, the microeconomic (firm) level approach is a major limitation. To overcome this limitation, the author proposes a macroeconomic approach and uses general assumptions to address more adequately and efficiently the issue of patent length and breadth. This approach is illustrated graphically and mathematically next.

## Graphic Demonstration of Unlikelihood of Finite Optimal Patent Length and Breadth

At the macro-level, the positive and negative impacts of patent protection can be illustrated using the graph of monopoly production. To avoid the complexity of the graphs, we assume a constant marginal cost

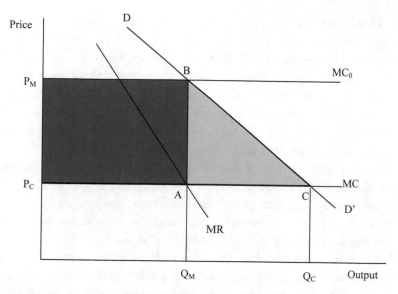

**Fig. 6.4**  Social benefit and social cost of a patent monopoly

of production. A more general case of increasing marginal cost of production will be discussed mathematically in section 'Mathematic Proof
of the Unlikelihood of Finite Optimal Patent Length and Breadth'.

Figure 6.4 illustrates the production and pricing strategy of a monopoly, as well as the social cost of patent monopoly and the incentives of
patent protection. With a marginal cost MC (assuming a constant MC
for simplification) and facing a demand curve DD', a competitive firm
will produce optimally at point C when supply equals demand, i.e. the
firm will produce quantity $Q_C$ and accept the market price $P_C$. If the
firm is a monopoly of the product, e.g. a monopoly protected by patent rights, the firm can maximize profit by producing output at point A
where marginal cost equals marginal revenue. As a result, the monopoly
produces less ($Q_M < Q_C$) and charges a higher price ($P_M > P_C$).

For a patent monopoly, the price difference ($P_M - P_C$) can be
viewed as a return on the patent, i.e. patent rent per unit of output, so
the line $P_M B$ or $MC_0$ can be viewed as the firm's marginal cost inclu-
sive of patent rent. Compared with production under a perfect competition scenario at point C, the monopoly achieves a super profit

(i.e. patent rent) of area $P_M BAP_C$ but leads to a social cost or dead-weight loss—the area of ABC—due to reduced production and consumption. The super profit $P_M BAP_C$ obtained by the patent holder can be viewed as an incentive to invent. Assuming the impact factor of $P_M BAP_C$ on future invention is $i$, the social benefit of patent protection is $i * P \, P_M BAP_C$. An optimal patent duration should maximize the net social benefit, namely $i * P_M BAP_C - ABC$. In relation to patent duration, we need to consider this net social benefit for each year.

First, we consider a simplified case: the demand curve is unchanged for each year. In this case, the net social benefit is the same for each year and the optimal patent duration will be at the two extreme ends: if $i * P_M BAP_C - ABC > 0$, the net social benefit is positive for each year, so the optimal patent duration is infinity; if $i * P_M BAP_C - ABC < 0$, the net social benefit is negative for each year, so the optimal patent duration is zero, i.e. abolishing the patent law. For a more general case, we can let all future values be discounted to the present value by a discount rate, but the conclusion will not change because the present values of these two parts will change by the same degree over time.

In the case of the linear demand curve shown in Fig. 6.4, the MR curve will always pass through the mid-point of the horizontal segment below the demand curve DD', i.e. $P_C A = AC$, or $Q_C = 2Q_M$, so $P_M BAP_C = 2 * ABC$. The ratio of the social benefit to the social cost of the invention can be calculated as: $R = i * P_M BAP_C / ABC = 2i$. As a result, optimal patent duration can be set based on the impact factor $i$: if $i = 1/2$, then $R = 1$, the patent system has no impact on net social welfare for each year, so any length of patent duration has the same result. If $i > 1/2$, then $R > 1$. The patent system has a positive social impact for each year, so an infinite patent duration is desirable. If $i < 1/2$, $R < 1$, the patent system has a negative social impact for each year, so a zero patent duration should be set, i.e. the item should not be patented.

Second, we consider the case of the demand curve changing over time. Since the variation of the demand curve is unpredictable, there is no general rule to determine the optimal patent duration for any changes in the demand curve. However, we can consider a simplified likely case: where the demand curve is linear, demand for the patent product will reduce over time because of market saturation, and the

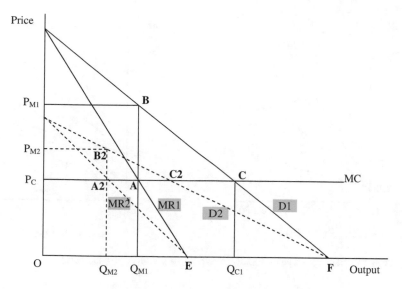

**Fig. 6.5** Impact of a patent monopoly with changing demand curve

demand curve will tend to become flatter in the long run (i.e. more elastic). Figure 6.5 demonstrates this case.

In Fig. 6.5, MC is the firm's marginal cost curve and $P_c$ is the price in a perfect competitive market. The demand curve and the marginal revenue curve for year 1 are $D_1$ and $MR_1$, respectively. For year 2, they are $D_2$ and $MR_2$. The monopoly output and price charged for year 1 are $Q_{M1}$ and $P_{M1}$ respectively, for year 2 they are $Q_{M2}$ and $P_{M2}$. Because the demand curves are linear here, it is easy to verify that the marginal curve passes through the mid-point of the horizontal segment between vertical axis and the demand curve. This geometry rule necessitates that $OE = EF$, $P_C A = AC$, $P_C A_2 = A_2 C_2$, so the area of $P_{M1} BAP_C$ is twice the area of $ABC$, and the area of $P_{M2} B_2 A_2 P_C$ is twice the area of $A_2 B_2 C_2$. As such, the ratio of social benefit to social cost is the same in year 1 and in year 2 because: $R_1 = i^* P_{M1} BAP_C / ABC = 2i$, $R2 = i^* P_{M2} B_2 A_2 P_C / A_2 B_2 C_2 = 2i$. Consequently, we reach the same conclusion as in the case where the demand curve does not change over time.

One can also change the assumption on the behaviour of the demand curve, the ratio of social benefit to social cost should be the same for each

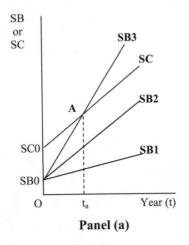
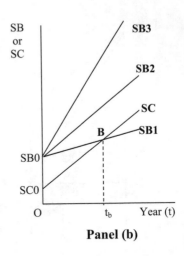

**Panel (a)**          **Panel (b)**

**Fig. 6.6** Impact of varying social benefit and social cost over time

year as long as the demand curve is linear. For the non-linear demand curve, the analysis becomes more complicated, but it can be shown that the social benefit and social cost go in the same direction and change by a similar degree.

Third, we allow various behavior of social cost and social benefit associated with patent protection. Figure 6.6 shows various setting of social cost (SC) and social benefit (SB). The slope of SC depends on how the demand for patented products changes over time.

In both panels of Fig. 6.6, the social cost (SC) is drawn for a case where SC increases over time, but this is not necessarily the only case in reality. The slope of SC depends on how the demand for patented products changes over time, but the change of the slope of SC will not affect our analysis.

The left panel of Fig. 6.6, Panel (a), shows the case where the initial social cost ($SC_0$) is greater than the initial social benefit ($SB_0$) of patent protection. If the pace of the increase in social cost is greater than or equal to that for social benefit, i.e. the slope of SC is not less than the slope of SB, e.g., $SB_1$ and $SB_2$, the social cost will be always greater than the social benefit and thus the optimal patent duration is zero. If the slope of SB is greater than SC, e.g., $SB_3$, SC intersects $SB_3$ at point A, the break-even point at which the social benefit equals social cost. Since the social benefit after year $t_a$ will be greater and greater than the social cost, so the optimal patent duration in this case is infinite. Consequently, if the initial social cost is greater than the social benefit, there is no case where optimal patent duration is positive and finite.

The right panel of Fig. 6.6, Panel (b), shows the case where the initial social cost is less than the social benefit. If the pace of the increase of SB is not less than that of SC, e.g., $SB_2$ and $SB_3$, social benefit will be greater than the social cost for any year, so the optimal patent duration is infinite. On the other hand, if the slope of SB is smaller than that of SC, e.g. $SB_1$, the net social benefit will be maximized at the intersection point B, so $t_b$ will be the optimal patent duration. However, no empirical evidence shows the growth of SC is faster than that of SB, so this finite optimal patent duration is not supported by evidence.

When we turn to patent breadth, a popular argument is worth noting, e.g. the wider the patent breadth, the more monopoly power and the higher the social cost, so the patent with wide breadth should be given shorter patent duration. This argument only considers the social cost and fails to see that the wider patent breadth also indicates higher social benefit. Moreover, higher monopoly power for the patent of wider breadth indicates the higher impact of patent protection in stimulating invention of wider application. Since the social cost and the social benefit go in the same direction as patent breadth varies, there is no general rule to vary patent duration based on patent breadth in order to maximize the net social benefit.

In short, since the total social benefit and total social cost tend to go in parallel, patent duration and patent breadth have little power in changing the balance between providing more protection and reducing deadweight loss, namely it is unlikely to have a positive finite optimal patent duration and an associated optimal patent breadth. Consequently, if the patent system has positive social effects, i.e. the net social benefit is positive in each year, an infinite patent duration can generate the highest level of net social benefit and thus should be most beneficial to society; otherwise, the patent system should be abolished.

## Mathematic Proof of the Unlikelihood of Finite Optimal Patent Length and Breadth

In this section, we will use mathematics as a tool to look into the patent length and breadth in a more accurate fashion. We will start with a simple case and then discuss more complicated general cases. Again, the monopoly production graph is our starting point. For generality, an increasing marginal production cost is assumed. The welfare effect of monopoly production is illustrated in Fig. 6.7.

As the patentee will monopolize production of the innovative product, he/she will establish the output and price so as to maximize his/her profit. The monopoly patentee sets the output at $Q_M$, so that the marginal cost equals the marginal revenue. Compared with production under a perfect competition scenario at point C, where marginal cost equals market demand, the monopoly produces less ($Q_M < Q_C$) and charges a higher price ($P_M > P_C$). As a result, the monopoly achieves a super profit $P_M BEP_C$ but leads to a deadweight loss—the area of ABC.

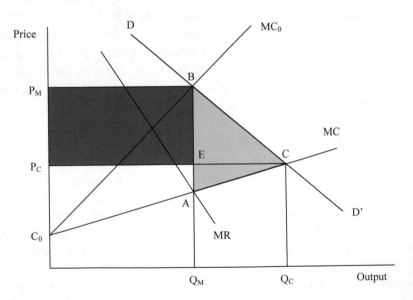

**Fig. 6.7** Net social benefit of a patent monopoly

1. A simplified case: all innovations are induced by patent protection

Assuming that the marginal cost of production before the innovation is $MC_0$, the total amount of research (i.e. social cost of research) for the innovation is SR, and the intertemporal discount rate is $\rho$. Assuming that the innovation leads to the reduction in production cost from $MC_0$ to MC, the social benefit of the innovation is the area $C_0AB$ during the patent protection and the area $C_0CB$ after the patent has expired. However, an important benefit of patent protection—stimulating future innovations—has also to be taken into account. It is reasonable to assume that the future innovation research is induced by the super profit obtained by the innovator, i.e. the area $P_MBEP_C$ in Fig. 6.7. Letting $q$ be the impact factor of super profit on the future innovation research, the innovation research induced by patent protection will be $q * P_MBEP_C$.

In summarizing all costs and benefits, the net social benefit (NSB) of a patented innovation can be expressed as follows:

$$NSB = \int_0^T C_0AB * e^{-\rho t}dt + \int_T^\infty C_0CB * e^{-\rho t}dt + q\int_0^T P_MBEP_C * e^{-\rho t}dt - SR$$

$$= \int_0^\infty C_0CB * e^{-\rho t}dt - \int_0^T ACB * e^{-\rho t}dt + q\int_0^T P_MBEP_C * e^{-\rho t}dt - SR$$

$$= C_0CB/\rho - ACB * (1 - e^{-\rho T})/\rho + q * P_MBEP_C * (1 - e^{-\rho T})/\rho - SR$$

To find the optimal patent duration, we find the derivative of net social benefit with respect to $T$:

$$\partial NSB/\partial T = (q * P_MBEP_C - ACB) * e^{-\rho T} \qquad (6.4)$$

Only two areas in Fig. 6.7 appear in the above equation, i.e. $P_MBEP_C$ and ACB. This is because other areas are not related to patent duration. The first-order condition for optimal patent duration requires $\partial NSB/\partial T = (q * P_MBEP_C - ACB) * e^{-\rho T} = 0$, or $q * P_MBEP_C = ACB$. Apparently, $T$ plays no role in this first-order condition for maximization of net social benefit.

If $q^* P_M BEP_C > ACB$, then $\partial NSB/\partial T > 0$. This means that the positive effect of patent protection (the incentive to innovate) always outweighs the negative effect (the deadweight loss) and thus the longer patent protection generates higher net social benefit. In this case, the optimal solution is to extend the duration of patent protection infinitely. On the other hand, if $q^* P_M BEP_C < ACB$, we have $\partial NSB/\partial T < 0$, i.e. patent protection has a negative net social effect all the time and thus, the optimal solution is no patent protection. In the case $q^* P_M BEP_C = ACB$, $\partial NSB/\partial T = 0$. This indicates that the positive and negative effects of patent protection are cancelled out. Hence, any length of patent protection has the same result. If we believe that the patent system has a positive impact overall, an infinite patent length should be desirable for society.

Next, we consider the case of patent breadth. As mentioned in the previous section, there are many definitions,[3] of which Denicolo (1996) provides a summary. In this book, a general and intuitive definition taken from Matutes et al. (1996) is used: the breadth of a patent is the number of products or procedures to which an innovation can apply and these applications are reserved for the patentee. Assuming a patent innovation has breadth of $n$, it can be applied to $n$ products and the innovator is able to monopolize production of $n$ products. Each product generates super profit and deadweight loss similar to those shown in Fig. 6.7. Adding up all super profits and deadweight loss for all $n$ products, the ratio of total super profit to total deadweight loss can be calculated. This ratio is constant over time because the super profit and deadweight loss for each product are the same each year. This constant ratio indicates that, even if patent breadth is considered, the trade-off between stimulating innovation and deadweight loss due to patent monopoly does not change over time. Consequently, there is no optimal patent duration for innovations of any patent breadth.

---

[3]The different definitions of patent breadth are largely due to the convenience to explain the models used by different researchers.

## 2. A general case: patent protection accelerates the advent of innovation

For all innovations that occurred each year in an economy, it is arguable that some innovation may have occurred naturally, with or without the patent system. In considering this argument, the effect of the patent system is actually to accelerate the advent of innovations. This is the starting point of the analysis for a general case. For the convenience of analysis and discussion, it is assumed initially that all innovations stimulated by the patent system have the same characteristics, i.e. the number of years advanced, and the amounts of social benefit and deadweight loss. We consider the impact of the patent length and of patent breadth in turn.

Let $N_0$ be the number of innovations occurring naturally, and $N$ be the number of innovations accelerated by the patent system. It is assumed that all innovations are accepted and protected by the patent system. For each innovation (either naturally occurring or induced by the patent system), let $i$ be the yearly discount rate, SS the social surplus (the area below the demand curve and above the marginal cost curve). The cost of innovation is assumed positively related to social benefit SS, i.e. $R = r*SS$ is, where $r$ is a positive parameter.[4] For each innovation stimulated by the patent system, $n$ is the number of years advanced by the patent protection, $t$ is the duration of patent right, which is assumed longer than the number of years advanced $n$ (otherwise, patent protection has no social cost and thus is definitely desirable), $w$ is the deadweight loss as the percentage of the total social surplus generated by each innovation. The variable $N$—the number of innovations which can be accelerated by the patent protection—is assumed as a positive function of $t$ (the longer patent duration leads to more profit to patentees and thus will stimulate more innovations), namely $N = N(t) > 0$ and $N(t)' > 0$. With this setting and in considering the innovation in the current year

---

[4] This assumption is convenient but not essential for the analysis. The cost of innovation can be set as a constant or proportional to an exogenous variable and the conclusion is not affected.

and the innovations delayed to year $n$, the total NSB without a patent system—BNP—can be expressed as follows:

$$\text{BNP} = N(t) * (\text{SS} - r * \text{SS}) * \left[(1 - i)^n + (1 - i)^{n+1} + \cdots\right]$$
$$+ N_0 * (\text{SS} - r * \text{SS}) * \left[1 + (1 - i) + (1 - i)^2 + \cdots\right];$$

Assuming all innovations—naturally occurred or patent-stimulated—are patented with a patent length of $t$, the NSB with a patent system—BP—can be expressed as:

$$\begin{aligned}
\text{BP} = {} & (N(t) + N_0) * (\text{SS} - r * \text{SS}) * (1 - w) * \left[1 + (1 - i) + \cdots + (1 - i)^{t-1}\right] \\
& + (N(t) + N_0) * (\text{SS} - r * \text{SS}) * \left[(1 - i)^t + (1 - i)^{t+1} + \cdots\right] \\
= {} & N(t) * (\text{SS} - r * \text{SS}) * \left[1 + (1 - i) + \cdots + (1 - i)^{t-1}\right] \\
& - N(t) * (\text{SS} - r * \text{SS}) * w * \left[1 + (1 - i) + \cdots + (1 - i)^{t-1}\right] \\
& + N_0 * (\text{SS} - r * \text{SS}) * \left[1 + (1 - i) + \cdots + (1 - i)^{t-1}\right] \\
& - N_0 * (\text{SS} - r * \text{SS}) * w * \left[1 + (1 - i) + \cdots + (1 - i)^{t-1}\right] \\
& + N(t) * (\text{SS} - r * \text{SS}) * \left[(1 - i)^t + (1 - i)^{t+1} + \cdots\right] \\
& + N_0 * (\text{SS} - r * \text{SS}) * \left[(1 - i)^t + (1 - i)^{t+1} + \cdots\right] \\
= {} & N(t) * (\text{SS} - r * \text{SS}) * \left[1 + (1 - i) + \cdots + (1 - i)^{t-1} + \cdots\right] \\
& - N(t) * (\text{SS} - r * \text{SS}) * w * \left[1 + (1 - i) + \cdots + (1 - i)^{t-1}\right] \\
& + N_0 * (\text{SS} - r * \text{SS}) * \left[1 + (1 - i) + \cdots + (1 - i)^{t-1} + \cdots\right] \\
& - N_0 * (\text{SS} - r * \text{SS}) * w * \left[1 + (1 - i) + \cdots + (1 - i)^{t-1}\right];
\end{aligned}$$

So, the overall effect of patent protection EP can be calculated as:

$$
\begin{aligned}
\text{EP} =\text{BP} - \text{BNP} = & N(t) * (\text{SS} - r * \text{SS}) * \left[1 + (1 - i) + \cdots + (1 - i)^{n-1}\right] \\
& - N(t) * (\text{SS} - r * \text{SS}) * w * \left[1 + (1 - i) + \cdots + (1 - i)^{t-1}\right] \\
& - N_0 * (\text{SS} - r * \text{SS}) * w * \left[1 + (1 - i) + \cdots + (1 - i)^{t-1}\right] \\
= & N(t) * \text{SS} * (1 - r) * \left[1 - (1 - i)^n\right]/i - N(t) * \text{SS} * (1 - r) * w* \\
& \left[1 - (1 - i)^t\right]/i - N_0 * \text{SS} * (1 - r) * w * \left[1 - (1 - i)^t\right]/i \\
= & N(t) * \text{SS} * (1 - r) * \left\{\left[1 - (1 - i)^n\right]/i - w\left[1 - (1 - i)^t\right]/i\right\} \\
& - N_0 * \text{SS} * (1 - r) * w * \left[1 - (1 - i)^t\right]/i \\
= & \text{SS} * (1 - r)/i * \left\{N(t) * \left[1 - (1 - i)^n - w + w(1 - i)^t\right]\right. \\
& \left. - N_0 * w * \left[1 - (1 - i)^t\right]\right\}
\end{aligned}
$$

For convenience of discussion, we move the constant positive item to the end of the equation and thus have:

$$
\begin{aligned}
\text{EP} = & \left\{N(t) * \left[1 - (1 - i)^n - w + w(1 - i)^t\right]\right. \\
& \left. - N_0 * w * \left[1 - (1 - i)^t\right]\right\} * [\text{SS}(1 - r)]/i
\end{aligned} \tag{6.5}
$$

Equation (6.5) simply states that the effect of patent protection is its net effect on stimulated innovations (the benefit of accelerated years after deducting the deadweight loss caused by patent protection) minus the deadweight loss due to patent protection related to the naturally occurred innovation. As $t$ increases, $(1 - i)^t$ decreases, the last term in the curly brackets in Eq. (6.5) becomes more negative, i.e. the negative impact of the patent protection related to natural innovation increase. The impact of an increasing t on the first term in the curly brackets in Eq. (6.5) is more complex but more important. Generally speaking, the number of naturally occurred innovations $N_0$ is small due to the high risk of innovation investment (if naturally occurred innovations were abundant, we would not be here to discuss whether or not we should have a patent system), and thus $N_0$ will be much smaller than stimulated innovation $N(t)$. As a result, the analysis will ignore the second term in the curly brackets in Eq. (6.5) and focus on the first term.

As $t$ increases, $N(t)$ increases but $w(1-i)^t$ decreases. Since $w$ and $(1-i)$ are between 0 and 1, the term $w(1-i)^t$ is a small number, so the impact of $N(t)$ dominates. Is an increasing $N(t)$ good or bad for the economy? It depends on the sign of the term $[1-(1-i)^n-w+w(1-i)^t]$.

If $[1-(1-i)^n-w]$ is positive, $[1-(1-i)^n-w+w(1-i)^t]$ must be positive because $w(1-i)^t$ is also positive, so an increase in t will benefit the economy more, thanks to the increase in $N(t)$. Consequently, the best outcome in this case would be an infinite patent right duration. On the other hand, if $[1-(1-i)^n-w]$ is significantly negative, we can infer that the whole term of $[1-(1-i)^n-w+w(1-i)^t]$ will be negative because $w(1-i)^t$ is only a small positive number. Thus, the patent system will have negative impact on the economy, and a longer patent duration would magnify this negative impact. The best outcome in this case would be a zero patent right duration, i.e. no patent protection. There is a small chance that, when $[1-(1-i)^n-w]$ is negative and in the vicinity of zero, a decrease in $t$ may lead to an increase in the positive value of $w(1-i)^t$ and thus change the sign of $[1-(1-i)^n-w+w(1-i)^t]$ to positive. In this case, a weak patent protection can avoid the marginal negative effect and achieve a marginal positive result. Only in this case, there exists a finite positive optimal patent duration.

The impact of patent breadth in reality is complicated by the difference between the patent breadth claimed and the true breadth of a patent technology. A true patent breadth represents the importance of an invention: a widely applicable invention will have a large positive impact on the economy. The claimed patent breadth, on the other hand, is generally positively related to the amount of super profit to the patentee and thus has a positive impact of stimulating innovation but, in the meantime, it is also positively related to the amount of deadweight loss. If the breadth claimed is less than the true breadth of the patent (that is, some applications of the invention are not claimed in the patent), there will be a smaller deadweight loss than when the breadth of invention is fully claimed; but this level of patent protection will deliver less income to the inventor and thus has less power to stimulate innovation in the future. As a result, it is not desirable to have a claimed patent breadth less than the true patent breadth.

If the breadth claimed is greater than the true breadth of the patent (that is, some claims in the patent are not associated with the innovation in practice), there will be a greater deadweight loss to society and greater unjustifiable super profit to the patentee—the part of super profit arising from the over-claimed patent protection. Since the over-claimed part of patent breadth is not truly innovative, patent protection of this part does not reward innovation, plays no role in stimulating further innovation and thus is not desirable. By this reasoning, the desirable maximum breadth of patent protection should be the true breadth of a patented innovation.

Let $b$ be the patent breadth claimed, and $m$ the true breadth of the patented technology. Since the true patent breadth $m$ represents the importance of an invention, a greater $m$ indicates a widely applicable invention that has a larger positive impact on the economy. So, SS (social surplus) should be a positive function of $m$: namely $SS = SS(m)$. On the other hand, the claimed patent breadth $b$ may have a positive impact in stimulating innovation, so $N$ (the number of innovations induced by the patent system) should be a positive function of the part of $b$ that is not greater than $m$. Since zero patent protection will be a result of either a zero duration of patent right or a zero patent breadth claimed by the patentee, we assume the function of stimulating innovation is positively related to $t*b$, namely $N = N(t*b)$.[5] The size of a claimed patent breadth relative to the size of a true breadth of patent innovation is positively related to the size of the deadweight loss relative to total social welfare, so we assume the deadweight loss is a positive function of $b/m$, namely $w = w(b/m)$. As such, the overall effect of patent protection EP can be obtained by extending Eq. (6.5) and be expressed as:

$$EP = N(t*b) * SS(m) * (1-r) * \left[1 - (1-i)^n - w(b/m) + w(b/m) * (1-i)^t\right]/i$$

$$- N_0 * SS(m) * (1-r) * w(b/m) * \left[1 - (1-i)^t\right]/i \qquad (6.6)$$

---

[5]More accurately, only the part of $b$ which is less than or equal to m stimulates innovation, so the function for innovation stimulation should be something like $N = N(t*(2mb - b^2)^{0.5})$. For simplicity, a simple function is used in the paper, but this will not affect the discussion.

Again, since $N(t, b)$ is assumed generally much bigger than $N_0$, we focus on the first term at the right-hand side of the Eq. (6.6). Both patent duration and the breadth of patent protection should have a larger impact on $N(t*b)$ than on $w(b/m)*(1-i)^t$; however, the impact on the latter is also important as it may alter the sign of EP. If $[1-(1-i)^n - w(b/m) + w(b/m)*(1-i)^t]$ is positive, a maximum breadth of patent protection should be $b=m$ because, as stated previously, over-claimed patent breadth has no effect on stimulating innovation but will cause deadweight loss. On the other hand, if $[1-(1-i)^n - w(b/m) + w(b/m)*(1-i)^t]$ is negative, the best solution will be zero patent protection (zero breadth or zero duration) so as to minimize the negative impact of patent protection. The third situation is the case of a small $[1-(1-i)^n]$. That is, when a small discount rate $i$ is used and when the number of years advanced $n$ is small, $1-(1-i)^n$ may be insignificantly positive. In this case, a change in breadth and/or duration of patent protection may enlarge the size of $-w(b/m)$ and thus reverse the sign of EP. Consequently, there may be an optimal breadth and length of patent protection to maximize EP. In this case, a short patent duration would be preferred as it can maximize the positive term $w(b/m)*(1-i)^t$. The desirable breadth of patent $b$ should also be small because its impact on the negative term $-w(b/m)$ is greater than its impact on the positive term $w(b/m)*(1-i)^t$.

In short, there is generally neither an optimal patent duration nor an optimal breadth of patent protection. If the economic impact of a patent system is positive, both patent duration and patent breadth should be maximized in order to maximize the contribution of the patent system to the economy. On the other hand, if the impact of a patent system is negative, the system should be abolished. When a patent system has only a marginal positive economic impact, an optimal patent duration with an optimal breadth of patent may exist, and thus weak patent protection is preferred. However, the likelihood of the patent system having only a marginal economic impact is generally not high. Moreover, in this unlikely case, the impact of the patent system would not be significant, and thus finding an optimal combination of patent duration and breadth would be of little significance.

# References

Aghion, P., Bloom, N., Blundell, R., Griffith, R., & Howitt, P. (2005). Competition and Innovation: An Inverted-U Relationship. *Quarterly Journal of Economics, 120*(2), 701–728.

Arora, A., Ceccagnoli, M., & Cohen, W. (2003). *R&D and the Patent Premium* (NBER Working Paper No. 9431).

Arrow, K. J. (1962). Economic Welfare and the Allocation of Resources for Invention. In R. R. Nelson (Ed.), *The Rate and Direction of Inventive Activity* (pp. 609–624). Princeton, NJ: Princeton University Press.

Arundel, A., & Kabla, I. (1998). What Percentage of Innovations Are Patented? *Empirical Estimates for European Firms, Research Policy, 27*(2), 127–141.

Barker, D. G. (2005, April 15). Troll or No Troll? Policing Patent Usage with an Open Post-grant Review. *Duke Law and Technology Review, 9,* 1–17.

Barney, J. (2002). A Study of Patent Mortality Rates: Using Statistical Survival Analysis to Rate and Value Patent Assets. *AIPLA Quarterly Journal, 30*(3), 317–352.

Baudry, M., & Dumont, B. (2006). Patent Renewals as Options: Improving the Mechanism for Weeding Out Lousy Patents. *Review of Industrial Organization, 28*(1), 41–62.

Baumol, W. J. (2002). *The Free Market Innovation Machine*. Princeton: Princeton University Press.

Bessen, J. (2006a). *The Value of US Patents by Owner and Patent Characteristics* (Boston University School of Law Working Paper, 06-46).

Bessen, J. (2006b). *Estimates of Firms' Patent Rents from Firm Market Value* (Boston University School of Law Working Paper No. 06-14).

Bessen, J., & Hunt, R. (2004). The Software Patent Experiment. *Federal Reserve Bank of Philadelphia Business Review, 2004*(Q3), 22–32.

Bessen, J., & Meurer, M. (2007). *The Private Costs of Patent Litigation* (Boston University School of Law Working Paper No. 07-08).

Bessen, J., & Meurer, M. (2008). *Patent Failure: How Judges, Bureaucrats, and Lawyers Put Innovations at Risk*. Princeton and Oxford: Princeton University Press.

Blair, J. M. (1972). *Economic Concentration—Structure, Behaviour and Public Policy*. New York: Harcourt Brace Jovanovich.

Block, R. (1991). *Bringing the Market to Bear on Research* (Report of the Task Force on the Commercialization of Research). Canberra: AGPS.

Blonder, G. (2005, December 19). Cutting Through the Patent Thicket. *Business Week*.

Bloom, N., & van Reenen, J. (2002). Patents, Real Options and Firm Performance. *Economic Journal, 112*, 97–116.

Boldrin, M., & Levine, D. (2004). The Case Against Intellectual Monopoly. *International Economic Review, 45*(2), 327–350.

Boldrin, M., & Levine, D. (2008). *Against Intellectual Monopoly*. Cambridge University Press.

Boldrin, M., & Levine, D. (2012). *The Case Against Patents*. Federal Reserve Bank of St. Louis (Working Paper Series 2012-035A).

Boldrin, M., & Levine, D. (2013). The Case Against Patents. *Journal of Economic Perspectives, 27*(1), 3–22.

Bonatti, L., & Comino, S. (2011). The Inefficiency of Patents When R&D Projects Are Imperfectly Correlated and Imitation Takes Time. *Journal of Institutional and Theoretical Economics, 167*(2), 327–342.

Bright, A. (1949). *The Electric-Lamp Industry: Technological Change and Economic Development from 1800 to 1947*. New York: Macmillan.

Carlson, S. C. (1999). Patent Pools and the Antitrust Dilemma. *Yale Journal on Regulation, 16*, 359–399.

Challu, P. (1995). Effects of the Monopolistic Patenting of Medicine in Italy Since 1978. *International Journal of Technology Management, 10*(2), 237–251.

Coase, R. H. (1960). The Problem of Social Cost. *Journal of Law and Economics, 3*(1), 1–44.

Cockburn, I., & Griliches, Z. (1988). Industry Effects and Appropriability Measures in the Stock Market Valuation of R&D and Patents. *American Economic Review, 78*(2), 419–423.

Cohen, J., & Lemley, M. (2001). Patent Scope and Innovation in the Software Industry. *California Law Review, 89*(1), 1–57.

Cohen, W. M., Nelson, R. R., & Walsh, J. P. (2000). *Protecting Their Intellectual Assets: Appropriability Conditions and Why U.S. Manufacturing Firms Patent (or not)*. Cambridge, MA: National Bureau of Economic Research (NBER Working Paper 7552).

Comanor, W. S., & Leibenstein, H. (1969). Allocative Efficiency, X-inefficiency, and the Measurement of Welfare Losses. *Economica, 36*, 392–415.

Dasgupta, P., & Stiglitz, J. (1988). Learning-by-Doing, Market Structure and Industrial and Trade Policies. *Oxford Economic Papers, 40*, 246–268.

Denicolo, V. (1996). Patent Race and Optimal Patent Breadth and Length. *Journal of Industrial Economics, 44,* 249–265.

Desrochers, P. (1998). On the Abuse of Patents as Economic Indicators. *Quarterly Journal of Austrian Economics, 1*(4), 51–74.

Domar, E., & Stiglitz, J. (1969). Discussion. *The American Economic Review, 59*(2), 44–49. Retrieved from http://www.jstor.org/stable/1823652.

Dutton, H. (1984). *The Patent System and Inventive Activity During the Industrial Revolution 1750–1852.* Manchester: Manchester University Press.

Encaoua, D., Guellec, D., & Martinez, C. (2006). Patent Systems for Encouraging Innovation: Lessons from Economic Analysis. *Research Policy, 35*(9), 1423–1440.

Epstein, C. (1934). *Industrial Profits in the United States.* New York: National Bureau of Economic Research.

Ferrill, F. D. (2005). Patent Investment Trust: Let's Build a Pit to Catch the Patent Trolls. *North Carolina Journal of Law and Technology, 6,* 367.

Firestone, O. (1971). *Economic Implications of Patents.* Ottawa: University of Ottawa Press.

Foster, N., & Breitwieser, A. (2012). *Intellectual Property Rights, Innovation and Technology Transfer: A Survey* (wiiw Working Papers 88). The Vienna Institute for International Economic Studies, wiiw.

Freeman, C., & Soete, L. (1997). *The Economics of Industrial Innovation.* London: Pinter.

Fudenberg, D., Gilbert, R., Stiglitz, J., & Tirole, J. (1983). Preemption, Leapfrogging and Competition in Patent Races. *European Economic Review, 22*(1), 3–31.

Gallini, N. (1992). Patent Policy and Costly Imitation. *RAND Journal of Economics, 23*(1), 52–63.

Gambardella, A., Giuri, P., & Luzzi, A. (2006). *The Market for Patents in Europe.* https://www.econstor.eu/bitstream/10419/89449/1/512141126.pdf.

Gambardella, A., Harhoff, D., & Verspagen, B. (2005). *The Value of Patents.* ftp://zinc.zew.de/pub/zew-docs/veranstaltungen/inno_patenting_conf/ GambardellaHarhoffVerspagen.pdf.

Geroski, P. A., & Pomroy, R. (1990). Innovation and the Evolution of Market Structure. *Journal of Industrial Economics, 38,* 299–314.

Gilbert, R. (2002). Antitrust for Patent Pools: Lessons from Recent U.S. Patent Reform. *Journal of Economic Perspectives, 16,* 131–154.

Gilbert, R., & Shapiro, C. (1990). Optimal Patent Length and Breadth. *The RAND Journal of Economics, 21,* 106–112.

Ginarte, J., & Park, W. (1997). Determinants of Patent Rights: A Cross-National Study. *Research Policy, 26*(3), 283–301.

Gort, M., & Klepper, S. (1982). Time Paths in the Diffusion of Product Innovations. *Economic Journal, 92*(367), 630–653.

Gould, D., & Gruben, W. (1996). The Role of Intellectual Property Rights in Economics Growth. *Journal of Development Economics, 48*(2), 323–350.

Granstrand, O. (1999). *The Economics and Management of Intellectual Property—Towards Intellectual Capitalism.* Cheltenham: Edward Elgar.

Griliches, Z. (1990). Patent Statistics as Economic Indicators: A Survey. *Journal of Economic Literature, 28,* 1661–1707.

Gustafsson, C. (2005). *Private Value of Patents in a Small Economy: Evidence from Finland* (Working Paper).

Hall, B., Jaffe, A., & Trajtenberg, M. (2005). Market Value and Patent Citations. *RAND Journal of Economics, 36*(1), 16–38.

Hall, B., Thoma, G., & Torrisi, S. (2007). *The Market Values of Patents and R&D: Evidence from European Firms* (NBER Working Paper 13426).

Harberger, A. (1954). Monopoly and Resource Allocation. *American Economic Review, 44,* 77–87.

Harhoff, D., Scherer, F., & Vopel, K. (1997). *Exploring the Tail of Patented Invention Value Distributions.* ftp://zew.de/pub/zew-docs/dp/dp3097.pdf.

Hart, P. E., & Prais, S. J. (1956). The Analysis of Business Concentration: A Statistical Approach. *Journal of the Royal Statistical Society* (Series A), *119,* 150–181.

Horowitz, A. W., & Lai, E. L. C. (1996). Patent Length and the Rate of Innovation. *International Economic Review, 37,* 785–801.

Jaffe, A. (2000). The US Patent System in Transition: Policy Innovation and the Innovation Process. *Research Policy, 29*(4–5), 531–557.

Jaffe, A., & Lerner, J. (2004). *Innovation and Its Discountents: How Our Broken Patents System Is Endangering Innovation and Progress, and What to Do About It.* Princeton: Princeton University Press.

Jenny, F., & Weber, A. (1983). Aggregate Welfare Loss Due to Monopoly Power in the French Economy: Some Tentative Estimates. *Journal of Industrial Economics, 32,* 113–130.

Kamerschen, D. (1966). An Estimation of the Welfare Losses from Monopoly in the American Economy. *Western Economic Journal, 4,* 221–236.

Kanwar, S., & Evenson, R. (2003). Does Intellectual Property Protection Spur Technological Change? *Oxford Economic Papers, 55,* 235–264.

Katari, R., & Baker, D. (2015). *Patent Monopolies and the Costs of Mismarketing Drugs*. Washington, DC: Center for Economic and Policy Research. http://cepr.net/documents/publications/mismarketing-drugs-2015–04.pdf.

Khan, B., & Sokoloff, K. L. (1993). Schemes of Practical Utility: Entrepreneurship and Innovation Among Great Inventors in the United States 1790–1865. *Journal of Economic History, 53*(2), 289–307.

Kitch, E. W. (1977). The Nature and Function of the Patent System. *Journal of Law and Economics, 20*(2), 265–290.

Klemperer, P. (1990). How Broad Should the Scope of Patent Protection Be? *RAND Journal of Economics, 21*, 113–130.

Landes, W., & Posner, R. (2004). *The Political Economy of Intellectual Property Law*. Washington, DC: AEI-Brookings Joint Center for Regulatory Studies.

Lanjouw, J. (1998). Patent Protection in the Shadow of Infringement: Simulation Estimations of Patent Value. *Review of Economic Studies, 65*, 671–710.

Lanjouw, J., & Schankerman, M. (2004). Protecting Intellectual Property Rights: Are Small Firms Handicapped. *Journal of Law and Economics, 47*(1), 45–74.

Lanjouw, J., Pakes, A., & Putnam, J. (1996). *How to Count Patents and Value Intellectual Property: Uses of Patent Renewal and Application Data* (NBER Working Paper No. 5741).

Lemley, M. A. (2001). Rational Ignorance at the Patent Office. *Northwestern University Law Review, 95*(4), 1495–1532.

Lerner, J. (2002). 150 Years of Patent Protection. *American Economic Association Papers and Proceedings, 92*(2), 221–225.

Lerner, J. (2009). The Empirical Impact of Intellectual Property Rights on Innovation: Puzzles and Clues. *American Economic Review Papers & Proceedings, 99*(2), 343–348.

Lerner, J., & Zhu, F. (2007). What Is the Impact of Software Patent Shifts? Evidence from Lotus v. Borland. *International Journal of Industrial Organization, 25*, 511–529.

Lerner, J., Strojwas, M., & Tirole, J. (2003). *The Structure and Performance of Patent Pools: Empirical Evidence* (Working Paper). IDEI – University of Toulouse, France.

Levin, R., Klevorick, A., Nelson, R., & Winter, S. (1987). Appropriating the Returns from Industrial Research and Development. *Brookings Papers on Economic Activity, Special Issue on Microeconomics, 3*, 783–831.

Levin, R., Merrill, S., & Myers, M. (2004). *A Patent System for the 21st Century*. Washington, DC: National Academies Press.

Machlup, F. (1958). *An Economic Review of the Patent System*, Study No. 15 of Committee on Judiciary, Subcommittee on Patents, Trademarks, and Copyrights, 85th Cong., 2d Sess. Washington, DC: U.S. Government Printing Office.

Lunney, G. (2001). E-obviousness. *Michigan Telecommunications Technology Law Review, 7*, 363–422.

Mandiville, T. D., Lamberton, D. M., & Bishop, E. J. (1982). *Supporting Papers for the Economic Effects of the Australian Patent System*. Canberra: Australian Government Publishing Service.

Mandeville, T. (1996). *Understanding Novelty: Information, Technological Change, and the Patent System*. Norwood, NJ: Ablex.

Mansfield, E. (1986). Patents and Innovation: An Empirical Study. *Management Science, 32*(2), 173–181.

Mansfield, E., Schwartz, M., & Wagner, S. (1981). Imitation Costs and Patents: An Empirical Study. *The Economic Journal, 91*(364), 907–918.

Matutes, C., Regibeau, P., & Rockett, K. (1996). Optimal Patent Design and the Diffusion of Innovations. *RAND Journal of Economics, 27*, 60–83.

McGee, J. S. (1966). Patent Exploitation: Some Economic and Legal Problems. *Journal of Law and Economics, 9*, 135–162.

Megna, P., & Klock, M. (1993). The Impact of Intangible Capital on Tobin's Q in the Semiconductor Industry. *American Economic Association Papers and Proceedings, 83*(2), 265–269.

Merges, R. P. (1999). *Institutions for Intellectual Property Transactions: The Case of Patent Pools* (Working paper). University of California at Berkeley.

Miller, S. P. (2009). Is There a Relationship between Industry Concentration and Patent Activity? Available at SSRN: http://ssrn.com/abstract=1531761.

Moir, H. (2008, October). *What Are the Costs and Benefits of Patent Systems* (Working Paper). Center for Governance of Knowledge and Development, the Australian National University.

Moir, H. (2013). *Patent Policy and Innovation: Do Legal Rules Deliver Effective Economic Outcomes*. Cheltenham: Edward Elgar.

Moser, P. (2002). *How Do Patent Laws Influence Innovation? Evidence from Nineteen-Century World Fairs* (NBER Working Paper No. 9909).

Moser, P. (2005). How Much Do Patent Laws Influence Innovation? Evidence from Nineteenth Century World Fairs. *American Economic Review, 95*(4), 1214–1236.

Murphy, F. H. (2002). The Occasional Observer: A New Source for What Is Happening in Operations Research Practice. *Interfaces, 32*(3), 26–29.

Needham, D. (1975). Market Structure and Firms' R&D Behaviour. *Journal of Industrial Economics, XXIII*(4), 241–255.

Nelson, R. R. (1990). Capitalism as an Engine of Progress. *Research Policy, 19,* 193–214.

Nelson, R. R., & Winter, S. (1982). *An Evolutionary Theory of Economic Change.* Cambridge: Harvard University Press.

Nielsen, C. M., & Samardzija, M. R. (2007). Compulsory Patent Licensing: Is It a Viable Solution in the United States? *Michigan Telecommunications and Technology Law Review, 13,* 509–539.

Nordhaus, W. D. (1967). *The Optimal Life of a Patent* (Cowles Foundation Discussion Paper No. 241). New Haven.

Nordhaus, W. D. (1972). The Optimum Life of a Patent: Reply. *The American Economic Review, 62*(3), 428–431.

Oxfam. (2001). *Patent Injustice: How World Trade Rules Threaten the Health of Poor People* (Oxfam Briefing Paper).

Pakes, A., & Schankerman, M. (1984). The Rate of Obsolescence of Patents, Research Gestation Lags, and the Private Rate of Return to Research Resources. In Z. Griliches (Ed.), *R&D, Patents and Productivity.* Chicago: University of Chicago Press/NBER.

Pakes, A. (1986). Patents as Options: Some Estimates of the Value of Holding European Patent Stocks. *Econometrica, 54*(4), 755–784.

Park, W. G., & Ginarte, J. C. (1997). Intellectual Property Rights and Economic Growth. *Contemporary Economic Policy, 15,* 51–61.

Parker, R., & Connor, J. (1979). Estimates of Consumer Losses Due to Monopoly in the U.S. Food Manufacturing Industries. *American Journal of Agricultural Economics, 61,* 626–639.

Pretnar, B. (2003). The Economic Impact of Patents in a Knowledge-based Market Economy. *International Review of Intellectual Property and Competition Law, 34*(3), 887–906.

Priest, G. L. (1977). Cartel and Patent Licensing Arrangements. *Journal of Law and Economics, 20,* 309–377.

Putnam, J. (1996). *The Value of International Patent Protection.* Ph.D. thesis, Yale University.

Qian, Y. (2007). Do National Patent Laws Stimulate Innovation in a Global Patenting Environment? A Cross-Country Analysis of Pharmaceutical Patent Protection, 1978–2002. *Review of Economics and Statistics, 89,* 436–453.

Reid, B. C. (1993). *A Practical Guide to Patent Law* (2nd ed.). London: Sweet and Maxwell.

Reinganum, J. (1982). A Dynamic Game of R and D: Patent Protection and Competitive Behaviour. *Econometrica, 50*(3), 671–688.

Rosenberg N. (2004). *Innovation and Economic Growth* (OECD Working Paper). https://www.oecd.org/cfe/tourism/34267902.pdf.

Sakakibara, M., & Branstetter, L. (2001). Do Stronger Patents Induce More Innovation? Evidence from the 1988 Japanese Patent Law Reforms. *RAND Journal of Economics, 32*(1), 77–100.

Saunders, K. M. (2002). Patent Nonuse and the Role of Public Interest as A Deterrent to Technology Suppression. *Harvard Journal of Law and Technology, 15*(2). http://jolt.law.harvard.edu/articles/pdf/v15/15Harv-JLTech389.pdf.

Schacht, W. H., & Thomas, J. R. (2005). *Patent Reform: Innovation Issues* (CRS Report for Congress). Congressional Research Service.

Schankerman, M. (1998). How Valuable Is Patent Protection: Estimates by Technology Fields. *RAND Journal of Economics, 29*(1), 77–107.

Scherer, F. M. (1970). *Industrial Market Structure and Economic Performance.* Chicago: Rand McNally.

Scherer, F. M. (2009). The Political Economy of Patent Policy Reform. *Journal of Telecommunication and High Technology, 7,* 167–216.

Scherer, F. M., & Ross, D. (1990). *Industrial Market Structure and Economic Performance.* Boston: Houghton Mifflin Company.

Scherer, F. M., & Weisburst, S. (1995). Economic Effects of Strengthening Pharmaceutical Patent Protection in Italy. *International Review of Industrial Property and Copyright Law, 26,* 1009–1024.

Scherer, F. M., Herzstein, S. E., Dreyfoos, A., Whitney, W., Bachmann, O., Pesek, P., et al. (1959). *Patents and the Corporation: A Report on Industrial Technology Under Changing Public Policy.* Boston: Privately Published.

Scotchmer, S. (1991). Standing on the Shoulders of Giants: Cumulative Research and the Patent Law. *Journal of Economic Perspectives, 5*(1), 29–41.

Serrano, C. (2005). *The Market for Intellectual Property: Evidence from the Transfer of Patents* (University of Minnesota Working Paper).

Shapiro, C. (2000). Navigating the Patent Thicket: Cross Licenses, Patent Pools, and Standard Setting. *Innovation Policy and the Economy, 1,* 119–150.

Shapiro, R., & Hassett, K. (2005). *The Economic Value of Intellectual Property* (USA for Innovation Working Paper).

Singh, K. (2001). *Patents vs. Patients: AIDS, TNCs and Drug Price Wars.* Public Interest Research Centre.

Tandon, P. (1982). Optimal Patents with Compulsory Licensing. *Journal of Political Economy, 90,* 470–486.

Taylor, C. T., & Silberston, Z. A. (1973). *The Economic Impact of the Patent System: A Study of the British Experience.* Cambridge: Cambridge University Press.

Thomas, P. (1999). The Effect of Technological Impact Upon Patent Renewal Decisions. *Technology Analysis & Strategic Management, 11*(2), 181–197.

Thomas, J. R. (2001). *Collusion and Collective Action in the Patent System: A Proposal for Patent Bounties* (p. 305). University of Illinois Law Review.

Toivanen, O., & Vaananen, L. (2012). Returns to Inventors. *The Review of Economics and Statistics, 94*(4), 1173–1190.

Varchaver, N. (2001, May 14). The Patent King. *Fortune.*

Winter, S. G. (1993). Patents and Welfare in an Evolutionary Model. *Industrial and Corporate Change, 2*(2), 211–231.

Woellert, L. (2005, July 4). A Patent War Is Breaking Out on the Hill. *Business Week,* p. 45.

# 7

# The Future of Our Economy and Society

What are the economic and social impacts of the proposed new patent system? Based on logical reasoning on the design of the new patent system, this chapter provides some projections about the economy and society into the future. It is true that projection into future is always speculative and often turns out to be incorrect, e.g. many quantitative projections from econometric models are rejected by reality. However, projections based on rigorous logical reasoning and containing plausible mechanisms should have a chance to be correct. Successful projections of this type include the prediction of the return of Halley's Comet and of the existence of Neptune; both are based on Newton's universal law of gravity. The projection here is based on the refined reward mechanism established by the new patent system, so it should contain some elements of truth.

## 7.1 New Pattern of Economic Growth

Economic history shows that our economic growth is featured by recessions, stagnations, and bubbles. During a recession, massive unemployment and idle capital are accompanied by stagnation of

© The Author(s) 2019
S. Meng, *Patentism Replacing Capitalism*,
https://doi.org/10.1007/978-3-030-12247-8_7

commodity sales. What would be the economic growth pattern under the new patent system? The author's projection is that the economy will grow relatively smoothly and at the highest speed possible. The well-functioning patent market brought about by the new patent system will play a key role in the new pattern of economic growth.

## 7.1.1 Abundance of Innovations

According to our previous discussion, the repeated occurrence of economic recessions suggests that our society lacks innovations and the shortage of innovation is due to the high risk of invention investment and the low return due to imitation activities. We are unable to reduce this risk of innovation investment, but we can increase the return to invention so that the high risk can be balanced out by high return. The high return to invention can be realized by reducing or even eliminating the externality of invention. The current patent system works in this direction, but it has a conflicting purpose and provides only weak stimulus to innovation activities.

The proposed new patent system removes the impediments on patent protection and gives inventors the maximum amount of reward. On the one hand, the new patent system will extend patent duration infinitely and abolish the compulsory licence rule. This makes sure that the patent holders can enjoy monopoly power over the natural life of their patents. On the other hand, the new patent system will forbid both exclusive patent licences and assignments of patents. These ensure that the patent monopoly power always stays with the patent holder and that all profits resulting from the patent monopoly power go back to the patent holder.

The strong monopoly power awarded to the patent holder will establish for the patent holder the monopoly position in the patent market. Subject to market demand, the patent holder will utilize its monopoly position to maximize profit. As such, some patents of high market demand will earn extremely high wealth for the patent holder. This would substantially inspire innovation activity. The success stories of Microsoft, Google and Facebook have inspired many to become successful innovative entrepreneurs such as Bill Gates, Larry Page, Sergey

Brin and Mark Zuckerberg. The success stories under the new patent system will do the same to inspire many to become innovators. As a result, a large number of people will be attracted to innovation activities and inspired to work harder in order to achieve innovations. Meanwhile, the high return to patents of high market demand will also attract investors so more resources will be devoted to innovation activity. The abundance of resources and manpower in innovation activity can more than offset the high risk of innovation failure. As a result, society can enjoy abundant innovations.

## 7.1.2  Synergy of Markets

Despite market failure in some circumstances, market economy has been enormously successful. The commodity market brings desirable goods and services to consumers at lowest cost, the labour market helps to put the right people in the right working place, and the capital market helps to allocate capital to the most efficient use. All these markets work together to ensure our resources are allocated efficiently in order to achieve desirable social welfare. However, our market system is not completed yet. The patent market is still in its infancy—it fails to bring enough resources into innovation activities.

The major obstacles to an effective patent market are limited patent duration and complicated patent transactions. The former necessitates a limited patent life which is a great distortion to the patent market. The latter leads to high transaction costs and thus makes the patent market inefficient.

The proposed new patent system overcomes the two obstacles and will lead to a thicker, simpler and less-distorted market. As discussed earlier, the new patent system enhances the monopoly power of the patent holders and thus can stimulate more invention and patents. This leads to an increasing supply of patents. The infinitely prolonged patent duration will remove the distortion to the patent market and thus will help to form a sustainable patent market. Forbidding the assignment of patents as well as exclusive patent licensing can eliminate sleeping patents due to cross-licensing, reduce the price of patent right, preserve demands for patent holders and thus form a thicker patent market. Standardizing patent licensing agreements will reduce the patent transaction cost. Separating patent licence agreements and technical support agreements will prevent the extension of patent monopoly power to patent technical support services and thus reduce the problems caused by information asymmetry. The new patent system may also give birth to new agencies such as patent licensing agents and help improve the function of patent pools. This can overcome the problem of incremental patent licensing. Once the obstacles on the patent market are removed, the patent market is expected to become efficient and effective.

Once a mature patent market is established, the synergy of capital and patent markets will allocate resources efficiently to both production activity and innovation activity. If invention activity is not enough, the return rate of innovation activity will be higher than that of production activity, and the market mechanism will ensure more funds will flow into invention activity. On the other hand, if production activity is underfunded, the return rate of production activity will be higher, so more funds will flow into production activity. As such, we will not see the situations where we have too many innovations but too little funds for production, and vice versa. Consequently, the synergy of both capital and patent markets will make sure the optimal allocation of funds to both activities and thus achieve a desirable balance between these two activities.

### 7.1.3 Realization of Economic Growth Potential

It is common wisdom that economic growth is often interrupted by economic recessions. The cause of economic recession is contentious. Keynesian economists blame the deficiency of effective demand, which causes overproduction in the economy (i.e. general gluts). Classical economists think economic recession is caused by a mismatch between demand and supply (partial gluts). The long durations of large economic recessions (e.g. 4 years for the Great Depression starting from 1929 and 2 years for the global financial crisis occurring in 2008) indicate that the mismatch theory is implausible. If there are products which are in high demand (or undersupply), producers would be able to find the mismatch and rebalance supply and demand fairly quickly and thus the economy would be back on an even footing very soon. However, reality shows otherwise. The long economic recessions indicate that the highly demanded commodity claimed by neoclassical economists must not exist or, in other words, have not been invented yet. In this reasoning, the long economic recessions are caused by the lack of new products or, to be more accurate, the lack of innovations.

This reasoning is also consistent with our experience. Experience tells that new products have market potential and old products tend to cause market saturation. If there are enough new products in an economy, a developed economy has sufficient resources (e.g. capital and labour) and technology to satisfy the demand, so the economy will keep growing. Thus, re-occurring economic recessions with abundant commodities must imply that inventions or new products are scarce in a modern economy. The lack of innovation indicates that the resource allocation is inefficient for production and innovation activities: too many resources are allocated to production activity and too little for innovation activity. Consequently, innovation scarcity not only causes economic recession but also prevents economic growth from reaching its potential.

With the new patent system, invention will be greatly encouraged. The new patent system will foster an efficient patent market to guide innovation activities. With an efficient patent market channelling funds to invention activity and an efficient capital market channelling funds to production activity, the synergy of these two markets can balance invention activity and production activity. Thus, invention shortage can be avoided and producers can always find and produce new products which have high market demand. Consequently, an economic recession will not occur. Moreover, the synergy of both capital and patent markets will ensure that innovation speed is matched by production capacity, so the economy can grow at the highest speed allowed by the resources available. As long as the two markets can effectively coordinate both innovation and production activities, no resource will be misallocated or wasted and thus economic growth will reach its potential. In other words, a smoother and faster economic growth is guaranteed by this market mechanism.

With this new pattern of economic growth, there will be little unemployment and idle capital because innovation activity needs a lot of resources. One may doubt: How can you expect all unemployed factory workers to find a job in innovation activities? Many people (e.g. Harari 2015) are even concerned that more innovation may cause more unemployment given the fact that the progress of technology (e.g. driverless cars, highly automated machines and personalized robots) is going to replace many jobs currently performed by manpower. Are these worries warranted?

The history of technological progress shows that new technology will generally bring more jobs, higher productivity and better living standards. Some industries may be outdated by new technology, and all jobs in these industries will be lost. However, the unemployed labour force and idle capital will be absorbed by new industries brought about by new technology. This is proved in history by industrial structural changes caused by innovations. With the new patent system in place, more people will be attracted to innovation activities, but there will still be the need for factories workers because the new products invented need to be produced to satisfy consumers. The high-tech machines equipped with artificial intelligence may replace the repetitive or low-intelligence labour work but people will still be needed to manage the factories as well as to control and maintain the automated machines. Although the overall requirement of skill and knowledge will be higher in the future, there will be both high-skill and low-skill jobs available.

## 7.2    The Patentist Society

The occurrence of economic recessions spells the end of capitalism and the beginning of patentism. This projection is based on the benign monopoly empowered by the new patent system. The reader may argue that there is no benign monopoly—all monopolies are inefficient, ugly and against democracy. However, the desirable monopoly under the new patent system—exclusive rights to grant non-exclusive patent licences—should be an exception. The right patent monopoly is efficient because it brings much-needed innovations to society at the least cost; it is fair because everyone has an equal chance to develop a patent and thus to have a patent monopoly right; it will not lead us to autocracy because the patent monopoly is restrained by natural factors and because the new patent system will bring about numerous patent monopolies (patent holders). The more patent monopolies society has, the more benefit and enjoyment for society.

## 7.2.1 The Desirable Patent Monopoly

Based on the previous discussion, new products are shown to be the key to sustaining economic growth and social welfare. To encourage inventors to invent sufficient new products, they must be given enough power over, or returns to, their inventions. This power or returns is guaranteed by the new patent system.

The new patent system enhances the monopoly power for the patent holder but also confines the monopoly power to the patent holder only. The infinitely prolonged patent length ensures the patent holder can enjoy the monopoly power over the natural life of the patent. The ban on exclusive licences and assignments of the patent strips the patent monopoly power from the producers. This gives the patent holder an absolute monopoly power over the producers. The monopoly of the patent holder over the producer is the most desirable monopoly because it ensures all profits from the monopoly power go back to the patent holder to stimulate innovation and thus can minimize the social cost of patent monopoly.

This desirable patent monopoly has important social implications—the patent holder will overpower the producers. On the one hand, patentists have monopoly power in setting the price of patent licences to maximize profit according to market demand. On the other hand, if the producer does not accept the price offered by the patent holder, the markets will be saturated with old products, so the producers will have plenty of resources and high production capacity but face the diminishing opportunities of making a profit. This eventually forces the producers to turn to inventors or patent holders to obtain patent permission to produce new products. The capitalists have to accept the price of the patent licences given by patentists because the new products are the only way to make a profit. The monopoly position of the patentist makes them superior to capitalist, and thus, the patent holder (patentist) will overtake producers (or capitalists) to become the richest group in society. As a result, patentism will replace capitalism.

## 7.2.2 Fashionable Innovators

The absolute monopoly power of the patent holders will improve the awareness of the importance of innovation and help patent holders to make super profits and to become the top class of society. Since the possibility of becoming super rich by conducting innovation is higher than other occupations, this will inevitably make innovation an attractive occupation. As a result, more people will swarm into innovation occupation and society will become an innovative one.

However, not everyone conducting innovation will become rich because of the high possibility of innovation failure. Even a successful innovator may not become rich when there is a lack of market demand for his/her innovations. This is an unfortunate situation. However, our economy and society need people to take the risk to innovate. Since the new patent system balances the high risk of innovation failure with the high return to the successful innovation, it is a fair game for innovators. Moreover, all people in society (including the unsuccessful innovators) will benefit from the patent system because its reward mechanism ensures the efficient allocation of resources to different innovation projects and also ensures the innovator will use the resources efficiently to increase the chance of success.

## 7.2.3 Restrainers on the Patent Monopoly

Economics theory tells us that monopoly is bad for a society because a monopoly tries to gain super profit at the net social cost of deadweight loss, so it is the responsibility of the government to restrain monopoly power. Examples include government regulation on natural monopolies like electricity generators, insurance companies, banks and transportation companies. Should the government do the same thing to patents?

The answer is negative for two reasons. One is that it is impractical to regulate the price of patent licences. There will be a numerous and constantly increasing number of patents, and the innovation costs and market demand are vastly different for each patent. The other reason is that, unlike other monopolies, patent monopoly generates social benefit—stimulating innovation, which is vital for economic growth and the welfare of the society. The more profit the patent monopoly obtains, the more innovations will be stimulated for society (contributed by both the patent monopoly and other innovators). Since all profits from the patent monopoly power will go back to the patent holder under the new patent system, the cost of the higher price paid by consumers will be more than compensated by the innovation stimulated. From this point of view, patent monopoly without government intervention will benefit society most.

Even without a government regulation of patent prices, the patent monopoly will not be out of control thanks to two natural constraints. The first is the market demand. The purpose of patent licensing is to make high profits. To maximize profits, the patent monopoly has to price its licences according to market demand—it is not reasonable to set an outrageous price but have no buyers. One may argue that the patent monopoly may set an outrageous price to ensure its monopoly position in production. It seems that this argument has provided a rationale for excessive pricing, but the argument itself is irrational: if one can make more profit by selling patent licences than by implementing patent technology, why will the patent holder bother to spend a lot of time and energy to produce the patented products? As stated earlier, the irrational behaviour of excess pricing with a few or no buyers will occur when the new patent system is first in place, but people will learn from mistakes and become rational over time. As a result, the patent monopoly will be under the control of the market demand.

The new innovations which will replace the existing patented technology will be the other natural restrainers of patent monopoly. The patent monopoly power will stimulate the efforts to out-innovate the patent. This has been shown in the history of many industries; the electrical lamp industry is a good example. As Thomas Edison successfully invented and patented his vacuum bulb carbon-filament electrical lamp, many other companies using this invention were forced out of business by court decisions. Meanwhile, tremendous efforts were devoted to finding new filaments and other ways of providing electrical lighting, e.g. to find cheaper and more energy efficient filaments for electrical lamps, to develop new ways of electrical lighting alternative to the vacuum bulb or even an alternative to incandescent lamps. These efforts eventually led to the discovery of ductile tungsten as a filament, the advent of nitrogen-filled bulbs and fluorescent lighting. Although many other factors may have induced people to try new methods, the patent monopoly power also played an important role.

## 7.3   Beyond Patentism

This is a projection about the distant future, so it is highly likely that the author risks projection accuracy here. However, the projection is based on the fact that scientific discoveries are the source and foundation of innovations. The logical line leads us to this projection, so it should have some elements of truth. If you are a reader after the author has long gone, you may give the author an objective mark on this projection.

### 7.3.1  The Source and Foundation of Innovations

Although the new patent system can stimulate innovation to the maximum extent, the speed and the successful rate of innovation are ultimately constrained by the advancement of knowledge in different fields. For example, without the knowledge of DNA, the invention of medicine and treatment targeting defective DNA is impossible; due to the lack of knowledge of the force of air flow, Leonardo da Vinci's flying machine remained only on paper; the fluorescent electrical lamp was invented only after the discovery that ultraviolet light can induce fluorescence and that the latter can be generated by very hot bodies. Since fundamental discoveries contribute greatly to humanity's knowledge, they are the source and foundation of inventions.

It is common wisdom that scientific discoveries and theoretical breakthroughs have widespread indirect influence on the economy, but the discoverers can obtain no economic benefit from their discoveries because they are not saleable products. This makes discoveries public goods of tremendous positive externality. Currently, neither patent laws nor copyright laws protect scientific discoveries and theoretical breakthroughs, so the vast difference between social and private return on scientific discoveries is severely underfunded, especially for theoretical and fundamental research. Few private companies are interested in it due to its inability to bring a profit. As a result, scientific research heavily depends on government funding.

## 7.3.2 The Law of Discoverers' Right

Since our economy is dependent on the speed of innovation and the latter is in turn dependent on the speed of discovery, discovery activity must be stimulated to satisfy the needs of our society and our economic growth. The author envisions a discoverers' right law which can solve the problem by imposing a non-exclusive licence on commercial users. The majority of knowledge licencees would be innovators and producers. They should pay because discoveries are their ultimate source of profits. The law should confine knowledge licences to commercial users also because, practically, the law can be enforced only for commercial users. Like in the case of patent licences, only commercial users can be tracked and also have the ability to pay.

The discoverers' right is very important both to discovers and to society as a whole, so it should be identified by an application and registration procedure similar to that of the patent system. The discovery must be proven beyond doubt with the recognition of discovers' right being based on evidence such as publication, conferences and research reports. The length of discoverers' right should also be infinite so as to give the discoverers the maximum amount of encouragement for their discovery activities. This absolute monopoly right on knowledge may worry some people but, as will be shown in the next section, carefully defined knowledge rights may not obstruct knowledge diffusion.

## 7.3.3 Knowledge Economy and the Discoverist Society

Since discoveries are the source of innovations, the former are far more important than the latter. This importance will be materialized by the law of discoverers' right; thus, knowledge licences will be a key feature of the economy in the far future. With the law of discoverers' right in place, the discoverers will have a monopoly power over the inventors and the producers. The monopoly position of discoverers will enable them to set a knowledge licence price according to market demand and thus make a super profit. Consequently, the discoverers will become the richest group. As a result, the discoverers' right law would highlight the importance of discoveries and discoverers, and thus transform the coming patentist society to the discoverist society in the far future.

One may be sceptical, or even fearful, about knowledge monopoly. This is indicated in the arguments about the patent system. For example, Stiglitz (2013) thought patenting knowledge makes a public good private so it is necessary to fight a patent in order to create a public good (i.e. open access). A more vivid illustration is provided by Tuccille (1971). In his popular satirical book *It Usually Begins with Ayn Rand*, Tuccille (1971) ridiculed the idea of Galambosianism.

Around this time I met the Galambosian. 'I am a Galambosian,' he said...

'What ... is Galambosian?'

'There are five legitimate functions of government,' said the Galambosian.

'No kidding. What are they?'

'I am not at liberty to say. The theory was originated by Andy Galambos and it is his primary property. ... If the rest of us were free to discuss his ideas,' said the Galambosian, 'there is no question in my mind that Galambosianism would spread throughout the world like wildfire'.

Since a Galambosian believes that the inventor or originator of an idea should have absolute and eternal rights over that idea and all the profits derived from it, diffusion of ideas needs permission and involves fees, so the idea of Galambosianism could not be popularized. The conversation highlights the conflict between the knowledge right and knowledge diffusion. However, this conflict will not exist in the discoverist society as the knowledge licences are required only for distribution or commercial use. The new knowledge due to discoveries is free to be used, but if one uses the new knowledge to obtain a profit, one needs to obtain a licence from the discoverers. In this way, a law of discoverers' right in the future can achieve a benign cycle for both knowledge creation and diffusion.

# References

Harari, Y. N. (2015). *Homo Deus: A Brief History of Tomorrow*. London: Harvill Secker.

Stiglitz, J. (2013). Institutional Design for China' Innovation System: Implications for Intellectual Property Rights. In D. Kennedy & J. Stiglitz (Eds.), *Law and Economic Development with Chinese Characteristics: Institutions for the 21st Century*. New York and Oxford: Oxford University Press. http://slideplayer.com/slide/5710070/.

Tuccille, J. (1971). *It Usually Begins with Ayn Rand*. New York: Stein and Day.

# Index

© The Editor(s) (if applicable) and The Author(s) 2019
S. Meng, *Patentism Replacing Capitalism*,
https://doi.org/10.1007/978-3-030-12247-8

Printed in the United States
By Bookmasters